Nonequilibrium Physics at Short Time Scales

Springer
Berlin
Heidelberg
New York
Hong Kong
London
Milan
Paris
Tokyo

Klaus Morawetz (Ed.)

Nonequilibrium Physics at Short Time Scales

Formation of Correlations

With 173 Figures

Springer

Dr. Klaus Morawetz
Institut für Physik
Technische Universität Chemnitz
Reichenhainer Str. 70
09107 Chemnitz, Germany
Email: klaus.morawetz@physik.tu-chemnitz.de

Library of Congress Cataloging-in-Publication Data

Nonequilibrium physics at short time scales : formation of correlations / Klaus Morawetz (ed.).
p.cm.
Includes bibliographical references and index.
ISBN 3-540-20031-2 (acid-free paper)
1. Nonequilibrium statistical mechanics. 2. Many-body problem. 3. Electron configuration. I. Morawetz, Klaus, 1963-

QC174.86.N65N66 2004
530.13--dc22 2003061496

ISBN 3-540-20031-2 Springer-Verlag Berlin Heidelberg New York

Springer-Verlag is a part of Springer Science+Business Media

springeronline.com

Typesetting: Authors and LE-TEX GbR, Leipzig using a Springer LATEX macro package
Production: LE-TEX Jelonek, Schmidt & Vöckler GbR, Leipzig
Cover design: *design & production* GmbH, Heidelberg

Printed on acid-free paper SPIN 10906255 57/3141/YL 5 4 3 2 1 0

Preface

This book is the result of the many discussions and collaborations that developed from the seven previous workshops held on this topic. This ongoing series of interdisciplinary workshops provided an opportunity for the presentation and exchange of results describing nonequilibrium phenomena at short time scales. The important questions concern the correlation and memory effects in dense interacting systems currently under study in various areas of physics. Experiments on very short time scales are especially characterized by the strong correlations in systems far from equilibrium and by their nonlinear dynamics. In this regard, conventionally applied theoretical techniques are critically reviewed, and new theoretical concepts are discussed. Possible signals of nonequilibrium effects are extracted from experiments using powerful techniques and skills. The exchange of views and techniques from different areas of physics helped to stimulate such developments.

The impressive success of experiments using short laser pulses to probe the properties of matter, as well as of the new methods of analysis of the early phases of heavy ion reactions, initiated a review of the available many body theoretical methods. These statistical methods describing strong nonequilibrium situations were known, at least partially, for a long time. However, recent developments in computing power have provided the possibility of accurately comparing *ab initio* and approximate methods. Encouraged by discussions during the meeting at Bad Honnef in 2002, which was supported as the 279th WE-Heraus-Seminar, a decision was reached to summarize and to organize the different experiences obtained regarding correlations on short time scales found in different subfields of physics, focusing upon their common themes.

One might at first question the notion that such different areas of physics, as the interaction of lasers with solids and of nuclear physics, might have important aspects in common. However, it turns out that the descriptions of correlation processes at very short time scales subsequent to a perturbation are similar in these and in other applications, although the time scales involved are different. Three common topics were found to be important. These are the description of the growth of correlation effects after a perturbation, the role of finite size effects on the observable quantities, and the nature of the signals that describe the correlations. Accordingly, the book is divided into two parts. While the first part reviews the theoretical concepts used to describe correlations in equilibrium and nonequilibrium, the focus of the second part is on actual applications and experiments. In the first chapter, modern developments in many body theory are reviewed. The second chapter is devoted to nonequilibrium kinetic approaches which are unavoidable in the description of short time dynamics. This should serve as a guide to a description of the time dependent correlations that arise in the subsequent chapters. The third chapter then describes experimental and theoretical progress on the interaction of pulsed lasers with matter. Here the focus is on the development in time of correlation effects, such as in the formation of excitons and in the time dependence of the response to very short laser pulses. The fourth chapter discusses the short time behavior of finite systems such as nanoparticles and clusters. Here the dynamics of superconductors, which is a subject that is currently hotly debated, has its place. These chapters may be useful in providing an understanding of the experimental and theoretical results on heavy ion reactions which are addressed in the subsequent contributions. Since there is no direct access to the short time correlation effects in heavy ion reactions, analogies with the results presented in the preceding chapters may be useful in providing a guide toward understanding the cluster formation observed in the heavy ion collisions at energies near to the Fermi energy.

The aim of this book is therefore to provide an introduction to the experimental and theoretical probes of the short time behaviors of systems perturbed far from equilibrium. By studying the time dependence of the correlations in such perturbed systems, one hopes to obtain a better insight into the meaning of such correlations. Each contribution to this book presents material from introductory level up to current research results, which is designed to be appropriate for graduate students as well as to researchers in other fields, in order that they might become acquainted with this exciting field.

In preparing this book the advice of Richard Klemm was very helpful and the hospitality of the Max-Planck Institute for the Physics of Complex Systems is gratefully acknowledged.

Chemnitz, November 2003 *Klaus Morawetz*

Contents

List of Contributors

Karsten Andrae
Universität Erlangen-Nürnberg
Institut für Theoretische Physik
Staudtstr. 7
91058 Erlangen, Germany

Arnaud Arbouet
Université Bordeaux I
CPMOH, 351
Cours de la Libération
33405 Talence, France
aarbouet@cpmoh.u-bordeaux.fr

Ladislaus Bányai
J.W. Goethe Universität
Institut für Theoretische Physik
60054 Frankfurt a.M., Germany

Mohamed Belkacem
Université Paul Sabatier
Laboratoire de Physique Quantique
118, route de Narbonne
31062 Toulouse, France

Markus Betz
TU München
Physik-Department E11
James-Franck-Str.
85748 Garching, Germany
mbetz@ph.tum.de

Rolf Binder
University of Arizona
Optical Sciences Center
Meinel Building
1630 East University Boulevard
Tucson, Arizona 85721, USA

Thomas Bornath
Universität Rostock
Fachbereich Physik
Universitätsplatz 3
18051 Rostock, Germany
thomas.bornath@
physik.uni-rostock.de

Robert Botet
CNRS/Université Paris Sud
Lab. de Physique des Solides
Bât. 510
Centre d'Orsay
91405 Orsay, France
botet@lps.u-psud.fr

Andreas Brodschelm
TU München
Physik-Department E11
85748 Garching, Germany

Wolfgang Cassing
Universität Giessen
Institut für Theor. Physik
Heinrich-Buff-Ring 16
35392 Giessen, Germany
Wolfgang.Cassing@
theo.physik.uni-giessen.de

Daniel S. Chemla
Department of Physics
University of California at Berkeley
Berkeley
Ca 94720, USA
and
Lawrence Berkeley National Laboratory
1 Cyclotron Road
Berkeley
Ca 94720, USA
DSChemla@lbl.gov

Dimitris Christofilos
School of Technology
Aristotle University of Thessaloniki
54006 Thessaloniki, Greece
christof@eng.auth.gr

Sung Chung
Western Michigan University
Dept. of Physics
Kalamazoo MI 49008-5252, USA
chung@wmich.edu

Natalia Del Fatti
Université Bordeaux I
CPMOH, 351
Cours de la Libération
33405 Talence, France
delfatti@cpmoh.u-bordeaux.fr

Thi Phuong Mai Dinh
Université Paul Sabatier
Laboratoire de Physique Quantique
118, route de Narbonne
31062 Toulouse, France

Massimo Di Toro
LNS/INFN, Theory44
Via S. Sofia
95123 Catania, Italy
ditoro@lns.infn.it

Jens Förstner
Technische Universität Berlin
Institut für Theoretische Physik
Nichtlineare Optik
und Quantenelektronik
10623 Berlin, Germany

Neil A. Fromer
Lawrence Berkeley National Laboratory
1 Cyclotron Road
Berkeley Ca 94720, USA
NAFromer@cal.berkeley.edu

Christian Fuchs
Universität Tübingen
Institut für Theor. Physik
Auf der Morgenstelle 14
72076 Tübingen, Germany
christian.fuchs@uni-tuebingen.de

Paul Gartner
Universität Bremen
Institut für Theoretische Physik
28334 Bremen, Germany
and
National Institute
for Materials Physics
Bucharest-Magurele, Romania

Eric Giglio
Université Paul Sabatier
Laboratoire de Physique Quantique
118, route de Narbonne
31062 Toulouse, France
and
CIRIL/GANIL
Rue Claude Bloch
Caen, France

Dieter Gross
Hahn-Meitner-Institut
Bereich SF5
Glienicker Str. 100
14109 Berlin, Germany
gross@hmi.de

Cyril Guillon
Université Bordeaux I
CPMOH, 351
Cours de la Libération
33405 Talence, France
cguillon@cpmoh.u-bordeaux.fr

Hartmut Haug
J.W. Goethe Universität
Institut für theoretische Physik
60054 Frankfurt a.M., Germany

Peter A. Henning
MediaLab Karlsruhe
University of Applied Sciences
Moltkestrasse 30
76133 Karlsruhe, Germany
P.Henning@FH-Karlsruhe.de

Paul Hilse
Ernst-Moritz-Arndt-Universität
Greifswald
Institut für Physik
Domstrasse 10a
17487 Greifswald, Germany
hilse@
physik.uni-greifswald.de

Walter Hoyer
Universität Marburg
Institut für Physik
Renthof 5
35032 Marburg, Germany
Walter.Hoyer@
Physik.Uni-Marburg.de

Rupert Huber
TU München
Physik-Department E11
85748 Garching, Germany

Remus-Amilcar Ionescu
Nat. Institute for Physics
and Nuclear Engineering
Heavy Ion Physics Dept.
P.O. Box MG-6
Bucharest, Romania
amilcar@tandem.nipne.ro

Sascha Juchem
Universität Giessen
Institut für Theoretische Physik
Heinrich-Buff-Ring 16
35392 Giessen, Germany

Mackillo Kira
Universität Marburg
Institut für Physik
Renthof 5
35032 Marburg, Germany
Mackillo.Kira
@Physik.Uni-Marburg.de

Stephan W. Koch
Universität Marburg
Institut für Physik
Renthof 5
35032 Marburg, Germany
Stephan.W.Koch
@Physik.Uni-Marburg.de

Dietrich Kremp
Universität Rostock
Fachbereich Physik
Universitätsplatz 3
18051 Rostock, Germany
dietrich.kremp@
physik.uni-rostock.de

Pierre Langot
Université Bordeaux I
CPMOH, 351
Cours de la Libération
33405 Talence, France
plangot@cpmoh.u-bordeaux.fr

Yoshiro Kakehashi
Max-Planck-Institut
für Physik komplexer Systeme
Nöthnitzer Str. 38
01187 Dresden, Germany
yok@mpipks-dresden.mpg.de

Anděla Kalvová
Academy of Sciences
of the Czech Republic
Institute of Physics
Na Slovance 2
182 21 Praha 8, Czech Republic
kalvova@fzu.cz

Richard Klemm
Max-Planck-Institut
für Physik komplexer Systeme
Nöthnitzer Str. 38
01187 Dresden, Germany
rklemm@mpipks-dresden.mpg.de

Andreas Knorr
Technische Universität Berlin
Institut für Theor. Physik
Hardenbergstr. 36
10623 Berlin, Germany
andreas.knorr@
physik.tu-berlin.de

Jan Koláček
Academy of Sciences
of the Czech Republic
Institute of Phys.
Cukrovarnická 10
162 53 Praha 6, Czech Republic
kolacek@fzu.cz

Nai-Hang Kwong
University of Arizona
Optical Science Dept.
Tucson AZ 85721, USA
kwong@physics.arizona.edu

Alfred Leitenstorfer
TU München
Physik-Department E11
James-Franck-Str.
85748 Garching, Germany
aleitens@ph.tum.de

Pavel Lipavský
Academy of Sciences
of the Czech Republic
Institute of Phys.
Cukrovarnická 10
162 53 Praha 6, Czech Republic
lipavsky@fzu.cz

Ming Ma
Université Paul Sabatier
Laboratoire de Physique Quantique
118, route de Narbonne
31062 Toulouse, France

Fabien Mégi
Université Paul Sabatier
Laboratoire de Physique Quantique
118, route de Narbonne
31062 Toulouse, France

Detlef Mentrup
Philips GmbH
Forschungslaboratorien
Abteilung Technische Systeme
Hamburg
Röntgenstrasse 24–26
22335 Hamburg, Germany
Detlef.Mentrup@philips.com
and
Universität Osnabrück
Fachbereich Physik
Barbarastr. 7
49069 Osnabrück, Germany

Klaus Morawetz
Institute of Physics
Chemnitz University of Technology
09107 Chemnitz, Germany
and
Max-Planck-Institut
für Physik komplexer Systeme
Nöthnitzer Str. 38
01187 Dresden, Germany
klaus.morawetz@
physik.tu-chemnitz.de

Jacek Okołowicz
Grand Accélérateur National
d'Ions Lourds (GANIL)
CEA/DSM – CNRS/IN2P3
BP 5027
14076 Caen Cedex 05, France
and
Institute of Nuclear Physics
Radzikowskiego 152
31342 Kraków, Poland

Ilias E. Perakis
University of Crete
Department of Physics
P.O. Box 2208
71003 Heraklion, Crete, Greece
ilias.e.perakis@vanderbilt.edu

Marek Płoszajczak
Grand Accélérateur National
d'Ions Lourds (GANIL)
CEA/DSM – CNRS/IN2P3
BP 5027
14076 Caen Cedex 05, France

Andreas Pohl
Universität Erlangen-Nürnberg
Institut für Theoretische Physik
Germany
and
Université Paul Sabatier
Laboratoire Physique Quantique
Toulouse, France

Paul-Gerhard Reinhard
Universität Erlangen-Nürnberg
Institut für Theoretische Physik
Staudtstr. 7
91058 Erlangen, Germany
reinhard@theorie2.physik.
uni-erlangen.de

Ingrid Rotter
Max Planck Institute for the Physics
of Complex Systems
Nöthnitzer Str. 38
01187 Dresden, Germany
rotter@mpipks-dresden.mpg.de

Ilia Rumyantsev
University of Arizona
Optical Sciences Center
Meinel Building
1630 East University Boulevard
Tucson, Arizona 85721, USA

Jürgen Schnack
Universität Osnabrück
Fachbereich Physik
Barbarastr. 7
49069 Osnabrück, Germany
jschnack@uos.de

Manfred Schlanges
Ernst-Moritz-Arndt-Universität
Greifswald
Institut für Physik
Domstr. 10a
17487 Greifswald, Germany
schlanges@
physik.uni-greifswald.de

W. Udo Schröder
University of Rochester
Dept. of Chemistry
Wilson Blvd.
Rochester NY 14627, USA
schroeder@chem.rochester.edu

Christian Schüller
Universität Hamburg
Institut für Angewandte Physik
Jungiusstr. 11
20355 Hamburg, Germany
schueller@
physnet.uni-hamburg.de
and
University of California at Berkeley
Department of Physics
and Materials Sciences Division
Lawrence Berkeley National
Laboratory
Berkeley, California 94720, USA

Eric Suraud
Université Paul Sabatier
Laboratoire de Physique Quantique
118, route de Narbonne
31062 Toulouse, France
suraud@irsamc2.ups-tlse.fr

Ryu Takayama
ACT-JST
Japan Science and Technology
Corporation
University of Tokyo
Department of Applied Physics
7-3-1 Hongo, Bunkyo-ku
Tokyo 113-8656, Japan

Florian Tauser
TU München
Physik-Department E11
85748 Garching, Germany

Jan Tõke
University of Rochester
Dept. of Chemistry
Wilson Blvd.
Rochester NY 14627, USA

Fabrice Vallée
Université Bordeaux I
CPMOH, 351
Cours de la Libération
33405 Talence, France
vallee@cpmoh.u-bordeaux.fr

Bedřich Velický
Academy of Sciences
of the Czech Republic
Institute of Physics
Na Slovance 2
182 21 Praha 8, Czech Republic
velicky@karlov.mff.cuni.cz
and
Charles University
Faculty of Mathematics and Physics
Ke Karlovu 2
121 16 Praha 2, Czech Republic

Inès Waldmüller
Technische Universität Berlin
Institut für Theoretische Physik
Hardenbergstr. 36
10623 Berlin, Germany
ines@itp.physik.tu-berlin.de

Zhen-Shan Yang
University of Arizona
Optical Sciences Center
Meinel Building
1630 East University Boulevard
Tucson, Arizona 85721, USA

Part I

Quantum Statistical Approaches to Correlations

Quantum Many-Body Theory and Coherent Potential Approximation

Yoshiro Kakehashi

Abstract. The coherent potential approximation (CPA) is a useful method for describing the electron correlations as well as the effects of disorder on electrons. Among the many-body theories using the CPA, the dynamical CPA, the many-body CPA, and the dynamical mean-field theory are reviewed to clarify how these theories use the CPA concept for the description of the electron correlations. The theories characterized by the momentum independent self-energy are shown to interpolate between the weak and strong Coulomb interaction limits, and therefore describe the basic properties of magnetism from metals to insulators, the metal-insulator transition, and the single particle excitations from the Fermi liquid to the insulator. The relation among various theories are clarified. In particular, it is shown that the dynamical CPA, the many-body CPA, and the dynamical mean-field theory are equivalent to each other, so that the theories of itinerant magnetism and those of the strongly correlated electron systems are unified within the single-site approximation. The nonlocal effects on the selfenergy are also discussed beyond the single-site approximation.

1 Introduction

The notion of the effective field is well-known to be useful for describing the electronic structure and phase transition in solids. The complex interactions such as the random potentials and the electron-electron interactions are treated there by means of some average potential or effective field. Essential physics is described by such an effective field in many phenomena, and the selfconsistency of the effective field allows us to describe best the physical properties and the phase transition [1].

In the random alloys, for example, such a method of describing the electronic structure is known as the coherent potential approximation (CPA) [2–8]. Consider the substitutional binary alloys A_xB_{1-x} with the concentration x of atom A. The electrons hop from site i to site j via transfer integral t_{ij} in the random potential $\{\epsilon_i\}$ (ϵ_A or ϵ_B). The eigen-value problem in this system is not trivial because of the lack of translational symmetry. The CPA approximates the random potential $\{\epsilon_i\}$ by means of an energy-dependent coherent potential $\Sigma(z)$ in the calculation of one electron Green function defined by $[(z-\mathbf{H})^{-1}]_{ij}$, where $(\mathbf{H})_{ij} = \epsilon_i\delta_{ij} + t_{ij}$. The coherent potential recovers the translational symmetry. To obtain the coherent potential, we introduce the Green function $G_A(z)$ for an impurity A embedded in the effective medium $\Sigma(z)$, and $G_B(z)$ for an impurity B embedded in the same

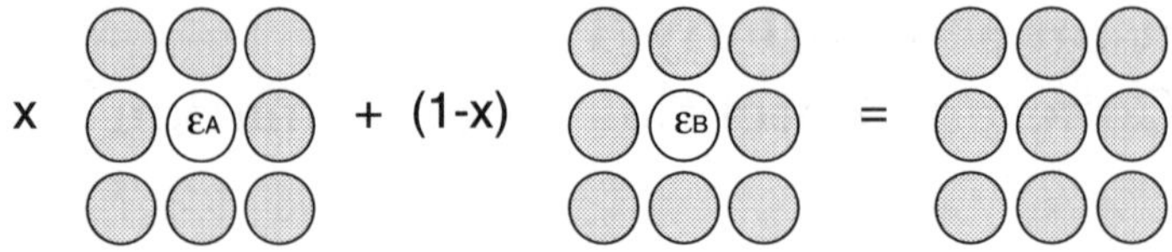

Fig. 1. Schematic picture showing the selfconsistent condition for determining the coherent potential in the binary alloy A_xB_{1-x}. The *left-hand-side* expresses the average of the impurity states embedded in the medium, while the *right-hand-side* expresses the coherent medium

medium. The coherent potential $\Sigma(z)$ is determined selfconsistently so that the averaged Green function $\langle G(z) \rangle = xG_A(z) + (1-x)G_B(z)$ equals the coherent Green function $F(z)$ for the effective medium only, which is defined by $F(z) = [(z - \tilde{\mathbf{H}})^{-1}]_{ii}$ and $(\tilde{\mathbf{H}})_{ij} = \Sigma(z)\delta_{ij} + t_{ij}$ [9]. (See Fig. 1.)

The concept of the CPA is also useful for describing the correlated electrons. Assume that electrons move from site to site with an on-site repulsive Coulomb interaction U. When the Coulomb interactions are weak, the electrons with spin σ may independently move over the crystal with an average Hartree–Fock potential $U\langle n_{i-\sigma} \rangle$. Here $\langle n_{i-\sigma} \rangle$ is the average electron occupation number with spin $-\sigma$ on site i. In the strongly correlated electron system, the motion of electrons becomes slow, and therefore the electrons with spin σ, for example, may feel an additional potential U when the opposite-spin electron is on the same site. Hubbard regarded this system as an alloy with different random potentials $\epsilon_0 + U$ and ϵ_0 having the concentration $\langle n_{i-\sigma} \rangle$ and $1 - \langle n_{i-\sigma} \rangle$, respectively. This is called the alloy analogy approximation in the electron correlations [10]. The retarded Green function for the single particle excitations is then obtained as an average of impurity Green function embedded in the effective medium according to the CPA. The CPA theory is the first which explained the metal-insulator transition in solids [4].

In this lecture note, we review the recent development of quantum many-body theories using the CPA. The topics include the dynamical CPA, the many-body CPA, the dynamical mean-field theory (DMFT), and the mutual relation among them.

The dynamical CPA [11,12] is based on the functional integral method [13,14]. The method transforms an interacting electron system into an independent particle system with the time dependent random potential. It allows us to treat the electron correlations by means of the alloy analogy approximation using the coherent potential. The dynamical CPA is an extension of the single-site spin fluctuation theory (SSF) developed by Cyrot [15], Hubbard [16], and Hasegawa [17] to describe the local-moment behavior as well as the itinerant-electron behavior in metallic magnetism.

The many-body CPA, which was first proposed by Hirooka and Shimizu [18], is an extension of the one-body CPA for random alloys to the many-body case using the temperature Green function. The method indicates how to include the surrounding electron correlations in solids solving the CPA equation.

The DMFT was developed in the past decade by many investigators [19–24] in the infinite dimensional system. The key point here is that the self-energy of the interacting electrons becomes independent of momentum in infinite dimensions [19,25], so that one can determine the "coherent" self-energy by using the concept of the CPA.

The theories mentioned above are similar to each other in a sense that the coherent potential (or self-energy) is determined by the similar self-consistent condition. It is also one of the aims here to show that these theories are in fact equivalent to each other [26].

In the following section, we elucidate the theoretical framework of the dynamical CPA based on the functional integral method. Various approximation schemes to solve the CPA equation are presented in Sect. 3. In the past decade, the numerical investigations to the dynamical effects of charge and spin fluctuations on various quantities have been made. We discuss the dynamical effects obtained from these numerical calculations in Sect. 4. In Sect. 5, we introduce the many-body CPA based on the temperature Green function and show its equivalence to the dynamical CPA. In Sect. 6, the DMFT is presented on the basis of the path integral method. It is shown that the DMFT is equivalent to the many-body CPA. Section 7 is devoted to the summary and discussion of the future problem, in particular, the nonlocal effects on the self-energy.

2 Dynamical CPA

Hubbard's idea of alloy analogy approximation and the CPA to the many-body problem was developed by Cyrot [15] on the basis of the functional integral method. He derived the CPA equation for the half-filled band within the static and saddle-point approximation to the functional integral method, and obtained the phase diagram showing the metal-insulator transition as well as the antiferromagnetic transition on the $T-U$ plane, T being temperature. Hubbard [16] and Hasegawa [17] established the interpolation theory between the weak and strong Coulomb interaction limits within the high temperature approximation, which is called the single-site spin fluctuation theory (SSF). The SSF reduces to the Hartree–Fock approximation at the ground state. Therefore it does not describe the quantum many-body effects at low temperatures. Kakehashi [11] finally proposed the dynamical CPA which completely takes into account the quantum spin and charge fluctuations within the single-site approximation.

Let us consider the correlated electrons in solids described by a narrow-band model with on-site Coulomb interaction U, whose Hamiltonian is given by

$$\hat{H} = \hat{H}_0 + \hat{H}_I \ , \tag{1}$$

$$\hat{H}_0 = \sum_{i,\sigma} (\epsilon_0 - \mu)\, \hat{n}_{i\sigma} + \sum_{i,j,\sigma} t_{ij}\, \hat{a}^{\dagger}_{i\sigma} \hat{a}_{j\sigma} \ , \tag{2}$$

$$\hat{H}_I = \sum_i U\, \hat{n}_{i\uparrow} \hat{n}_{i\downarrow} \ . \tag{3}$$

Here $\hat{H}_0$ is the noninteracting Hamiltonian. $\hat{H}_I$ describes the intra atomic Coulomb interactions on each site. $\hat{a}^{\dagger}_{i\sigma}$ ($\hat{a}_{i\sigma}$) is the creation (annihilation) operator for an electron with spin σ on site i, and $\hat{n}_{i\sigma} = \hat{a}^{\dagger}_{i\sigma}\hat{a}_{i\sigma}$. Note that the chemical potential μ has been introduced for convenience into (2).

The functional integral method transforms the interacting Hamiltonian into an independent particle system in the time-dependent random fictitious fields by using the Hubbard–Stratonovich transformation [13,14,27].

$$\mathrm{e}^{A\hat{O}^2} = \sqrt{\frac{A}{\pi}} \int d\xi \mathrm{e}^{-A\xi^2 + 2A\hat{O}\xi} \ . \tag{4}$$

Here $\hat{O}$ stands for $i\hat{n}_i(\tau)$ or $\hat{m}_i(\tau)$ in the interaction representation, and $A = \beta U_i/(4N')$ denotes the interaction strength when the imaginary time τ between 0 and β has been discretized into N' intervals in the interaction representation. Here β is the inverse temperature.

The free energy $\mathcal{F}$ is then expressed by a functional integral with respect to the time-dependent exchange and charge fields $\{\xi_i(\tau)\}$ and $\{\eta_i(\tau)\}$ as

$$e^{-\beta\mathcal{F}} = \int \left[\prod_{i=1}^{N} \delta\xi_i \delta\eta_i \right] \exp\left(-\beta E[\xi, \eta] \right) \ , \tag{5}$$

$$E[\xi, \eta] = -\frac{1}{\beta} \ln \mathrm{Tr} \left(\mathcal{T} \exp \left[-\int_0^{\beta} \hat{H}_1(\xi, \eta, \tau) d\tau \right] \right) + \frac{1}{4\beta} \sum_i U \int_0^{\beta} \left[\xi_i(\tau)^2 + \eta_i(\tau)^2 \right] d\tau \ . \tag{6}$$

The functional integral in (5) is defined by

$$\int \delta\xi_i = \lim_{N' \to \infty} \int \left[\prod_{n=1}^{N'} \sqrt{\frac{\beta U_i}{4\pi N'}} d\xi_i(\tau_n) \right] \ . \tag{7}$$

Here τ_n is the n-th imaginary time. $\mathcal{T}$ in (6) denotes the time ordered product. $\hat{H}_1(\xi, \eta, \tau)$ is the time-dependent Hamiltonian defined by

$$\hat{H}_1(\xi, \eta, \tau) = \hat{H}_0 + \sum_{i,\sigma} v_{i\sigma}(\xi_i, \eta_i, \tau)\, \hat{n}_{i\sigma}(\tau) \ . \tag{8}$$

Here $v_{i\sigma}(\xi_i, \eta_i, \tau)$ is the time dependent random potential defined by

$$v_{i\sigma}(\xi_i, \eta_i, \tau) = -\frac{1}{2} U \left(i\eta_i(\tau) + \xi_i(\tau)\sigma \right) . \tag{9}$$

The form (8) indicates that the electrons move independently in the time dependent random potential $v_{i\sigma}(\xi, \eta, \tau)$, therefore the alloy analogy approximation (*i.e.* the CPA) is available in this system. In the dynamical CPA [11], we approximate the time-dependent random potential $\{v_{i\sigma}\}$ by means of the time dependent coherent potential $\Sigma_\sigma(\tau - \tau')$, so that the Hamiltonian (8) is replaced by the effective Hamiltonian.

$$\tilde{H}(\tau) = \hat{H}_0 + \sum_{i\sigma} \int_0^\beta d\tau' \hat{a}_{i\sigma}^\dagger(\tau) \Sigma_\sigma(\tau - \tau') \hat{a}_{i\sigma}(\tau') . \tag{10}$$

The coherent potential $\Sigma_\sigma(\tau - \tau')$ is determined in the same way as in Fig. 1. We consider an impurity system with the dynamical potential $v_{i\sigma}$ embedded in the coherent potential (see the left-hand-side of Fig. 2), whose Hamiltonian is given by

$$\begin{aligned} \hat{H}^{(i)}(\xi, \eta, \tau) &= \tilde{H}(\tau) \\ &+ \sum_\sigma \int_0^\beta d\tau' \hat{a}_{i\sigma}^\dagger(\tau) [v_{i\sigma}(\xi, \eta, \tau)\delta(\tau - \tau') - \Sigma_\sigma(\tau - \tau')] \hat{a}_{i\sigma}(\tau') . \end{aligned} \tag{11}$$

The impurity Green function to the time-dependent Hamiltonian (11) is given by

$$G_{i\sigma}^{(i)}(\xi, \eta, \tau, \tau') = -\frac{\mathrm{Tr}\left(\mathcal{T} \hat{a}_{i\sigma}(\tau) \hat{a}_{i\sigma}^\dagger(\tau') \exp\left(-\int_0^\beta \hat{H}^{(i)}(\xi, \eta, \tau'') d\tau'' \right) \right)}{\mathrm{Tr}\left(\mathcal{T} \exp\left(-\int_0^\beta \hat{H}^{(i)}(\xi, \eta, \tau'') d\tau'' \right) \right)} . \tag{12}$$

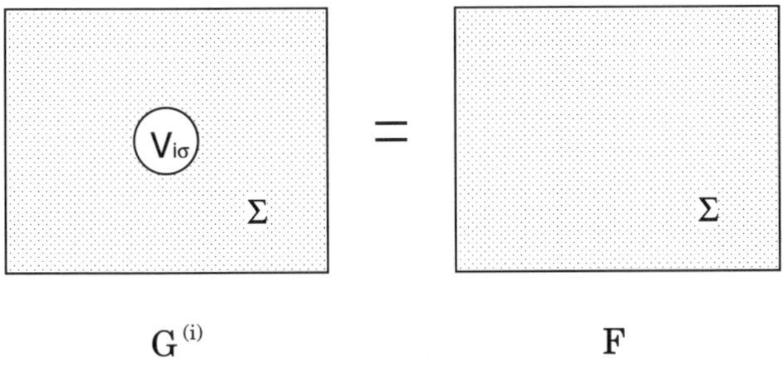

Fig. 2. Schematic picture showing the dynamical coherent potential approximation. The *left-hand-side* shows an impurity state with a dynamical potential $v_{i\sigma}$ on site i in the surrounding effective potential Σ_σ. The *right-hand-side* shows a uniform state with the effective potential

By solving the Dyson equation, it is expressed as

$$G^{(i)}_{i\sigma}(\xi,\eta,\tau,\tau') = \left[\left(F_i^{-1} - v_i + \Sigma\right)^{-1}\right]_{i\sigma i\sigma}(\tau,\tau') \ . \tag{13}$$

Here F_i is the site-diagonal coherent Green function to the Hamiltonian (10), which is expressed as $(F_i)_{j\sigma k\sigma}(\tau - \tau') = F_\sigma(\tau - \tau')\delta_{ji}\delta_{ki}\delta_{\sigma\sigma'}$, and $v_i - \Sigma$ is defined by the matrix element in the second term at the right-hand-side (r.h.s.) of (11).

According to the alloy analogy approximation, the Green function (13) should be identical with what was obtained from $\tilde{H}(\tau)$ when the average is taken over the random potential. (See Fig. 2.)

$$\left\langle G^{(i)}_{i\sigma}(\xi,\eta,\tau,\tau')\right\rangle = F_\sigma(\tau - \tau') \ . \tag{14}$$

This is the CPA equation in the dynamical CPA which determines the coherent potential. The average $\langle\ \rangle$ at the left-hand-side (l.h.s.) is taken on the impurity site with use of the weight $\exp(-\beta E^{(i)}[\xi,\eta])$. The impurity energy functional $E^{(i)}[\xi,\eta]$ is defined by (6) in which $\hat{H}_1(\xi,\eta,\tau)$ has been replaced by the impurity Hamiltonian (11). The Green function F_σ at the r.h.s. of (14), called the coherent Green function, is defined by (12) in which the impurity Hamiltonian $H^{(i)}(\xi,\eta,\tau'')$ has been replaced by the coherent one $\tilde{H}(\tau)$ (i.e. (10)). It is given in the Fourier representation as

$$F_\sigma(i\omega_l) = \int \frac{\rho(\epsilon)d\epsilon}{i\omega_l + \mu - \Sigma_\sigma(i\omega_l) - \epsilon} \ . \tag{15}$$

Here ω_l is the Matsubara frequency for Fermion, and $\rho(\epsilon)$ denotes the density of state for noninteracting Hamiltonian.

The free energy is expanded with respect to the deviation from the medium. Within the single-site approximation, it is given by

$$\mathcal{F}_{\mathrm{CPA}} = \tilde{\mathcal{F}} - \sum_i \beta^{-1}\ln \int \delta\xi_i\delta\eta_i \ \mathrm{e}^{-\beta E^{(i)}[\xi_i,\eta_i]} \ . \tag{16}$$

Here $\tilde{\mathcal{F}}$ is the free energy which depends on the effective medium only. $E^{(i)}[\xi_i,\eta_i]$ is the impurity energy functional in the effective medium.

3 Approximation Scheme to the Dynamical CPA

Solving the dynamical CPA equation (14) is not an easy problem because the time dependent Green function is given by an inverse of the infinite dimensional matrix as seen in (13), and secondly because there are still functional integrals to be implemented on the impurity site as seen in (14).

The simplest approximation to the dynamical CPA is to neglect the time dependence of the charge and spin field variables, which is valid in the high temperature limit because the time interval $[0,\beta]$ becomes zero in the limit. This is called the static approximation [28]. The impurity Green function then recovers the translational symmetry in time. The CPA equation (14) reduces in the frequency representation to

$$\left\langle G^{(i)}_{i\sigma}(\xi,\eta,i\omega_l)\right\rangle_{\mathrm{st}} = F_\sigma(i\omega_i) \;, \tag{17}$$

$$G^{(i)}_{i\sigma}(\xi,\eta,i\omega_l) = \left[F_\sigma(i\omega_l)^{-1} - v_{i\sigma}(\xi,\eta) + \Sigma_\sigma(i\omega_l)\right]^{-1} \;. \tag{18}$$

Here ξ (η) is defined by $\xi(i\omega_l=0)$ ($\eta(i\omega_l=0)$), and $v_{i\sigma}(\xi,\eta)$ is defined by the static value of $v_{i\sigma}$. The average $\langle\ \rangle_{\mathrm{st}}$ is taken with respect to the static energy $E^{(i)}_{\mathrm{st}}(\xi,\eta)$.

$$\langle(\sim)\rangle_{\mathrm{st}} = \frac{\displaystyle\int d\xi_i d\eta_i \;(\sim)\; \mathrm{e}^{-\beta E^{(i)}_{\mathrm{st}}(\xi_i,\eta_i)}}{\displaystyle\int d\xi_i d\eta_i \; \mathrm{e}^{-\beta E^{(i)}_{\mathrm{st}}(\xi_i,\eta_i)}} \;. \tag{19}$$

Equation (17) is identical to the CPA equation in the SSF by Hubbard [16] and
Hasegawa [17].

The static approximation is a high temperature approximation; it becomes worse at lower temperatures. The free energy reduces to the Hartree–Fock approximation at the ground state. This means that the electron correlations are not taken into account at the ground state. Physically the local electron correlations persist even at finite temperatures because the associated energy is higher than the typical energy scale of phase transitions in solids.

We can take into account the electron correlations missing in the static approximation using the variational approach [29]. Note that the free energy $\mathcal{F}_{\mathrm{CPA}}$ (16) is expressed by the effective potentials $E^{(i)}_{\mathrm{eff}}(\xi_i)$ projected onto the static exchange field $\{\xi_i\}$ as

$$\mathcal{F}_{\mathrm{CPA}} = \tilde{\mathcal{F}} - \beta^{-1}\sum_i \ln\int\sqrt{\frac{\beta U}{4\pi}}d\xi_i \,\exp\left[-\beta E^{(i)}_{\mathrm{eff}}(\xi_i)\right] \;. \tag{20}$$

Here the correction to the static energy $(E^{(i)}_{\mathrm{st}}(\xi_i))$, which is defined by $\langle E^{(i)}_{\mathrm{c}}(\xi_i)\rangle = \langle E^{(i)}_{\mathrm{eff}}(\xi_i)\rangle - \langle E^{(i)}_{\mathrm{st}}(\xi_i)\rangle$, reduces to the correlation energy at the ground state. Therefore one may adopt for it an analytic form of the ground state correlation energy $E^{(i)}_{\mathrm{c}}(\xi_i,\lambda)$. The variational parameter λ may be determined from the Feynman inequality [30] given by

$$\mathcal{F}_{\mathrm{CPA}} \le \mathcal{F}_{\mathrm{t}} + \sum_i \left\langle E^{(i)}_{\mathrm{eff}}(\xi_i) - E^{(i)}_{\mathrm{t}}(\xi_i,\lambda)\right\rangle_{\mathrm{t}} \;. \tag{21}$$

Here the trial free energy $\mathcal{F}_{\mathrm{t}}$ is given by (20) in which the effective potential has been replaced by the trial one $E^{(i)}_{\mathrm{t}}(\xi_i,\lambda) = E^{(i)}_{\mathrm{st}}(\xi_i) + E^{(i)}_{\mathrm{c}}(\xi_i,\lambda)$. The average $\langle\ \rangle_{\rm t}$ in (21) is defined by (19) in which $E_{\rm st}^{(i)}(\xi_i)$ has been replaced by $E_{\rm t}^{(i)}(\xi_i, \lambda)$. The coherent potential should be determined from the stationarity condition

$$\frac{\delta \mathcal{F}_{\rm t}}{\delta \Sigma_\sigma(\tau - \tau')} = 0 \ , \tag{22}$$

because the original CPA equation (14) is obtained from the stationarity condition to $\mathcal{F}_{\rm CPA}$.

The theory mentioned above is called the variational approach (VA) [29]. Apparently, the VA is an adiabatic approximation because we assumed that the form $E_{\rm c}^{(i)}(\xi_i, \lambda)$ does not change with increasing temperature. It does not necessarily improve the entropy at finite temperatures, though it does the energy.

To go beyond the adiabatic approximation, one has to take into account directly the dynamical effects. The free energy is expressed in general by means of the effective potential $E(\xi, \eta)$, which is projected onto the static fields ξ and η, as

$$\mathcal{F}_{\rm CPA} = \tilde{\mathcal{F}} - \beta^{-1} \ln \int \sqrt{\frac{\beta U}{4\pi}} d\xi \sqrt{\frac{\beta U}{4\pi}} d\eta \ e^{-\beta E(\xi,\eta)} \ . \tag{23}$$

Here we assumed for brevity that all the sites are equivalent, so that $\mathcal{F}_{\rm CPA}$ and $\tilde{\mathcal{F}}$ stand for the free energy per site.

The effective potential $E(\xi, \eta)$ consists of the static part $E_{\rm st}(\xi, \eta)$ and the dynamical part $E_{\rm dyn}(\xi, \eta)$. The latter is given by the Gaussian average of the determinant D_σ for the dynamical scattering matrix as follows

$$e^{-\beta E_{\rm dyn}(\xi,\eta)} = \int \left[\prod_{l=1}^{\infty} \frac{\beta U}{2\pi} d^2\xi_l \frac{\beta U}{2\pi} d^2\eta_l \right] D_\uparrow D_\downarrow \exp\left[-\frac{\beta U}{2} \sum_{l=1}^{\infty} (|\eta_l|^2 + |\xi_l|^2) \right] \ , \tag{24}$$

$$D_\sigma = \det\left[\delta_{lm} - \tilde{v}_\sigma(i\omega_l - i\omega_m)\tilde{g}_\sigma(i\omega_m)\right] \ . \tag{25}$$

Here $\xi_l(\eta_l)$ is defined by $\xi(i\omega_l)(\eta(i\omega_l))$. $\tilde{v}_\sigma(i\omega_l)$ is the dynamical potential in which the static potential $v_\sigma(0)\delta_{l0}$ has been subtracted from $v_\sigma(i\omega_l)$. $\tilde{g}_\sigma(i\omega_m)$ is the diagonal Green function in the static approximation.

The Monte-Carlo technique allows us to calculate statistically the dynamical part via (24) and to solve the CPA equation numerically. It is called the Monte-Carlo dynamical CPA [11]. The Monte-Carlo method is a useful method for examining an overall feature of the theory, varying the Coulomb interaction at finite temperatures. At low temperatures, accurate calculations become more difficult because the range of imaginary time $[0, \beta]$ becomes wider.

A more analytic way which does not rely on the statistical method is to adopt the harmonic approximation [31,32] to the calculation of the dynamical part [12]. In the harmonic approximation, we expand the determinant of the dynamical scattering matrix with respect to the frequency modes of the dynamical potential, and take into account the lowest order correction, which is expressed by the sum of the individual frequency contributions

$$D_\sigma = 1 + \sum_\nu (D_{\nu\sigma} - 1) + \cdots \ , \tag{26}$$

$$D_{\nu\sigma} = \det \left[\delta_{lm} - (\tilde{v}_\sigma(i\omega_\nu)\delta_{l-m,\nu} + \tilde{v}_\sigma(i\omega_{-\nu})\delta_{l-m,-\nu})\tilde{g}_\sigma(i\omega_m) \right] . \tag{27}$$

The harmonic approximation interpolates between the weak and strong Coulomb interaction limits, and allows us to perform the numerical calculations over the wide ranges of the occupation number, the Coulomb interaction, and the temperature.

4 Dynamical Effects on Various Properties

The development of the dynamical CPA in the past decade has allowed us to examine the effects of the dynamical charge and spin fluctuations. Numerical investigations revealed that the dynamical effects play an important role on various properties in the intermediate region of the Coulomb interaction. In particular, it has been clarified that the dynamical CPA explains the localized as well as the delocalized features of charge and spins in solids on the same footing.

Figure 3 shows the momentum distributions for a half-filled narrow band model calculated by both the dynamical and static CPA at finite temperatures. The static approximation causes a random static exchange potential called the disordered local moments in the paramagnetic state at high temperatures. The random potential causes a broad momentum distribution. The quantum charge and spin fluctuations due to the dynamical potentials, however, suppress the disordered local moment behavior in the static approximation and produce the Fermi-liquid like curve with a steep slope at the Fermi level, as seen in Fig. 3.

The dynamical effects also influence the magnetic properties. Figure 4 shows the result of the model calculations for the temperature variation of magnetic moments and susceptibility for Fe. Calculated magnetization temperature curves are close to the Brillouin curve of the local moment model in agreement with the experiment [33]. Both the static and dynamical CPA lead to the Curie Weiss susceptibility. The Curie temperature, however, is reduced by a factor of two due to the dynamical spin and charge fluctuations.

The dynamical CPA describes the many-body effects in the single-particle excitations. In Fig. 5, we show the results of the model calculations for Ni. The static approximation broadens the band width because of thermal spin fluctuations. The dynamical effects suppress the thermal spin fluctuations, and shrink the main band near the Fermi level by about 15% as compared

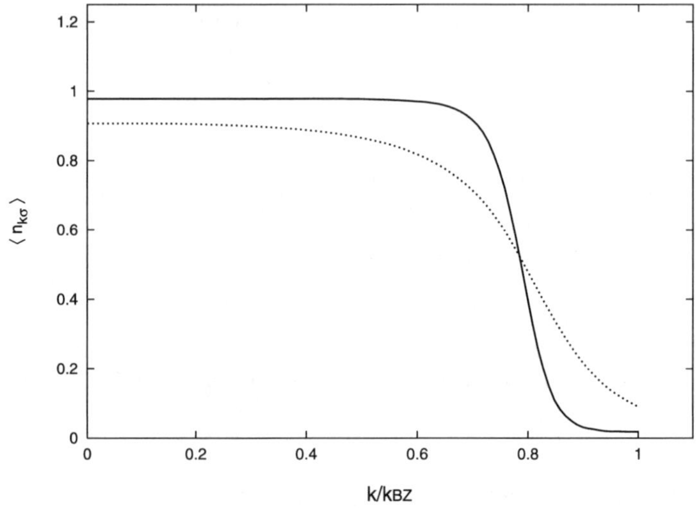

Fig. 3. Momentum distributions for the half-filled band Hubbard model calculated by the Monte-Carlo dynamical CPA (*solid curve*) and the static approximation (*dotted curve*) at $U/W = 0.4$ and $T/W = 0.05$ [11]. The rectangular density of states with the band width W is adopted for the noninteracting system

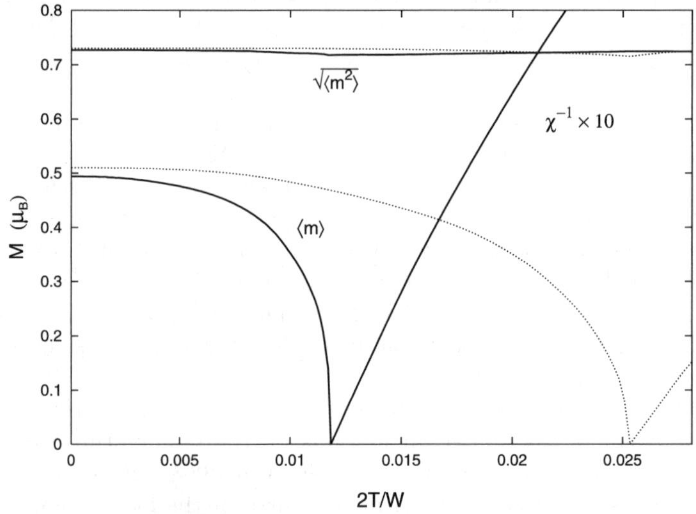

Fig. 4. Magnetization vs temperature curves, the inverse susceptibility curves, and the curves of amplitude of local moment obtained by the dynamical CPA (*solid curves*) and the static approximation (*dotted curves*) [12]

with that of the noninteracting system. The dynamical effects also create a satellite peak below the main band as seen in Fig. 5. The results are consistent with those obtained from the ground-state theories [34–37]. The satellite

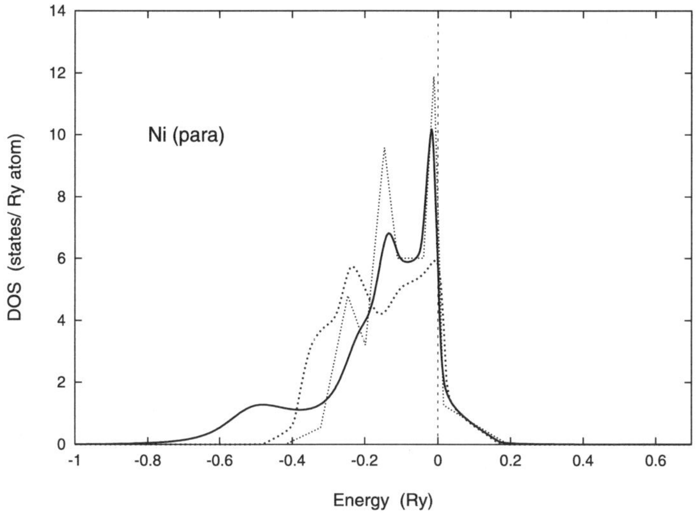

Fig. 5. Calculated densities of states in the dynamical CPA (*solid curve*) and the static approximation (*dashed curve*) for Ni. The noninteracting density of states is shown by the *dotted curve* [12]

peak corresponds to the 6 eV satellite in the photoemission experiment of Ni [38–40]. These results indicate that the dynamical spin and charge fluctuations suppress the thermal fluctuations so that the dynamical CPA describes the quantum many-body aspects on the electronic and magnetic properties in the intermediate region of the Coulomb interaction strength.

5 Many-Body CPA

The CPA to the binary alloy problem, as has been mentioned in the introduction, was extended to the many-body case by using the temperature Green function by Hirooka and Shimizu [18]. It is called the many-body CPA. The selfconsistent CPA equation to the coherent potential is nontrivial even for the pure metal when the electron correlations are taken into account, because the frequency dependent coherent potential describes the local electron correlations on the surrounding sites self-consistently.

Let us consider a pure metal with the intra-atomic Coulomb interactions to clarify the basic feature of the many-body CPA. The Green function for the system is given in the interaction representation as

$$G_{i\sigma}(\tau - \tau') = -Z^{-1}\,\mathrm{Tr}\left[\mathcal{T}\hat{a}_{i\sigma}(\tau)\hat{a}^{\dagger}_{i\sigma}(\tau')\,\mathrm{e}^{-\int_0^\beta \hat{H}(\tau'')d\tau''}\right] . \tag{28}$$

Here Z denotes the partition function of the system.

In the many-body CPA, the Hamiltonian $\hat{H}(\tau)$ in the interaction representation in (28) is now approximated by the effective Hamiltonian (10) with the time dependent coherent potential. An approximate Green function is then given by the coherent Green function $F_\sigma(\tau - \tau')$ (see (15)) according to the Dyson equation. (Note that the Dyson equation for the Hamiltonian (10) has the same form as obtained from the Green function (28) after the single-site approximation.) To choose the best coherent potential, we consider an impurity Hamiltonian embedded in the effective medium

$$\hat{H}^{(i)}(\tau) = \tilde{H}(\tau) - \int_0^\beta d\tau' \sum_\sigma \hat{a}^\dagger_{i\sigma}(\tau) \Sigma_\sigma(\tau - \tau') \hat{a}_{i\sigma}(\tau') + U\, \hat{n}_{i\uparrow}(\tau) \hat{n}_{i\downarrow}(\tau) \ . \tag{29}$$

Here the second and the third terms at the r.h.s. means that the coherent potential on the impurity site i has been replaced by the real Coulomb interaction $U\, \hat{n}_{i\uparrow}(\tau)\hat{n}_{i\downarrow}(\tau)$. Note that (29) corresponds to (11) in the dynamical CPA, but the time dependent random potential is now replaced by the real interaction in (29).

Solving the Dyson equation to the impurity Hamiltonian (29) we obtain the Green function

$$G^{(i)}_{i\sigma}(i\omega_l) = \left[F_\sigma(i\omega_l)^{-1} - \Lambda^{(i)}_\sigma(i\omega_l) + \Sigma_\sigma(i\omega_l) \right]^{-1} \ . \tag{30}$$

Here $\Lambda^{(i)}_\sigma(i\omega_l)$ is the self-energy due to the Coulomb interaction on the impurity site i. The coherent potential in the many-body CPA is determined so that the diagonal impurity Green function agrees with the coherent Green function for the effective medium

$$G^{(i)}_{i\sigma}(i\omega_l) = F_\sigma(i\omega_l) \ . \tag{31}$$

Note that the self-energy $\Lambda^{(i)}_\sigma(i\omega_l)$ has to be obtained separately by using one of the many-body theories. Hirooka and Shimizu adopted the self-energy in the low density approximation.

The many-body CPA is equivalent to the dynamical CPA [26]. In fact, applying the Hubbard–Stratonovich transformation (4) directly to the Green function of the impurity Hamiltonian (29), we obtain the relation

$$G^{(i)}_{i\sigma}(\tau - \tau') = \left\langle G^{(i)}_{i\sigma}(\xi, \eta, \tau, \tau') \right\rangle \ . \tag{32}$$

The above relation manifests that the CPA equation (14) in the dynamical CPA is identical with that in the many-body CPA, i.e. (31).

6 Dynamical Mean-Field Theory

In the last decade, the many-body theory in infinite dimensions has been significantly developed [41]. There, the hopping integrals $|t_{ij}|$ for example on the hyper-cubic lattice are scaled as $|t|/\sqrt{2d}$ so that the band width becomes finite for any dimensions d. Accordingly, the off-diagonal Green function in the noninteracting system becomes of order of $1/\sqrt{2d}$. Since the Feynman diagrams to the off-diagonal self-energy have at least three electron lines between the different sites, the contribution from all the off-diagonal self-energy diagrams to the Dyson equation becomes of order of $2d(1/\sqrt{2d})^3 = 1/\sqrt{2d}$. Therefore the contribution from the off-diagonal self-energy $\Sigma_{ij\sigma}(i\omega_l)(i \neq j)$ vanishes in the limit $d = \infty$.

The dynamical mean-field theory (DMFT) determines the site-diagonal self-energy (or the momentum independent self-energy) $\Sigma_\sigma(i\omega_l)$ as follows. The diagonal Green function in the path integral representation is assumed to be given by an impurity action S_i with an effective field and the local Coulomb interaction [41]

$$G^{(i)}_{i\sigma}(\tau - \tau') = -Z^{(i)-1} \int \left[\prod_\sigma \mathcal{D}a^*_{i\sigma} \mathcal{D}a_{i\sigma} \right] \mathrm{e}^{-S_i} a_{i\sigma}(\tau) a^*_{i\sigma}(\tau') \;, \tag{33}$$

$$\begin{aligned} S_i = &- \int_0^\beta d\tau \int_0^\beta d\tau' \sum_\sigma a^*_{i\sigma}(\tau) (F^{(i)-1})_\sigma(\tau - \tau') a_{i\sigma}(\tau) \\ &+ \int_0^\beta d\tau \, U n_{i\uparrow}(\tau) n_{i\downarrow}(\tau) \;. \end{aligned} \tag{34}$$

Here $Z^{(i)}$ denotes the partition function to the action S_i. $a^*_{i\sigma}(\tau)$ and $a_{i\sigma}(\tau)$ are the Grassmann variables being conjugate to the creation and annihilation operators, $\hat{a}^\dagger_{i\sigma}(\tau)$ and $\hat{a}_{i\sigma}(\tau)$, and $n_{i\sigma}(\tau) = a^*_{i\sigma}(\tau) a_{i\sigma}(\tau)$. $\mathcal{D}a^*_{i\sigma}\mathcal{D}a_{i\sigma}$ denotes the path integrals for these variables. $F^{(i)}_\sigma(\tau - \tau')$ in (34) is called the Weiss field function. According to the Dyson equation to the Green function (33) it is given by

$$F^{(i)}_\sigma(i\omega_l)^{-1} = \Sigma_\sigma(i\omega_l) + G^{(i)}_{i\sigma}(i\omega_l)^{-1} \;. \tag{35}$$

Since the site-diagonal Green function $G^{(i)}_{i\sigma}(i\omega_l)$ should be equal to the average of the Green function of the momentum representation $(1/(i\omega_l + \mu - \Sigma_\sigma(i\omega_l) - \epsilon_k))$, we have the relation,

$$G^{(i)}_{i\sigma}(\tau - \tau') = \int \frac{\rho(\epsilon) d\epsilon}{i\omega_l + \mu - \Sigma_\sigma(i\omega_l) - \epsilon} \;. \tag{36}$$

This means that

$$i\omega_l + \mu - \Sigma_\sigma(i\omega_l) = R\left[G^{(i)}_\sigma(i\omega_l)\right] \;. \tag{37}$$

Here R denotes the reciprocal function to the Hilbert transform. Eliminating $\Sigma_\sigma(i\omega_l)$ from (35) and (37), we obtain the expression of the Weiss function by means of the impurity Green function as

$$F^{(i)}_\sigma(i\omega_l)^{-1} = i\omega_l + \mu + G^{(i)}_\sigma(i\omega_l)^{-1} - R\left[G^{(i)}_\sigma(i\omega_l)\right] \;. \tag{38}$$

Equations (33), (34), and (38) form the selfconsistent equation in the DMFT. Solving the impurity problem ((33) and (34)) and the Weiss effective field selfconsistently via (38), the metal-insulator transition in the $d = \infty$ limit has been clarified [42,43]. One of the new features is that the quasiparticle peak at the Fermi level remains until the metal-insulator transition occurs with increasing Coulomb interaction (see Fig. 6).

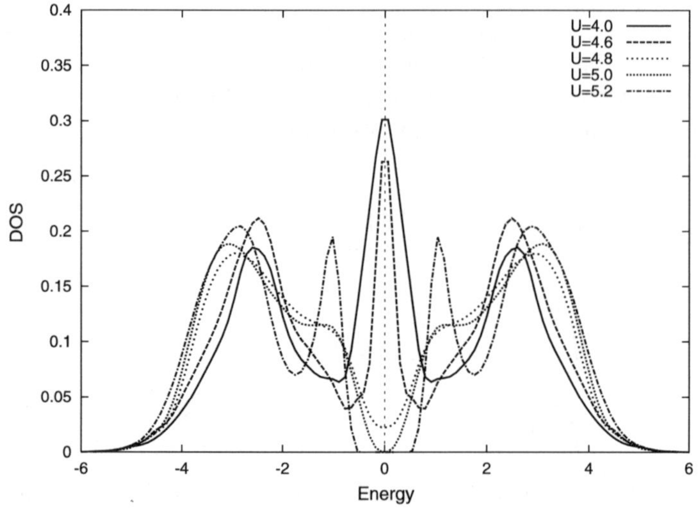

Fig. 6. Densities of states for the half-filled band Hubbard model on the Bethe lattice in the infinite dimensions at $T = 1/20$ [42]

Detailed analysis at the ground state shows that the self-energy follows the Fermi-liquid theory [41]; the self-energy (*i.e.* the coherent potential) near the Fermi level is expressed for the half-filled band as

$$\Sigma_\sigma(\omega + i\delta) = \frac{1}{2}U + \left(1 - \overline{Z}^{-1}\right)\omega - i\gamma\omega^2 + \dots \; . \tag{39}$$

The quasiparticle weight $\overline{Z}$ defined by $(1 - \partial \mathrm{Re}\Sigma_\sigma(0^+)/\partial\omega)^{-1}$ vanishes at the critical Coulomb interaction above which the insulator gap appears. It should be noted that Hubbard's theory for the metal-insulator transition does not describe the Fermi-liquid behavior in the metallic region [10].

The DMFT is equivalent to the many-body CPA [26]. To see this fact, we start from the temperature Green function expressed by the path integral.

$$G_{i\sigma}(\tau - \tau') = -Z^{-1}\int \left[\prod_{j\sigma} \mathcal{D}a^*_{j\sigma}\mathcal{D}a_{j\sigma}\right] \mathrm{e}^{-S} a_{i\sigma}(\tau)a^*_{i\sigma}(\tau) \; , \tag{40}$$

$$S = \int_0^\beta d\tau \left[\sum_{i\sigma} a_{i\sigma}^*(\tau) \left(\frac{\partial}{\partial \tau} \right) a_{i\sigma}(\tau) + H(\{a^*\}\{a\}) \right] . \tag{41}$$

Here the Hamiltonian operator $H(\{a^*\}\{a\})$ in the action has the same form as the Hamiltonian in the interaction representation (see (28)). Therefore we can construct the many-body CPA using the path integral method taking the same steps as in Sect. 5.

The CPA equation in the path integral formulation is given by

$$G_\sigma^{(i)}(i\omega_l) = F_\sigma(i\omega_l) . \tag{42}$$

The r.h.s. in (42) is the coherent Green function defined by

$$F_\sigma(\tau - \tau') = -\tilde{Z}^{-1} \int \left[\prod_{j\sigma} \mathcal{D}a_{j\sigma}^* \mathcal{D}a_{j\sigma} \right] \mathrm{e}^{-\tilde{S}} a_{i\sigma}(\tau) a_{i\sigma}^*(\tau') . \tag{43}$$

Here $\tilde{Z}$ is the partition function to the coherent action $\tilde{S}$ described by the coherent Hamiltonian operator $\tilde{H}(\{a^*\}\{a\})$. It is defined by

$$\tilde{H}(\{a^*\}\{a\}) = H_0(\{a^*(\tau)\}\{a(\tau)\}) + \sum_{i\sigma} \int_0^\beta d\tau' a_{i\sigma}^*(\tau) \Sigma_{i\sigma}(\tau - \tau') a_{i\sigma}(\tau') . \tag{44}$$

The coherent Green function given by (43) is identical with that in the many-body CPA (15) because both of them satisfy the same Dyson equations.

The impurity Green function at the l.h.s. in (42) is

$$G_\sigma^{(i)}(\tau - \tau') = -Z^{(i)-1} \int \left[\prod_{j\sigma} \mathcal{D}a_{j\sigma}^* \mathcal{D}a_{j\sigma} \right] \mathrm{e}^{-S^{(i)}} a_{i\sigma}(\tau) a_{i\sigma}^*(\tau') . \tag{45}$$

Here $Z^{(i)}$ is the partition function to the action $S^{(i)}$ which is described by the impurity Hamiltonian operator given by

$$\begin{aligned} H^{(i)}(\{a^*\}\{a\}) = \tilde{H}(\{a^*\}\{a\}) &- \int_0^\beta d\tau' \sum_\sigma a_{i\sigma}^*(\tau) \Sigma_\sigma(\tau - \tau') a_{i\sigma}(\tau') \\ &+ U\, a_{i\uparrow}^*(\tau) a_{i\downarrow}^*(\tau) a_{i\downarrow}(\tau) a_{i\uparrow}(\tau) . \end{aligned} \tag{46}$$

Because the impurity Green function (45) follows the same Feynman diagram and the Dyson equation as in the many-body CPA, it is identical with the impurity Green function in the many-body CPA (30). This means that the CPA equation (42) is identical with that in the many-body CPA (31).

We can derive the impurity Green function in the DMFT (33) from (45) performing the functional integrals at all the sites except the impurity one. The Weiss field function $F_\sigma^{(i)}$ in (33) is then obtained as

$$F_\sigma^{(i)}(i\omega_l)^{-1} = F_\sigma(i\omega_l)^{-1} + \Sigma_\sigma(i\omega_l) . \tag{47}$$

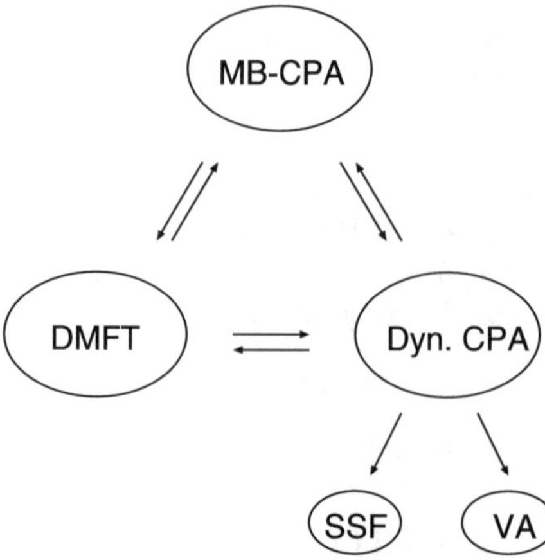

Fig. 7. Schematic diagram showing the equivalence among the many-body CPA (MB-CPA), the dynamical CPA (Dyn. CPA), and the dynamical mean-field theory (DMFT). The single-site spin fluctuation theory (SSF) [16,17] is a high-temperature approximation to the dynamical CPA, and the variational approach (VA) [29] is an adiabatic approximation to the dynamical CPA

Using the reciprocal representation of (15) and the CPA equation (42), we can derive the selfconsistent condition (38) in the DMFT from (47), which completes the proof of the equivalence between the DMFT and the many-body CPA.

Figure 7 shows a summary of the relationship among various many-body theories using the CPA. The dynamical CPA, the many-body CPA, and the DMFT are equivalent to each other. The SSF is obtained from the dynamical CPA as a high temperature approximation, and the VA is an adiabatic approximation to the dynamical CPA. These relations indicate that the interpolation theories developed in the magnetism and those developed in the strongly correlated electron system are unified within the single-site approximation.

7 Summary and Discussion

We have surveyed how the concept of the coherent potential approximation (CPA) has been used in the many-body problem in solids. In the dynamical CPA, the correlated motion of electrons is described by the time dependent random charge and exchange potentials for the independent electron system. The random potentials are replaced by the coherent potential in the spirit of the alloy analogy approximation. The theory was developed in the early stage within the high temperature approximation (*i.e.* the static approximation). It is called the single-site spin fluctuation theory (SSF). The theory including the dynamical effects was developed in the past decade.

The dynamical effects were investigated numerically by using the Monte-Carlo method and the harmonic approximation. One of the important features is that the disordered local moment behavior caused by the thermal spin fluctuations is suppressed by the dynamical spin and charge fluctuations and it is replaced by the characteristic behavior of the Fermi liquid and the quantum spin fluctuations in the correlated electron system. Furthermore it was demonstrated that the dynamical CPA describes both the itinerant and the localized features of electrons on the same footing because of its interpolative nature.

The many-body CPA approximates the real Coulomb interactions by means of the single-site coherent potential in the calculation of the temperature Green function. The latter is self-consistently obtained so that the many-body Green function for an impurity embedded in the effective medium is identical with the coherent Green function.

In the dynamical mean-field theory (DMFT), the momentum independent self-energy (*i.e.* the coherent potential) is identified with the impurity self-energy calculated from an impurity action with the Weiss effective field. The latter is determined so that the impurity Green function calculated from the impurity action should be identical with that obtained by averaging the Green function in the **k** representation.

We have clarified that the dynamical CPA, the many-body CPA, and the DMFT are equivalent to each other. The frequency-dependent coherent potential in these theories takes into account best the local electron correlations on the surrounding sites solving a many-body impurity problem embedded in the effective medium and the CPA equation. Such a self-consistency is significant in solids, in particular, for the description of the phase transitions caused by the Coulomb interactions (e.g. magnetism and the metal-insulator transition).

The theories which we discussed here are mostly based on the temperature Green function. It is also possible to construct a many-body CPA on the basis of the retarded Green function. Hubbard's original CPA theory is the first along this line. The development of the CPA theory based on the projection operator method [1] using the retarded Green function might be of value towards realistic calculations, because the method allows us to perform the first principles calculations on the excitation problem.

It is obvious that the many-body theories based on the CPA neglecting the momentum dependence of the self-energy, do not describe the nonlocal correlations of electrons. An extension of the coherent potential $\Sigma_\sigma(i\omega_l)$ to the cluster $\{\Sigma_{ij\sigma}(i\omega_l)\}$, is called the cluster CPA, and it enables us to describe the short-range intersite electron correlations. A large scale computer calculation has been performed recently for the 2 dimensional system [44]. The application to the 3 dimensional system, however, is not practical because of more difficulty in the numerical calculations. The incremental method [1] may be more efficient for taking into account the inter-site correlations. The method allows us to take into account the pair-site correlations, the triple-site correlations, and the higher order cluster correlations successively starting from the single-site approximation. Examining the range and the size of correlations at each step, one can describe the intersite correlations efficiently. The application of the method combined with the CPA is left for the future investigations.

Acknowledgements

The author would like to thank Professor P. Fulde for valuable discussions on the present subject.

References

1. See for example, P. Fulde: *Electron Correlations in Molecules and Solids* (Springer, Berlin, 1995)
2. P. Soven: Phys. Rev. **156**, 809 (1967); **178**, 1136 (1969)
3. D.W. Taylor: Phys. Rev. **156**, 1017 (1967)
4. B. Velický, S. Kirkpatrick, and H. Ehrenreich: Phys. Rev. **175**, 747 (1968)
5. F. Yonezawa: Prog. Theor. Phys. **40**, 734 (1968)
6. R.J. Elliott, J.A. Krumhansl, and P.L. Leath: Rev. Mod. Phys. **46**, 465 (1974)
7. H. Ehrenreich and L.M. Schwartz: Solid State Physics, edited by H. Ehrenreich, F. Seitz, and D. Turnbull (Academic, New York, 1980), Vol. 30
8. M. Jarrell and H.R. Krishnamurthy: Phys. Rev. B **63**, 125102 (2001)
9. H. Shiba: Prog. Theor. Phys. **46**, 77 (1971)
10. J. Hubbard: Proc. Roy. Soc. (London) **A281**, 401 (1964)
11. Y. Kakehashi: Phys. Rev. B **45**, 7196 (1992); J. Magn. Magn. Mater. **104–107**, 677 (1992)
12. Y. Kakehashi: Phys. Rev. B **65**, 184420 (2002)
13. J. Hubbard: Phys. Rev. Lett. **3**, 77 (1959)
14. R.L. Stratonovich: Dokl. Akad. Nauk. SSSR **115**, 1097 (1958) [Sov. Phys. – Dokl. **2**, 416 (1958)]
15. M. Cyrot: J. Phys. (Paris) **33**, 25 (1972)
16. J. Hubbard: Phys. Rev. B **19**, 2626 (1979); **20**, 4584 (1979); **23**, 5974 (1981)
17. H. Hasegawa: J. Phys. Soc. Jpn. **46**, 1504 (1979); **49**, 178 (1980)
18. S. Hirooka and M. Shimizu: J. Phys. Soc. Jpn. **43**, 70 (1977)
19. E. Müller-Hartmann: Z. Phys. B **74**, 507 (1989)
20. U. Brandt and C. Mielsch: Z. Phys. B **75**, 365 (1989); **79**, 295 (1991); **82**, 37 (1991)
21. V. Janis: Phys. Rev. B **40**, 11331 (1989); Z. Phys. B **83**, 227 (1991)
22. M. Jarrell: Phys. Rev. Lett. **69**, 168 (1992)
23. A. Georges and G. Kotliar: Phys. Rev. B **45**, 6479 (1992)
24. F.J. Ohkawa: Phys. Rev. B **46**, 9016 (1992)
25. W. Metzner and D. Vollhardt: Phys. Rev. Lett. **62**, 324 (1989)
26. Y. Kakehashi: Phys. Rev. B **66**, 104428 (2002)
27. G. Morandi, E. Galleani D'Agliano, F. Napoli, C.F. Ratto: Adv. Phys. **23**, 867 (1974)

28. S.Q. Wang, W.E. Evanson, and J.R. Schrieffer: Phys. Rev. Lett. **23**, 92 (1969); J. Appl. Phys. **41**, 1199 (1970)
29. Y. Kakehashi and P. Fulde: Phys. Rev. B **32**, 1595 (1985)
30. R.P. Feynman: *Statistical Mechanics* (W.A. Benjamin, Inc., London 1972), Chap. 8
31. D.J. Amit and C.M. Bender: Phys. Rev. B **4**, 3115 (1971)
32. D.J. Amit and H. Keiter: J. Low Temp. Phys. **11**, 603 (1973)
33. R.M. Bozorth: *Ferromagnetism* (Van Nostrand, Princeton, 1968)
34. D.R. Penn: Phys. Rev. Lett. **42**, 921 (1979)
35. A. Liebsch: Phys. Rev. Lett. **43**, 1431 (1979)
36. R.H. Victora and L.M. Falicov: Phys. Rev. Lett. **55**, 1140 (1985)
37. P. Unger, J. Igarashi, and P. Fulde: Phys. Rev. B **50**, 10485 (1994)
38. D.E. Eastman, F.J. Himpsel, and J.A. Knapp, Phys. Rev. Lett. **40**, 1514 (1978); F.J. Himpsel, J.A. Knapp, and D.E. Eastman: Phys. Rev. B **19**, 2919 (1979)
39. W. Eberhardt and E.W. Plummer: Phys. Rev. B **21**, 3245 (1980)
40. H. Martensson and P.O. Nilsson: Phys. Rev. B **30**, 3047 (1984)
41. A. Georges, G. Kotliar, W. Krauth, and M.J. Rosenberg: Rev. Mod. Phys. **68**, 13 (1996)
42. J. Schlipf, M. Jarrell, P.G. van Dongen, N. Blümer, S. Kehrein, Th. Pruschke, and D. Vollhardt: Phys. Rev. Lett. **82**, 4890 (1999)
43. R. Bulla: Phys. Rev. Lett. **83**, 136 (1999)
44. M. Jarrell, Th. Maier, C. Huscroft, and S. Moukouri: Phys. Rev. B. **64**, 195130 (2001)

Cluster Transfer Matrix Method: A Novel Many-Body Technique in Nanoelectronics

Sung Chung

Abstract. The cluster transfer matrix method is introduced for the single electron box connected by a tunnel barrier to the environmental lead. After integrating out electron degrees of freedom of background electrons, the effective action for a phase contains the kinetic term due to the Coulomb energy and the long-range interaction in the temperature space, representing dissipation due to electron tunneling through the barrier. The path-integral expression for the partition function is exactly calculated by the newly developed cluster transfer matrix method. We demonstrate that the method is as powerful as Monte Carlo and renormalization group methods. We particularly emphasize that the method is systematic and exact. Application of the new method to the single Josephson junction, particularly the superconductor-insulator transition, will then be discussed. Finally we will briefly discuss its potential generalization to dynamics, i.e. to calculate the exact current-voltage characteristics in nano-structures such as single electron transistor and quantum dot.

1 Introduction

The study of low-dimensional systems and nanostructures has been exploding. Its technological importance is certainly a driving force, but it is also quite challenging theoretically. For theorists it is an ideal playground for studying correlations and cooperative phenomena of many-particle systems. Since the discovery of high temperature superconductors, the strong correlation has become quite a practical issue, not just an interest of a few theorists. A recent discovery of the metal-insulator transition in a clean two-dimensional electron gas is clear evidence that we still know very little about correlations in many-particle systems.

The very definition of strong correlation to the author is that the standard perturbative or mean-field approximations are questionable there. The good quantum basis based on which we confidently apply perturbation theories and mean-field theories emerges only after we solve the problem. Novel many-body formalism armed with powerful computers are certainly one of the directions theorists are pursuing.

This paper introduces a novel nonperturbative method, cluster transfer matrix (CTM) for nanostructures such as single electron transistor (SET), quantum dot (QD) and Josephson junction (JJ) arrays. This method has met a success in the analysis of single electron box (SEB) and proved to be as powerful as Monte Carlo and real-time renormalization group (RG) method. Concerning this new method, two points may be worth emphasizing. First, it is systematic and exact as far as it converges with the cluster size. Second, the method provides a means to drastically accelerate the convergence, enabling us to reach a fairly wide parameter space. In the next section, the CTM method for thermodynamic, imaginary time formalism, will be presented for the SEB.

Nanostructures involving superconductors such as JJ arrays are equally technologically important, but an additional theoretical challenge is to calculate, in mechanical analogy, the escape rate of a particle out of a metastable potential well, a long-standing theoretical issue called quantum Kramers rate. We have recently proposed a Landauer-like formula for the escape rate which well outperformed the quantum transient state theory. In Sect. 3, the CTM method will be applied to the single JJ, and we particularly discuss the superconductor-insulator transition. With this Landauer-like formula, one can calculate the IV characteristics of single JJ from its thermodynamics. The experimental results on single JJ at temperature around 80 mK agree well with this work. But a highly nontrivial prediction of this work is that the single JJ becomes all superconducting at sufficiently low temperature as long as dissipation is present. On the other hand, two-dimensional JJ arrays clearly undergo the superconductor-insulator (SI) transition. An important future work in JJ arrays will then be to see how correlation develops and cooperative phenomenon emerges with the array size.

An obvious challenge to the newly developed CTM method is its possible generalization to dynamics such as to calculate the IV characteristics of SET and QD. The existing methods are perturbative, real-time RG, and Monte Carlo in imaginary time with analytic continuation to real time by means such as maximum entropy method and SVD (singular-value-decomposition). However, success so far along these lines is limited, and a successful generalization of the CTM method to dynamics would give a definitive answer to the issue. In Sect. 4, we will discuss some initial success along this line. A conclusion follows in Sect. 5.

2 CTM Method

The CTM method has been developed for the SEB and applied recently to a single JJ. An account of these works follows in this section and in Sect. 3.

The SEB is the most elementary device [1]. It consists of a metallic island connected to an outside lead by a tunnel junction and coupled to a gate voltage. Gate voltage induces a continuous background electrical charge n_x in the units of electronic charge e. Now if n electrons occupy the metallic ball, the associated electro-static energy is proportional to $(n - n_x)^2$. When this energy dominates the situation, then discreteness of electron numbers 0, 1, 2, ... will show up in experimental observations. This phenomenon is called the Coulomb blockade. The controlling parameter of the system is n_x, temperature T, and g which characterizes the strength of electron transfer through the tunnel junction connecting the metallic ball to the lead. Theoretical studies so far have been based on perturbation theories [2], RG [3], the instanton method [4], scaling theories [5], and Monte Carlo simulations [6,7]. However, these theories either contain uncontrolled assumptions or their accuracies are not quite clear. Also the two Monte Carlo calculations are with not so small error bars and disagree with each other. The CTM method is for arbitrary temperature T, g, and n_x [8].

The Hamiltonian is

$$H = \frac{1}{2}(n - n_x)^2 + \sum_{ab} t_{ab} C_a^+ C_b + h.c. + H_0 \ , \tag{1}$$

where the Coulomb energy is chosen to be an energy unit, $e^2/C = 1$, t_{ab} is the transfer integral for electron transfer between the box and the lead, and H_0 describes noninteracting reservoir fermions. Following the standard path-integral formalism [9], the fermion degrees of freedom can be integrated out, and the partition function is written as a path integral over a phase variable ϕ conjugate to the electron number n,

$$Z(n_x) = \sum_{m=-\infty}^{\infty} e^{2\pi i m n_x} \int_{-\infty}^{+\infty} d\phi_0 \int_{\phi_0}^{\phi_0 + 2\pi m} \mathcal{D}\phi$$
$$\exp\left[- \int_0^\beta d\tau \frac{1}{2}\dot{\phi}^2 - \frac{2g}{\pi^2} \int_0^\beta \int_0^\beta d\tau d\tau' \frac{\cos\left(\phi(\tau) - \phi(\tau')\right)}{\sin^2(\tau - \tau')} \right] \ , \tag{2}$$

where the winding number m runs over integers, β is inverse temperature and $g = R_Q/R_t$, with the quantum resistance $R_Q = h/4e^2$ and a junction resistance R_t, $1/R_t = 4\pi e^2/\hbar |t_{ab}|^2 N_l N_I$ (Ns are the density of states at the Fermi energies in the lead and the metallic island), is the dimensionless tunneling conductance.

The path-integral (2) can be evaluated exactly by CTM as follows. Let us divide the interval $(0, \beta)$ into N islands and each island contains M discrete points denoted by M dimensional vectors $\phi_i, i = 1, 2, \ldots N$. We write the integrand in (2) as

$$\exp(\ldots) = \prod_{i=1}^{N} K(\phi_i, \phi_{i+1}) \times \mathrm{Rm} \ , \tag{3}$$

where Rm contains the contributions from the double integral

$$\iint d\tau d\tau' \quad \text{with} \quad |\tau - \tau'| > 2M\Delta \ , \tag{4}$$

where $\Delta = \beta/MN$. The $K(\phi_i, \phi_{i+1})$ contains the remaining action. Our procedure is a cluster generalization of the transfer matrix method [10]. We thus consider a CTM equation

$$\int d\phi K(\phi, \phi')\psi_p(\phi) = \lambda_p \psi_p(\phi') \; . \tag{5}$$

The CTM equation is essentially a quantum M body problem with a free center of mass motion. In fact, the invariance of the CTM operator K, for arbitrary simultaneous translations of the center of mass of ϕ and ϕ', indicates that the CTM eigenstates ψ_p are also eigenstates for the center of mass translation operator. This is a generalization of the Bloch–Floquet theorem. The quantum number p is thus decomposed into l for the relative motion and k for the center of mass motion. The CTM equation is reduced to a matrix diagonalization problem. An important note here is that the resulting matrix is complex-nonsymmetric. As a result, we need both the right ψ_p^R and left ψ_p^L eigenstates. Let us now write the integral in (2) as

$$\int_{-\infty}^{+\infty} d\phi_0 \int_{\phi_0}^{\phi_0+2\pi m} \mathcal{D}\phi = \prod_{i=1}^{N+1} \int d\phi_\mathbf{i} \delta\left(\phi_\mathbf{N+1} - \phi_\mathbf{1} - 2\pi m\right) \; . \tag{6}$$

The δ-function in (6) is expressed as

$$\delta\left(\phi_\mathbf{N+1} - \phi_\mathbf{1} - 2\pi m\right) = \sum_p \psi_p^R(\phi - 2\pi m)\psi_p^L(\phi_\mathbf{1}) \; . \tag{7}$$

Substituting (3), (6) and (7) into (2) and repeatedly using the CTM equation (5), we arrive at

$$Z(n_x) \;=\; \mathrm{Rm} \sum_{l,m} \lambda_{l,n_x+m}^N \; . \tag{8}$$

Our remaining task is to calculate Rm. We do this by cumulant approximation as

$$\mathrm{Rm} = \exp\left(\frac{2g}{\pi^2} I\right) \; , \tag{9}$$

where I is the double integral in (4) with the $\cos(\phi(\tau) - \phi(\tau'))$ being replaced by its expectation value.

With the cluster size M, both the perturbative treatment of Rm and its cumulant evaluation, thereby neglecting higher order correlations, will become increasingly correct.

The procedure for calculating the correlation function in (9) is the same as in [10]. We have

$$\langle \cos\phi(\tau)\cos\phi(\tau') \rangle = \frac{1}{\sum_{m,n} \lambda_{m,n_x+n}^N} \sum_{m,m',k',n} \lambda_{m,n_x+n}^{N-(q-p)} \lambda_{m',k'}^{q-p} W(m, m', k', n) \; , \tag{10}$$

where, we have assumed that τ is the i-th lattice point in the p-th island, and τ' j-th lattice point in the q-th island, and

$$W(m, m', k', n) = \langle \psi^R_{m',k'} | \cos\phi(\tau) | \psi^L_{m,n_x+n} \rangle \langle \psi^R_{m,n_x+n} | \cos\phi(\tau') | \psi^L_{m',k'} \rangle \ . \tag{11}$$

Using (10) and (11), we can calculate (9) and hence the remnant Rm from the eigenvalues and eigenstates of the CTM equation. Once the partition function and hence the free energy f is obtained, then the average electron number in the box is calculated by

$$\langle n \rangle = n_x - \frac{\partial f}{\partial n_x} \ . \tag{12}$$

For $n_x = 0$, a quantity which directly measures the degree of Coulomb blockade is the slope $\chi = \partial \langle n \rangle / \partial n_x$. $\chi = 0$ means perfect blockade, while $\chi = 1$ its absence. There are two important parameters in the theory, the lattice constant Δ and the cluster size M. The larger the Δ, the faster the convergence of physical quantities such as χ with increasing M. The theory contains a means to find maximum efficient Δ when other parameters are specified. That is to calculate χ for a Δ by some M and for $\Delta' = \Delta M/M'$ by a larger M'. If the obtained χ and χ' agree within an allowable relative error, we are not missing important degrees of freedom by choosing that Δ.

Let us first consider the case, $T = 0$ and $n_x = 0$, and calculate χ. We found that $\beta = 300$ practically corresponds to $T = 0$. Figure 1 shows the maximum efficient Δ vs g, determined by comparing χs for $M = 2$ and 3, setting allowable relative error to be less than 1%. Figure 2 shows calculated χ using this Δ for $M = 2, 3, 4$ and 5 for g up to 10. It is noted that the discrepancy between the $M = 4$ and 5 results is less than 1% for $g \leq 6$, indicating convergence and hence the exactness of the final result for this g. Note, as is clear in (2) that, the partition function is maxim at $n_x = 0$ as a function of n_x, and hence the free energy is minim there. This means from

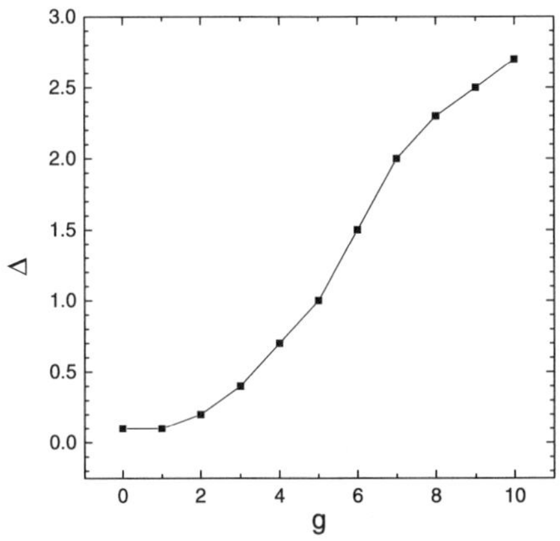

Fig. 1. The maximum efficient Δ vs g

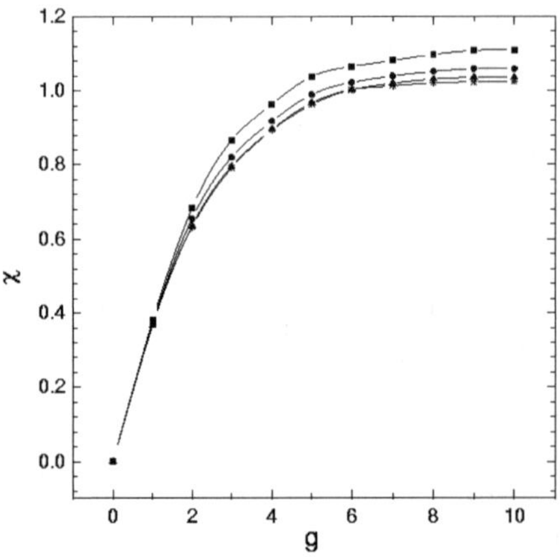

Fig. 2. The slope χ vs. g for the cluster sizes $M = 2$ (*square*), 3 (*circle*), 4 (*triangle*) and 5 (*star*). For $M = 5$, calculation was done only for $g \geq 5$

(12) that $\chi \leq 1$. One thus expects $\chi \to 1$ asymptotically for large g. The 2% inaccuracy, $\chi = 1.02$ at $g = 10$, may be due to inaccuracy in multidimensional integrals.

Figure 3 is $1 - \chi$ vs g for $g \leq 6$ where converged and hence exact result is obtained together with some recent results due to other methods. Figure 4 shows the $M = 4$ result for χ vs g with $n_x = 0$ and $T = 0.003$ (square), 0.1 (circle) and 0.2 (triangle). It is noted that our $M = 2$ cluster calculation applied to the $g = 0$ case perfectly reproduces the exact result for arbitrary temperature. Figure 5 shows the average electron number n vs n_x at $T = 0$ and $g = 1$ (square), 3 (circle) and 5 (triangle). Except for the $n_x = 0.5$

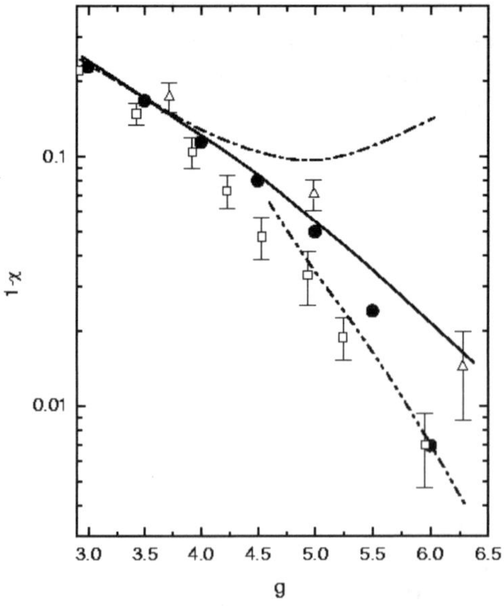

Fig. 3. $1-\chi$ vs g up to $g = 6$. *Solid circle*: CTM method [8] *Dash-dot line*: perturbation theory [2]. *Solid line*: RG [3]. *Dash-dot-dot line*: Instanton [4]. The *open triangles* are MC data from [6], whereas the *open squares* are MC data from [7]

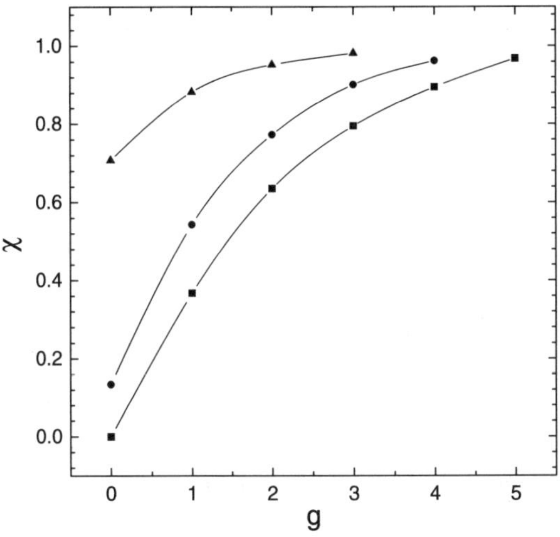

Fig. 4. The slodpe χ vs g with $n_x = 0$ and $M = 4$ for $T = 0.003$ (*square*), 0.1 (*circle*) and 0.2 (*triangle*)

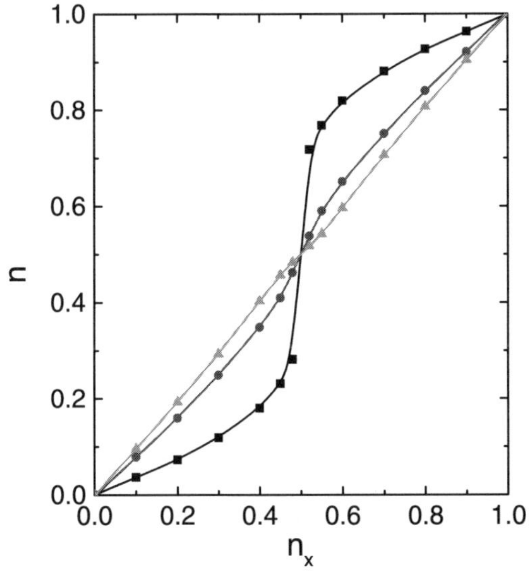

Fig. 5. n vs n_x at $T = 0$ and for $g = 1$ (*square*), 3 (*circle*) and 5 (*triangle*)

vicinity for $g = 5$ where our result is slightly (1%) too large compared to the expected straight line $n = n_x$, the result is in excellent agreement with the recent RG theory [3].

To summarize up to this point, the most notable point of the CTM method is that it is systematic and exact as far as it converges with the cluster size. In fact, convergence in M means that a part of the remnant calculated perturbatively gives the same result when treated exactly through the CTM equation, indicating the exactness of the CTM calculation of the partition function. Moreover it contains a means to determine the maximum efficient lattice constant, which drastically accelerates the convergence, enabling us to reach a fairly wide parameter space.

3 Single JJ

A Josephson junction (JJ) is composed of two superconductors separated by a thin oxide layer. The oxide layer is characterized by the tunnel conductance g as in the SEB, whereas the superconductors are characterized by the Josephson energy E_J. Experimentalists let the current I flow through the single JJ and measure the voltage V between the two superconductors, and thus the differential electrical resistance $R = dV/dI$. The controlling parameters here are temperature T, g and E_J/E_C where the Coulomb energy E_C is a measure of the size of the oxide layer. The Coulomb energy increases with the decrease of the size of the oxide layer. We have recently proposed a Landauer-like formula for the voltage which reduces the problem to a CTM evaluation of the path-integral similar to (2) in the above. Explicitly, the path-integral expression of the partition function is [11],

$$Z_0 = \int \mathcal{D}\phi \, \exp(-S/\hbar) \ , \tag{13}$$

$$\begin{aligned} S = & \int_0^{\beta\hbar} d\tau \left[\frac{C}{2} \left(\frac{\hbar}{2e} \dot{\phi} \right)^2 + U(\phi) \right] \\ & - \int_0^{\beta\hbar} \int_0^{\beta\hbar} d\tau d\tau' \alpha(\tau - \tau') \cos \left(\frac{\phi(\tau) - \phi(\tau')}{2} \right) \ , \end{aligned} \tag{14}$$

where C is the capacitance and

$$\begin{aligned} U(\phi) &= -E_J \cos\phi - \frac{I\hbar}{2e}\phi \\ \alpha(\tau) &= \frac{2g}{\pi^2} \frac{(\pi/\beta\hbar)^2}{\sin^2(\pi\tau/\beta\hbar)} \ . \end{aligned} \tag{15}$$

The new term $U(\phi)$, the tilted washboard potential, contains the effect of superconductor and the bias current.

The first question concerning the phase dynamics is whether it is *coherent* or *incoherent.* A band theory assumes a coherent motion of phase over many local minima of the washboard potential. In band picture, the voltage calculation is easy because it is proportional to the group velocity of the quasi-charge [12]. On the other hand, a non-band theory is appropriate for incoherent motion, and the problem is reduced to the calculation of the escape rate of particle out of a dissipative metastable well, the quantum Kramers rate.

Watanabe and Haviland [13] discussed their recent experiment on the Coulomb blockade at $g \gg 1$ in light of the band theory. However, other experiments [14,15] have not found the negative differential resistance in the SI transition regime $g < 1$. This indicates that the phase dynamics near the SI transition is dominated by *incoherent* tunneling. Indeed a typical experimental setup is, $E_C = 1\,\mathrm{K}$, $E_J/E_C = 0.1 \sim 20$, $T/E_C = 0.1 \sim 1$ and $g < 1$. And most importantly, the minimum voltage experimentally observable is of the order of $0.5\,\mu\mathrm{V}$ [14] which translates to the temperature 10 mK. So when the energy splitting due to tunneling exceeds this value, the corresponding state is experimentally recognized as insulating. However, 10 mK is much smaller than the potential barrier E_J at the SI transition. Near the SI transition at $g = 0$, therefore, the effect of many wells is expected to be unimportant. Switching on dissipation, the SI transition shifts to a smaller critical E_J/E_C, but then dissipation will suppress the effect of increased tunneling strength. One thus expects that the phase dynamics essentially at a single well would determine the SI transition.

We thus propose a non-band theory for the SI transition. Two methods have been employed to calculate the quantum Kramers rate along this line. One is the imaginary free energy method with an instanton technique [16,17]. The other is the quantum transition state (QTS) theory long known in chemistry [18,19]. Here we develop a new approach combining a Landauer-like formula for the quantum Kramers rate [20] and the CTM method [8]. The Landauer formula for electrical conductance has proven to be quite useful in mesoscopic systems. In the quantum Kramers rate problem, there is a formally exact formula [21] analogous to the Green–Kubo–Nakano formula for electrical conductivity. The currently available Landauer-like formula is the quantum transient state (QTS) theory which tries to calculate the Kramers rate by summing up all the probabilities of finding particles at the top of the barrier which have positive, outgoing momentum. The QTS theory, however, has often been exercised with a semiclassical approximation [19,22] and its drawbacks do not show up this way. As soon as one tries to exactly evaluate the path-integrals, however, one encounters difficulties. First, the contribution to the escape rate is no longer local. The phase motion at different places than at the barrier top contributes. Moreover, the non-local contribution could become negative, as discussed in [23]. This undesirable nonlocality and negativity originates from the Weyle quantization procedure [17,18]. The original idea to collect all the outgoing, positive contribution *at* the barrier site, is thus spoiled.

We have recently proposed a new Landauer-like formula for the quantum Kramers rate. Applied to the voltage calculation here, it reads [20,23]

$$\frac{V}{e/2C} = [1 - \exp(-\pi I \hbar \beta / e)] \times \frac{1}{2}\sqrt{\langle \delta(\phi - \phi_0) P^2 / L \rangle}\ , \tag{16}$$

where $P \equiv -\partial_\phi$, ϕ_0 is the barrier height in the potential $U(\phi)$, L is the system size, and $\langle \& \rangle$ is a thermal average. The first factor in (16) takes into account the backward flux through detailed balance, whereas the second term is essentially the square root of the average of the kinetic energy, in a mechanical analogy, at the barrier top with the factor 1/2 only taking into account the right-going contribution. In contrast, the QTS formula is obtained from (16) by replacing $\sqrt{\langle\delta(\phi-\phi_0)P^2/L\rangle} \to \langle\delta(\phi-\phi_0)|P|\rangle$. The difference between formula (16) and the QTS theory is thus a fluctuation which is precisely the undesirable non-local, possibly negative contributions in the QTS theory described above. After some manipulation, one can express the average in (16) as

$$\langle\delta(\phi-\phi_0)P^2\rangle = Z/Z_0 \ , \tag{17}$$

$$Z = -\frac{1}{4}\partial_\phi^2 W(\phi, 2\phi_0 - \phi)|_{\phi\,=\,\phi_0} \ , \tag{18}$$

$$Z_0 = \int d\phi W(\phi,\phi) \ , \tag{19}$$

in terms of the path integral:

$$W(\phi,\phi') = \int_{\phi\to\phi'} \mathcal{D}\tilde{\phi}\, \exp(-S/\hbar) \ , \tag{20}$$

where $\tilde{\phi}(0) = \phi$ and $\tilde{\phi}(\beta\hbar) = \phi'$.

The path-integral (20) with the action S given by (14) can be evaluated precisely by the cluster TM method [20]. In the present problem, however, the dimensionless conductance is at $g < 1$, and from the study of the single electron box, which has a similar action as (14), this regime of g can be accurately handled by the 1-cluster TM method.

Figure 6 shows the temperature dependence of the resistance at $V = 0.5\,\mu\mathrm{V}$ for $g = 0.4$. The small E_J/E_C case shows an insulator-like temperature dependence $\frac{dR}{dT} < 0$, while the large E_J/E_C case shows a superconductor-like

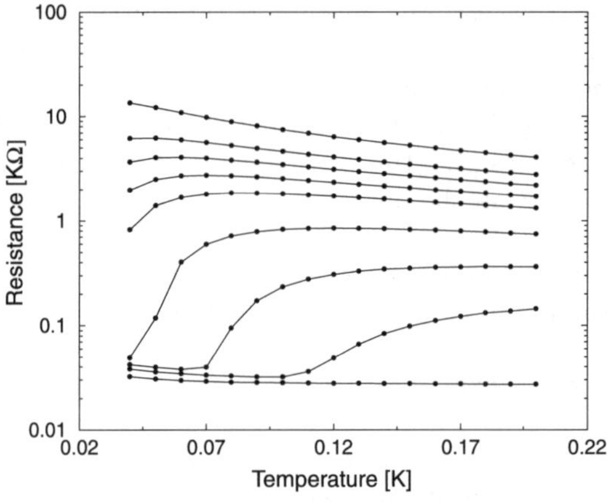

Fig. 6. Temperature dependence of the "zero-bias" resistance at $g = 0.4$. From *top* to *bottom*, $E_J/E_C = 0.07$, 0.56, 0.84, 1.12, 1.4, 1.96, 2.52, 3.0 and 4.0

behavior $\frac{dR}{dT} > 0$. Repeating the calculations for different g, one can draw a phase diagram. Figure 7 is a phase diagram at $T = 80\,\mathrm{mK}$. The experimental results [14,15] are *open circles* (superconductor-like) and *solid circles* (insulator-like) at $T \approx 80\,\mathrm{mK}$. Our result is denoted by *open diamonds.* The band theory result, *thick solid line*, is obtained by simply replacing the minimum measurement energy eV_{min} by the thermal energy $k_{\mathrm{B}}T$ [14]. The QTS theory result, *triangles*, is also plotted for comparison. A major disagreement between theory and experiment is for some data points near $g = 2.8$ [15]. However a similar phase diagram experimentally found for the 2D JJ arrays with similar parameter ranges for $g, E_J/E_C$ and T is bounded, $E_J/E_C < 0.5$, and $g < 0.5$ (cf. Fig. 3 in [24]). The above data near $g = 2.8$ is currently mysterious.

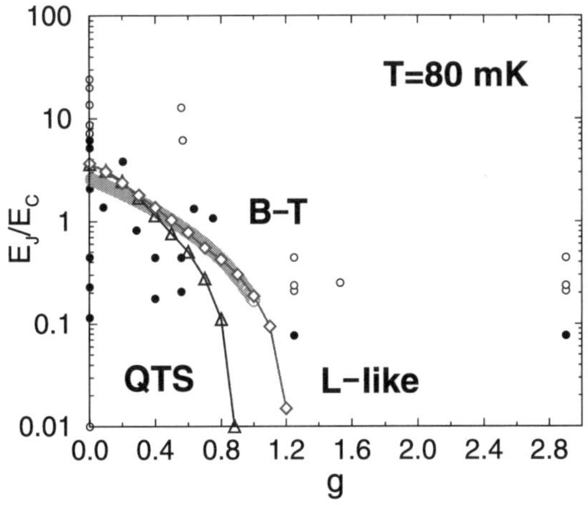

Fig. 7. Phase diagram of shunted Josephson junction at $T = 80\,\mathrm{mK}$. The phase boundary lies between the insulator-like (*solid circles*) and superconductor-like (*open circles*) samples experimentally found in [14,15]. The *thick line* is the band theory (B-T), the *triangle* the QTS theory and the *diamond* is due to the present theory (L-like)

There are two notable points in the obtained temperature dependence of resistance. First it becomes flat for large E_J/E_C irrespective of g. This is consistent with experimental findings [14,15]. Theoretically, when the ratio E_J/E_C is sufficiently large, the quantum Kramers rate would become less dependent on T and g. Note a sharp contrast between the single JJ and the 2D JJ arrays [24]. The temperature dependence of resistance in the latter case showed a *true* phase transition behavior, namely resistance decreases or increases roughly *exponentially* with temperature. Interestingly, the temperature dependence of resistance in 1D JJ arrays also becomes flat deep inside the superconducting phase [25]. The second notable point is that the SI phase boundary defined by vanishing dR/dT actually depends on temperature. In Fig. 8, we have plotted the SI phase boundary for $T = 80, 60, 40, 20, 10$ and 5 mK. Clearly, the insulator phase diminishes with decreasing temperature. We thus reach a conclusion that, *as long as dissipation is present, the single JJ becomes all superconducting at absolute zero temperature.* Note that, in contrast, the band theory predicts an essentially temperature independent phase boundary. It is interesting to see if our finding can be observed experimentally.

The single Josephson junction for which the CTM method was tested is quite general: It contains quantum fluctuation, dissipation and external bias. The reasonable outcome of the new method in comparison with experiments may be encouraging for a further testing of the method, Landauer-like formula + CTM, in a variety of physical systems.

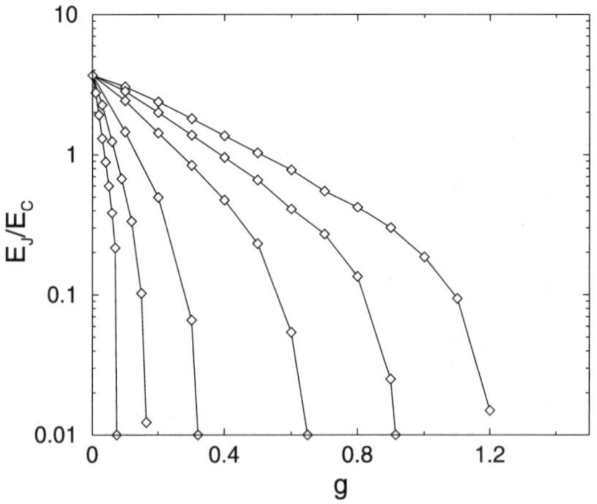

Fig. 8. Temperature dependence of the superconductor-insulator phase boundary in the $E_J/E_C - g$ plane. From *right* to *left*, $T = 80, 60, 40, 20, 10$, and 5 mK

4 CTM for Dynamics

The CTM method has been successfully developed for the SEB and the single JJ. Now a challenging question is if the CTM works for dynamical cases such as SET and QD. The SEB is in fact replaced in reality by SET or QD, thereby letting the current flow through the system and various experimental observations are made.

SET is a series connection of the left lead, metal island, and the right lead, with a bias voltage applied across the left-right lead. Recent theoretical studies of the problem are based on a diagrammatic method [26], a quasi-classical Langevin equation approach [27] and the Monte Carlo for imaginary time path-integral combined with a SVD (singular value decomposition) method to analytically continue to real time [28]. The first is applicable to the regime $g < 1$, the second to not so low temperatures and the last gives a good result for $g \leq 2.5$.

Here the CTM method proceeds as follows. First express the current as a path-integral. Obviously the resulting action is in real time and contains the backward as well as the forward time integrals involving now the two phase variables in the two time branches. The current as a function of applied voltage can be expressed as integral over the correlation function

$$\left\langle e^{\mathrm{i}\phi(t)} e^{-\mathrm{i}\phi(t')} \right\rangle , \tag{21}$$

where $\mathrm{e}^{\mathrm{i}\phi(t)}$ creates an electron in the central metallic island. The job of the CTM is thus to evaluate the path integral precisely.

We have done a preliminary calculation of the single cluster. Figure 9 shows the comparison of CTM and QMC for the conductance as a function of the gate voltage n_x at $T = 0.048$ in units of Coulomb energy and $g = 1.187$. The lowest order CTM is of course not exact, but already captures an essential feature of the conductance. Figure 10 shows the same conductance as a function of time in units of $\beta\hbar/\pi$ which is in the order of 1 nano-second in typical experiments. To our knowledge, none of the existing theories can calculate such transient phenomenon. A nice demonstration of a quantum beat at small g may deserve a special note.

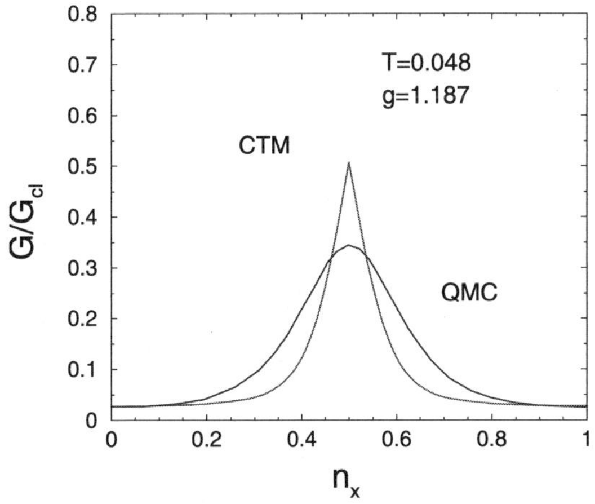

Fig. 9. Conductance vs n_x for $T = 0.048$ and $g = 1.187$. Comparison between CTM and Monte Carlo

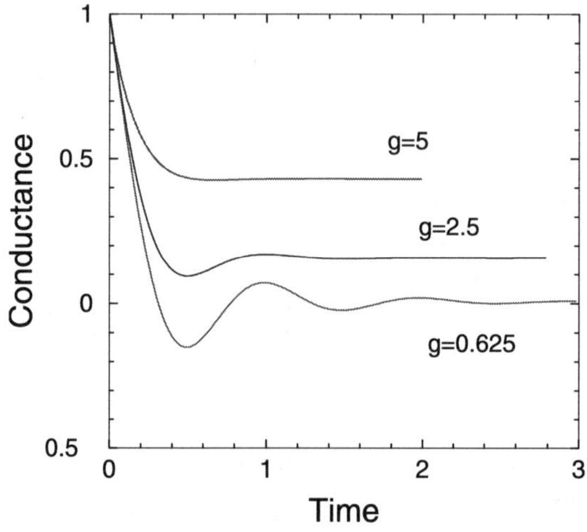

Fig. 10. Conductance vs time in the unit of $\beta\hbar/\pi$ for $n_x = 0$ and $T = 0.05$. A quantum beat is clearly seen for the $g = 0.625$ case

5 Conclusion

We have discussed the CTM method as a novel many-body technique in nanoelectronics. Its potential for nonlinear, nonequilibrium and transient phenomena as demonstrated in the conductance fluctuation in the SET is quite significant, urging a further effort to implement a higher order calculation to see if the CTM really converges with the cluster size.

Acknowledgements

I thank Gianaurelio Cuniberti for enlightening discussions. I also thank the MPI-PKS where a part of the present work was conducted for their hospitality and support. This work was also partially supported by NSF under DMR 990002N and utilized the SGI/CRAY Origin2000 at the National Center for Supercomputing Applications, University of Illinois at Urbana-Champaign.

References

1. D.V. Averin and K.K. Likharev: *Mesoscopic Phenomena in Solids*, ed. B.L. Altshuler, P.A. Lee and R.A. Webb (Elsevier, Amsterdam, 1991) p. 173
2. G. Göppert and H. Gravert: cond-mat/9802248
3. J. König and H. Schoeller: Phys. Rev. Lett. **81**, 3511 (1998)
4. S.V. Panyukov and A.D. Zaikin: Phys. Rev. Lett. **67**, 3168 (1991)
5. G. Falci, G. Schön, and G.T. Zimanyi: Phys. Rev. Lett. **74**, 3257 (1995)
6. X. Wang and H. Gravert: Phys. Rev. B **53**, 12621 (1996)

7. C.P. Herrero, G. Schön and A.D. Zaikin: cond-mat/9807112
8. S.G. Chung: J. Korea Phys. Soc. **33**, S25 (1998); J. Phys. Soc. Japan **68**, 1778 (1999)
9. G. Schön and A.D. Zaikin: Phys. Rep. **198**, 237 (1990)
10. S.G. Chung, D.R.A. Johlen, and J. Kastrup: Phys. Rev. **B48**, 5049 (1993); S.G. Chung: J. Phys. Condens. Matter **9**, L219 (1997)
11. U. Eckern, G. Schön, and V. Ambegaokar: Phys. Rev. B **30**, 6419 (1984)
12. K.K. Likharev and A.B. Zorin: J. Low Temp. Phys. **59**, 347 (1985)
13. M. Watanabe and D. Haviland: cond-mat/0103502, Phys. Rev. Lett. **86**, 5120 (2001)
14. J.S. Penttilä, Ü. Parts, P.J. Hakonen, M.A. Paalanen, and E.B. Sonin: Phys. Rev. Lett. **82**, 1004 (1999)
15. R. Yagi, S. Kobayashi, and Y. Otuka: J. Phys. Soc. Japan **66**, 3722 (1997)
16. H. Grabert and U. Weiss: Phys. Rev. Lett. **53**, 1787 (1984); U. Weiss, H. Gravert, P. Hänggi, and P. Riseborough: Phys. Rev. B **35**, 9535 (1987)
17. U. Weiss: *Quantum Dissipative Systems*, 2-nd ed., Series in Modern Condensed Matter Physics, Vol. 10 (World Scientific, Singapore, 1998)
18. W.H. Miller: J. Chem. Phys. **61**, 1823 (1974); **62**, 1899 (1975)
19. P. Hänggi, P. Talkner, and M. Borkovec: Rev. Mod. Phys. **62**, 251 (1990)
20. S.G. Chung: Physica E **12**, 931 (2002); Phys. Rev. **B66**, 12503 (2002)
21. T. Yamamoto: J. Chem. Phys., **33**, 281 (1960); W.H. Miller, S.D. Schwartz and J.W. Tromp: *ibid.* **79**, 4889 (1983)
22. V.I. Mel'nikov: Phys. Rep. **209**, 1 (1991)
23. S.G. Chung: cond-mat/0202164
24. T. Yamaguchi, R. Yagi, A. Kanda, Y. Ootuka, and S. Kobayashi: Phys. Rev. Lett. **85**, 1974 (2000)
25. D.B. Haviland, K. Anderson, and P. Argen: cond-mat/0001143; W. Kuo, J.-H. Shyu, Y.-J. Chien, D.-S. Chung, M.-L. Jeng, and C. Chen: cond-mat/0007355; W. Kuo and C. Chen: cond-mat/0103245
26. J. König, H. Schoeller, and G. Schön: Phys. Rev. B **58**, 7882 (1998)
27. D. Chouvaev, L.S. Kuzmin, D.S. Goluvev, and A.D. Zaikin: cond-mat/9803015
28. C. Wallisser, B. Limbach, P. vom Stein, R. Schäfer, C. Theis, G. Göppert, and H. Grabert: cond-mat/0205220

Dynamics of Open Quantum Systems

Jacek Okołowicz, Marek Płoszajczak, Ingrid Rotter

Abstract. The coupling between the states of a system and the continuum into which it is embedded, induces correlations that are especially at short time scales. These correlations cannot be calculated by using a statistical or perturbational approach. They are, however, involved in an approach describing structure and reaction aspects in a unified manner. Such a model is the SMEC (shell model embedded in the continuum). Some characteristic results obtained from SMEC as well as some aspects of the correlations induced by the coupling to the continuum are discussed.

1 Introduction

Most states of a nucleus are embedded in the continuum of decay channels due to which they get a finite lifetime. That means: the discrete states of a nucleus shade off into resonance states with complex energies $\mathcal{E}_k = E_k - \frac{i}{2}\,\Gamma_k$. The E_k give the positions in energy of the resonance states while the widths Γ_k are characteristics of their lifetimes. The E_k may be different from the energies of the discrete states, and the widths Γ_k may be large corresponding to a short lifetime. Nevertheless, there is a well defined relation between the discrete states characterizing the closed system, and the resonance states appearing in the open system. The main difference in the theoretical description of quantum systems with and without coupling to an environment is that the function space of the system is supposed to be complete in the first case while this is not so in the second case. Accordingly, the Hamiltonian operator is Hermitian in the first case, and the eigenvalues are discrete. The resonance states, however, characterize a subsystem described by a non-Hermitian Hamiltonian operator with complex eigenvalues. The function space containing everything consists, in the second case, of system plus environment.

The mathematical formulation of this problem goes back to Feshbach [1] who introduced the two subspaces Q and P, with $Q + P = 1$, containing the discrete and scattering states, respectively. Feshbach was able to formulate a unified description of nuclear reactions with both direct processes at the short-time scale and compound nucleus processes at the long-time scale. Due to the high excitation energy and high level density in compound nuclei, he introduced statistical approximations in order to describe the discrete states of the Q subspace. A unified description of nuclear structure and nuclear reaction aspects is much more complicated and became possible only at the end of the last century (see [2] for a recent review). In this formulation, the states of both subspaces are described with the same accuracy. All the coupling matrix elements between different discrete states, different scattering states as well as between discrete and scattering states are calculated in order to get results that can be compared with experimental data. This method has been applied to the description of light nuclei by using the shell model approach for the discrete many-particle states of the Q subspace [2].

In the unified description of structure and reaction aspects, the system is described by an effective Hamiltonian $\mathcal{H}$ that consists of two terms: the Hamiltonian matrix H of the closed system with discrete eigenstates, and the coupling matrix between system and environment. The last term is responsible for the finite lifetime of the resonance states. The eigenvalues of $\mathcal{H}$ are complex and give the poles of the S matrix.

The dynamics of quantum systems is determined by the S matrix, more exactly by its poles and the postulation of unitarity. The unitarity is involved in the continuum shell model calculations [2], but is in conflict with the statistical assumptions when calculations in the overlapping regime are performed [4].

Characteristical for the motion of the poles of the S matrix as a function of a certain parameter are the following generic results obtained for very different systems [5–8]: in the overlapping regime, the trajectories of the S matrix poles avoid crossing with the only exception of exact crossing when the S matrix has a double pole. At the avoided crossing, either level repulsion or level attraction occurs. The first case is caused by a predominantly real interaction between the crossing states and is accompanied by the tendency to form a uniform time scale of the system. Level attraction occurs, however, when the interaction is dominated by its imaginary part arising from the coupling via the continuum. It is accompanied by the formation of different time scales in the system: while some of the states decouple more or less completely from the continuum and become long-lived (trapped), a few of the states become short-lived and wrap the long-lived ones in the cross section. The dynamics of quantum systems at high level density is determined by the interplay of these two opposite tendencies. For a more detailed discussion see [2].

At large overall coupling strength, quick direct reaction processes may appear from slow resonance processes by means of the resonance trapping phenomenon. In recent calculations on microwave cavities with large fixed overall coupling strength, the details of resonance trapping are shown to depend on the position of the attached leads [8]. In microwave cavities of the Bunimovich type, that are chaotic when closed, coherent whispering gallery and bouncing ball modes may be strongly enhanced by trapping other (incoherent) modes. Since the coherent modes determine the value of the conductivity, resonance trapping may cause observable effects that are not small. It is interesting to remark that the trapped long-lived states can be described by random matrix theory, as a shot-noise analysis of the numerical results has shown. The enhanced conductivity, however, is related to the short-lived whispering gallery modes that are regular and dominant when the leads are attached to the cavity in a suitable manner [9].

Meanwhile, the phenomenon of resonance trapping has been proven experimentally on a microwave cavity as a function of the degree of opening of the cavity to an attached lead [10]. In this experiment, the parameter varied is the overall coupling strength between discrete and scattering states. Resonance trapping may appear, however, as function of any parameter [2].

In the following, we will discuss the interplay of the different time scales in nuclei. Most interesting is the mechanism of formation of short-lived states in open quantum systems. In Sect. 2, the effective Hamiltonian and the S matrix are written down for a quantum system embedded in a continuum while in Sect. 3, the basic relations for the spectroscopic information are discussed. Characteristic features of the different approaches to the system and to the environment are sketched in Sect. 4. In Sect. 5, some results obtained from calculations with unified description of structure and reaction aspects are shown. The relation between lifetimes and decay widths of resonance states in the overlapping region is discussed in Sect. 6 while in Sect. 7, properties of the system in the short time scale are illustrated. In any case, the correlations induced by the coupling to the continuum are large. The last section contains some concluding remarks.

2 Effective Hamiltonian and S Matrix for a Quantum System Embedded in a Continuum

In the unified description of structure and reaction aspects of quantum systems, the Schrödinger equation

$$(H - E)\Psi = 0 \tag{1}$$

is solved in a function space containing everything, i.e. discrete as well as continuous states. The Hamiltonian operator H is Hermitian, the wave functions Ψ depend on energy as well as on the decay channels and all the resonance states of the system. Knowing the wave functions Ψ, an expression for the S matrix can be derived that holds true also in the overlapping regime, see the recent review [2]. In the continuum shell model, it reads

$$S_{cc'} = e^{i(\delta_c - \delta_{c'})} \left[\delta_{cc'} - S^{(1)}_{cc'} - S^{(2)}_{cc'} \right] \, , \tag{2}$$

where $S^{(1)}_{cc'}$ is the smooth direct reaction part related to the short-time scale, and

$$S^{(2)}_{cc'} = i \sum_{k=1}^{N} \frac{\tilde{\gamma}_k^c \, \tilde{\gamma}_k^{c'}}{E - \tilde{E}_k + \frac{i}{2}\tilde{\Gamma}_k} \tag{3}$$

is the resonance reaction part related to the long-time scale. Here, the $\tilde{\mathcal{E}}_k = \tilde{E}_k - \frac{i}{2}\,\tilde{\Gamma}_k$ are the complex eigenvalues of the non-Hermitian Hamiltonian operator

$$\mathcal{H}_{QQ} = H_{QQ} + H_{QP}\,G_P^{(+)}\,H_{PQ} \tag{4}$$

appearing effectively in the system (Q subspace) after embedding it into the continuum (P subspace). They are energy dependent functions and determine the positions $E_k = \tilde{E}_k\,(E = E_k)$ and widths $\Gamma_k = \tilde{\Gamma}_k\,(E = E_k)$ of the resonance states k [2]. The $G_P^{(+)}$ in (4) are the Green functions in the P subspace. The $\tilde{\gamma}_k^c$ are the coupling matrix elements between the resonance states and the scattering states. They are also energy dependent functions. The wave functions $\tilde{\Omega}_k$ of the resonance states are related to the eigenfunctions $\tilde{\Phi}_k$ of $\mathcal{H}_{QQ}$ by a Lippmann Schwinger like relation [2],

$$\tilde{\Omega}_k = (1 + G_P^{(+)}H_{PQ})\,\tilde{\Phi}_k\ . \tag{5}$$

The eigenfunctions of $\mathcal{H}_{QQ}$ are bi-orthogonal,

$$\langle\tilde{\Phi}_l^*|\tilde{\Phi}_k\rangle = \delta_{kl}\ , \tag{6}$$

so that

$$\langle\tilde{\Phi}_k|\tilde{\Phi}_k\rangle = \mathrm{Re}(\langle\tilde{\Phi}_k|\tilde{\Phi}_k\rangle) \quad ; \quad A_k \equiv \langle\tilde{\Phi}_k|\tilde{\Phi}_k\rangle \geq 1 \tag{7}$$

$$\langle\tilde{\Phi}_k|\tilde{\Phi}_{l\neq k}\rangle = i\ \mathrm{Im}(\langle\tilde{\Phi}_k|\tilde{\Phi}_{l\neq k}\rangle) = -\langle\tilde{\Phi}_{l\neq k}|\tilde{\Phi}_k\rangle \ ; \ B_k^{l\neq k} \equiv |\langle\tilde{\Phi}_k|\tilde{\Phi}_{l\neq k}\rangle| \geq 0\ . \tag{8}$$

As a consequence of (7), it holds [2]

$$\tilde{\Gamma}_k = \frac{\sum_c |\tilde{\gamma}_k^c|^2}{A_k} \ \leq \ \sum_c |\tilde{\gamma}_k^c|^2\ . \tag{9}$$

The main difference to the standard theory is that the $\tilde{\Gamma}_k$, $\tilde{\gamma}_k^c$ and $\tilde{E}_k$ are not numbers but energy dependent functions [2]. The energy dependence of $\mathrm{Im}\{\tilde{\mathcal{E}}_k\} = -\frac{1}{2}\,\tilde{\Gamma}_k$ is large near to the threshold for opening the first decay channel. This causes not only deviations from the Breit Wigner line shape of isolated resonances lying near to the threshold, but also an interference with the above-threshold "tail" of bound states, see Sect. 5.2 for an example. Also an inelastic threshold may have an influence on the line shape of a resonance when the resonance lies near to the threshold and is coupled strongly to the channel which opens [11]. Also in this case, $\tilde{\Gamma}_k$ depends strongly on energy. In the cross section, a cusp may appear in the cross section instead of a resonance of Breit Wigner shape. Both types of threshold effects in the line shape of resonances can explain experimental data known in nuclear physics [2]. They can not be simulated by a parameter in the S matrix.

In the numerical calculations in the framework of the continuum shell model, the coupling matrix elements $\tilde{\gamma}_k^c$ between resonance states and continuum are obtained by representing the eigenfunctions $\tilde{\Phi}_k$ of the effective non-Hermitian Hamiltonian operator $\tilde{\mathcal{H}}_{QQ}$ in the set of eigenfunctions $\{\Phi_k\}$ of the Hermitian Hamiltonian operator H_{QQ},

$$\tilde{\Phi}_k = \sum_l b_{kl}\,\Phi_l \; . \tag{10}$$

The Φ_k are real, while the $\tilde{\Phi}_k$ are complex and energy dependent. The coefficients b_{kl} and the $(\tilde{\gamma}_k^c)^2$ are complex and energy dependent, too. The $(\tilde{\gamma}_k^c)^2$ characterizing the coupling of the resonance state k to the continuum, are related to the width of this state. In the overlapping regime, their sum over all channels is, however, not equal to the width even in the one-channel case, (9). Both functions, $(\tilde{\gamma}_k^c)^2$ and $\tilde{\Gamma}_k$, may show a different energy dependence. An example is shown in Sect. 5.3.

3 Spectroscopy of Resonance States

3.1 Isolated Resonance States

The energies and widths of the resonance states follow from the solutions of the fixed-point equations:

$$E_k = \tilde{E}_k(E = E_k) \tag{11}$$

and

$$\Gamma_k = \tilde{\Gamma}_k(E = E_k) \quad , \tag{12}$$

on condition that the two subspaces are defined adequately [2]. The values E_k and Γ_k correspond to the standard spectroscopic observables. The functions $\tilde{E}_k(E)$ and $\tilde{\Gamma}_k(E)$ follow from the eigenvalues $\tilde{\mathcal{E}}_k$ of $\tilde{\mathcal{H}}_{QQ}$. The wave functions of the resonance states are defined by the functions $\tilde{\Omega}_k$, (5), at the energy $E = E_k$. The partial widths are related to the coupling matrix elements $(\tilde{\gamma}_k^c)^2$ that are calculated independently by means of the eigenfunctions $\tilde{\Phi}_k$ of $\tilde{\mathcal{H}}_{QQ}$. For isolated resonances, $A_k = 1$ according to (7) and $(\tilde{\gamma}_k^c)^2 = |\gamma_k^c|^2$. In this case the standard relation $\Gamma_k = \sum_c |\gamma_k^c|^2$ follows from (9).

It should be underlined that different $\tilde{\Phi}_k(E = E_k)$ are neither strictly orthogonal nor bi-orthogonal since the bi-orthogonality relation (6) holds only when the energies of both states k and l are equal. The spectroscopic studies on resonance states are performed, therefore, with the wave functions being only approximately bi-orthogonal. The deviations from the bi-orthogonality relation (6) are small, however, since the $\tilde{\Phi}_k$ depend only weakly on the energy.

This drawback of the spectroscopic studies of resonance states has to be contrasted with the advantage it has for the study of observable values: the S matrix and therefore the cross section is calculated with the resonance wave functions being strictly bi-orthogonal at every energy E of the system. Furthermore, the full energy dependence of $\tilde{E}_k, \tilde{\Gamma}_k$ and, above all, of the coupling matrix elements $\tilde{\gamma}_k^c$ is taken into account in the S matrix and therefore in all calculations for observable values.

As a result of the formalism sketched in Sect. 2 for describing the nucleus as an open quantum system, the influence of the continuum of scattering states on the spectroscopic values consists mainly in the following: there is (i) an additional shift in energy of the states and (ii) an additional mixing of the states through the continuum of decay channels.

For isolated resonances, the additional shift is usually taken into account by simulating $\mathrm{Re}\,(\mathcal{H}_{QQ}) = H_{QQ} + \mathrm{Re}\,(W)$ (see (4)) by $H_0 + V'$, where V' contains the two-body effective residual forces and $W \equiv H_{PQ}G_P^{(+)}H_{PQ}$. Furthermore, the widths of isolated states are not calculated from $\mathrm{Im}\,(W)$, but from the sum of the partial widths. The amplitudes of the partial widths are the coupling matrix elements between the discrete states of the Q subspace and the scattering wave functions of the P subspace. The additional mixing of the states via the continuum is neglected in the standard calculations.

It should be mentioned, however, that $\mathrm{Re}\,(W)$ can not be completely simulated by an additional contribution to the residual two-body interaction since it contains many-body effects, as follows from the analytical structure of W. $\mathrm{Re}\,(W)$ is an integral over energy and depends explicitly on the energies ϵ_c at which the channels c open. As a matter of fact, the thresholds for neutron and proton channels in nuclei open at different energies. Therefore, $\mathrm{Re}\,(W)$ causes some charge dependence of the effective nuclear forces in spite of the charge symmetry of the Hamiltonian H_{QQ}. It arises as a many-body effect depending on shell closures, and is directly related to the different binding energies of neutrons and protons in nuclei [2,11].

Since only a few data on isolated resonances are sensitive to the many-body effects involved in $\mathrm{Re}\,(W)$, the standard calculations performed by using a Hermitian operator are mostly justified. However, the standard calculations can not be justified for closely-lying levels which are coupled via the continuum of decay channels, as well as for well isolated levels in the neighborhood of thresholds where new decay channels open.

3.2 Correlations Induced by the Coupling via the Continuum

The coupling of the resonance states via the continuum induces correlations between the states that are described by the term $H_{QP}G_P^{(+)}H_{PQ} \equiv W$ of the effective Hamiltonian $\mathcal{H}_{QQ}$, (4). W is complex and energy dependent [2]. The real part $\mathrm{Re}\,(W)$ causes level repulsion in energy and is accompanied by the tendency to form a uniform time scale in the system. In contrast to this behavior, the imaginary part $\mathrm{Im}\,(W)$ causes different time scales in the system and is accompanied by level attraction in energy. That means, the formation of correlations at short-time scales is essentially influenced by $\mathrm{Im}\,(W)$.

In the overlapping regime, many calculations have shown the phenomenon of resonance trapping caused by Im (W),

$$\sum_{k=1}^{N} \tilde{\Gamma}_k \approx \sum_{K=1}^{K} \tilde{\Gamma}_k \quad ; \qquad \sum_{k=K+1}^{N} \tilde{\Gamma}_k \approx 0 \, . \tag{13}$$

It means almost complete decoupling of $N - K$ resonance states from the continuum while K of them become short-lived. Usually, $K \ll N - K$. The long-lived resonance states in the overlapping regime appear often to be well isolated from one another. The few short-lived resonance states determine the evolution of the system (short time scale).

The formation of different time scales in an open quantum system that is accompanied by level attraction and also by the appearance of a non-trivial energy dependence of W [3]. This energy dependence can directly be expressed by non-linear terms appearing in the overlapping regime. As a consequence, the use of an effective Hamiltonian in describing scattering processes is meaningful only when, at the same time, the energy dependence of W is considered.

In the framework of statistical approaches, the coupling matrix elements between resonance states and continuum are assumed to be energy independent parameters. Also in the different versions of R matrix approaches, the correlations induced by W cannot be studied. The interplay between the different time scales of open quantum systems at high level density can be studied only microscopically, without any statistical assumptions on the level distribution or perturbation theory approaches.

3.3 Overlapping Resonance States

The solutions E_k and Γ_k of the fixed point equations (11) and (12) are basic for spectroscopic studies not only of isolated but also of overlapping resonances since the energy dependence of the eigenvalues $\tilde{\mathcal{E}}_k = \tilde{E}_k - i/2\; \tilde{\Gamma}_k$ of the effective Hamiltonian $\mathcal{H}_{QQ}$ is smooth everywhere. The E_k and Γ_k are therefore well defined and it makes sense to use them for spectroscopic studies. The coupling coefficients $\tilde{\gamma}_k^c$ are however less well defined since the wave functions $\tilde{\Phi}_k(E = E_k)$ are bi-orthogonal. The bi-orthogonality relations (7) and (8) become important at the avoided level crossings where $A_k > 1$. In approaching a double pole of the S matrix, $A_k \to \infty$. The same holds for the modulus square of the coupling coefficients: $|\tilde{\gamma}_k^c|^2 \to \infty$, in accordance with the relation (9).

The numerator of the resonance part of the S matrix (3) is

$$\langle \tilde{\Phi}_k^* | \hat{W}_{cc'} | \tilde{\Phi}_k \rangle = 2\pi \langle \tilde{\Phi}_k^* | V^\dagger | \xi_E^c \rangle \langle \xi_E^{c'} | V | \tilde{\Phi}_k \rangle = \tilde{\gamma}_k^c \tilde{\gamma}_k^{c'} \quad . \tag{14}$$

For $c = c'$, this is $(\tilde{\gamma}_k^c)^2$ and not $|\tilde{\gamma}_k^c|^2$ as often assumed [12]. Expression (14) remains meaningful also in approaching the double pole of the S matrix [2], and the S matrix (2) with (3) is unitary also in the overlapping regime. When the energy difference $\Delta E = |E_k - E_l|$ between two neighboring resonance states is smaller than their widths, higher-order terms in the S matrix that are related to the bi-orthogonality of the eigenfunctions of the non-Hermitian Hamiltonian operator $\mathcal{H}_{QQ}$, can not be neglected. At a double pole of the S matrix, $(\tilde{\gamma}_k^c)^2 \to -(\tilde{\gamma}_l^c)^2$ corresponding to $\tilde{\Phi}_k \to \pm i \tilde{\Phi}_l$ [2]. Here, the two resonance terms cancel, and the system decouples from the continuum at the energy of the double pole. The same relations hold when the two states avoid crossing in the complex plane by varying a certain parameter [6]. The point is, however, that in such a case the transition $\tilde{\Phi}_k \to \pm i \tilde{\Phi}_l$ influences the wave functions not only at the critical point but in a certain region around the critical value of the parameter [6]. At high level density, this fact will cause deviations from the relation $\tilde{\Gamma}_k = \sum_c (\tilde{\gamma}_k^c)^2$. For numerical results on the relation between $\tilde{\Gamma}_k$ and $(\tilde{\gamma}_k^c)^2$, see Sect. 5.3.

Furthermore, the energies and widths of overlapping resonance states are given by the values E_k and Γ_k ((11) and (12)), at which the S matrix has poles. However, the positions of the maxima in the cross section do, generally, not appear at the energies E_k when the resonance states overlap [2].

The relation between $\tilde{\Gamma}_k = -2 \text{ Im } \{\langle \tilde{\Phi}_k^* | \mathcal{H}_{QQ} | \tilde{\Phi}_k \rangle\}$ and the sum of the coupling coefficients $\sum_c (\tilde{\gamma}_k^c)^2$ is, in general, more complicated than for isolated resonances due to the avoidance of level crossings in the complex plane [2]. The S matrix behaves smoothly in the neighborhood of a double pole. The same is true for measurable values due to their relation to the S matrix. The value $|\tilde{\gamma}_k^c|^2$ loses its physical meaning in the overlapping regime.

4 Different Approaches

4.1 Statistical Approach to the System

More than 40 years ago, the *unified theory of nuclear reactions* has been formulated by Feshbach [1]. Feshbach introduced the projection operator technique in order to make possible the concurrent numerical solution of equations with discrete and scattering states in spite of their very different mathematical properties. By means of the projection operator technique, the whole function space is divided into the subspace of discrete states (Q subspace) and the subspace of scattering states (P subspace). Then, the problem in the P subspace is solved numerically by coupled-channel methods while the problem in the Q subspace is not solved directly. Here, statistical assumptions are introduced by which the mean properties of the discrete states are described. Also the coupling matrix elements between discrete and scattering states are determined statistically and characterized by their mean values.

The advantage of using different approximations in the two subspaces consists, above all, in the possibility of solving the coupled-channel problem with high accuracy. Since the P subspace is constructed from all open decay channels, it changes with energy since new channels open in passing the corresponding thresholds. Furthermore, the inclusion of, e.g., α decay channels into the P subspace is not a problem. The method is applied successfully to the description of nuclear reactions in energy regions with high level density of the excited nucleus which makes it possible for a statistical treatment of the discrete states of the Q subspace. It represents the standard method in analyzing nuclear reaction data on medium and heavy nuclei at low energy.

The *shell model approach to nuclear reactions* [12] is formulated by Mahaux and Weidenmüller. Also in this approach, the whole function space is divided into the two subspaces. However, the P subspace contains open as well as closed decay channels and, therefore, does not change with energy. The inclusion of more than one particle in the continuum becomes a principal problem. The bi-orthogonality of the eigenfunctions of the effective Hamiltonian is not considered, what causes problems with the unitarity of the S matrix in the overlapping regime due to $\Gamma_k < \sum_c |\gamma_k^c|^2$ [12]. Eventually, the states of the Q subspace are treated by means of statistical methods in the same manner as in the Feshbach formulation [1]. The restrictions in the applicability of both treatments are therefore the same: as long as the (long-lived) resonance states are isolated from each other and their individual properties can be neglected to a good approximation, the method gives reliable results.

The formation of different time scales in a realistic system cannot be studied by using a statistical description of the states, since the interplay between the real and imaginary parts of the interaction in the effective Hamiltonian $\mathcal{H}_{QQ}$ is not taken into account.

4.2 *R* Matrix Approach

In contrast to the Feshbach unified theory of nuclear reactions, different approaches for the description of decaying states are worked out by starting from well established nuclear structure models. These approaches are based on the R matrix theory of nuclear reactions that is justified at low level density [13]. Here, the resonance levels are assumed to be isolated, i.e. the influence of resonance overlapping on the nuclear structure is not considered.

The advantage of these studies consists, above all, in the integration of proven nuclear structure models into the calculations. That means, the wave functions of the Q subspace are realistic. The coupling to the supplementary P subspace (continuum of decay channels) is described in a straightforward manner. The feedback from the continuum of decay channels on the nuclear structure is however hidden, if at all taken into account, in the results of the numerical studies. When the resonances overlap, some averaging over many levels is performed in the R matrix theory of nuclear reactions [13].

The formation of different time scales in the system cannot be studied since it arises from the feedback from the continuum to the states of the system that is not taken into account in the R matrix approach.

4.3 Shell Model Approach to the System

In reactions on light nuclei and in studying nuclei near to the drip line, the level density is low and the individual properties of the nuclear states can not be neglected. In these nuclei, the restriction to a description of the mean properties of the states is not justified. The problem in the Q subspace has to be solved with a higher accuracy.

The spectroscopic properties of light nuclei are described successfully in the framework of the shell model. It is therefore reasonable to identify the Q subspace with the function space of the shell model used in performing numerical calculations for these nuclei. Two different approaches have been developed: (i) the CSM-FDP approach (continuum shell model with finite depth potential), that generates the single particle basis states in a Woods Saxon potential [11,14], and has been used mainly for a description of giant resonances in light nuclei, and (ii) the SMEC (shell model embedded in the continuum) which uses the shell model effective interaction in the Q subspace and provides, in particular, a realistic description of resonance phenomena near particle decay thresholds [15]. Common to both approaches is that (1) is solved numerically by using similar approximations in the two subspaces. The bi-orthogonality of the eigenfunctions of the effective Hamiltonian ((6) to (8)) is taken into account in both approaches. As a consequence, the unitarity of the S matrix is also ensured in the overlapping regime. These calculations provide a *unified description of nuclear structure and nuclear reaction* aspects.

In the SMEC, the nuclear shell model is involved what makes it possible for a realistic description of the nuclear structure, as in the models based on the R matrix approach. However, in contrast to these models, the feedback from the continuum of decay channels on the nuclear structure is explicitly taken into account. Therefore, the formation of different time scales in the system can be studied by means of SMEC.

5 Some Results Obtained for ^{24}Mg in SMEC

5.1 The ^{24}Mg Nucleus

Let us consider ^{24}Mg with the inner core ^{16}O and the phenomenological sd-shell interaction among the valence nucleons. Within this configuration space, the ^{24}Mg nucleus has 325 states with $J^{\pi} = 0^{+}$, $T = 0$. These states can couple to a number of open channels which correspond to excited states in the neighboring $(A-1)$ nucleus. For details see [2,16].

For illustration, we show in Fig. 1 the dependence of energies $\tilde{E}_k$ and widths $\tilde{\Gamma}_k$ of the ten lowest 0^+ states of ^{24}Mg on the energy E of the particle in the continuum, as well as the eigenvalue picture with the energy E parametrically varied. The number of channels is one (the l.h.s plot) and two (the r.h.s. plot). We can see the non-random features occurring at this edge

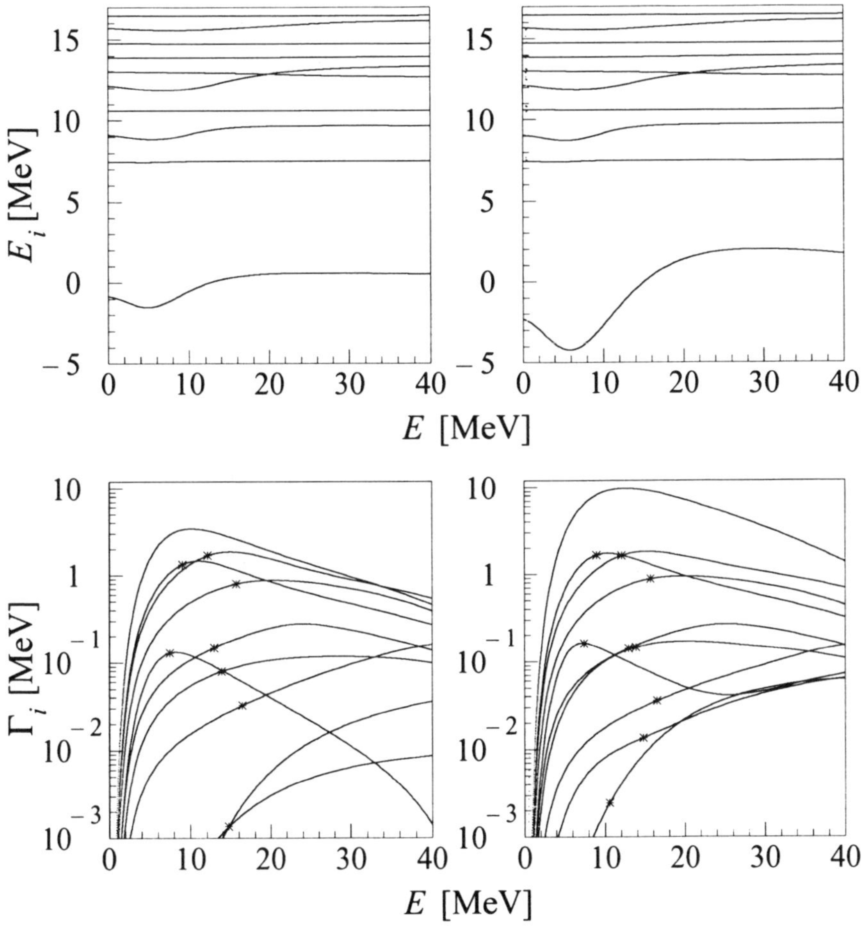

Fig. 1. Energy dependence of the positions $\tilde{E}_i$ (*upper row*) and widths $\tilde{\Gamma}_i$ (*lower row*) of the ten lowest 0^+ states of ^{24}Mg as a function of the energy E of the particle in the continuum. In the first column, the calculations include coupling to only one channel in ^{23}Mg. In the other column, two channels are taken into account. The stars at the trajectories mark the fixed-point solutions

of the spectrum. The coupling between the channels reduces the differences between the widths of the different states. It has almost no influence onto their positions.

The positions $\tilde{E}_k$ of the resonance states are almost independent of a variation of the energy of the system (Fig. 1). The widths $\tilde{\Gamma}_k$ however depend on energy: they rise at low energies above the particle decay threshold and decrease again at energies beyond the positions E_k of the states. Most of the resonances have therefore a tail at the high energy side. This feature is well pronounced especially for the lowest-lying state which is bound. Due to its large width, it can contribute to the cross section in the threshold region.

5.2 Near-threshold Behavior of the Cross Section

In Fig. 2, we summarize the generic features of cross sections near thresholds. As an example we show the cross section for the reaction $\mathrm{n}+{}^{23}\mathrm{Mg}\,(1/2^+) \longrightarrow {}^{24}\mathrm{Mg}(0^+)$ (one open channel, the s-wave scattering). The upper part shows the cross section with only one excited (resonance) state 0_2^+ of ^{24}Mg. The minimum in the cross section is an effect of destructive interference of the resonance ($E^* = 2.23\,\mathrm{MeV}$, $\Gamma = 1.76\,\mathrm{MeV}$) with the background of the potential scattering (the direct part of the reaction cross section), denoted by the dashed line. In the middle part, the cross section with only the ground state (bound state) 0_1^+ in ^{24}Mg is shown. The cross section exhibits a strong increase for $E \to 0$. This is caused by the bound state for which $\tilde{\Gamma}_k \neq 0$ at $E > 0$ in spite of $\tilde{\Gamma}_k = 0$ at $E = \tilde{E}_k < 0$. Finally, the cross section with both ground state 0_1^+ and resonance state 0_2^+ is shown in the bottom part of the figure. The interference picture of these two states shows level repulsion accompanied by a decrease of the width of the higher-lying state ($E^* = 2.40\,\mathrm{MeV}$, $\Gamma = 0.47\,\mathrm{MeV}$). The line shape of the resonance resembles a typical interference picture for overlapping resonance states in spite of the fact that the calculation is performed with only one resonance state while the other state is bound.

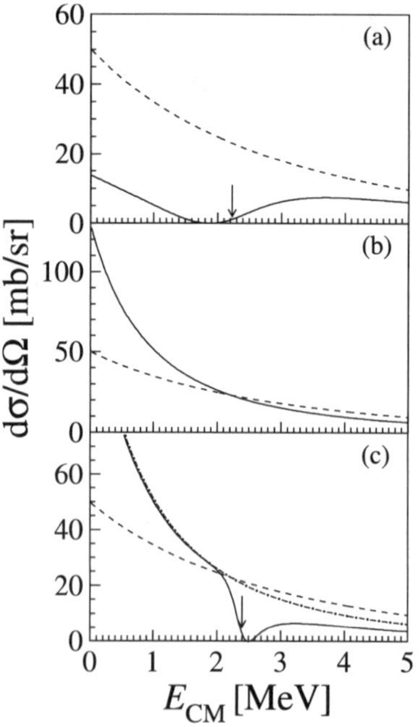

Fig. 2. Cross section for the reaction $\mathrm{n} + {}^{23}\mathrm{Mg} \to {}^{24}\mathrm{Mg}$ calculated for one open neutron channel and (a) the resonance state 0_2^+ of ^{24}Mg, (b) the ground state 0_1^+ of ^{24}Mg, (c) the bound 0_1^+ and the resonance state 0_2^+ of ^{24}Mg. The *dashed lines* show the direct reaction part of the cross section. The *arrows* denote the position of the resonances

5.3 Relation Between Total and Partial Widths

The resonance states shown in Fig. 1 do not overlap strongly. Nevertheless, the relation between their widths $\tilde{\Gamma}_k$ and the coupling matrix elements $(\tilde{\gamma}_k^c)^2$ is far from being both well defined and energy independent, even in the one-channel case. In Figs. 3 and 4, we show the total widths $\tilde{\Gamma}_k$ and the real and imaginary parts of the coupling matrix elements $(\tilde{\gamma}_k^c)^2$ for six of these states and, additionally, the partial widths $|\gamma_k^c|^2$ (the unambiguous identification of the label for different states can be obtained from Fig. 5). All results are for the one-channel case and therefore $\Gamma_k = |\gamma_k^c|^2$ is assumed in the R matrix theory.

The states '1' and '8' are well isolated from the other ones due to the large distance in energy (state '1') and small width (state '8'), respectively. For these two states the relation $\tilde{\Gamma}_k \approx \mathrm{Re}(\tilde{\gamma}_k^c)^2$ holds in the whole energy region considered. The values $\tilde{\Gamma}_k$ and $|\gamma_k^c|^2$ differ from each other, but show a similar energy dependence (Fig. 4).

States '5' and '6' are coming near to one another at an energy higher than their position. As a consequence, the total widths $\tilde{\Gamma}_k$ are different from the $\mathrm{Re}\,(\tilde{\gamma}_k^c)^2$ that are the real parts of the coupling matrix elements of the resonance states to the continuum. They differ also from the $|\gamma_k^c|^2$ that are the coupling matrix elements of the discrete states to the continuum. The differences are noticeable in the whole energy region considered, and not only at their nearest distance in energy (Fig. 3). This happens for the states '7' and '10' in a similar manner (Fig. 4) as for the states '5' and '6' although the distance to the neighboring states is, in these cases, much larger than in the case of two states '5' and '6'.

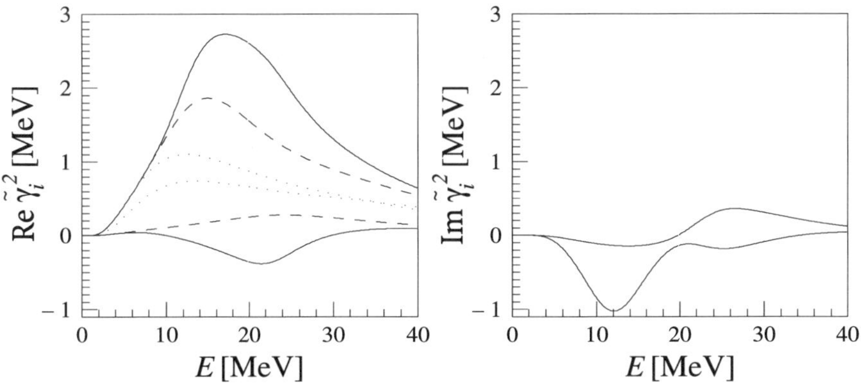

Fig. 3. Energy dependence of the coupling coefficients $(\tilde{\gamma}_i)^2$ (*solid lines*) for crossing resonances in ^{24}Mg (resonances 5 and 6 in Fig. 1, one-channel case). The real parts are shown on the *left-hand side* and the imaginary parts on the *right-hand side*. In addition, the figure on the *left-hand side* shows the dependence of the widths $\tilde{\Gamma}_i$ (*dashed lines*) and $|\gamma_i|^2$ (*dotted lines*) on the energy of the particle in the continuum

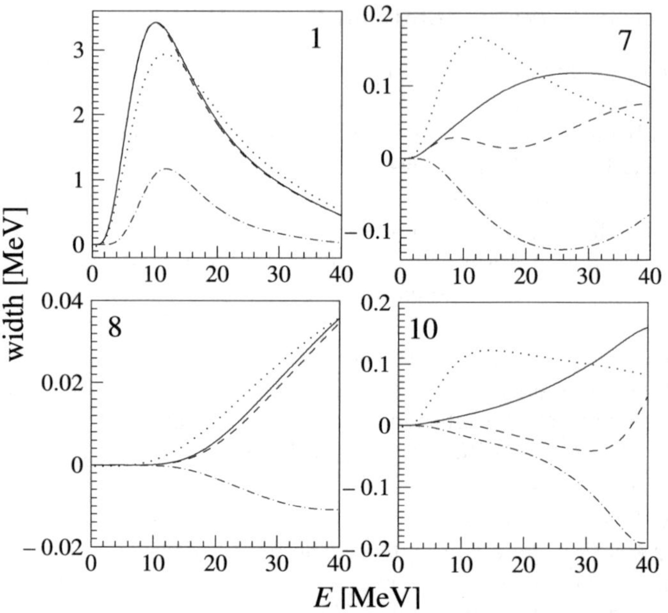

Fig. 4. Energy dependence of various 'widths' for resonances '1', '7', '8', '10' (see Fig. 1) in ^{24}Mg. The different lines denote: $\tilde{\Gamma}_i$ (*solid line*), $\mathrm{Re}\,(\tilde{\gamma}_i)^2$ (*dashed line*), $\mathrm{Im}\,(\tilde{\gamma}_i)^2$ (*dashed-dotted line*) and $|\gamma_i|^2$ (*dotted line*)

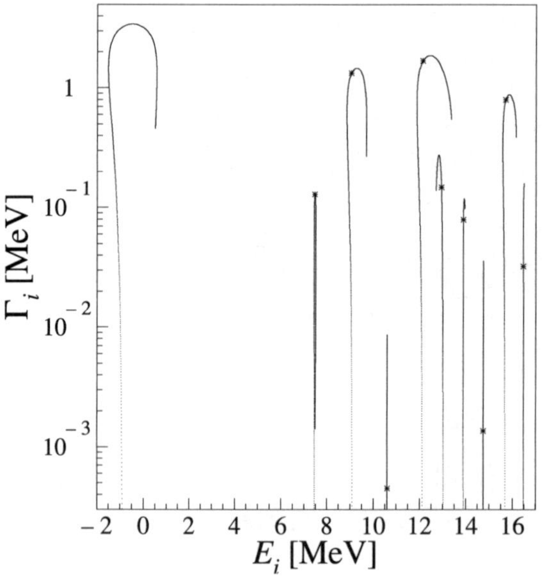

Fig. 5. The eigenvalue picture with the energy E of the particle in the continuum parametrically varied for ten lowest 0^+ states of ^{24}Mg

Figures 3 and 4 illustrate that the standard one-level formula for the cross section (the Breit Wigner representation) can be applied only for well isolated resonances. Only in such a case, unitarity of the S matrix provides a clear relation between the total widths $\tilde{\Gamma}_k$ and the coupling coefficients between system and environment. Well isolated resonances appear, however, seldom in realistic situations. Therefore, a generic relation of the type $\tilde{\Gamma}_k = \sum |\gamma_k^c|^2$ (or $\tilde{\Gamma}_k = \sum(\tilde{\gamma}_k^c)^2$) does not exist, even in the one-channel case. The partial widths $|\gamma_k^c|^2$ of the state k relative to the channels c lose their physical meaning when the resonance states are not well isolated. Furthermore, the $\tilde{\Gamma}_k$ and even the $|\gamma_k^c|^2$ are energy dependent.

These numerical results show that, in general, the influence of the different states onto the properties of the system can not be restricted to the small energy region that is determined by their energies and widths. This restriction being one of the basic approximations of R matrix approaches, is justified neither for bound states lying just below the first particle decay threshold nor for resonance states at high level density.

Thus, even though the SMEC and the different R matrix approaches start both from a reliable nuclear structure model, the coupling of the resonance states via the continuum of decay channels is taken into account correctly only in the SMEC.

5.4 Statistical Versus Dynamical Aspects of Resonance States at High Level Density

The statistical properties of ^{24}Mg are studied in [16]. As a result, the dynamics of the ^{24}Mg nucleus in the short-time scale is determined by the states at the edges of the spectra of the parent and daughter nuclei. These states are strongly related to each other with the result that the corresponding resonance states have short lifetimes. Randomness in an open quantum system can be found only in the long-time scale and, even here, only in the one-channel case. Since the short-lived and long-lived states are created together at avoided level crossings, both time scales exist simultaneously in the nucleus. This statement is in agreement with experimental results on different nuclei of the *sd*-shell, including ^{24}Mg. The experimental data show the interplay of various reaction times, ranging from the lifetime of the compound nucleus to the time associated with shape resonances in the ion-ion potentials [17]. For a more detailed discussion see [2].

6 Relation Between Lifetimes and Decay Widths

The phenomenon of resonance trapping has been discussed also in quantum chemistry for unimolecular reactions. For illustration, let us consider here the unimolecular decay processes in the regime of overlapping resonance states with the goal to elucidate how unimolecular reaction rates depend on resonance widths [18]. Using the definition

$$k^{\text{eff}} = -\frac{d}{dt} \ln\langle\phi(t)|\phi(t)\rangle \tag{15}$$

for the decay rate, and

$$\langle\Gamma\rangle = \frac{1}{N} \sum_{k=1}^{N} \Gamma_k \ , \tag{16}$$

where the sum runs over all N resonance states in the energy region considered, the result is as follows [18]: in all studied cases, the dependence of the average decay rate on $\langle\Gamma\rangle$ for a given energy interval is characterized by a saturation curve. In other words: in the regime of nonoverlapping resonances (the weak coupling regime), the standard relation between decay rate and $\langle\Gamma\rangle$ holds, i.e. the unimolecular decay rate is equal to the resonance width divided by $\hbar$. When, however, the resonance overlap increases (the strong coupling regime), the decay rate saturates as a function of increasing $\langle\Gamma\rangle$. Identifying the average resonance width Γ^{av} with $\langle\Gamma\rangle$, it was claimed [18] that the fundamental quantum mechanical relation between the average decay rate and the average resonance width does not hold in the strong overlapping limit.

This conclusion is, however, justified only under the assumption of uniform level broadening that makes it possible to identify $\langle\Gamma\rangle$ with Γ^{av} [19]. According to the phenomenon of resonance trapping, (13), the levels are, however, broadened non-uniformly in the overlapping regime due to the reordering processes taking place under the influence of the environment into which the system is embedded. As a consequence

$$\sum_{k=1}^{M} \Gamma_k \gg \sum_{k=M+1}^{N} \Gamma_k \ , \tag{17}$$

and Γ^{av} is different from $\langle\Gamma\rangle$ in the overlapping region.

A meaningful definition of the average width of the long-lived states is [20]

$$\Gamma^{\text{av}} = \frac{1}{N-M} \sum_{k=M+1}^{N} \Gamma_k \ . \tag{18}$$

The sum in (18) runs over the $N-M$ long-lived (trapped) states only. These states do not overlap and the value Γ^{av} saturates in the long-time scale.

The saturation of the decay rate in the overlapping regime may be related also to the broadening of the widths distribution occurring in this regime [18]. This result is *not* in contradiction with the conclusion that the saturation is related to resonance trapping. The point is that resonance trapping creates differences in the transmission coefficients for the different states that cause a broadening of the widths distribution [20]. Consequently both, the broadening of the widths distribution and the saturation of the decay widths in the overlapping regime, can be traced back to the same origin, i.e. to resonance trapping or, more generally, to avoided level crossings in the complex plane.

It is possible to invert the discussion: it is not the standard relation between decay rate and average decay width which ceases to hold in the overlapping regime. The saturation of the decay rate is rather a proof of the formation of different time scales. A uniform level broadening does not take place in the system at high level density, and the unimolecular decay rate in the long-time scale is equal to the average resonance width $\Gamma^{\rm av}$ divided by $\hbar$. It should be mentioned here, that also in atomic nuclei a similar saturation effect is known: the spreading width obtained from an analysis of experimental data on isobaric analogue resonances in different nuclei saturates [21].

We conclude that the standard relation between decay rate and average decay width $\Gamma^{\rm av}$ holds in the regime of overlapping resonance states in the long-time scale. The point is that different time scales exist in this regime that are caused by resonance trapping, i.e. by the bifurcation of the widths at the avoided level crossings.

7 Properties of the Broadened States

7.1 Dynamical Localization

An answer to the question whether resonance trapping is accompanied by dynamical changes of the shape of the system, can not be found directly from studies on microwave cavities, since their shape is fixed from outside. Nevertheless, a *dynamical localization* of the wave function density inside the cavity may occur. A study of different wave functions in open microwave cavities showed, indeed, that the localization of the probability density for short-lived and long-lived states inside the cavity is different. While the short-lived states are localized along particular short paths related to the position(s) of the attached wave guide(s), all the long-lived trapped states have pronounced nodal structure that is distributed over the whole cavity [7,8]. The long-lived states can be described well by random matrix theory [9].

A similar result has been obtained in calculations for nuclei [22]. The numerical results obtained for the radial profile of partial widths of 1^- resonance states with 2p-2h structure in ^{16}O show that, also in this case, resonance trapping is accompanied by a dynamical localization of the short-lived states. In other words: *structures in space and time* are created that are characterized by a small radial extension, a short lifetime and a small information entropy [22,23].

7.2 Classical Description

As has been discussed in Sect. 3.2, resonance trapping is accompanied by the broadening of some states when the system is opened to the continuum. An example are the whispering gallery modes that appear in microwave cavities and may give an important contribution to the conductance when the cavity is opened by attaching wave guides to it [8].

From a physical point of view, most interesting is the following fact. Special states of a certain type are characterized by their structure that is similar for all the states of this type. For example, all whispering gallery modes are spatially localized in different groups parallel to each other and near to the convex boundary of the cavity. The states of each group differ by the number of nodes, but not by the localization region. They couple therefore coherently to the continuum of decay channels when the leads are attached in a suitable manner. As a consequence, special states existing in closed systems among other states, may become dominant by opening the system to a small number of decay channels [8]. The widths of these states may increase strongly by trapping other incoherent or less coherent resonance states lying in the same energy region. The special states align with the channels while the trapped states decouple more or less completely from the environment. Eventually, the properties of the system as a whole are determined at the short time scale mainly by the special states. Since these states are localized, they partly lose their wave character, and it is even possible to describe some properties of the system by using the methods of classical physics.

The conductance of the cavity is determined by the non-diagonal terms of the S matrix. The transmission coefficients between channel n and m may be represented by a Fourier transform in order to get the length spectra of the quantum mechanical calculations [8]. They are compared to the histograms of trajectories calculated classically as a function of the length L of the path for the same cavity. In the classical calculations, the trajectories are obtained from paths of different lengths corresponding to a different number of bouncings of the particle at the convex boundary. The results display a remarkable and surprisingly good agreement between the quantum mechanical results of the Fourier analysis and the classical results in spite of the small value of the wave vector of the propagating waves [8].

The correspondence between the quantum mechanical and purely classical results holds not only in the length scale but also in the time scale: varying parametrically the length L of the paths causes corresponding changes in the widths of the special states that agree with the changes of the times for transmission of a classical particle through the cavity [8]. Obviously, the correspondence is related to the spatial localization of the whispering gallery modes due to which the wave properties of the states are somewhat suppressed.

These results illustrate the strong correlations which may be induced in the system due to its coupling to the continuum. The correlations appear in the long time scale and, above all, in the short time scale. While the long-lived trapped states can be described well by random matrix theory, the short-lived coherent modes are regular [9].

8 Concluding Remarks

All the studies of open quantum systems have shown that the coupling of the system to the environment may change the properties of the system. The changes are small as long as the coupling strength between system and environment is smaller than the distance between the individual states of the unperturbed system, i.e. smaller than the distance between the eigenstates of the Hamiltonian H. The changes can, however, not be neglected when the coupling to the continuum is of the same order of magnitude as the level distance or larger. In such a case, the changes can be described neither by perturbation theory nor by introducing statistical assumptions for the level distribution. The point is that non-linear effects become important which cause a redistribution of the spectroscopic properties of the system and, consequently, changes of its properties. Under the influence of the coupling to the continuum, level repulsion as well as level attraction may appear that are accompanied by the tendency to form a uniform time scale for the system in the first case, but different time scales in the second case.

The resonance phenomena are described well by two ingredients also at high level density. The first ingredient is the effective Hamiltonian $\mathcal{H}$ that contains all the basic structure information involved in the Hamiltonian H, i.e. in the Hamiltonian of the corresponding closed system with discrete eigenstates. Moreover, $\mathcal{H}$ contains the coupling matrix elements between discrete and continuous states that account for the changes of the system under the influence of its coupling to the continuum. These matrix elements are responsible for the non-Hermiticity of $\mathcal{H}$ and its complex eigenvalues which determine not only the positions of the resonance states but also their (finite) lifetimes.

The second ingredient is the unitarity of the S matrix that has to be fulfilled in all calculations of resonance phenomena. It is taken into account in the unified description of structure and reaction aspects since any statistical or perturbative assumptions are avoided in solving the basic equation (1).

The studies within the formalism of a unified description of structure and reaction phenomena show that the coupling of the states via the continuum induces correlations that are not small. The correlations are important especially in the short time scale, but appear also in the long time scale. The short-lived states, involving information on the environment, characterize the system far from equilibrium. The long-lived states, however, are described well by random matrix theory. They are more or less decoupled from the environment.

References

1. H. Feshbach, Ann. Phys. (N.Y.) **5**, 357 (1958) and **19**, 287 (1962)
2. J. Okołowicz, M. Płoszajczak, and I. Rotter, Phys. Reports **374**, 271 (2003)
3. I. Rotter, Phys. Rev. E **68**, 016211 (2003)

4. T. Guhr, A. Müller-Groeling, and H.A. Weidenmüller, Phys. Rep. **299**, 190 (1998)
5. A.I. Magunov, I. Rotter, S.I. Strakhova, J. Phys. B **323**, 1669 (1999); J. Phys. B **34**, 29 (2001)
6. I. Rotter, Phys. Rev. C **64**, 034301 (2001); Phys. Rev. E **64**, 036213 (2001)
7. E. Persson, K. Pichugin, I. Rotter, and P. Šeba, Phys. Rev. E **58**, 8001 (1998); P. Šeba, I. Rotter, M. Müller, E. Persson, and K. Pichugin, Phys. Rev. E **61**, 66 (2000); I. Rotter, E. Persson, K. Pichugin, and P. Šeba, Phys. Rev. E **62**, 450 (2000)
8. R.G. Nazmitdinov, K.N. Pichugin, I. Rotter, and P. Seba, Phys. Rev. E **64**, 056214 (2001); R.G. Nazmitdinov, K.N. Pichugin, I. Rotter, and P. Seba, Phys. Rev. B **66**, 085322 (2002)
9. R.G. Nazmitdinov, H.S. Sim, H. Schomerus, and I. Rotter, Phys. Rev. B **66**, 241302(R) (2002)
10. E. Persson, I. Rotter, H.J. Stöckmann, and M. Barth, Phys. Rev. Lett. **85**, 2478 (2000); H.J. Stöckmann, E. Persson, Y.H. Kim, M. Barth, U. Kuhl, and I. Rotter, Phys. Rev. E **65**, 066211 (2002)
11. I. Rotter, Rep. Prog. Phys **54**, 635 (1991)
12. C. Mahaux and H.A. Weidenmüller, *Shell model approach to nuclear reactions* (North-Holland, Amsterdam, 1969)
13. A.M. Lane and R.G. Thomas, Rev. Mod. Phys. **30**, 257 (1958)
14. H.W. Barz, I. Rotter, and J. Höhn, Nucl. Phys. A **275**, 111 (1977)
15. K. Bennaceur, F. Nowacki, J. Okołowicz, and M. Płoszajczak, J. Phys. G **24**, 1631 (1998); Nucl. Phys. A **651**, 289 (1999); K. Bennaceur, J. Dobaczewski, and M. Ploszajczak, Phys. Rev. C **60**, 034308 (1999); R. Shyam, K. Bennaceur, J. Okolowicz, and M. Ploszajczak, Nucl. Phys. A **669**, 65 (2000); K. Bennaceur, F. Nowacki, J. Okolowicz, and M. Ploszajczak, Nucl. Phys. A **671**, 203 (2000); K. Bennaceur, N. Michel, F. Nowacki, J. Okolowicz, and M. Ploszajczak, Phys. Letters B **488**, 75 (2000); N. Michel, J. Okolowicz, F. Nowacki, and M. Ploszajczak, Nucl. Phys. A **703**, 202 (2002)
16. S. Drożdż, J. Okołowicz, M. Płoszajczak, and I. Rotter, Phys. Rev. C **62**, 4313 (2000)
17. P. Braun-Munzinger and J. Barrette, Phys. Rep. **87**, 209 (1982)
18. U. Peskin, H. Reisler, and W.H. Miller, J. Chem. Phys. **101**, 9672 (1994); ibid. **106**, 4812 (1997)
19. I. Rotter, J. Chem. Phys. **106**, 4810 (1997)
20. E. Persson, T. Gorin, and I. Rotter, Phys. Rev. E **54**, 3339 (1996); Phys. Rev. E **58**, 1334 (1998)
21. H.L. Harney, A. Richter, and H.A. Weidenmüller, Rev. Mod. Phys. **58**, 607 (1986); J. Reiter and H.L. Harney, Z. Phys. **337**, 121 (1990)
22. W. Iskra, M. Müller, and I. Rotter, Phys. Rev. C **51**, 1842 (1995)
23. W. Iskra, M. Müller, and I. Rotter, J. Phys. G **19**, 2045 (1993); ibid. **20**, 775 (1994)

Variational Approaches to the Dynamics of Mixed Quantum States

Remus-Amilcar Ionescu

Abstract. The time-dependent Schrödinger equation could be derived starting from the Dirac–Frenkel variational principle for the Dirac action $\int dt \langle \Psi(t)|i\hbar\frac{\partial}{\partial t} -H|\Psi(t)\rangle$. Restricting the variational space one can obtain different approximations. For instance one can obtain the time-dependent-Hartree–Fock approximation (TDHF) if the variational space is restricted to states which can be written as single Slater determinants. We discuss an approach based on the quantum variational principle which works also for mixed quantum states, states which are described by a density operator. We use the formalism of thermo field dynamics proposed by Umezawa to describe the mixed quantum states as pure states in a larger Hilbert space and we apply the Dirac–Frenkel variational principle in this space. In this way, using restricted variational states which have the form of vacuum states for Bogoliubov quasiparticles in the full Hilbert space (which mix not only creation and annihilation operators for physical degrees of freedom, but also creation and annihilation operators for unphysical degrees of freedom introduced in the formalism of thermo field dynamics) one can derive time-dependent-Hartree–Fock–Bogoliubov type of equation for mixed states. We illustrate this approach in detail for a single bosonic degree of freedom and we comment on the possibilities to use it for description of systems with an infinite number of degrees of freedom.

1 Introduction

A transport description of many-body non-equilibrium systems is a very convenient approximation used with success in the study of heavy ion collisions in the last two decades [1,2]. It is also used in the solid state physics to describe the transport properties like electric conductivity, thermal conductivity, etc, for more than 40 years. In this approach the one-body phase-space distribution function satisfies a semi-classical transport equation of Boltzmann type. The validity of the transport description can be established by a rigorous derivation from quantum field theory [3].

The dynamical evolution of a many-body system can be described, depending on the problem, either by classical mechanics or by quantum mechanics (quantum field theory being the quantum mechanics of classical fields). The classical mechanics for mixed states, neglecting the two-particle correlations, results in the celebrated Boltzmann equation for the phase-space one-particle distribution function. The conditions to use classical mechanics are not fulfilled for nucleons in a nucleus. Nevertheless, the Boltzmann-like equation was proven to be useful in the description of heavy ion collisions.

There are two different ways to derive useful approximations for quantum systems. One way is to work in the Heisenberg picture (all the time dependence is included in time-dependent operators). Then, the standard perturbation expansion results into the Dyson equation for the one-particle Green functions. Further approximations result into a transport equation of Boltzmann type [3] (Boltzmann–Uehling–Uhlenbeck equation with collision term which takes into account the Pauli exclusion principle for fermions).

The other way is to work in the Schrödinger picture of quantum mechanics and to use judiciously chosen restricted trial states in the Dirac–Frenkel variational principle equivalent to the time-dependent Schrödinger equation. The problem is that this variational principle works only for pure states and we expect that in practice the state of the system would be more correctly approximated by a mixed state. Let us mention that there is a variational principle for the time evolution of a quantum density matrix in the literature [4], but the resulting equations for the approximate dynamical evolution of the system between time t_1 and time t_2 depend on the observable we want to calculate and they should be solved in general with both initial (at t_1) and final (at t_2) conditions and not as simple evolution equations with initial conditions.

In what follows we use the thermo field dynamics [5] to describe the mixed states as pure states in an enlarged Hilbert space. In this way the machinery of the Dirac–Frenkel variational principle could be used to derive approximations also for the time-evolution of mixed states, which are important for all the physical systems far from equilibrium.

The paper is organized as follows. First we will present the basic results of the time-dependent variational principle. Then we will discuss the description of mixed states as pure ones as it is realized in the formalism of thermo field dynamics. These results are used to construct a Lagrangian description for the parameters of some trial states; the trial states are chosen as vacuum states for Bogoliubov quasiparticles in the full Hilbert space of thermo field dynamics. Then we will discuss how the Lagrangian dynamics can be reformulated in the Hamiltonian form with a noncanonical Poisson bracket. In the next section we particularize the results to a simple example and we end with a discussion about the application of this treatment to problems in field theory.

2 Time-Dependent Variational Principle for Pure Quantum States

In the Schrödinger picture of quantum mechanics, in which the state (a vector in the Hilbert space of the system) depends on time and the observables (linear operators on the Hilbert space) are time independent, the time evolution of a pure state is given by the time-dependent Schrödinger equation

$$i\frac{d}{dt}\left|\Psi(t)\right\rangle = H\left|\Psi(t)\right\rangle \ , \tag{1}$$

where H is the Hamiltonian and $|\Psi(t)\rangle$ is the normalized many-body state which describes the physical system. This equation of motion can be obtained from the variation of the action [6–9]

$$A = \int_{t1}^{t2} dt \, \langle\Psi(t)| \, i\frac{d}{dt} - H \, |\Psi(t)\rangle \ . \tag{2}$$

This is the Dirac–Frenkel variational principle. In general, the set $|\Psi(t)\rangle$ contains infinitely many dynamical degrees of freedom, for example the complex coefficients of an orthonormal basis in the Hilbert space.

The advantage of the reformulation of the Schrödinger equation as a variational principle is the possibility that a suitably chosen restriction in the dynamical variables will lead to an useful approximation of the full problem. For instance, in the case of n fermions, one can obtain the time-dependent-Hartree–Fock approximation (TDHF) if the variational n-body space is restricted to states which can be written as single Slater determinants formed with n orthonormal one-body states [10], or one can obtain the Fermionic Molecular Dynamics (FMD) if the many-body trial state is a Slater determinant formed from n (nonorthogonal) one-body Gaussian wave packets [11]. In TDHF the dynamical variables are the n one-body states and in FMD the dynamical variables are the mean positions, mean momentums and complex widths of the n Gaussian wave packets.

From the action given by Eq. (2), and some restriction of the dynamical variables one can see that the time evolution of the system is a classical one for a system with the Lagrangian

$$\mathcal{L} = \langle\Psi(t)| \, i\frac{d}{dt} \, |\Psi(t)\rangle - \mathcal{H} \ , \tag{3}$$

where $\mathcal{H}$ is the expectation value of the quantum Hamiltonian

$$\mathcal{H} = \langle\Psi(t)| \, H \, |\Psi(t)\rangle \ . \tag{4}$$

The Lagrangian is a function of the usually complex dynamical variables $\Psi(t)$, so that $\Psi(t)$ and $\Psi^*(t)$ may be regarded as independent variables.

Before to discuss how the variational principle works with some particular trial states, we want to present a way to describe the mixed quantum states (states given by a density operator) as pure states. In this way we can use the Dirac–Frenkel variational principle to obtain variational approximations for systems in mixed states, too.

3 Mixed Quantum States and Thermofield Dynamics

In the formalism of thermo field dynamics [5,2] a mixed state is described as a pure state, the price to be paid for that being a doubling of the number of degrees of freedom. The idea is to consider the density operator, ρ, as a vector in the Liouville space (the Liouville space is the vector space of linear operators acting on the Hilbert space of quantum system). Therefore, the Liouville space is the direct product of two copies of the Hilbert space of the system. We consider that the algebra of operators acting on the Hilbert space is generated by the creation and annihilation operators $a^\dagger$, a satisfying the commutation or anticommutation relations

$$\begin{aligned} a\,a^\dagger \mp a^\dagger a &= 1\,, \\ a\,a \mp a\,a &= 0\,, \\ a^\dagger a^\dagger \mp a^\dagger a^\dagger &= 0\,, \end{aligned} \tag{5}$$

depending on the bosonic or fermionic character of the considered system. In this case the algebra of operators acting on the Liouville space is generated by two commuting (for bosons) or anticommuting (for fermions) copies of the physical creation and annihilation operators, $a^\dagger$, a, denoted $a^\dagger$, a and $\tilde{a}^\dagger$, $\tilde{a}$. The vector in the Liuoville space is denoted $||\Psi\rangle\rangle$ and one of the main results of thermo field dynamics is that the time evolution of the state $||\Psi\rangle\rangle$ is given by a Schrödinger equation

$$i\frac{d}{dt}\,||\Psi(t)\rangle\rangle = \underline{H}\,||\Psi(t)\rangle\rangle\ , \tag{6}$$

with

$$\underline{H} = H(a^\dagger, a) - H(\tilde{a}^\dagger, \tilde{a})\ , \tag{7}$$

$H(a^\dagger, a)$ being the Hamiltonian of the system expressed in terms of the creation and annihilation operators. Therefore, instead to work with the density operator, $\rho(t)$, we can work equivalently with the pure state $||\Psi(t)\rangle\rangle$. The above evolution equation is equivalent to the quantum Liouville equation for the density matrix, ρ, and the expectation value of an operator O is given by its matrix element

$$\langle O\rangle = \langle\langle\Psi||\,O\,||\Psi\rangle\rangle \tag{8}$$

instead of $\mathrm{Tr}(\rho O)$. To illustrate the correspondence between the description by density matrix and the pure state description in Liouville space we give the correspondence between the density matrix for thermally equilibrated systems with one degree of freedom and the pure state in thermo field dynamics. For a non-interacting bosonic system the density matrix (not normalized) is

$$\rho \sim \mathrm{e}^{\frac{1}{k_B T} a^\dagger a}\ , \tag{9}$$

where k_B is the Boltzmann constant and T is the temperature of the system and the corresponding pure state in the thermo field dynamics is [5]

$$||\Psi\rangle\rangle \sim \mathrm{e}^{\operatorname{arcsinh}\left(\mathrm{e}^{\frac{1}{k_B T}}-1\right)^{-\frac{1}{2}} a^\dagger \tilde{a}^\dagger}\,||0\rangle\rangle\ , \tag{10}$$

where $||0\rangle\rangle$ is the vacuum state in the Liouville space annihilated by the operators a and $\tilde{a}$. In the case of a non-interacting fermionic system with one degree of freedom we have the same density matrix, Eq. (9), (now the creation and annihilation operators will satisfy anticommutation relations) and the corresponding pure state is [5]

$$||\Psi\rangle\rangle \sim \mathrm{e}^{\arcsin\left(\mathrm{e}^{\frac{1}{k_B T}}+1\right)^{-\frac{1}{2}} a^\dagger \tilde{a}^\dagger} ||0\rangle\rangle \ . \tag{11}$$

We remark that the above states are annihilated by certain linear combinations of a and $\tilde{a}^\dagger$ and respectively $\tilde{a}$ and $a^\dagger$. In this sense the thermal equilibrated states are vacuum states for some special Bogoliubov quasiparticles named "thermal quasiparticles". The Bogoliubov transformation depend on temperature and does not mix only the creation and annihilation operators for physical particles, $a^\dagger$ and a, but also the creation and annihilation operators for unphysical particles, $\tilde{a}^\dagger$ and $\tilde{a}$, introduced in thermo field dynamics.

The above picture can be extended to a finite number, M, of fermionic or bosonic operators $a_\alpha^\dagger$ and a_α, $\alpha = 1, \ldots, M$ by introducing the operators $\tilde{a}_\alpha^\dagger$ and $\tilde{a}_\alpha$, $\alpha = 1, \ldots, M$.

To summarize this section, the description of a mixed state can be realized in the same way as for a pure state if we use $N = 2M$ creation and annihilation operators denoted by $A_i^\dagger$, A_i, $i = 1, \ldots, N$, where the set of operators A_i include a_α and $\tilde{a}_\alpha$.

Inspired by the form of thermal states, Eqs. (10, 11), we will use trial states which are annihilated by arbitrary linear combinations of the operators A_i and $A_i^\dagger$ (see Eq. (13) below).

4 Trial States and Classical Lagrange Dynamics

Let us consider a system described in terms of an arbitrary finite number, N, of creation and annihilation operators $A_i^\dagger$, A_i satisfying the commutation or anticommutation relations

$$\begin{aligned} A_i A_j^\dagger \mp A_j^\dagger A_i &= \delta_{ij} \ , \\ A_i A_j \mp A_j A_i &= 0 \ , \\ A_i^\dagger A_j^\dagger \mp A_j^\dagger A_i^\dagger &= 0 \ , \end{aligned} \tag{12}$$

where the upper sign refers to bosons and the lower one to fermions.

We choose a restricted class of variational states which have the form of a coherent superposition of $0, 2, 4, \cdots$ particle number components

$$|Z\rangle = \mathrm{e}^{\frac{1}{2} Z_{ij} A_i^\dagger A_j^\dagger} |0\rangle \ , \tag{13}$$

where $|0\rangle$ is the normalized vacuum state for fermionic or bosonic operators satisfying

$$A_i |0\rangle = 0 \ , \tag{14}$$

and Z is a skew symmetric complex matrix for fermions and a symmetric complex one for bosons and it parameterizes the trial state. The Einstein convention of summation over repeated indices is used. Our goal is to obtain equations for the time evolution of this variable. As the system was assumed to have an arbitrary finite number of degrees of freedom, N, Z is a $N \times N$ matrix and only at the end we allow N to become infinity.

When we want to describe pure states only, the operators $A_i^\dagger$ and A_i are the physical creation and annihilation operators of the system. The generalization to mixed states is simply done by doubling the dimension of the matrix Z which parameterizes the trial state.

The trial state given by Eq. (13) is the most general state which is not orthogonal to the vacuum in the fermionic case [10] (Thouless theorem). It is annihilated by certain linear combination of creation and annihilation operators and in this sense it can be considered as a vacuum for Bogoliubov quasiparticles. It is extensively used in time-independent problems in solid state and nuclear physics, mainly in the pure state (zero temperature) approximation, but it was not used in the time-dependent case.

Although it is not an eigenstate of the particle number operator and the Hamiltonian in the fermionic case conserves the number of particles, this is not a problem in practice for time-independent problems (e.g. BCS or Hartree–Fock–Bogoliubov calculations for nuclear structure) because, imposing a mean particle number, the particle number fluctuations are proven to be small; they are completely negligible in condensed matter physics. Moreover, when the particle number is small and one cannot neglect the unphysical particle number fluctuations introduced by such a trial state there are methods to retain only the component with the correct particle number by projecting the above trial state on a subspace of the Hilbert space [10,12–14]. We will not discuss such projection techniques in what follows. We want to stress that, as the above trial state is a superposition of components with $0, 2, 4, \cdots$ particles, it is useful for the description of systems with an even number of particles; the systems with an odd number of particles could be described using 1-particle excitations of the considered quasiparticle vacuum [10,15].

In the bosonic case there is a possibility to use a coherent state as trial state, too; it will be also a vacuum for some quasiparticles [10], but we restrict ourselves in the following to the above form.

The quantum state given by Eq. (13) is not normalized to unity and, before we can proceed further, we need to calculate its norm. Let us consider the complex valued function $\mathcal{F}(\xi)$ of a real variable ξ given by

$$\mathcal{F}(\xi) = \langle \xi \cdot Z_1 | Z_2 \rangle \ , \tag{15}$$

which reduces to $\langle Z_1 | Z_2 \rangle$ for $\xi = 1$ and when $Z_1 = Z_2 = Z$ it gives the norm of the state $|Z\rangle$. Taking into account the form of the trial state given by Eq. (13) one can see that $\mathcal{F}(0) = 1$ due to the normalization of the vacuum to unity. To calculate $\mathcal{F}(1)$ we use a trick to obtain a first order differential equation for the function $\mathcal{F}(\xi)$. The solution of this differential equation which satisfies $\mathcal{F}(0) = 1$ will results into the overlap between two considered trial states, in particular into the norm of the state. Using the form of the hermitian adjoint of the state $|Z\rangle$

$$\langle Z| = \langle 0| \, e^{\frac{1}{2} Z^*_{ij} A_j A_i} = \langle 0| \, e^{\frac{1}{2} (Z^\dagger)_{ij} A_i A_j} \,, \tag{16}$$

where $*$ denotes complex conjugation and $\dagger$ denotes the adjoint of a matrix (transpose plus complex conjugation), one can write

$$\frac{d\mathcal{F}(\xi)}{d\xi} = \frac{1}{2} (Z_1^\dagger)_{ij} \langle \xi \cdot Z_1 | \, A_i A_j \, | Z_2 \rangle \ . \tag{17}$$

Using the commutation or anticommutation relations for creation and annihilation operators, Eq. (12), and the properties of the vacuum, Eq. (14), one obtains

$$A_j \, |Z\rangle = Z_{jk} A_k^\dagger \, |Z\rangle \ , \tag{18}$$

$$\langle Z| \, A_k^\dagger = \langle Z| \, A_l (Z^\dagger)_{lk} \ . \tag{19}$$

From the above two relations and the commutation relations for creation and annihilation operators, Eq. (12), we get

$$\begin{aligned} \pm \langle Z_1| \, A_j A_i \, |Z_2\rangle &= \langle Z_1| \, A_i A_j \, |Z_2\rangle \\ &= (Z_2)_{jk} \langle Z_1| \, A_i A_k^\dagger \, |Z_2\rangle \\ &= (Z_2)_{ji} \langle Z_1|Z_2\rangle \pm (Z_2)_{jk} \langle Z_1| \, A_k^\dagger A_i \, |Z_2\rangle \\ &= (Z_2)_{ji} \langle Z_1|Z_2\rangle \pm (Z_2)_{jk} (Z_1^\dagger)_{lk} \langle Z_1| \, A_l A_i \, |Z_2\rangle \\ &= (Z_2)_{ji} \langle Z_1|Z_2\rangle + (Z_2 Z_1^\dagger)_{jl} \langle Z_1| \, A_l A_i \, |Z_2\rangle \ , \end{aligned}$$

where the upper sign refers to bosons and the lower one to fermions and in the last line the symmetry or skew symmetry of the Z matrix was used. Therefore, we have

$$(\mathbf{1} \mp Z_2 Z_1^\dagger)_{jl} \langle Z_1| \, A_l A_i \, |Z_2\rangle = \pm (Z_2)_{ji} \langle Z_1|Z_2\rangle \ ,$$

from which, multiplying by $(\mathbf{1} \mp Z_2 Z_1^\dagger)_{kj}$ and replacing the indices k, i by i, j, we obtain

$$\langle Z_1| \, A_i A_j \, |Z_2\rangle = \pm \left((\mathbf{1} \mp Z_2 Z_1^\dagger)^{-1} Z_2 \right)_{ij} \langle Z_1|Z_2\rangle \ . \tag{20}$$

The above relation for Z_1 replaced by $\xi \cdot Z_1$ is now used in Eq. (17) and with the appropriate symmetry of the matrix Z we obtains the following first order differential equation

$$\frac{d\mathcal{F}(\xi)}{d\xi} = \frac{1}{2} \mathrm{Tr} \left[Z_2 Z_1^\dagger \left(\mathbf{1} \mp \xi Z_2 Z_1^\dagger \right)^{-1} \right] \mathcal{F}(\xi) \,, \tag{21}$$

where Tr denotes the trace of a matrix and the cyclic permutation invariance of the trace was used.

The solution of this differential equation satisfying $\mathcal{F}(0) = 1$ is given by a simple integration

$$\begin{aligned}\mathcal{F}(\xi) &= \mathrm{e}^{\mp\frac{1}{2}\mathrm{Tr}\left[\ln\left(\mathbf{1}\mp\xi Z_2 Z_1^\dagger\right)\right]} \\ &= \mathrm{e}^{\mp\frac{1}{2}\mathrm{Tr}\left[\ln\left(\mathbf{1}\mp\xi Z_1^\dagger Z_2\right)\right]} .\end{aligned} \tag{22}$$

The equivalence of the above two forms can be established by using the power expansion of the logarithmic function and the invariance of the trace of a product with respect to a cyclic permutation of the factors. Therefore, the overlap of two trial states is

$$\langle Z_1 | Z_2 \rangle = \mathrm{e}^{\mp\frac{1}{2}\mathrm{Tr}\left[\ln\left(\mathbf{1}\mp Z_1^\dagger Z_2\right)\right]} , \tag{23}$$

which in the fermionic case is known as Onishi formula [10,16]. The norm of the state given by Eq. (13) is therefore

$$\langle Z | Z \rangle = e^{\mp\frac{1}{2}\mathrm{Tr}\left[\ln\left(\mathbf{1}\mp Z^\dagger Z\right)\right]} . \tag{24}$$

Using Eq. (20) we have the following expectation values

$$\frac{\langle Z| A_i A_j |Z\rangle}{\langle Z|Z\rangle} = \pm \left(Z(\mathbf{1} \mp Z^\dagger Z)^{-1}\right)_{ij} , \tag{25}$$

Taking the hermitian conjugate of the above relation we get

$$\frac{\langle Z| A_i^\dagger A_j^\dagger |Z\rangle}{\langle Z|Z\rangle} = \pm \left((\mathbf{1} \mp Z^\dagger Z)^{-1} Z^\dagger\right)_{ij} , \tag{26}$$

and using Eq. (18) it is a simple exercise to obtain

$$\frac{\langle Z| A_i^\dagger A_j |Z\rangle}{\langle Z|Z\rangle} = \pm \left[(\mathbf{1} \mp Z^\dagger Z)^{-1} - \mathbf{1}\right]_{ij} . \tag{27}$$

The expectation values of products of an even number, larger than two, of creation and annihilation operators can be obtained by taking the partial derivatives of the Eq. (23) with respect to the elements of matrix Z_1 or/and Z_2, using the commutation relations, Eq. (12), and properties of the trial state, Eq. (18), and at the end the result is written for $Z_1 = Z_2 = Z$. The expectation value of products of an odd number of creation and annihilation operators vanishes for trial state (13). In this way we can calculate $\mathcal{H}(Z, Z^\dagger)$ as the Hamiltonian of the system is given in terms of products of creation and annihilation operators.

To calculate the Lagrangian given by Eq. (3) with the trial states of the form written in Eq. (13) we need to use normalized states and the first term in Eq. (3) will be

$$\langle Z| \frac{1}{\langle Z|Z\rangle^{1/2}} i \frac{d}{dt} \frac{1}{\langle Z|Z\rangle^{1/2}} |Z\rangle ,$$

with $\langle Z|Z\rangle$ given by Eq. (24). The time derivative of the norm will result in a term which is a total time derivative and is irrelevant in the Lagrangian, so the contribution of the above term to the Lagrangian can be written as

$$\frac{1}{\langle Z|Z\rangle}\langle Z|\,i\frac{d}{dt}\,|Z\rangle \ .$$

Now we can use the form of the trial state, Eq. (13), and the above contribution can be expressed as

$$\frac{1}{2}\frac{dZ_{ij}}{dt}\frac{\langle Z|\,A_i^\dagger A_j^\dagger\,|Z\rangle}{\langle Z|Z\rangle} \ .$$

With the expectation value given in Eq. (26) we can write the Lagrangian in the form

$$\begin{aligned}\mathcal{L} &= \pm i\frac{1}{2}\left((\mathbf{1}\mp Z^\dagger Z)^{-1}\right)_{ik} Z_{kj}^\dagger \frac{dZ_{ij}}{dt} - \mathcal{H}(Z,Z^\dagger)\\ &= i\frac{1}{2}\mathrm{Tr}\left[(\mathbf{1}\mp Z^\dagger Z)^{-1} Z^\dagger \frac{dZ}{dt}\right] - \mathcal{H}(Z,Z^\dagger)\ .\end{aligned} \tag{28}$$

Therefore, the quantum dynamical problem was reformulated as classical Lagrange dynamics for the parameters of the trial state. We want to stress that this approach works for pure states as well as for mixed states, but the dimension of the configuration space is larger in the second case.

5 Phase Space, Poisson Bracket and Hamilton Dynamics

We restrict in this section to the fermionic case, as the bosonic case can be studied along the same lines. First we remark that the Lagrangian (28) is first order in the time derivatives and the standard definitions of the conjugate momenta $P_{ij} = \frac{\partial \mathcal{L}}{\partial \frac{dZ_{ij}}{dt}}$ result into constraint equations in the associated phase space. This means that the phase space can be identified with the configuration space, but the Poisson brackets will be not in the canonical form. To approach this problem we can use the general method of Dirac to take into account the constraints by using them to replace the canonical Poisson bracket by a noncanonical one. In our particular case we can obtain directly the Poisson bracket as follows: the independent coordinates in the configuration space are Z_{ij}, for $j > i$ and their conjugate momenta are

$$P_{ij} = i\left[(\mathbf{1}+Z^\dagger Z)^{-1}Z^\dagger\right]_{ij} \ . \tag{29}$$

We mention that the other matrix elements of Z are not independent since $Z_{ij} = -Z_{ji}$. Therefore, we have the following Poisson bracket between coordinates and associated momenta

$$\{Z_{mn}, P_{ij}\} = \{Z_{mn}, i(\mathbf{1}+Z^\dagger Z)^{-1}_{ik} Z^\dagger_{kj}\} = \delta_{mi}\delta_{nj} \tag{30}$$

for $j > i$ and $n > m$. Using the skew symmetry of the matrix Z, we can write a relation which is true for every i, j, m, n

$$\{Z_{mn}, P_{ij}\} = \{Z_{mn}, i(\mathbf{1} + Z^\dagger Z)^{-1}_{ik} Z^\dagger_{kj}\} = \delta_{mi}\delta_{nj} - \delta_{ni}\delta_{mj} \ . \qquad (31)$$

From the above relation it is a straightforward but tedious exercise to obtain the Poisson bracket between the components of the matrices Z and $Z^\dagger$

$$\{Z_{mn}, Z^\dagger_{ij}\} = i\big[(\mathbf{1} + Z^\dagger Z)_{mj}(\mathbf{1} + Z^\dagger Z)_{in} - (\mathbf{1} + Z^\dagger Z)_{nj}(\mathbf{1} + Z^\dagger Z)_{im}\big] \ . \qquad (32)$$

The components of the matrix Z (or $Z^\dagger$) have zero Poisson bracket with themselves as they are coordinates of the configuration space in the Lagrange formulation.

Therefore, with the Poisson bracket, Eq. (32), we can use the classical Hamiltonian $\mathcal{H}(Z, Z^\dagger)$ (the expectation value of the quantum Hamiltonian of the system for pure states, or the thermal Hamiltonian, Eq. (7), for mixed states) together with the Poisson bracket, Eq. (32), to describe the system. In this way the quantum dynamics is approximated by a classical Hamiltonian dynamics for the parameters of trial states. The Poisson bracket is not in the canonical form, but, due to Darboux theorem, there is a local system of coordinates in which it has the canonical form. Nevertheless, we do not need to find the canonical system of coordinates in practice.

6 Discussion

The variational formalism with the trial states of the form given in Eq. (13) was developed for an arbitrary finite number, N, of fermionic or bosonic operators. We are now at the position to discuss the limit $N \to \infty$, relevant for quantum field theories.

For a field theory (we have in mind the nonrelativistic Schrödinger field theory as the simplest example) there are an infinite number of creation and annihilation operators. The indices i can be taken as the coordinates of the space in which the field theory lives plus the spin indices and the sums will be replaced by integrals. In this way the parameters of the trial state, Eq. (13), become an infinite matrix Z_{x_1,x_2} which is a complex two-point function. Therefore, the finite number of creation/annihilation operators case can be generalized to a classical Hamiltonian field theory for the complex function $Z(x_1, x_2)$. The classical Hamiltonian will be obviously the expectation value of the quantum Hamiltonian on the trial states, but the Poisson bracket will be highly non-trivial being non-local. It is of interest to see how to apply further approximations to this non-local Poisson bracket. This subject is under investigation.

The formalism presented is general. To have a taste of its power we collect below some results obtained in one of the simplest imaginable case, namely the one-dimensional quantum system [18]. For this case we have the bosonic creation and annihilation operators, a, $a^\dagger$, we restrict ourselves to pure state, and, slightly generalizing the trial state, Eq. (13), we use the following trial state

$$|Z\eta\rangle = \mathrm{e}^{\frac{1}{2}Za^\dagger a^\dagger + \eta a^\dagger}\,|0\rangle\ , \tag{33}$$

where Z and η are two complex parameters. The above trial state is a Gaussian and the parameter η is related to its mean position and the parameter Z to its width. One can obtain, using simple but long algebraic manipulations as above, the following expectations values:

$$\frac{\langle Z\eta|\,a\,|Z\eta\rangle}{\langle Z\eta|Z\eta\rangle} = \frac{\eta+\eta^* Z}{1-ZZ^*}\,, \tag{34}$$

$$\frac{\langle Z\eta|\,a\ a\,|Z\eta\rangle}{\langle Z\eta|Z\eta\rangle} = \left(\frac{\eta+\eta^* Z}{1-ZZ^*}\right)^2 + \frac{Z}{1-ZZ^*}\,, \tag{35}$$

$$\frac{\langle Z\eta|\,a^\dagger a\,|Z\eta\rangle}{\langle Z\eta|Z\eta\rangle} = \frac{(\eta+\eta^* Z)(\eta^*+\eta Z^*)}{(1-ZZ^*)^2} + \frac{ZZ^*}{1-ZZ^*}\,. \tag{36}$$

The expectation value of $a^\dagger$ is the complex conjugate of Eq. (34) and the expectation value of $a^\dagger a^\dagger$ is the complex conjugate of Eq. (35). The norm of the considered trial state is proven to be

$$\ln\langle Z\eta|Z\eta\rangle = -\frac{1}{2}\ln(1-ZZ^*) + \frac{\eta\eta^*}{1-ZZ^*} + \frac{1}{2}\frac{\eta^2 Z^* + \eta^{*2} Z}{1-ZZ^*}\,. \tag{37}$$

If we take a Hamiltonian, H, expressed in terms of the operators $a^\dagger$, a, it is only a problem of algebra to calculate $\mathcal{H}(Z,\eta,Z^*,\eta^*)$ as its expectation value on the trial state. The classical dynamics in the space of the complex trial state parameters is given by the Lagrangian

$$\mathcal{L} = \frac{i}{2}\left[\left(\frac{\eta^*+\eta Z^*}{1-ZZ^*}\right)^2 + \frac{Z^*}{1-ZZ^*}\right]\frac{dZ}{dt} + i\frac{\eta^*+\eta Z^*}{1-ZZ^*}\frac{d\eta}{dt} - \mathcal{H}(Z,\eta,Z^*,\eta^*) \tag{38}$$

or, alternatively, by the classical Hamiltonian, $\mathcal{H}(Z,\eta,Z^*,\eta^*)$, with the non-vanishing Poisson brackets

$$\begin{aligned}
\{Z,Z^*\} &= -2i(1-ZZ^*)^2\\
\{Z,\eta^*\} &= 2i(\eta^*+\eta Z^*)(1-ZZ^*)\\
\{\eta,\eta^*\} &= -2i(\eta^*+\eta Z^*)(\eta+\eta^* Z) - i(1-ZZ^*)
\end{aligned} \tag{39}$$

together with their complex conjugates. In this way the quantum dynamics of the one-dimensional problem is approximated by the classical Hamiltonian dynamics in a two dimensional complex space, or a four dimensional real space. It is obvious that, despite of its classical character, the dynamics retains some quantum effects through the increased dimensionality of the phase space, and the above Poisson brackets. We want to stress that the above Poisson brackets reduce to the usual Poisson bracket

$$\{\eta, \eta^*\} = -i \; , \tag{40}$$

if we impose the constraints $Z = 0$ and $Z^* = 0$ (they are second class constraints in the terminology of Dirac), as it should be (to impose the constraints it is not enough to put $Z = Z^*$ in the last of Eqs. (39), we need to follow the Dirac prescription).

Alternatively, we can impose the constraints $\eta = \eta^* = 0$. They are also second class constraints and we get

$$\{Z, Z^*\} = -2i(1 - ZZ^*)^2 \; . \tag{41}$$

This corresponds to our initial trial state, Eq. (13), for this simple one-dimensional case (it happens that the above Poisson bracket coincides with the first Eqs. (39)!).

7 Conclusions

We have presented a classical description for the dynamics of the parameters of a trial states. The formalism works in the same way for pure and mixed state descriptions of a system, the only difference being the number of dimensions of the configuration space.

This description can be extended to infinite dimensional quantum system where it results into a classical field theory for the parameters of the trial state; the fields are two-point complex functions and further approximations might be introduced in the same way as in the Green functions formalism as the one-particle Green functions are also two-point functions. The difference is that the formalism is completely deterministic and there are no problems with the nonlocality in time of some of the quantities involved.

The present description includes some quantum effects despite the fact that the equation of motion looks classically. It could be an alternative to the nonequilibrium Green function formalism for description of time evolution of systems with important quantum effects, as is the case for heavy ion collisions, especially in the initial stage.

The methods developed in this work were particularized to the simple case of a one-dimensional quantum system. The Lagrangian and the Poisson brackets for the Hamiltonian formalism has been given.

The present approach can be used to treat a system of bosons, or a coupled fermion-boson system, as quantum hadrodynamics [17] which is the modern way for describing the nuclear many-body problem in the framework of effective field theories.

Acknowledgments

The author would like to thank Professor H.H. Wolter for interesting discussions on the present subject.

References

1. W. Botermans, R. Malfliet: Phys. Rep. **198** (1990) 115
2. P.A. Henning: Phys. Rep. **253** (1995) 235
3. L.P. Kadanoff, G. Baym: *Quantum statistical theory* (Benjamin/Cummings, Menlo Park, CA, 1962)
4. R. Balian, M. Vénéroni: Phys. Rev. Lett. **47** (1981) 1353; Ann. Phys. **164** (1985) 334
5. Y. Takahashi, H. Umezawa: Coll. Phenomena **2** (1975) 55
6. P.A.M. Dirac: Proc. Camb. Phil. Soc. **26** (1930) 376
7. J. Frenkel: *Wave mechanics* (Oxford University Press, 1934)
8. A.K. Kerman, S.E. Koonin: Ann. Phys. **100** (1976) 332
9. P. Kramer, M. Saraceno: *Geometry of the time-dependent variational principle in quantum mechanics*, Lecture Notes in Physics **140** (Springer, Berlin, 1981)
10. P. Ring, P. Schuck: *The nuclear many-body problem* (Springer, Berlin, 1980)
11. H. Feldmeier, J. Schnack: Rev. Mod. Phys. **72** (2000) 655
12. R.E. Peierls, Y. Yoccoz: Proc. Phys. Soc. **A70** (1957) 381
13. H.A. Flocard, N. Onishi: Ann. Phys. **254** (1997) 275
14. J.A. Sheikh, P. Ring: Nucl. Phys. **A665** (2000) 71 (*nucl-th*/9907065)
15. V.G. Soloviev: *Theory of atomic nuclei: quasiparticles and phonons* (Institute of Physics Publishing, Bristol, 1992)
16. N. Onishi, S. Yoshida: Nucl. Phys. **80** (1966) 367
17. B.D. Serot, J.D. Walecka: Int. J. Mod. Phys. **E6** (1997) 515 (*nucl-th*/9701058)
18. R.A. Ionescu: *Simple quantum systems and time-dependent variational approximations*, in preparation

Thermal Non-analyticity in the Damping Rate of a Massive Fermion

Peter A. Henning

Abstract. In a hot system every excitation acquires a finite lifetime, manifesting itself in a non-zero spectral width. Ordinary damping as well as quantum memory effects arise from this nontrivial spectral function. This report presents an alternative method for the self-consistent calculation of the spectral width of a fermion coupled to massless bosons: The self-consistent summation of the corresponding Fock diagram eliminates all infrared divergences although the bosons are not screened at all. The solutions for the fermion damping rate are analytical in the coupling constant g, but not analytical in the temperature parameter, i.e., $\gamma \propto g^2 T + \mathcal{O}(g^4 T \log(T/M))$.

1 Introduction

The *damping rate* of a fermion moving through a hot system, i.e., the width of its spectral function, is an interesting quantity for non-equilibrium physics: It allows one to estimate the relaxation time in collisions as well as the quantum memory time [1] and radiation properties [2].

The damping rate of *massive* fermions interacting with *massless* bosons is an especially interesting problem in this context. From the phenomenological point of view this importance is due to the fact that all fundamental forces of nature involve massless gauge bosons. From the theoretical standpoint a similarly big interest lies in the infrared divergence associated with this particular problem.

At zero temperature such an infrared divergence occurs in QED as well as in QCD, and its removal (or regularization) has become a standard textbook content. At non-zero temperature however, the question of the gauge-independent, causality-preserving and therefore *physical* removal of the infrared divergence is far from settled. Many schemes involve momentum scale cutoff factors or quasi-particles with "magnetic mass" [3]. As will be argued in the next section, such schemes violate several fundamental rules of quantum theory.

The present paper attempts a new solution to the infrared problem in hot systems, motivated by one of the few mathematically rigorous theorems of finite temperature quantum theory. This NRT theorem due to Narnhofer and Thirring states that particle-like excitations with infinite lifetime can exist in a hot system only if they do not interact [4].

The use of particle-like excitations in a perturbative scheme, which are asymptotically stable and only have a temperature dependent "mass", is therefore unjustified *a priori* and can only be obtained as a limiting case at the very end of all calculations.

The deeper mathematical reason is that the symmetry group of space and time in thermal states is not the Poincaré group as in the vacuum state, but rather a product of SO(3) and the four-dimensional translation group [5]. Its irreducible representations can be interpreted as having a continuous mass spectrum, i.e., the concept of a mass-shell (and thereby of stable asymptotic states) is not well-defined at non-zero temperature.

The solution of this problem has been pointed out by Landsman [6], using earlier work of Licht [7] and Wightman [8] and leading to a perturbative expansion in terms of *generalized free fields* without a mass-shell. To account for the proper temporal boundary conditions in a thermal system, these generalized free fields have to be embedded in a description with doubled Hilbert space, i.e., the single-particle propagators are 2×2 matrices [9]. Two flavors exist for such a formalism, the Schwinger–Keldysh (or closed time-path) method [10] and the method called thermo field dynamics (TFD) [11]. Within the latter this merger of different aspects of finite temperature field theory has been discussed in [12], which also introduces the notation used in the following.

Readers not interested in the computational details are referred to Sect. 6, where the results are assembled to obtain the approximate full fermion propagator calculated with a one-loop self-energy function at finite temperature. More computational details may be found in [13].

2 Spectral Functions

Appropriate for a description of dynamical phenomena in a thermal system is a formalism with 2×2 matrix valued Green's functions. In TFD as well when using the Schwinger–Keldysh method the full propagator matrix is

$$S^{(ab)}(x,x') = -\mathrm{i} \begin{pmatrix} \left\langle \mathrm{T}\left[\psi_x \overline{\psi}_{x'}\right]\right\rangle & -\left\langle \overline{\psi}_{x'} \psi_x \right\rangle \\ \left\langle \psi_x \overline{\psi}_{x'} \right\rangle & \left\langle \widetilde{\mathrm{T}}\left[\psi_x \overline{\psi}_{x'}\right]\right\rangle \end{pmatrix} , \tag{1}$$

where $\langle \cdot \rangle$ denotes the statistical average, $\mathrm{T}[\cdot]$ the time ordered and $\widetilde{\mathrm{T}}[\cdot]$ the "anti-time ordered" product (see [12,14] for detailed discussions).

By construction the propagator matrix obeys the linear relation $S^{11} + S^{22} - S^{12} - S^{21} = 0$. Furthermore, in an equilibrium state it has to fulfill the Kubo–Martin–Schwinger condition [15], i.e., an anti-periodic boundary condition in the imaginary time direction. In momentum space the KMS condition then reads [9,12,14]

$$\left(1 - n_F(p_0)\right) S^{12}(p_0, \mathbf{p}) + n_F(p_0) S^{21}(p_0, \mathbf{p}) = 0 \; . \tag{2}$$

S^{12} and S^{21} are the Green's functions without time ordering (Wigner functions) and $n_F(E)$ is the fermion equilibrium distribution function at a given temperature, the Fermi–Dirac function. An almost identical KMS relation holds for bosons, but here the fields are periodic in the imaginary time direction. Consequently, occupation number factors are given as Bose–Einstein function $n_B(E)$. These functions are used here with zero chemical potential as

$$n_{F,B}(E) = \frac{1}{e^{\beta E} \pm 1} \, . \tag{3}$$

In interacting systems, the KMS boundary conditions lead to a *causal* propagator which has a cut along the real energy axis. Its analytical structure therefore is not easily understood, and especially when combining several of such propagators in a perturbative scheme one has to implement more or less complicated cutting rules to understand pieces of diagrams *physically* [16,17].

It is therefore generally a safe method to use only retarded and advanced propagators, and to condense the matrix structure of (1) into the vertices where such propagators join. This amounts to a *diagonalization scheme* for the matrix (1), which has been described in several publications [12,14].

2.1 Retarded and Advanced Propagator

The retarded and advanced propagators are, by definition, analytical functions of the energy parameter in the upper or lower complex energy half plane. Analytical functions obey the Kramers–Kronig relation, and this implies that the retarded propagator is known completely if only its imaginary part (or spectral function) $\mathcal{A}_F$ is known along the real axis. Hence, for arbitrary complex E

$$S^{R,A}(E,\mathbf{p}) = \int_{-\infty}^{\infty} dE' \, \mathcal{A}_F(E',\mathbf{p}) \, \frac{1}{E - E' \pm i\epsilon} \, . \tag{4}$$

Trivially, the spectral function is recovered for real E,

$$\mathcal{A}_F(E,\mathbf{p}) = \mp\frac{1}{\pi} \operatorname{Im}\left[S^{R,A}(E,\mathbf{p})\right] = \frac{1}{2\pi i}\left(S^A(E,\mathbf{p}) - S^R(E,\mathbf{p})\right) \, . \tag{5}$$

The diagrammatic rules to combine these propagators in the calculation of physical quantities are well established. In terms of the spectral function $\mathcal{A}_F(E,\mathbf{p})$, the matrix valued propagator of (1) can be expressed as [12,14]

$$S^{(ab)}(p_0,\mathbf{p}) = \int_{-\infty}^{\infty} dE \, \mathcal{A}_F(E,\mathbf{p})$$

$$\times \tau_3 \left(\mathcal{B}(n_F(E))\right)^{-1} \begin{pmatrix} \dfrac{1}{p_0 - E + i\epsilon} & \\ & \dfrac{1}{p_0 - E - i\epsilon} \end{pmatrix} \mathcal{B}(n_F(E)) \, . \tag{6}$$

τ_3 is the diagonal Pauli matrix, and the transformation matrix $\mathcal{B}$ is

$$\mathcal{B}(n_F(E)) = \begin{pmatrix} (1 - n_F(E)) & -n_F(E) \\ 1 & 1 \end{pmatrix} . \tag{7}$$

A similar relation holds for the boson case, see [12,14] for details of the corresponding $\mathcal{B}$. For both cases, the transformation matrices $\mathcal{B}$ have a special meaning in the TFD formalism, where they play the role of a thermal Bogoliubov transformation. However, the explicit form (4) of the propagator is the same in the Schwinger–Keldysh method.

2.2 Normalization, Locality and Ghost Poles

The spectral function has two features which are intimately related to fundamental requirements of quantum field theory. Here they are commented on only for the fermionic case, and refer to [18] for the case of bosons. Firstly, the quantization rules for fields,

$$\left\{\psi(t, \mathbf{x}), \psi^\dagger(t, \mathbf{y})\right\} = \delta^3(\mathbf{x} - \mathbf{y}) \tag{8}$$

require that the spectral function is *normalized*

$$\int_0^\infty dE \, \mathrm{Tr}\left[\gamma^0 \mathcal{A}_F(E, \mathbf{p})\right] = 2 . \tag{9}$$

The second important feature of the spectral function is that its four dimensional Fourier transform into coordinate space must vanish for space-like arguments. This is equivalent to the Wightman axiom of *locality*, i.e., field operators must (anti-)commute for space-like separations in Minkowski space [19]. The function

$$\begin{aligned} C_F(x, y) &= \left\langle \left\{\psi(x), \psi^\dagger(y)\right\} \right\rangle \\ &= \int \frac{dE\, d^3\mathbf{p}}{(2\pi)^3} \, \mathrm{e}^{-\mathrm{i}(E(x_0 - y_0) - \mathbf{p}(\mathbf{x} - \mathbf{y}))} \, \mathcal{A}_F(E, \mathbf{p}) \end{aligned}$$

must vanish if $x - y$ is space-like. In an interacting many-body system, one may very well expect non-locality in a causal sense. The locality requirement ensures that this influence does not occur over space-like separations, i.e., faster than a physical signal can propagate. Thus, to distinguish between the *causal* non-locality and the violation of the locality axiom, the latter will henceforth be denoted a violation of *causality*.

It is easy to establish that normalization and locality axiom are satisfied for free fermions of mass M. However, for a general spectral function, locality is not automatically guaranteed, a careful check is necessary in any application involving spectral functions.

For an interacting system, the full fermion propagator and thus also the spectral function is defined in terms of a self energy function Σ:

$$\mathcal{A}_F(p_0, \mathbf{p}) = \mp\frac{1}{\pi}\,\mathrm{Im}\left[\left(p_\mu\gamma^\mu - M - \Sigma^{R,A}(p_0, \mathbf{p})\right)^{-1}\right] . \tag{10}$$

Note, that also the self-energy function is Dirac matrix valued, i.e., this equation contains the inversion of a 4×4 matrix.

However, the real part of the full propagator is *not* given by the real part of the quantity in square brackets of (10). The reason is, that according to Weinberg's theorem [20] the self-energy function in a relativistic theory is more than linearly divergent. This implies, in general, the presence of unwanted poles when the interaction is not asymptotically free. For fermions, these Landau ghost poles appear at four points in the complex energy plane [21, p. 636].

It was pointed out in [22], that they are unphysical and the only consistent way for their removal is to use (4) to calculate the propagator. This amounts to a *non-perturbative* correction to the perturbative propagator, in the sense that the correction term is not an analytical function of the coupling constant. In general, it will be of order $\exp(-1/g^2)$ instead (see [22] for details).

The calculation of the propagator by dispersion integral gives a function which is free of poles on both sides of the real axis. As a matter of fact it has been shown very long ago in quantum field theory, that this is the only scenario compatible with causality, i.e., retarded and advanced propagator have a common analytical continuation [23].

2.3 Approximate Spectral Functions

An ansatz for the spectral function for bosons, which is compatible with all the above requirements reads

$$\begin{aligned}\mathcal{A}_B(E, \mathbf{k}) &= \frac{1}{2\omega_k}\left(\mathcal{A}_0(E, \mathbf{k}) - \mathcal{A}_0(-E, \mathbf{k})\right)\\ &= \frac{1}{\pi}\,\frac{2E\gamma_k}{\left(E^2 - \Omega_k^2\right)^2 + 4E^2\gamma_k^2} ,\end{aligned} \tag{11}$$

with $\Omega_k^2 = \omega_k^2 + \gamma_k^2$. This spectral function is normalized according to the canonical field commutation relations, i.e.,

$$\int_0^\infty dE\, E\, \mathcal{A}_B(E, \mathbf{k}) = \frac{1}{2} . \tag{12}$$

Note, that in this ansatz the dependence of γ_k on the momentum of the relativistic boson is completely arbitrary, and one may also introduce a general relationship between ω_k and momentum.

The approximation of the full propagator by such an approximate spectral function amounts to the approximation of the denominator of a relativistic retarded boson propagator along the real axis as

$$k_\mu k^\mu - m^2 - \Pi^R(k) = (k_0 - (\omega_k - \mathrm{i}\gamma_k))\,(k_0 + (\omega_k + \mathrm{i}\gamma_k))\;, \tag{13}$$

where Π is the boson self-energy function, and solving this equation for ω_k, γ_k. An example for such an approximation has been given in reference [12], and there the momentum dependence of ω_k as well as of γ_k was quite strong for the case of a pseudoscalar coupling between bosons and fermions.

Note at this point, that a *general* momentum dependence of the γ_k-parameter does not guarantee the causality of the model, i.e., it may violate the locality axiom as discussed above. We are therefore considering an ω_k^2 which is quadratic in $\mathbf{k}$ and a constant γ_k.

This ansatz may be extended to the fermionic case, where one has to use a more complicated spectral function,

$$\begin{aligned}\mathcal{A}_F(E,\mathbf{p}) &= \frac{\gamma_p}{\pi}\frac{\gamma^0\left(E^2+\omega_p^2+\gamma_p^2\right)+2E\gamma\mathbf{p}+2EM}{\left(E^2-\omega_p^2-\gamma_p^2\right)^2+4E^2\gamma_p^2}\\ &= \frac{1}{4\pi\mathrm{i}\omega_p}\left(\frac{\omega_p\gamma^0+\mathbf{p}\gamma+M}{E-\omega_p-\mathrm{i}\gamma_p}-\frac{-\omega_p\gamma^0+\mathbf{p}\gamma+M}{E+\omega_p-\mathrm{i}\gamma_p}\right.\\ &\qquad\left.-\frac{\omega_p\gamma^0+\mathbf{p}\gamma+M}{E-\omega_p+\mathrm{i}\gamma_p}+\frac{-\omega_p\gamma^0+\mathbf{p}\gamma+M}{E+\omega_p+\mathrm{i}\gamma_p}\right)\end{aligned} \tag{14}$$

where in principle γ_p may be momentum dependent, ${\Omega_p}^2 = {\omega_p}^2 + {\gamma_p}^2$ and M is a constant which may be different from the physical fermion mass in the vacuum state.

To conform to the basic axioms of quantum field theory outlined above, we use within this spectral function the approximations

$$\begin{aligned}\gamma_p \quad &\equiv \gamma &&= \text{const.}\\ \omega_p^2 \quad &\equiv \omega^2 &&= \mathbf{p}^2 + M^2\\ \Rightarrow (\omega(\mathbf{p}+\mathbf{k}))^2 &&&= \omega^2 + 2|\mathbf{p}|\,|\mathbf{k}|\,\eta + \mathbf{k}^2\;,\end{aligned} \tag{15}$$

where η is the cosine of the angle between the two momenta $\mathbf{p}$ and $\mathbf{k}$.

3 Fermion Damping Rate

The next step is to use spectral functions, together with the definition of the complete propagator by dispersion integral, in a “perturbative” expansion at finite temperature. One finds, that “perturbative” is certainly not the correct label for such a skeleton expansion, because in general due to the properties of the dispersion integral the results will *not* be expandable into power series in the coupling parameters (see Sect. 2.3). Indeed up to correlation diagrams and vertex corrections, a perturbative expansion in terms of *full* propagators is exact already at the one-loop level [21, pp. 476].

For simplicity, first consider a simple scalar coupling of a boson field having the spectral function $\mathcal{A}_B$ to a fermion field having the spectral function $\mathcal{A}_F$.

The "one-loop" diagram for the fermion self-energy, the Fock diagram, is given by the integral

$$\Sigma^R(p_0, \mathbf{p}) = g^2 \int \frac{d^3\mathbf{k}}{(2\pi)^3} \int_{-\infty}^{\infty} dE dE' \, \mathcal{A}_F(E, \mathbf{p}+\mathbf{k}) \, \mathcal{A}_B(E', \mathbf{k}) \times \left(\frac{n_B(E') + n_F(E)}{p_0 + E' - E + \mathrm{i}\epsilon} \right) . \tag{16}$$

This function is split into real and imaginary part as $\Sigma^R = \mathrm{Re}\Sigma - \mathrm{i}\pi\Gamma$, and for the latter one obtains

$$\Gamma(p_0, \mathbf{p}) = g^2 \int \frac{d^3\mathbf{k}}{(2\pi)^3} \int_{-\infty}^{\infty} dE \, \mathcal{A}_F(E + p_0, \mathbf{p}+\mathbf{k}) \, \mathcal{A}_B(E, \mathbf{k}) \times \left(n_B(E) + n_F(E + p_0) \right) . \tag{17}$$

To fulfill the Kubo–Martin–Schwinger boundary condition (2) it is *absolutely essential* that both propagators in the above expression have the same equilibrium temperature. In other words, one may *not* disregard the fermion distribution function $n_F(E + p_0)$ in the above expression in any case.

In principle, one could introduce a self-consistent calculation scheme: The expressions (16) and the corresponding self-energy function for bosons are calculated with an approximate spectral function, then used as input to (10) and a similar equation for the full boson propagator in terms of its self-energy. This gives new spectral functions, which may be used again to determine self energies. Such a scheme has been used in [25,26], and based on these papers one might hope that a few iterations are enough to determine the spectral function numerically quite well.

However, as pointed out, there is no way to guarantee that in such a scheme causality (in the representation of the locality axiom) is preserved. Moreover, in an entirely numerical scheme it would be impossible to point out the path to the solution of the infrared problem.

We will therefore use a simple causality-preserving parametrization of the spectral functions. The parameters are then determined self-consistently in a closed scheme. Apart from the possibility to achieve analytical approximations, this strategy also has the virtue that one can use it as an input to the first one.

3.1 Approximations to the Spectral Functions

The full boson propagator may have isolated poles only on the real energy axis, but not away from it on the physical Riemann sheet [23]. Furthermore it may exhibit cuts along the real energy axis due to self-energy corrections with a continuous non-zero imaginary part.

In the absence of condensation phenomena these self-energy corrections have zero imaginary part when the boson energy parameter is equal to the chemical potential. Consequently, the product of a Bose–Einstein distribution function (3) and a continuous part of the boson spectral function is always *infrared finite.*

The only possible source for an infrared divergence therefore are poles on the real axis, corresponding to particles with infinite lifetime. As argued above, such poles may not exist in a finite temperature system [4], and thus it is on one hand clear that infrared divergences in finite temperature field theory are not present in the full theory but a mere artifact of perturbation theory.

Introducing a modified (i.e., *less* singular) spectral function for the bosons of course diminishes the result for Γ because it leads to a "smearing" of the Bose pole with some distribution. Consequently, one may consider the values obtained in the following as an upper bound on the full calculation with screened massles bosons.

Suppose, one really would approximate the full spectral function according to (10) by the simple parametrization of (14). This then would amount to equating the denominators on the real energy axis as

$$\begin{aligned}&\left(p_\mu - V_\mu(p) + \mathrm{i}\pi\Gamma^v_\mu(p)\right)^2 - \left(M + S(p) - \mathrm{i}\pi\Gamma^s(p)\right)^2\\ &= \left(p_0 - (\omega_p - \mathrm{i}\gamma_p)\right)\left(p_0 + (\omega_p + \mathrm{i}\gamma_p)\right)\ . \end{aligned} \tag{18}$$

The functions of momentum appearing here are the components of the retarded self-energy function, split into real and imaginary part according to

$$\begin{aligned}\Sigma^R(p) \;&= \mathrm{Re}\Sigma(p) - \mathrm{i}\pi\Gamma(p)\\ \mathrm{Re}\Sigma(p) &= S(p) + V_\mu(p)\gamma^\mu\\ \Gamma(p) \;&= \Gamma^s(p) + \Gamma^v_\mu(p)\gamma^\mu\ . \end{aligned} \tag{19}$$

In the above expressions, the self-energy functions have been split into a Lorentz scalar and a Lorentz vector piece, henceforth simply abbreviated as scalar and vector part.

It was stated above, that in general this may give rise to a momentum dependent γ_p as well as to a non-quadratic dependence of ω_p on $\mathbf{p}$. Consequently, checking for the locality axiom, i.e., the preservation of causality in the interactions of the system, requires great numerical effort and one would lose all the virtues that come with a simple parametrization of $\mathcal{A}_F$. It is therefore useful to first study how the choice of spectral function (14) affects the imaginary part of the self-energy (17). Clearly it has only two independent components, since its Lorentz scalar and vector part are

$$\begin{aligned} \Gamma_0^v(p) &= \Gamma^I(p_0, \mathbf{p}, \gamma) \\ \Gamma_i^v(p) &= \frac{\mathbf{p}_i}{M}\,\Gamma^{II}(p_0, \mathbf{p}, \gamma) \\ \Gamma^s(p) &= \Gamma^{II}(p_0, \mathbf{p}, \gamma)\;. \end{aligned} \tag{20}$$

To study a slow *massive* fermion on its effective mass-shell, i.e., for $p_0 = \omega_p = \sqrt{\mathbf{p}^2 + M^2}$, one may therefore make the approximation that the real part of the self energy function may be absorbed into the mass parameter M.

For the scalar coupling considered here, the self-consistency equation obtained from (18) reads

$$\gamma_S = \pi\Gamma^I(\omega, \mathbf{p}, \gamma_S) + \pi\Gamma^{II}(\omega, \mathbf{p}, \gamma_S)\left(1 - \frac{\mathbf{p}^2}{M^2}\right)\frac{M}{\omega}\;. \tag{21}$$

According to the assumption, the left side does not depend on the momentum $\mathbf{p}$. While one certainly cannot hope that the momentum dependence on the right side cancels completely, one may nevertheless expect that it possesses an expansion of the type

$$\gamma_S = \nu_0 + \nu_1\frac{\mathbf{p}^2}{M^2} + \nu_2\frac{\mathbf{p}^4}{M^4} + \ldots\;. \tag{22}$$

Consequently, the approximations made here are reasonable if the coefficients in this expansion are at least of the same magnitude – and they must be considered a failure when e.g. $|\nu_0| \ll |\nu_1|$. To check this condition therefore will provide an a-posteriori test of the assumptions.

It is quite simple to generalize this scheme to the exchange of a massless vector boson, if the boson propagator is taken in Feynman gauge [21, p. 329]: The inclusion of Dirac matrices γ^μ, γ_μ at the vertices of the Fock diagram simply multiplies Γ^v by a factor -2 and Γ^s by a factor 4. The resulting *vector boson* self-consistency relation is

$$\gamma_V = -2\pi\Gamma^I(\omega, \mathbf{p}, \gamma_V) + 4\pi\Gamma^{II}(\omega, \mathbf{p}, \gamma_V)\left(1 + \frac{\mathbf{p}^2}{2M^2}\right)\frac{M}{\omega}\;. \tag{23}$$

It is worthwhile to note, that the ansatz of a momentum independent imaginary part of the fermion self-energy ensures the gauge invariance of this self-consistent damping rate in the limit $\mathbf{p} \to 0$. Gauge invariance even of the first momentum dependent correction term requires to introduce a vertex correction in (17) to satisfy the correct Ward identity [27].

The result for massless *pseudoscalar bosons* is also physically interesting, see [13].

3.2 Angular Integrations

The integration over the angle between the two momenta **k** and **p** may be done analytically. Using the above spectral functions, one obtains as the γ^0-proportional imaginary part of the self-energy

$$\begin{aligned}\Gamma^I(\omega,\mathbf{p},\gamma) = \frac{g^2\gamma^0\gamma}{4\pi^3}\int\limits_0^\infty dk\,k\Bigg\{&\Big(\big((k+\omega)^2 - k\omega + \gamma^2/2\big)\;I_1(k+\omega)\\ &+pk\,I_2(k+\omega)\Big)\,(n_B(k)+n_F(k+p_0))\\ &-\Big(\big((k-\omega)^2 + k\omega + \gamma^2/2\big)\;I_1(k-\omega)\\ &+pk\,I_2(k-\omega)\Big)\,(n_B(-k)+n_F(-k+p_0))\Bigg\}\,. \end{aligned}\tag{24}$$

I_1 and I_2 are integrals over the angle between the momenta **p** and **k**, and $k = |\mathbf{k}|$, $p = |\mathbf{p}|$ (see appendix for technical details and the definition of the functions I_1 and I_2).

A similar expression can be found for the second (scalar) piece of the self-energy function,

$$\begin{aligned}\Gamma^{II}(\omega,\mathbf{p},\gamma) = \frac{g^2\,M\,\gamma}{4\pi^3}\int\limits_0^\infty dk\,k\Bigg\{&(k+\omega)\,I_1(k+\omega)\;(n_B(k)+n_F(k+p_0))\\ &+(k-\omega)\,I_1(k-\omega)\;(n_B(-k)+n_F(-k+p_0))\Bigg\}\,. \end{aligned}\tag{25}$$

The infrared problem addressed above finds its representation in the fact, that in the limit $\gamma \to 0$, I_1 and I_2 behave like a *negative* power ≤ -2 of k. This becomes obvious when expanding the integrand around the value $k = 0$.

To check for infrared finiteness, one may perform this expansion in the above expressions and obtains for the quantity in curly brackets of (24), at the point $p_0 = \omega$:

$$\lim_{k\to 0}\Big\{\ldots\Big\}_{p_0=\omega} = \gamma^0\frac{T}{\gamma}\,\frac{\omega^2+\gamma^2/2}{\omega^2+\gamma^2/4} + \mathcal{O}(k^2)\,. \tag{26}$$

Consequently, the infrared divergence of the fermion damping rate is removed by starting with a small non-zero γ.

4 Vacuum State

In this section, the temperature independent part of the Bose–Einstein and Fermi–Dirac distribution functions is used to calculate contributions to the self energy function of fermions. In this limit, $n^0_B(E) = -\Theta(-E)$ and $n^0_F(E) = \Theta(-E)$.

One may argue, that these are vacuum contributions, therefore not present in a properly renormalized theory. However, the propagators employed here are *not* the ordinary vacuum propagators for the fermion field, rather, they contain a non-zero damping rate as input. Only in the very end one may be able to find a self-consistent solution of zero damping rate in a "true" vacuum state.

Using the above step functions in the expression for the imaginary part of the self-energy leads to the simpler form

$$\Gamma^I_{\mathrm{vac}}(\omega, \mathbf{p}, \gamma) = \frac{g^2\gamma}{4\pi^3} \int\limits_0^\infty dk\, k \left[\left((k-\omega)^2 + k\omega + \gamma^2/2\right) I_1(k-\omega) + pk\, I_2(k-\omega) \right] \Theta(\omega - k)\,. \tag{27}$$

This part of the self-energy function has an asymptotic behavior according to Weinberg's theorem [20]: The above integral is ultraviolet finite, but the corresponding real part is divergent and has to be renormalized. It is connected to the above imaginary part only through a *subtracted* dispersion relation. This asymptotic behavior of the self-energy function for large values of p_0 implies, that the ghost problem discussed above becomes relevant.

Clearly, at $p_0 = \omega$, the imaginary part of the self-energy vanishes if γ is zero. This no longer holds once the integral in (27) is calculated with a non-zero positive γ as input. Instead one obtains a non-zero imaginary part also "on-shell". To put it bluntly: Zero input γ gives zero output, whereas non-zero input γ gives non-zero output Γ_{vac} even at $p_0 = \omega$.

One may now exploit the virtue of a simply parametrized spectral function for fermions, by expanding the continuous function Γ_{vac} around its value at $\gamma = 0$. The imaginary part of the self energy function with $p_0 = \omega$ is not an analytical function of γ: It has a cut in the complex γ-plane, starting at $\gamma = 0$. Thus, a Taylor expansion in γ is *not* possible, one has to perform an asymptotic expansion, explicitly taking into account the leading non-analyticity.

Various techniques exist for such an expansion, for the purpose of the present paper the leading terms were isolated by first substituting $k \to x\gamma$ and then expanding the integrand in powers of γ. It turns out, that for finite momentum p the scale for the expansion is $\omega^2 - p^2$ – which implies, that this asymptotic expansion is reliable only up to momenta $p \approx M$.

Within this limitation, the result for the vector imaginary part of the Fock self-energy function is

$$\Gamma^I_{\mathrm{vac}}(\omega, \mathbf{p}, \gamma) = -\frac{g^2}{4\pi^3}\left\{\frac{\gamma}{4}\frac{\omega^2}{\omega^2-p^2}\left(\log\left(\frac{\gamma^2}{\omega^2-p^2}\right) - 1 + \frac{p}{\omega}\log\left(\frac{\omega+p}{\omega-p}\right)\right)\right.$$
$$-\frac{\gamma^2}{8}\frac{\pi\,\omega(\omega^2+3p^2)}{(\omega^2-p^2)^2}$$
$$+\frac{\gamma^3}{24}\frac{1}{(\omega^2-p^2)^3}\left(\frac{1}{6}\left(15\omega^4+122\omega^2p^2-9p^4\right)\right.$$
$$-\left(9\omega^4+20\omega^2p^2+3p^4\right)\log\left(\frac{\gamma^2}{\omega^2-p^2}\right)$$
$$\left.\left.-\frac{\omega}{p}\left(2\omega^4+18\omega^2p^2+12p^4\right)\log\left(\frac{\omega+p}{\omega-p}\right)\right)\right\} + \mathcal{O}(\gamma^4)\,, \quad (28)$$

where $p = |\mathbf{p}|$ and $k = |\mathbf{k}|$. In Fig. 1, this expansion in γ is compared to the full numerical calculation of (27): With each additional order that is included, the quality of the approximation grows.

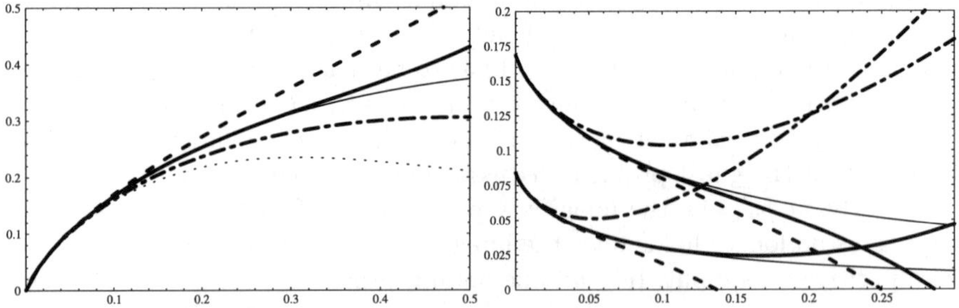

Fig. 1. Γ^I_{vac} (*left*) and Γ'_T (*right*) as function of γ (both in units of M, energy $p_0 = \omega$, momentum $p = 0.5M$, $g^2/4\pi^3$ set to one). Temperatures for the right panel are $T = 0.05M$ (*bottom group*) and $T = 0.1M$ (*top group*). *Continuous thin line*: Full numerical calculation. *Dash-Dotted thick line*: Expansion to order γ. *Dashed thick line*: Expansion to order γ^2. *Continuous thick line*: Expansion to order γ^3. *Dotted thin line*: Expansion to order γ^3, but *without* the term of order γ^2

An important aspect of this expansion is the occurrence of the quadratic term: In a naive view of the integral involved, it is not present because $I_1(k - p_0)$ as well as $I_2(k - p_0)$ are odd functions of γ. However, as pointed out above, the expanded function is not analytical in γ – and this effect causes the appearance of the quadratic contribution. To demonstrate this, Fig. 2 also contains a curve where this quadratic term has been left out. Obviously in this case the third order contribution makes the approximation worse instead of improving it.

In view of the above results one has to conclude, that for not too large values of γ the asymptotic expansion to *second order* is well under control.

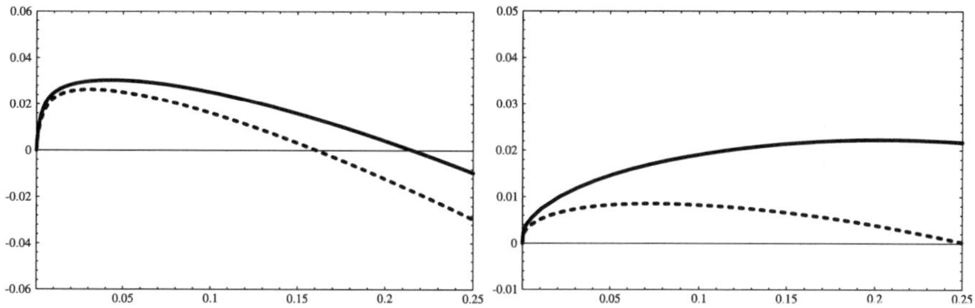

Fig. 2. $(\gamma_S - \alpha T)/M$ (*left*) and $(\gamma_V - \alpha T)/M$ (*right*) as function of T/M at strong coupling $\alpha = 1$. *Continuous lines*: $p = 0$; *dashed lines*: $p = 0.5M$

Henceforth, to keep the results simple, the discussion is restricted to this next-to-leading log order.

A similar expansion is performed for the second piece of the imaginary part of the fermion self-energy. According to (25) one obtains in the vacuum

$$\Gamma^{II}_{\text{vac}}(\omega, \mathbf{p}, \gamma) = -M\,\frac{g^2}{4\pi^3}\int_0^\infty dk\,k\,\gamma\,(k-\omega)\,I_1(k-\omega)\,\Theta(\omega-k)\,, \tag{29}$$

leading consequently to

$$\begin{aligned}\Gamma^{II}_{\text{vac}}(\omega, \mathbf{p}, \gamma) = -\frac{g^2}{4\pi^3}\Bigg\{ & \frac{\gamma}{4}\,\frac{M\omega}{\omega^2-p^2}\left(\log\left(\frac{\gamma^2}{\omega^2-p^2}\right)+\frac{\omega}{p}\log\left(\frac{\omega+p}{\omega-p}\right)\right)\\ & -\frac{\gamma^2\,\pi}{8}\,\frac{M(3\omega^2+p^2)}{(\omega^2-p^2)^2}\\ & +\frac{\gamma^3}{8}\,\frac{M}{\omega(\omega^2-p^2)^3}\left(\frac{1}{6}\left(21\omega^4+14\omega^2p^2-3p^4\right)\right.\\ & -4\omega^2(\omega^2+p^2)\,\log\left(\frac{\gamma^2}{\omega^2-p^2}\right)\\ & \left.-\frac{\omega}{p}\left(\omega^4+6\omega^2p^2+p^4\right)\,\log\left(\frac{\omega+p}{\omega-p}\right)\right)\Bigg\}+\mathcal{O}(\gamma^4)\,.\end{aligned} \tag{30}$$

Note again, that as a sign of the non-analyticity in the parameter γ, a quadratic as well as a $\gamma\log(\gamma)$ contribution appear. This asymptotic expansion was compared to the fully numerical calculation of the above integral, with three findings: 1. The approximation improves systematically from first to third order in γ, which proves that the non-analyticity was treated consistently. 2. The quadratic contribution (due to the non-analyticity of the integral) is necessary to provide a meaningful third order result. 3. For $\gamma/M \leq 0.1$, the use of a first order asymptotic expansion is sufficient.

5 Thermal State

In this section, the explicitly temperature dependent part of the distribution functions is used to perform a similar approximation as in the previous section. Contributions to the full imaginary part of the self-energy function according to (17) therefore contain an explicit temperature dependence.

For calculational convenience they are split of the complete expressions according to

$$\begin{aligned}\Gamma(p_0, \mathbf{p}, \gamma) &= \Gamma_{\mathrm{vac}}(p_0, \mathbf{p}, \gamma) + \Gamma_T(p_0\mathbf{p}, \gamma) \\ &= \Gamma_{\mathrm{vac}}(p_0, \mathbf{p}, \gamma) + \Gamma_T^I(p_0, \mathbf{p}, \gamma)\gamma^0 + \Gamma_T^{II}(p_0, \mathbf{p}, \gamma)\left(1 + \frac{\mathbf{p}\gamma}{M}\right) , \end{aligned} \quad (31)$$

similar to the decomposition for the vacuum part only in (20). For the moment it seems quite hopeless to obtain an *analytical* approximation to the full expression (24) valid for all temperatures. However, for not too high temperatures the Bose–Einstein and Fermi–Dirac distribution function may be expanded around their zero-temperature value,

$$\begin{aligned} & n_B(k) + n_F(p_0 + k) \\ & \quad \approx \mathrm{e}^{-k/T}\left(\frac{T}{k} + \frac{1}{2} + n_F(p_0) + \frac{k}{T}\left(\frac{1}{12} + n_F(p_0)^2\right)\right) \end{aligned}$$

$$\begin{aligned} & n_B(-k) + n_F(p_0 - k) \\ & \quad \approx -\mathrm{e}^{-k/T}\left(\frac{T}{k} + \frac{1}{2} + \frac{k}{T}\frac{1}{12}\right) \\ & \quad\quad -\mathrm{e}^{-k/T}\mathrm{e}^{p_0/T}\Theta(k - p_0)\left(n_F(p_0) + \frac{k}{T}n_F(p_0)(1 - n_F(p_0))\right) \\ & \quad\quad + \underbrace{\Theta(p_0 - k)\left(-1\right.}\left. + n_F(p_0) + \frac{k}{T}n_F(p_0)(1 - n_F(p_0))\right) . \end{aligned} \quad (32)$$

The term underlined with the curly brace constitutes the "vacuum" part as obtained in the previous section. The other terms are ordered according to their dominance, i.e., the contribution proportional to T/k is the largest due to the strongly peaked nature of the function $n_B(k)$. The terms including the fermionic distribution functions are negligible for $p_0 \gg T$. Note however, that the negative energy states of the fermionic distribution function are properly taken care of since $n_F(-p_0) = 1 - n_F(p_0) \approx 1$.

Similar to the vacuum case, one obtains a result for Γ_T which cannot be expanded in a Taylor series around the point $\gamma = 0$. The asymptotic expansion is obtained in the same way as above, and improved using the following trick: First, the momentum integral is calculated using only the leading term $T/k = 1/(\beta k)$ of the above expansion. The result is then improved by acting on it with a differential operator

$$\Gamma_T^{I'}(\omega, \mathbf{p}, \gamma) = \Gamma_T^I(\omega, \mathbf{p}, \gamma) - \left(\frac{1}{2}\frac{\partial}{\partial\beta} - \frac{1}{12T}\frac{\partial^2}{\partial\beta^2}\right)\frac{\Gamma_T^I(\omega, \mathbf{p}, \gamma)}{T}\ . \tag{33}$$

For both pieces of the self-energy function this is a quite laborious task, for our calculation we relied on symbolic computation using Mathematica™. For the first piece one obtains

$$\begin{aligned}
\Gamma_T^{I'}(\omega, \mathbf{p}, \gamma) &= \frac{g^2}{4\pi^3}\Bigg\{\frac{\pi T\omega}{4p}\log\left(\frac{\omega+p}{\omega-p}\right) \\
&+\frac{\gamma}{4}\frac{\omega^2}{\omega^2-p^2} \\
&\times\left(\log\left(\frac{\gamma^2\omega^2}{T^2(\omega^2-p^2)}\right)+\frac{\omega}{p}\log\left(\frac{\omega+p}{\omega-p}\right)+2C_\Gamma-\frac{17}{3}+\frac{20T^2}{3\omega^2}\right) \\
&-\frac{\gamma^2\pi}{24T}\frac{\omega^4+18\omega^2T^2+18T^2}{(\omega^2-p^2)^2} \\
&+\frac{\gamma^3}{(\omega^2-p^2)^2}\Bigg[\frac{(\omega-p)^5}{8p}\left(\log\left(\frac{\gamma\omega}{T(\omega+p)}\right)+C_\Gamma-\frac{19}{12}\right) \\
&-\frac{(\omega+p)^5}{8p}\left(\log\left(\frac{\gamma\omega}{T(\omega-p)}\right)+C_\Gamma-\frac{19}{12}\right) \\
&+\left(3\omega^2+p^2\right)\left(\frac{\omega^4}{216T^2}-\frac{5}{9}T^2\right)\Bigg]\Bigg\}+\mathcal{O}(\gamma^4)\ ,
\end{aligned} \tag{34}$$

where $C_\Gamma = 0.57721\ldots$ is Euler's constant. A comparison of this approximation to different orders in γ with the full numerical calculation of the temperature dependent contribution is also presented in Fig. 1. As before, it turns out that the results are already quite reliable in second order of γ, provided the temperature is not too high.

In the case of the vacuum parts of the self-energy function it was possible to gather all terms of a given order in γ in this asymptotic expansion. For the present temperature dependent pieces this is *not* possible, because of the above expansion: It automatically counts T and γ to be of the same order, hence it is a simultaneous expansion in these parameters. As a small reminder of this fact appears a term γT^2 in the above expansion, and the $\mathcal{O}(\gamma^4)$ must be replaced by a $\mathcal{O}(\gamma^4, \gamma T^3)$.

The net result is that in second order γ the asymptotic expansion is reliable up to $\gamma \approx T/2$, whereas it is good up to the temperature in third order. In anticipation of the following section, note that the quality of the approximation grows tremendously for very small temperatures $T \ll 0.1M$.

With similar quality one may also obtain the second piece, asymptotically given by

$$\Gamma_T^{II'}(\omega,\mathbf{p},\gamma) = \frac{g^2}{4\pi^3}\Bigg\{\frac{\pi T M}{4p}\,\log\left(\frac{\omega+p}{\omega-p}\right)$$
$$+\frac{\gamma}{4}\frac{M\omega}{\omega^2-p^2}\left(\log\left(\frac{\gamma^2\omega^2}{T^2(\omega^2-p^2)}\right)+\frac{\omega}{p}\log\left(\frac{\omega+p}{\omega-p}\right)+2C_\Gamma-\frac{17}{3}\right)$$
$$-\frac{\gamma^2\pi M\omega}{24T}\frac{\omega^2+12T^2}{(\omega^2-p^2)^2}$$
$$+\frac{\gamma^3 M}{(\omega^2-p^2)^3}\Bigg[\frac{(\omega-p)^4}{8p}\left(\log\left(\frac{\gamma\omega}{T(\omega+p)}\right)+C_\Gamma-\frac{19}{12}\right)$$
$$-\frac{(\omega-p)^4}{8p}\left(\log\left(\frac{\gamma\omega}{T(\omega-p)}\right)+C_\Gamma-\frac{19}{12}\right)$$
$$+\frac{\omega^3}{216T^2}\left(3\omega^2+p^2\right)-\frac{1}{16\omega}\left(\omega^4+6\omega^2p^2+p^4\right)\Bigg]\Bigg\}+\mathcal{O}(\gamma^4,\gamma T^3)\,. \quad (35)$$

To this expression one may apply the same considerations as before: Although there is no piece $\propto \gamma T^2$, one may trust this expansion in second order γ only up to $\gamma \approx T/2$.

The claims laid on the accuracy of our asymptotic expansions are supported by Fig. 1, where they are compared with a completely independent numerical calculation. The accuracy of the results was checked with various values for momentum and temperature, and found to persist up to momenta $p \approx M$.

To be completely sure of the findings, the corrections of orders γ^3 will be dropped henceforth: Unusual as it may seem for quantum field theory, one therefore has an approximate method with controlled accuracy in the next-to-leading-log order.

6 Fermion Damping Rate II

In this section, the results are assembled to obtain a consistent solution for the fermion damping rate in hot systems. This applies to the explicitly temperature dependent part as well as the "vacuum" part of the self-energy. Combining (28) with (34) and (30) with (35), *all the terms of order* $\gamma\log(\gamma)$ *cancel*!

The cancellation occurs for scalar and vector part of the self-energy function independently, hence holds for all types of massless bosons. As a consequence of this cancelation, the self-consistent damping rate for the fermion moving slowly through a hot medium is the solution of an *algebraic equation* of the form

$$f_0(\omega, \mathbf{p}, T) - \left(f_1(\omega, \mathbf{p}, T) + \frac{4\pi}{g^2} \right) \gamma + f_2(\omega, \mathbf{p}, T)\, \gamma^2 = 0\,. \tag{36}$$

Note however, that the results do not exclude the possibility of corrections of order $\gamma^3 \log(\gamma)$ to this equation.

Equation (36) always has two solutions for γ, but in general one of them is negative or very large. Hence it is obvious how to choose the physical solution, and in the following this is the only one discussed.

The coupling constant appears only at one point, the three functions f_0–f_2 do not depend on it. f_0 is proportional to the temperature T, f_1 contains terms which are not analytical around zero temperature, i.e., terms of order $T \log(T)$.

The solution γ of (36) therefore is a power series in the coupling constant – but still it includes a non-perturbative effect in form of the non-analytical behavior in the temperature, which is of order $g^4\, T \log(T)$.

One might argue, that these non-analytical terms are not significant because in order g^4 other two-loop diagrams also give a contribution that were neglected. However, as may be expected from the results obtained here, their non-analytical contribution involving $\log(T)$ is suppressed with respect to their leading order contribution by a factor γ, which may be translated into a factor $g^2 T$. This implies, that our results to order g^4 yield the *dominant* non-analyticity and are important for small temperatures.

Moreover, this chain of arguments also supports the conclusion that terms of order $g^2 T p^2 / M^2 \log(1/g)$ cannot appear: Terms logarithmic in the coupling constant wear at least a coefficient g^6 or $g^4 T^2 / M^2$, even if vertex corrections are introduced in the calculation of the self-energy function.

Figure 2 displays the self consistent solution of (36) for scalar bosons coupled to the fermions. For small temperatures, the deviation from the linear temperature dependence is very striking: Due to the $T \log(T)$ terms in the self-energy functions, $\gamma_S(T)$ rises sharply at very small temperature.

The self-consistent γ_V due to massless vector boson exchange in Feynman gauge is also shown in Fig. 2. The curves basically employ the same features as for the scalar bosons: A pronounced non-analyticity of the function $\gamma_V(T)$ at small temperatures.

Figure 2 also shows that the momentum dependence of the self-consistent solution is small up to quite high temperatures. In view of (22) this means $|\nu_0| \geq |\nu_1|$ for scalar and vector boson exchange. One may therefore conclude, that the ansatz of a momentum independent γ_S and γ_V is very well justified.

6.1 Analytical Approximations for the Damping Rate

It is not necessary to repeat the analytical expressions for the self-energy pieces in order to give the three functions f_0–f_2 in closed form. However, it is instructive to extract the dominant terms in the calculation of γ.

To this end, perform an expansion of the solution of (36) in powers of the coupling constant. Only even powers occur, hence the expansion parameter is

$$\alpha = \frac{g^2}{4\pi}\ . \tag{37}$$

The contributions of order T and of order $T\ \log(T)$ are, to second order in α

$$\begin{aligned}\gamma_S \approx\, & \alpha T \left(1 - \frac{2}{3}\frac{p^2}{M^2} + \frac{8}{15}\frac{p^4}{M^4}\right)\\ & - \alpha^2 \frac{T}{\pi}\left(\log\left[\frac{T}{M}\left(1 - \frac{11}{12}\frac{p^2}{M^2} + \frac{1061}{1440}\frac{p^4}{M^4}\right)\right] + \frac{25}{12} - C_\Gamma\right)\\ & \times \left(1 - \frac{2}{3}\frac{p^2}{M^2} + \frac{8}{15}\frac{p^4}{M^4}\right) + \mathcal{O}\left(\alpha\frac{p^6}{M^6}, \alpha^2\frac{T^2}{M^2}, \alpha^3\right)\ ,\end{aligned} \tag{38}$$

where $p = |\mathbf{p}|$. Naturally this last expansion is only good in the weak coupling limit. Performing the same expansion for the vector boson exchange yields

$$\begin{aligned}\gamma_V \approx\, & \alpha T \left(1 - \frac{2}{3}\frac{p^2}{M^2} + \frac{8}{15}\frac{p^4}{M^4}\right)\\ & - \alpha^2 \frac{T}{\pi}\left(\log\left[\frac{T}{M}\left(1 - \frac{1}{3}\frac{p^2}{M^2} + \frac{31}{180}\frac{p^4}{M^4}\right)\right] + \frac{13}{3} - C_\Gamma\right)\\ & \times \left(1 - \frac{2}{3}\frac{p^2}{M^2} + \frac{8}{15}\frac{p^4}{M^4}\right) + \mathcal{O}\left(\alpha\frac{p^6}{M^6}, \alpha^2\frac{T^2}{M^2}, \alpha^3\right)\ .\end{aligned} \tag{39}$$

It is a crucial aspect of these results, that they are analytical functions of the coupling constant, i.e., in contrast to other calculations it is *not* proportional to $\log(1/g)$.

During the discussion of the self-consistency criterion (23) it was already stated, that the vector boson result is gauge independent in the limit $\mathbf{p} \to 0$. A vertex correction is needed to ensure gauge invariance for the momentum dependent parts (which in principle also require a completely different calculation scheme). However, this will at most give a regular modification of order $\alpha^2 p^2/M^2$, logarithmic terms may appear only in even higher order α.

The contribution of order $T\ \log(T)$, apart from being non-analytical around $T = 0$, has the effect of destroying the linear relationship between γ and T: Neither for high temperatures nor for low temperatures can one approximate the self-consistent γ by a linear function. This functional piece constitutes the leading non-analyticity observed numerically in Fig. 2.

7 Conclusion

This report presents a self-consistent calculation of the damping rate (= spectral width parameter) of a slow massive fermion in a gas of hot massless bosons. The result is obtained as a series in the coupling constant, the momentum of the fermion and the temperature. Instead of a non-analyticity in the coupling constant, which is sometimes stated in the literature, the final results for the damping rate exhibit a non-analyticity in the temperature parameter arising from contributions of order $T \log(T)$.

To confirm and explain this discrepancy several careful investigations were made. First of all, to control thc quality of the approximations every asymptotic expansion was made to higher order than assembled in the final result. On the other hand, all approximations were also compared to numerical calculations. Therefore, the quality of the series expansion is very well under control and considered *mathematically* sound.

Furthermore there exists an independent *physical* support for this result, since it is in accordance with the symmetry of space and time: In the presence of temperature the Poincaré symmetry is broken to $\mathrm{SO}(3) \times \mathrm{T}_4$, and the symmetry restoration with temperature $T \to 0$ is expected to be singular. Consequently, on physical grounds one would *expect* a non-analyticity in the temperature parameter.

Another check to be made for *physical* reasons is, whether the most dominant contribution to the fermion damping rate was taken into account. The "diagram" used in the present paper is the imaginary part of the self-consistent Fock diagram. Cutting this apart yields the proper scattering amplitudes hidden in the result – in the present case, these contains the coulomb scattering of the slow massive fermion by other fermions in the hot medium.

The only effect not included here is a polarization of the medium, i.e., a modification of the boson spectral function. However, such a modification could only weaken the boson propagator singularity, and consequently would give contributions which are *not* logarithmic (see [13] for more details). Thus it must be concluded, that the use of a modified boson propagator does not change the leading order result for the fermionic spectral width, i.e., the non-analyticity in temperature.

By these careful investigations one is thus forced to conclude the correctness of the results at least to the accuracy given in the previous section: The damping rate of a slow massive fermion in a gas of moderately hot massless bosons is $\gamma \propto g^2 T + \mathcal{O}(g^4 T \log(T/M))$.

Another thread of this report is the investigation, how different mathematical or physical assumptions may produce damping rates $\gamma \propto g^2 T \log(1/g)$. It was shown, that generally the methods used to obtain such a result violate important axioms of quantum theory and therefore should be avoided. Moreover it was found, that non-analyticities in the coupling constant arise, if the vacuum contributions are neglected – a common error in many existing papers on the subject. Only a consistent treatment of negative energy states produces the subtle cancellations of logarithmic pieces in (34) and (35).

For our approximate fermion spectral function the summation of all nested Fock diagrams was carried out to infinite order. Apart from vertex corrections, such a summation constitutes the solution of the full one-body problem. As pointed out, such vertex corrections would lead to momentum dependent corrections to the damping rate – and therefore would not affect the momentum independent parts of our results.

Acknowledgements

The author owes many thanks to H. Weigelt and R. Sollacher.

Appendix: Angular integration

In the following the explicit form of the angular integrals over the approximate fermion spectral function is given, which occur in the calculation of (24) and (25). In this appendix, η is the cosine of the angle between the two momenta $\mathbf{k}$ (internal) and $\mathbf{p}$ (external), and will henceforth will be used $k = |\mathbf{k}|$, $p = |\mathbf{p}|$.

$$\begin{aligned} I_1(k \pm p_0) &= \int_{-1}^{1} d\eta \, \frac{1}{(c_1 - 2pk\eta)^2 + c_2^2} \\ &= \frac{-1}{4pk(k \pm p_0)\gamma} \arctan\left(\frac{(k \pm p_0)^2 - t^2 - \gamma^2}{2(k \pm p_0)\gamma} \right)\Bigg|_{\omega_{p-k}}^{\omega_{p+k}} \end{aligned} \tag{40}$$

and

$$\begin{aligned} I_2(k \pm p_0) &= \int_{-1}^{1} d\eta \, \frac{\eta}{(c_1 - 2pk\eta)^2 + c_2^2} \\ &= \frac{c_1}{2kp} I_1(k \pm p_0) \\ &+ \frac{1}{8p^2k^2} \log\left(\frac{\left(t^2 + \gamma^2 - (k \pm p_0)^2\right)^2 + 4\gamma^2 (k \pm p_0)^2}{\gamma^2} \right)\Bigg|_{\omega_{p-k}}^{\omega_{p+k}} \end{aligned} \tag{41}$$

with

$$\begin{aligned} c_1 &= \pm 2p_0 k + p_0^2 - \omega^2 - \gamma^2 \\ c_2 &= 2(k \pm p_0)\gamma \end{aligned}$$

and boundaries of the integration defined as

$$\omega_{p\pm k}^2 = \omega^2 \pm 2pk + k^2 \; . \tag{42}$$

An important aspect of the above analytical results is to select the proper Riemann sheet for the arctan-functions: I_1 and I_2 are continuous functions in each variable.

References

1. P.A. Henning, K. Nakamura and Y. Yamanaka, Int. J. Mod. Phys. **B 10** (1996) 1599
2. P.A. Henning and E. Quack, Phys. Rev. Lett. **75** (1995) 2811
3. E. Braaten and R.D. Pisarski, Nucl. Phys. **B 337** (1990) 569
4. H. Narnhofer, M. Requardt and W. Thirring, Commun. Math. Phys. **92** (1983) 247
5. H.J. Borchers, R.N. Sen, Commun. Math. Phys. **21** (1975) 101
6. N.P. Landsman, Phys. Rev. Lett. **60** (1988) 1990 and Ann. Phys. **186** (1988) 141
7. A.L. Licht, Ann. Phys. **34** (1965) 161
8. A.S. Wightman, *in: Cargèse Lectures in Theoretical Physics: High Energy Electromagnetic Interactions and Field Theory*, ed. M. Lévy (Gordon & Breach, New York 1967)
9. N.P. Landsman and Ch.G. van Weert, Phys. Rep. **145** (1987) 141
10. J. Schwinger, J. Math. Phys. **2** (1961) 407; L.V. Keldysh, Zh. Exsp. Teor. Fiz. **47** (1964) 1515 and JETP **20** (1965) 1018
11. H. Umezawa, *Advanced Field Theory: Micro, Macro and Thermal Physics* (American Institute of Physics, 1993)
12. P.A. Henning, Phys. Reports **253** (1995) 235
13. P.A. Henning, Cond. Mat. Phys. **3** (2000) 75
14. P.A. Henning and H. Umezawa, Nucl. Phys. **B 417** (1994) 463; Phys. Lett. **B 355** (1995) 241
15. R. Kubo, J. Phys. Soc. Japan **12** (1957) 570; C. Martin and J. Schwinger, Phys. Rev. **115** (1959) 1342
16. R.L. Kobes and G.W. Semenoff, Nucl. Phys. **B 260** (1985) 714; Nucl. Phys. **B 272** (1986) 329
17. J. Knoll and D. Voskresensky, Ann. Phys. **249** (1996) 532
18. P.A. Henning, E. Poliatchenko, T. Schilling and J. Bros, Phys.Rev. **D 54** (1996) 5239
19. J. Glimm and A. Jaffe, *Quantum Physics* (Springer, New York 1981)
20. S. Weinberg, Phys. Rev. **118** (1960) 848
21. C. Itzykson and J.B. Zuber, *Quantum Field Theory* (McGraw-Hill, New York 1980)
22. P.A. Henning, Nucl. Phys. **A 546** (1992) 653
23. S. Mandelstam, Nuovo Cim. **15** (1960) 658
24. J. Kapusta, P. Lichard and D. Seibert, Phys. Rev. **D 44** (1991) 2774
25. C.M. Korpa and R. Malfliet, Phys. Lett. **B 315** (1993) 209
26. A.F. Bielajew and B.D. Serot, Ann. Phys. **156** (1984) 215
27. P.A. Henning and M. Blasone, hep-ph 9507273

Second Law of Thermodynamics, but Without a Thermodynamic Limit

Dieter Gross

Abstract. A geometric foundation of thermo-statistics is presented with only the axiomatic assumption of Boltzmann's principle $S(E,N,V) = k \ln W$. This relates the entropy to the geometric area $e^{S(E,N,V)}$ of the manifold of constant energy in the (finite) N-body phase space. From this principle, all of thermodynamics and especially all phenomena of phase transitions and critical phenomena can be unambiguously identified for even small systems. Within Boltzmann's principle, Statistical Mechanics becomes a geometric theory addressing the whole ensemble or the manifold of all points in phase space which are consistent with the few macroscopic conserved control parameters. This interpretation leads to a straight derivation of irreversibility and the second law of thermodynamics out of the time-reversible microscopic mechanical dynamics. It is the whole ensemble that spreads irreversibly over the accessible phase space not the single N-body trajectory. This is all possible without invoking the thermodynamic limit, extensivity, or concavity of $S(E,N,V)$. Without the thermodynamic limit or at phase-transitions the systems are usually not self-averaging, i.e. do not have a single peaked distribution in phase space. The main obstacle against the second law, the conservation of the phase-space volume due to Liouville, is overcome by realizing that a macroscopic theory like Thermodynamics cannot distinguish a fractal distribution in phase space from its closure.

1 Introduction

Recently interest in the thermo-statistical behavior of non-extensive many-body systems, like atomic nuclei, atomic clusters, soft-matter, biological systems – and also self-gravitating astrophysical systems lead to consider thermo-statistics without using the thermodynamic limit. This is most safely done by going back to Boltzmann.

Einstein calls Boltzmann's definition of entropy as e.g. written on his famous epitaph

$$\boxed{S = k \cdot \ln W} \tag{1}$$

Boltzmann's principle [1] from which Boltzmann was able to deduce thermodynamics. Here W is the number of micro-states at a given energy E of the N-body system in the spatial volume V:

$$W(E,N,V) = tr[\epsilon_0 \delta(E - \hat{H}_N)] \tag{2}$$

$$tr[\delta(E - \hat{H}_N)] = \int_{\{q \in V\}} \frac{1}{N!} \left(\frac{d^3q\, d^3p}{(2\pi\hbar)^3}\right)^N \delta(E - \hat{H}_N), \tag{3}$$

ϵ_0 is a suitable energy constant to make W dimensionless, $\hat{H}_N$ is the N-particle Hamiltonian and the N positions q are restricted to the volume V, whereas the momenta p are unrestricted. In what follows, we remain at the level of classical mechanics. The only reminders of the underlying quantum mechanics are the measure of the phase space in units of $2\pi\hbar$ and the factor $1/N!$ which respects the indistinguishability of the particles (Gibbs paradox). In contrast to Boltzmann [2,3] who only used the principle for dilute gases and to Schrödinger [4], who thought Eq. (1) is otherwise useless, I take the principle as *the fundamental, generic definition of entropy.* In a recent book [5] cf. also [6,7] I demonstrated that this definition of thermo-statistics works well especially also at higher densities and at phase transitions without invoking the thermodynamic limit.

Before we proceed we must comment on Einstein's attitude to the principle [8]. Originally, Boltzmann called W the "Wahrscheinlichkeit" (probability), i.e. the relative time a system spends (along a time-dependent path) in a given region of $6N$-dimensional phase space. Our interpretation of W to be the number of "complexions" (Boltzmann's second interpretation) or quantum states (trace) with the same energy was criticized by Einstein [1] as artificial. It is exactly this criticized interpretation of W which I use here and which works so well [5]. In Sect. 4 I will come back to this fundamental point.

2 Aim of This Contribution

Boltzmann's principle defines the entropy and with it thermodynamics for an *equilibrium* system. It opens a transparent *geometric* interpretation of equilibrium statistics. In this *static* scenario time plays no role.

2.1 Equilibrium Statistics as a Topology of $S(E,\cdots)$

At phase separations the canonical and grand-canonical potentials become singular, the Yang-Lee singularities. In contrast the microcanonical entropy $S(E,\cdots)$ is multiply differentiable there. The microcanonical probability $P(E) = e^{S(E)-T_{tr}E}$ is *multi-modal.* Between the peaks which characterize the various mono-phases, $S(E,\cdots)$ has a convex intruder with positive curvature. In this region of energies the specific heat $C = -(\frac{\partial S}{\partial E})^2/\frac{\partial^2 S}{\partial E^2}$ becomes *negative.* This is impossible in any canonical ensemble. In view of the fact that Thermodynamics was invented in the 19-century to explain the working of steam engines just at phase-separation, the *geometric interpretation of Statistical Mechanics* only now does justice to the original task of thermodynamics [5].

Only in homogeneous phases of systems with short-range interactions do the three ensembles, microcanonical, canonical, and grand-canonical be equivalent, but only in the thermodynamic limit. Thus the celebrated equivalence of the various ensembles is expected to be violated. The question for the correct ensemble has to be decided.

Not only diluted systems can be described by microcanonical statistics as done by Boltzmann [3]. In contrast, the power of the *microcanonical* statistics as a geometrical theory is most beautifully seen for dense systems and at phase-separations.

It is a common belief that phase transitions can be identified only in very large systems close to the thermodynamic limit. Especially the distinction between first order and continuous transitions is rather difficult in the canonical (conventional) ensemble of a finite system. In a series of papers I pointed out that in contrast to the canonical ensemble, the existence and the type of a phase transition can be very well clarified in the *microcanonical* ensemble for relatively small systems of some hundreds of particles [9–11]. Moreover inhomogeneities, fluctuations of interfaces between different phases especially the surface entropy as the source of surface tension, can be calculated on the same level as the other transition parameters.

The geometric interpretation of statistical mechanics has one of its most important success in the calculation of the phase diagram of rotating self-gravitating systems. The gas-phase, the collapse to a mono-star, to binaries or even more exotic configurations depending on the total angular-momentum and energy are the various *equilibrium* phases in that phase-diagram [12].

2.2 Nonequilibrium Statistics

However, thermodynamics is more than equilibrium statistics. The second law describes how a non-equilibrium system approaches the equilibrium with an increase of its entropy. It is the central law of thermodynamics that explains why thermodynamics works at all.

This book addresses short-time behavior of complex many-body systems. Before investigating non-equilibrium thermo*dynamics* microscopically one should first derive the second law from microscopic time-reversal-invariant Hamiltonian dynamics. This, however, is a still debated but challenging problem of the most fundamental importance. We will present a straight *geometrical* derivation of the second law also for small mixing systems.

After successfully deducing equilibrium statistics including all phenomena of phase transitions from Boltzmann's principle even for "Small" systems, i.e. non-extensive many-body systems, it is challenging to explore how far this "most conservative and restrictive way to thermodynamics" [13] is able to describe also the *approach* of (eventually "Small") systems to equilibrium and the second law of thermodynamics. {With the name "Small" systems I think of nuclei, atomic clusters and stars or star systems. They are all of a size comparable to the range of their force and thus never in the thermodynamic limit.}

Thermodynamics describes the development of *macroscopic* features of many-body systems without specifying them microscopically in all details. Before we address the second law, we have to clarify what we mean with the label "macroscopic observable".

3 Measuring a Macroscopic Observable. The "EPS-formulation"

A single point $\{q_i(t), p_i(t)\}_{i=1\cdots N}$ in the N-body phase space corresponds to a detailed specification of the system with all degrees of freedom (d.o.f) completely fixed at time t (microscopic determination). Fixing only the total energy E of an N-body system leaves the other $(6N-1)$-degrees of freedom unspecified. A second system with the same energy is most likely not in the same microscopic state as the first, it will be at another point in phase space, the other d.o.f.s will be different. That is the measurement of the total energy $\hat{H}_N$, or any other macroscopic observable $\hat{M}$, determines a $(6N-1)$-dimensional *sub-manifold* $\mathcal{E}$ or $\mathcal{M}$ in phase space. All points in N-body phase space consistent with the given value of E and volume V, i.e. all points in the $(6N-1)$-dimensional sub-manifold $\mathcal{E}(N,V)$ of phase space are equally consistent with this measurement. $\mathcal{E}(N,V)$ is the microcanonical ensemble. This example tells us that *any macroscopic measurement is incomplete and defines a sub-manifold of points in phase space not a single point.* An additional measurement of another macroscopic quantity $\hat{B}\{q,p\}$ reduces $\mathcal{E}$ further to the cross-section $\mathcal{E}\cap\mathcal{B}$, a $(6N-2)$-dimensional subset of points in $\mathcal{E}$ with the volume:

$$W(B,E,N,V) = \frac{1}{N!}\int\left(\frac{d^3q\,d^3p}{(2\pi\hbar)^3}\right)^N \epsilon_0\delta(E-\hat{H}_N\{q,p\})\,\delta(B-\hat{B}\{q,p\})\,. \tag{4}$$

If $\hat{H}_N\{q,p\}$ as also $\hat{B}\{q,p\}$ are continuous differentiable functions of their arguments, which we assume in the following, $\mathcal{E}\cap\mathcal{B}$ is closed. In the following we use W for the Riemann or Liouville volume of a many-fold.

Microcanonical thermo*statics* gives the probability $P(B,E,N,V)$ of finding the N-body system in the sub-manifold $\mathcal{E}\cap\mathcal{B}(E,N,V)$:

$$P(B,E,N,V) = \frac{W(B,E,N,V)}{W(E,N,V)} = e^{\ln[W(B,E,N,V)]-S(E,N,V)}\,. \tag{5}$$

This is what Krylov seems to have had in mind [14] and what I will call the "ensemble probabilistic formulation of statistical mechanics (EPS)".

Similarly thermo*dynamics* describes the development of some macroscopic observable $\hat{B}\{q_t,p_t\}$ in time of a system which was specified at an earlier time t_0 by another macroscopic measurement $\hat{A}\{q_0,p_0\}$. It is related to the volume of the sub-manifold $\mathcal{M}(t) = \mathcal{A}(t_0) \cap \mathcal{B}(t) \cap \mathcal{E}$:

$$W(A, B, E, t) = \frac{1}{N!} \int \left(\frac{d^3 q_t \, d^3 p_t}{(2\pi\hbar)^3} \right)^N \delta(B - \hat{B}\{q_t, p_t\}) \, \delta(A - \hat{A}\{q_0, p_0\})$$
$$\epsilon_0 \delta(E - \hat{H}\{q_t, p_t\}) \,, \tag{6}$$

where $\{q_t\{q_0, p_0\}, p_t\{q_0, p_0\}\}$ is the set of trajectories solving the Hamilton Jacobi equations

$$\dot{q}_i = \frac{\partial \hat{H}}{\partial p_i}, \qquad \dot{p}_i = -\frac{\partial \hat{H}}{\partial q_i}, \qquad i = 1 \cdots N \tag{7}$$

with the initial conditions $\{q(t = t_0) = q_0;\ p(t = t_0) = p_0\}$. For a very large system with $N \sim 10^{23}$ the probability of finding a given value $B(T)$, $P(B(t))$, is usually sharply peaked as a function of B. Ordinary thermodynamics treats systems in the thermodynamic limit $N \to \infty$ and gives only $\langle B(t) \rangle$. However, here we are interested in formulating the second law for "Small" systems i.e. we are interested in the whole distribution $P(B(t))$ not only in its mean value $\langle B(t) \rangle$. Thermodynamics does *not* describe the temporal development of a *single* system (single point in the $6N$-dim phase space).

There is an important property of macroscopic measurements. Whereas the macroscopic constraint $\hat{A}\{q_0, p_0\}$ determines (usually) a compact region $\mathcal{A}(t_0)$ in $\{q_0, p_0\}$ this does not need to be the case at later times $t \gg t_0$. $\mathcal{A}(t)$ defined by $\mathcal{A}\{q_0\{q_t, p_t\}, p_0\{q_t, p_t\}\}$ might become a *fractal* i.e. "spaghetti-like" manifold as a function of $\{q_t, p_t\}$ in $\mathcal{E}$ at $t \to \infty$ and loose compactness.

This can be expressed in mathematical terms. There exist a series of points $\{a_n\} \in \mathcal{A}(t)$ which converge to a point a_∞ which is *not* in $\mathcal{A}(t)$. E.g. such points a_∞ may have intruded from the phase space complimentary to $\mathcal{A}(t_0)$. Illustrative examples for this evolution of an initially compact sub-manifold into a fractal set are the baker transformation discussed in this context by [15,16]. Then no macroscopic (incomplete) measurement at time t can resolve a_∞ from its immediate neighbors a_n in phase space with distances $|a_n - a_\infty|$ less then any arbitrary small δ. In other words, *at the time* $t \gg t_0$ *no macroscopic measurement with its incomplete information about* $\{q_t, p_t\}$ *can decide whether* $\{q_0\{q_t, p_t\}, p_0\{q_t, p_t\}\} \in \mathcal{A}(t_0)$ *or not.* That is any macroscopic theory like thermodynamics can only deal with the *closure* of $\mathcal{A}(t)$. If necessary, the sub-manifold $\mathcal{A}(t)$ must be artificially closed to $\overline{\mathcal{A}}(t)$ as developed further in Sect. 5. *Clearly, in this approach this is the physical origin of irreversibility.* We come back to this in Sect. 5.

4 On Einstein's Objections Against the EPS-probability

According to Abraham Pais "Subtle is the Lord" [8], Einstein was critical with regard to the definition of relative probabilities given by Eq. (5), Boltzmann's counting of "complexions". He considered it as artificial and not corresponding to the immediate picture of probability used in the actual problem: "The word probability is used in a sense that does not conform to its definition as given in the theory of probability. In particular, cases of equal probability are often hypothetically defined in instances where the theoretical pictures used are sufficiently definite to give a deduction rather than a hypothetical assertion" [1]. He preferred to define probability by the relative time a system (a trajectory of a single point moving with time in the N-body phase space) spends in a subset of the phase space.

However, is this really the immediate picture of probability used in statistical mechanics? This definition demands the ergodicity of the trajectory in phase space. As we discussed above, thermodynamics as any other macroscopic theory handles incomplete, macroscopic information of the N-body system. It handles, consequently, the temporal evolution of *finite sized submanifolds* – ensembles – not single points in phase space. The *typical* outcomes of macroscopic measurements are calculated. Nobody waits in a macroscopic measurement, e.g. of the temperature, long enough that an atom can cross the whole system.

In this respect, I think the EPS version of statistical mechanics is closer to the experimental situation than the duration-time of a single trajectory. Moreover, in an experiment on a small system like a nucleus, the excited nucleus, which then may fragment statistically later on, is produced by a multiple *repetition* of scattering events and statistical averages are taken. No ergodic covering of the whole phase space by a single trajectory in time is demanded. At the high excitations of the nuclei in the fragmentation region their life-time would be too short for that. This is analogous to the statistics of a falling ball on a Galton's nail-board where also a single trajectory is not touching all nails but is random. Only after many *unbiased* repetitions the smooth binomial distribution is established. As I am discussing here the second law in *finite* systems, this is the correct scenario, not the time average over a single ergodic trajectory. (In this context see the remarks on Einstein's objection by [17] or [18] where the necessity of "new super-statistics" is speculated.)

5 Fractal Distributions in Phase Space, Second Law

Here we will first describe a simple working-scheme (i.e. a sufficient method) which allows us to deduce mathematically the second law. Later, we will show how this method is necessarily implied by the reduced information obtainable by macroscopic measurements.

Let us examine the following Gedanken experiment: Suppose the probability of finding our system at points $\{q_t, p_t\}_1^N$ in phase space is uniformly distributed for times $t < t_0$ over the sub-manifold $\mathcal{E}(N, V_1)$ of the N-body phase space at energy E and spatial volume V_1. At time $t > t_0$ we allow the system to spread over the larger volume $V_2 > V_1$ without changing its energy. If the system is *dynamically mixing*, the majority of trajectories $\{q_t, p_t\}_1^N$ in phase space starting from points $\{q_0, p_0\}$ with $q_0 \in V_1$ at t_0 will now spread over the larger volume V_2. Of course the Liouvillean measure of the distribution $\mathcal{M}\{q_t, p_t\}$ in phase space at $t > t_0$ will remains the same ($= tr[\mathcal{E}(N, V_1)]$) [19]. (The label $\{q_0 \in V_1\}$ of the integral means that the positions $\{q_0\}_1^N$ are restricted to the volume V_1, the momenta $\{p_0\}_1^N$ are unrestricted.)

$$
\begin{aligned}
&tr[\mathcal{M}\{q_t\{q_0,p_0\},p_t\{q_0,p_0\}\}]|_{\{q_0\in V_1\}} \\
&= \int_{\{q_0\{q_t,p_t\}\in V_1\}} \frac{1}{N!}\left(\frac{d^3q_t\, d^3p_t}{(2\pi\hbar)^3}\right)^N \epsilon_0\delta(E - \hat{H}_N\{q_t, p_t\}) \\
&= \int_{\{q_0\in V_1\}} \frac{1}{N!}\left(\frac{d^3q_0\, d^3p_0}{(2\pi\hbar)^3}\right)^N \epsilon_0\delta(E - \hat{H}_N\{q_0, p_0\}) \,, \qquad (8)
\end{aligned}
$$

$$
\text{because of:} \ \frac{\partial\{q_t, p_t\}}{\partial\{q_0, p_0\}} = 1 \,. \qquad (9)
$$

But as already argued by Gibbs the distribution $\mathcal{M}\{q_t, p_t\}$ will be filamented like ink in water and will approach arbitrarily close to any point of $\mathcal{E}(N, V_2)$. $\mathcal{M}\{q_t, p_t\}$ becomes dense in the new, larger $\mathcal{E}(N, V_2)$ for times sufficiently larger than t_0. The closure $\overline{\mathcal{M}}$ becomes equal to $\mathcal{E}(N, V_2)$. This is clearly expressed by Lebowitz [20,21].

In order to express this fact mathematically, *we have to redefine Boltzmann's definition of entropy Eq. (1) and introduce the following fractal "measure" for integrals like Eqs. (3) or (4):*

$$
W(E, N, t \gg t_0) = \frac{1}{N!}\int_{\{q_0\{q_t,p_t\}\in V_1\}} \left(\frac{d^3q_t\, d^3p_t}{(2\pi\hbar)^3}\right)^N \epsilon_0\delta(E - \hat{H}_N\{q_t, p_t\}) \,. \qquad (10)
$$

With the transformation:

$$
\int \left(d^3q_t\, d^3p_t\right)^N \cdots = \int d\sigma_1 \cdots d\sigma_{6N} \cdots \qquad (11)
$$

$$
d\sigma_{6N} := \frac{1}{||\nabla\hat{H}||}\sum_i \left(\frac{\partial\hat{H}}{\partial q_i}dq_i + \frac{\partial\hat{H}}{\partial p_i}dp_i\right) = \frac{1}{||\nabla\hat{H}||}dE \,, \qquad (12)
$$

$$
||\nabla\hat{H}|| = \sqrt{\sum_i\left(\frac{\partial\hat{H}}{\partial q_i}\right)^2 + \sum_i\left(\frac{\partial\hat{H}}{\partial p_i}\right)^2} \,, \qquad (13)
$$

$$
W(E, N, t \gg t_0) = \frac{1}{N!(2\pi\hbar)^{3N}}\int_{\{q_0\{q_t,p_t\}\in V_1\}} d\sigma_1 \cdots d\sigma_{6N-1}\frac{\epsilon_0}{||\nabla\hat{H}||} \,, \qquad (14)
$$

we replace $\mathcal{M}$ by its closure $\overline{\mathcal{M}}$ and *define* now:

$$\begin{aligned} W(E,N,t \gg t_0) &\to M(E,N,t \gg t_0) \\ &:= \langle G(\mathcal{E}(N,V_2))\rangle * \mathrm{vol}_{\mathrm{box}}[\mathcal{M}(E,N,t \gg t_0)]\,, \end{aligned} \tag{15}$$

where $\langle G(\mathcal{E}(N,V_2))\rangle$ is the average of $\frac{\epsilon_0}{N!(2\pi\hbar)^{3N}||\nabla\hat{H}||}$ over the (larger) manifold $\mathcal{E}(N,V_2)$, and $\mathrm{vol}_{\mathrm{box}}[\mathcal{M}(E,N,t \gg t_0)]$ is the box-counting volume of $\mathcal{M}(E,N,t \gg t_0)$ which is the same as the volume of $\overline{\mathcal{M}}$, see below.

To obtain $\mathrm{vol}_{\mathrm{box}}[\mathcal{M}(E,N,t \gg t_0)]$ we cover the d-dim. sub-manifold $\mathcal{M}(t)$, here with $d = (6N-1)$, of the phase space by a grid with spacing δ and count the number $N_\delta \propto \delta^{-d}$ of boxes of size δ^{6N}, which contain points of $\mathcal{M}$. Then we determine

$$\mathrm{vol}_{\mathrm{box}}[\mathcal{M}(E,N,t \gg t_0)] := \underline{\lim}_{\delta\to 0}\delta^d N_\delta[\mathcal{M}(E,N,t \gg t_0)] \tag{16}$$

with $\underline{\lim}\,* = \inf[\lim *]$ or symbolically:

$$\begin{aligned} M(E,N,t \gg t_0) &=: \oint\!\!\!\!\!\!B_d{}_{\{q_0\{q_t,p_t\}\in V_1\}} \frac{1}{N!}\left(\frac{d^3q_t\,d^3p_t}{(2\pi\hbar)^3}\right)^N \epsilon_0\delta(E-\hat{H}_N) \\ &\to \frac{1}{N!}\int_{\{q_t\in V_2\}}\left(\frac{d^3q_t\,d^3p_t}{(2\pi\hbar)^3}\right)^N \epsilon_0\delta(E-\hat{H}_N\{q_t,p_t\}) \\ &= W(E,N,V_2) \geq W(E,N,V_1)\,, \end{aligned} \tag{17}$$

where $\oint\!\!\!\!\!\!B_d$ means that this integral should be evaluated via the box-counting volume Eq. (16) here with $d = 6N-1$.

This is illustrated by Fig. 1. With this extension of Eq. (3) Boltzmann's entropy (1) is at time $t \gg t_0$ equal to the logarithm of the *larger* phase space $W(E,N,V_2)$. *This is the second law of thermodynamics.* The box-counting is also used in the definition of the Kolmogorov entropy, the average rate of entropy gain [22,23]. Of course still at t_0 $\overline{\mathcal{M}(t_0)} = \mathcal{M}(t_0) = \mathcal{E}(N,V_1)$:

$$M(E,N,t_0) =: \oint\!\!\!\!\!\!B_d{}_{\{q_0\in V_1\}} \frac{1}{N!}\left(\frac{d^3q_0\,d^3p_0}{(2\pi\hbar)^3}\right)^N \epsilon_0\delta(E-\hat{H}_N) \tag{18}$$

$$\begin{aligned} &\equiv \int_{\{q_0\in V_1\}} \frac{1}{N!}\left(\frac{d^3q_0\,d^3p_0}{(2\pi\hbar)^3}\right)^N \epsilon_0\delta(E-\hat{H}_N) \\ &= W(E,N,V_1)\,. \end{aligned} \tag{19}$$

The box-counting volume is analogous to the standard method of determining the fractal dimension of a set of points [22] by the box-counting dimension:

$$\mathrm{dim}_{\mathrm{box}}[\mathcal{M}(E,N,t \gg t_0)] := \underline{\lim}_{\delta\to 0}\frac{\ln N_\delta[\mathcal{M}(E,N,t \gg t_0)]}{-\ln\delta}\,. \tag{20}$$

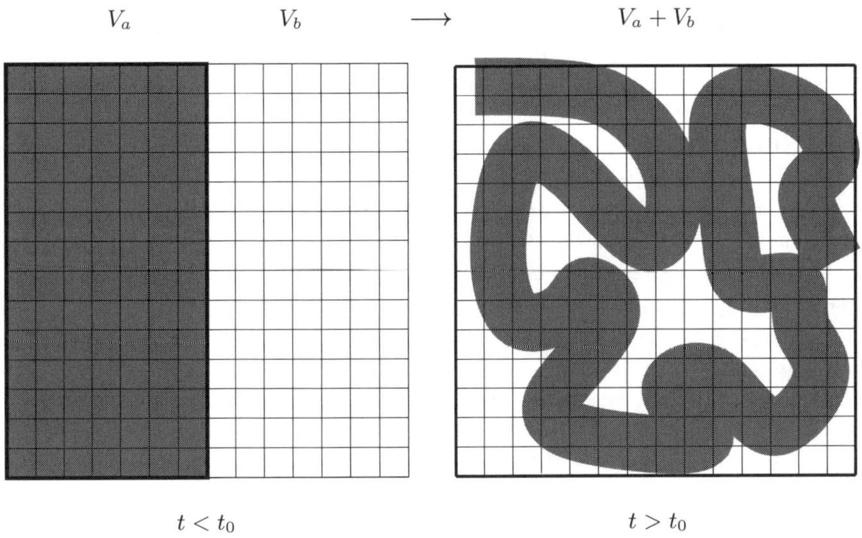

Fig. 1. The compact set $\mathcal{M}(t_0)$, *left side*, develops into an increasingly folded "spaghetti"-like distribution in phase-space with rising time t. This figure shows only the early form of the distribution. At much larger times it will become more and more fractal and finally dense in the new phase space. The grid illustrates the boxes of the box-counting method.All boxes which overlap with $\mathcal{M}(t)$ are counted in N_δ in Eq. (16)

Like the box-counting dimension, vol_{box} has the peculiarity that it is equal to the volume of the smallest *closed* covering set. E.g.: The box-counting volume of the set of rational numbers $\{\mathbf{Q}\}$ between 0 and 1, is $\text{vol}_{\text{box}}\{\mathbf{Q}\} = 1$, and thus equal to the measure of the *real* numbers, c.f. Falconer [22] Sect. 3.1. This is the reason why vol_{box} is not a measure in its mathematical definition because then we should have

$$\text{vol}_{\text{box}}\left[\sum_{i\in\{\mathbf{Q}\}}(\mathcal{M}_i)\right] = \sum_{i\in\{\mathbf{Q}\}}\text{vol}_{\text{box}}[\mathcal{M}_i] = 0\,, \tag{21}$$

therefore the quotation marks for the box-counting "measure" c.f. the appendix.

Coming back to the the end of Sect. 3, the volume $W(A, B, \cdots, t)$ of the relevant ensemble, the *closure* $\overline{\mathcal{M}(t)}$ must be "measured" by something like the box-counting "measure" Eqs. (16), (17) with the box-counting integral $\oint_{\mathcal{B}_d}$, which must replace the integral in Eq. (3). Due to the fact that the box-counting volume is equal to the volume of the smallest closed covering set, the new, extended, definition of the phase-space integral Eq. (17) is for compact sets like the equilibrium distribution $\mathcal{E}$ identical to the old one Eq. (3) and nothing changes for equilibrium statistics. Therefore, one can simply replace the old Boltzmann-definition of the number of complexions and with it the old entropy by the new one Eq. (17).

6 Conclusion

Macroscopic measurements $\hat{M}$ determine only very few of all the $6N$ d.o.f.s. Any macroscopic theory like thermodynamics deals with the *area* M of the corresponding closed sub-manifolds $\overline{\mathcal{M}}$ in the $6N$-dimensional phase space not with single points. The averaging over ensembles or finite sub-manifolds in phase space becomes especially important for the microcanonical ensemble of a *finite* system.

Because of this necessarily coarsed information, macroscopic measurements, and with it also macroscopic theories are unable to distinguish fractal sets $\mathcal{M}$ from their closures $\overline{\mathcal{M}}$. Therefore, I make the conjecture: the proper manifolds determined by a macroscopic theory like thermodynamics are the closed $\overline{\mathcal{M}}$. However, an initially closed subset of points at time t_0 does not necessarily evolve again into a closed subset at $t \gg t_0$. That is the closure operation and the $t \to \infty$ limit do not commute, and the macroscopic dynamics becomes irreversible.

Here is the origin of the misunderstanding by the famous reversibility paradoxes which were invented by Loschmidt [24] and Zermelo [25,26] and which bothered Boltzmann so much [27,28]. These paradoxes address to trajectories of *single points* in the N-body phase space which must return after Poincaré's recurrence time or which must run backwards if all momenta are exactly reversed. Therefore, Loschmidt and Zermelo concluded that the entropy should decrease as well as it was increasing before. The specification of a single point demands of course a *microscopic exact* specification and control in time of all $6N$ degrees of freedom not a determination of a few macroscopic degrees of freedom only. No entropy is defined for a single point.

This way various non-trivial limiting processes can be avoided. Neither does one invoke the thermodynamic limit of a homogeneous system with infinitely many particles nor does one rely on the ergodic hypothesis of the equivalence of (very long) time averages and ensemble averages. *The use of ensemble averages is justified directly by the very nature of macroscopic (incomplete) measurements.* Coarse-graining appears as a natural consequence of this. The box-counting method mirrors the averaging over the overwhelming number of non-determined degrees of freedom. Of course, a fully consistent theory must use this averaging explicitly. Then one would not depend on the order of the limits $\lim_{\delta\to 0}\lim_{t\to\infty}$ as it was tacitly assumed here. Presumably, the rise of the entropy can then be already seen at finite times when the fractality of the distribution in phase space is not yet fully developed. The coarse-graining is no longer a mathematical ad hoc assumption. Moreover the second law is in the EPS-formulation of statistical mechanics not linked to the thermodynamic limit as was thought up to now [20,21]. Also

Gaspard [29] derives the second law for an infinite system (thermodynamic limit). This is more or less the same as Boltzmann's answer to Zermelo's objection.

Acknowlegements

I thank A. de Martino for a careful reading of this manuscript and giving useful suggestions for its improvement.

Appendix

In the mathematical theory of fractals [22] one usually uses the Hausdorff measure or the Hausdorff dimension of the fractal [23]. This, however, would be wrong in statistical mechanics. Here I want to point out the difference between the box-counting "measure" and the proper Hausdorff measure of a manifold of points in phase space. Without going into too much mathematical detail we can make this clear again with the same example as above: The Hausdorff measure of the rational numbers $\in [0,1]$ is 0, whereas the Hausdorff measure of the real numbers $\in [0,1]$ is 1. Therefore, the Hausdorff measure of a set is a proper measure. The Hausdorff measure of the fractal distribution in phase space $\mathcal{M}(t \to \infty)$ is the same as that of $\mathcal{M}(t_0)$, $W(E, N, V_1)$. Measured by the Hausdorff measure the phase space volume of the fractal distribution $\mathcal{M}(t \to \infty)$ is conserved and Liouville's theorem applies. This would demand that thermodynamics could distinguish between any point inside the fractal from any point outside of it independent of how close it is. This, however, is impossible for any macroscopic theory that can only address macroscopic information where all unobserved degrees of freedom are averaged over. That is the deep reason why the box-counting "measure" must be taken and where irreversibility comes from in a fully time reversible microscopic dynamics.

References

1. A. Einstein. Über einen die Erzeugung und Verwandlung des Lichtes betreffenden heuristischen Gesichtspunkt. *Annalen der Physik*, 17:132, 1905.
2. L. Boltzmann. Über die Beziehung eines allgemeinen mechanischen Satzes zum Hauptsatz der Wärmelehre. *Sitzungsbericht der Akadamie der Wissenschaften, Wien*, II:67–73, 1877.
3. L. Boltzmann. Über die Begründung einer kinetischen Gastheorie auf anziehende Kräfte allein. *Wiener Berichte*, 89:714, 1884.
4. E. Schrödinger. *Statistical Thermodynamics, a Course of Seminar Lectures, delivered in January-March 1944 at the School of Theoretical Physics*. Cambridge University Press, London, 1946.

5. D.H.E. Gross. *Microcanonical thermodynamics: Phase transitions in "Small" systems*, volume 66 of *Lecture Notes in Physics.* World Scientific, Singapore, 2001.
6. D.H.E. Gross and E. Votyakov. Phase transitions in "small" systems. *Eur. Phys. J. B*, 15:115–126, (2000), http://arXiv.org/abs/cond-mat/?9911257.
7. D.H.E. Gross. Micro-canonical statistical mechanics of some non-extensive systems. *Chaos, Solitons, and Fractals*, 13:417–430, 2002; http://arXiv.org/abs/astro-ph/?cond-mat/0004268.
8. A. Pais. *Subtle is the Lord.* Oxford University Press, Oxford, 1982.
9. D.H.E. Gross and M.E. Madjet. Fragmentation phase transition in atomic clusters IV – the relation of the fragmentation phase transition to the bulk liquid-gas transition. *Z. Physik B*, 104:541–551, 1997; and http://xxx.lanl.gov/abs/cond-mat/9707100.
10. D.H.E. Gross and M.E. Madjet. Cluster fragmentation, a laboratory for thermodynamics and phase-transitions in particular. In Abe, Arai, Lee, and Yabana, editors, *Proceedings of "Similarities and Differences between Atomic Nuclei and Clusters"*, pages 203–214, Tsukuba, Japan 97, 1997. The American Institute of Physics.
11. D.H.E. Gross. Phase transitions without thermodynamic limit. In X. Campi J.P. Blaizot, M. Ploszaiczak, editor, *Proceedings of the Les Houches Workshop on Nuclear Matter in Different Phases and Transitions*, pages 31–42, Les Houches, France, 31.3-10.4.98, 1999; http://xxx.lanl.gov/abs/cond-mat/9812120. Kluwer Acad. Publ.
12. E.V. Votyakov, H.I. Hidmi, A. De Martino, and D.H.E. Gross. Microcanonical mean-field thermodynamics of self-gravitating and rotating systems. *Phys. Rev. Lett.*, 89:031101–1–4; http://arXiv.org/abs/cond–mat/0202140, (2002).
13. J. Bricmont. Science of chaos or chaos in science? *Physicalia Magazine, Proceedings of the New York Academy of Science, to apear*, pages 1–50, 2000.
14. N.S. Krylov. *Works on the Foundation of Statistical Physics.* Princeton University Press, Princeton, 1979.
15. R.F. Fox. Entropy evolution for the baker map. *Chaos*, 8:462–465, 1998.
16. T. Gilbert, J.R.Dorfman, and P.Gaspard. Entropy production, fractals, and relaxation to equilibrium. *Phys. Rev. Lett.*, 85:1606,nlin.CD/0003012, 2000.
17. C. Beck and E.G.D. Cohen. Superstatistics. *cond-mat/0205097*, 2002.
18. C. Tsallis. Entropic nonextensivity: a possible measure of complexity. *Chaos, Solitons, and Fractas*, page 13, 2002.
19. H. Goldstein. *Classical Mechanics.* Addison-Wesley, Reading, Mass, 1959.
20. J.L. Lebowitz. Microscopic origins of irreversible macroscopic behavior. *Physica A*, 263:516–527, 1999.
21. J.L. Lebowitz. Statistical mechanics: A selective review of two central issues. *Rev. Mod. Phys.*, 71:S346–S357, 1999.
22. Kenneth Falconer. *Fractal Geometry – Mathematical Foundations and Applications.* John Wiley & Sons, Chichester, New York, Brisbane, Toronto,Singapore, 1990.
23. E.W. Weisstein. *Concise Encyclopedia of Mathemetics.* CRC Press, London, New York, Washington D.C:, 1999.
24. J. Loschmidt. *Wiener Berichte*, 73:128, 1876.
25. E. Zermelo. *Wied. Ann.*, 57:778–784, 1896.
26. E. Zermelo. Über die mechanische Erklärung irreversiblen Vorgänge. *Wied. Ann.*, 60:392–398, 1897.

27. E.G.D. Cohen. Boltzmann and statistical mechanics. In *Boltzmann's Legacy, 150 Years after his Birth*, pages http://xxx.lanl.gov/abs/cond–mat/9608054, Rome, 1997. Atti dell Accademia dei Lincei.
28. E.G.D. Cohen. *Boltzmann and Statistical Mechanics*, volume 371 of *Dynamics: Models and Kinetic Methods for Nonequilibrium Many Body Systems*. Kluwer, Dordrecht, The Netherlands, e, j. karkheck edition, 2000.
29. P. Gaspard. Entropy production in open volume preserving systems. *J. Stat. Phys*, 88:1215–1240, 1997.

Part II

Description of Nonequilibrium Correlations

Isothermal Molecular Dynamics in Classical and Quantum Mechanics

Detlef Mentrup and Jürgen Schnack

Abstract. The typical problem in statistical physics is the determination of ensemble averages given in terms of phase-space integrals (classical case) or traces of the density matrix (quantum case). Especially for strongly correlated, i.e. interacting many-body systems, the direct evaluation of these averages is usually impossible. Therefore, a large number of different approaches has been developed. In this contribution, we will focus on finite temperature or *canonical ensemble* properties.

A specific approach due to Nosé is based on classical molecular dynamics (MD) and time averages [1]. It permits one to calculate canonical ensemble averages by averaging over an ergodic deterministic isothermal time evolution. The key idea is to add an additional degree of freedom to the original system in order to mimic the influence of the heat bath on the dynamics. Nowadays, the most reliable (ergodic) techniques are the so-called Nosé–Hoover chains and the demon method.

Static finite temperature properties of *quantum* systems have been determined very successfully with path integral Monte Carlo (MC) techniques in the past, e.g., for liquid ^{4}He [2]. A drawback of these methods is the fact that MC dynamics is highly artificial and non-physical. Therefore, the calculation of dynamical quantities is inhibited. Surprisingly, the calculation of path integrals using *classical* isothermal MD methods is also possible on the basis of the "classical isomorphism" due to Feynman [3], but this method does not shed light on the real-time quantum dynamics of the system either.

Hence, a more direct isothermal quantum molecular dynamics scheme is highly desirable in order to realistically describe the dynamics of a quantum system at finite temperature. As a first step, we are able to generalize the method of Nosé and Hoover to a genuine quantum system, the ubiquitous harmonic oscillator. The method is valid both for a single particle and for ideal Fermi and Bose gases. We discuss possibilities of further generalization.

In our contribution we review the various classical thermostat methods and sketch fields of applicability. We also discuss indirect (path integrals) and direct (our own work) applications of thermostat methods in quantum mechanical many-body problems.

1 Introduction

Classical MD simulations are performed by solving Hamilton's equations of motion numerically. Therefore, the internal energy of the system is conserved during time evolution. If the respective system is *ergodic*, a time average of a macroscopic observable is, under equilibrium conditions, equivalent to a microcanonical ensemble average. Hence, isoenergetic molecular dynamics naturally induces a method for the calculation of microcanonical ensemble averages.

In the canonical ensemble, however, the internal energy of the system is not constant, but free to fluctuate by thermal contact to an external heat bath. In order to adapt classical MD simulations to the problem of the calculation of canonical ensemble averages, i.e. to pass over from an isoenergetic to an isothermal time evolution, powerful methods have been developed since the 1980s, and they are commonly used nowadays [4]. Some of them, like the Nosé–Hoover technique or the Kusnezov–Bulgac–Bauer thermostat, are based on deterministic time-reversal equations of motion that are designed to fill the phase space of the system according to the canonical thermal weight during time evolution. In Sect. 2, it will be demonstrated that this allows the calculation of canonical ensemble averages by means of molecular dynamics.

Regarding correlations, a thermostated dynamics offers the general opportunity to model the rise, the stability, and the decay of correlations under the influence of a heat bath. For such investigations, which are not within the scope of the present contribution, one may as well apply other constraints than constant temperature, e.g., constant pressure involving barostats. Moreover, it is also possible to study the system in nonequilibrium situations, e.g. by varying the temperature of the thermostat, and thus investigate cooling processes as in glass transitions [5]. A good starting point for an overview of possible applications is the report by Nosé [1] that treats the classical thermostat methods in detail.

Quantum canonical ensemble averages are accessible via Feynman's path integral method and the so-called "classical isomorphism" that allows a mapping of a quantum statistical problem to a classical system, see Sect. 3.

In quantum mechanics, a number of approximate quantum molecular dynamics methods are available [6]. It is an open question whether they can be modified in a manner appropriate to permit the calculation of quantum canonical averages. In a first step towards an isothermal quantum dynamics scheme, we show that in the special case of a quantum particle in a harmonic oscillator potential, the framework of coherent states (Sect. 4) permits to set up equations of motion for an isothermal quantum dynamics (Sect. 5) [7]. Following the approach of Nosé–Hoover chains or the similar Kusnezov–Bulgac–Bauer-technique, time-dependent pseudofriction terms are added to the equations of motion for the parameters of coherent states. The dynamics of the pseudofriction coefficient is designed such that the desired thermal weight function is sampled in time if the system is ergodic. The case of identical quantum particles adds an important aspect to the problem of isothermal quantum dynamics, the principle of indistinguishability. The resulting (anti-)symmetry of the wave function leads to additional terms in the equations of motion that act like an attractive (bosons) or repulsive force (fermions) in the dynamics.

2 Classical Thermostat Methods

A classical statistical ensemble average $\langle\langle B \rangle\rangle$ of an observable $B(q,p)$ (which in classical mechanics is a function on phase space) is given by the phase space integral

$$\langle\langle\, B \,\rangle\rangle = \frac{1}{Z_{cl}} \int_\Gamma \mathrm{d}q\mathrm{d}p\, B(q,p) f(q,p) \;, \qquad \text{with } Z_{cl} = \int_\Gamma \mathrm{d}q\mathrm{d}p\, f(q,p) \;. \quad (1)$$

The basic idea of MD methods is to calculate ensemble averages (1) via time averaging over a dynamics adapted to the respective statistical ensemble characterized by $f(q,p)$, i.e. create temporal trajectories $(q(t),p(t))$ in phase space such that

$$\lim_{\mathcal{T}\to\infty} \frac{1}{\mathcal{T}} \int_0^{\mathcal{T}} \mathrm{d}t\, B(q(t),p(t)) = \frac{1}{Z_{cl}} \int_\Gamma \mathrm{d}q\mathrm{d}p\, B(q,p) f(q,p) \;. \quad (2)$$

For the microcanonical ensemble, Hamilton's equations of motion are directly applicable, since they conserve energy. The condition (2) is fulfilled for ergodic systems. In the canonical ensemble, however, the energy of the system is free to fluctuate via thermal contact to a heat bath. Therefore, suitable modifications of the equations of motion have to be studied.

The non-Hamiltonian set of equations of motion,

$$\begin{aligned} \frac{d}{dt} q_i &= \frac{p_i}{m} \;, \qquad \frac{d}{dt} p_i = -\frac{\partial V(q)}{\partial q_i} - p_i \frac{p_\eta}{Q} \;, \\ \frac{d}{dt} p_\eta &= \sum_{i=1}^{N} \frac{p_i^2}{m} - N k_B T \;, \\ \frac{d}{dt} \eta &= \frac{p_\eta}{Q} \;, \end{aligned} \quad (3)$$

defines Nosé–Hoover dynamics. It is obtained from Hamilton's equations of motion by adding the pseudofrictional term $-p_i p_\eta / Q$ to the equation of motion of the momenta. p_η is a time-dependent supplementary degree of freedom added to the system to induce the energy fluctuations required in the canonical ensemble. Its time dependence is driven by the deviation of the momentary value of the kinetic energy of the system from its canonical ensemble average. In more detail, it is derived from the postulate that the set (3) be a stationary solution of the following Liouville equation,

$$\frac{d}{dt} f = -f \left(\frac{\partial}{\partial x} \cdot \dot{x} \right) \;, \quad (4)$$

where $x = (q, p, p_\eta)$ denotes an element of the phase space of the enlarged system, and $f(x)$ is the distribution function

$$f(q,p,p_\eta) = \exp\left(-\beta \left(H(q,p) + \frac{p_\eta^2}{2Q} \right) \right) \;, \quad (5)$$

that contains Boltzmann's distribution function on the subspace of the original system as a marginal. Equation (4) is obtained from a continuity equation,

$$\frac{\partial f}{\partial t} + \mathrm{div}_x(f\dot{x}) = 0 \;, \tag{6}$$

and the expression for the total temporal derivative of f,

$$\frac{d}{dt}f = \frac{\partial f}{\partial t} + \dot{x} \cdot \frac{\partial f}{\partial x} \;. \tag{7}$$

The set of equations (3) conserves the quantity

$$H'(q,p,p_\eta,\eta) = H(q,p) + \frac{p_\eta^2}{2Q} + Nk_BT \int^t \mathrm{d}t' \, \frac{p_\eta}{Q} \;, \tag{8}$$

and it can be shown under the assumption of ergodicity that a microcanonical average on the phase space (q,p,p_η) involving the weight function $\delta(H'-E')$ corresponds to a canonical average in the subspace of the original system [8].

In summary, the essence of the Nosé–Hoover scheme is that a temporal average of a classical observable over the trajectories generated by (3) corresponds to a phase space average with the distribution function (5). This statement is true only if the system is *ergodic*, which implies that the system actually runs through all phase space points with the correct weight, independent of the initial conditions. In practice, the original Nosé–Hoover method (3) frequently fails to produce ergodic motion, in particular in the important case of the harmonic oscillator where Gaussian distributions need to be sampled. For this reason, Martyna et al. [9] have proposed the following scheme,

$$\begin{aligned}
\frac{d}{dt}q_i &= \frac{p_i}{m} \;, \qquad \frac{d}{dt}p_i = -\frac{\partial V(q)}{\partial q_i} - p_i\frac{p_{\eta_1}}{Q_1} \;, \\
\frac{d}{dt}p_{\eta_1} &= \left(\sum_{i=1}^{N}\frac{p_i^2}{m} - Nk_BT\right) - p_{\eta_1}\frac{p_{\eta_2}}{Q_2} \;, \\
\frac{d}{dt}p_{\eta_j} &= \left(\frac{p_{\eta_{j-1}}^2}{Q_{j-1}} - k_BT\right) - p_{\eta_j}\frac{p_{\eta_{j+1}}}{Q_{j+1}} \;, \\
\frac{d}{dt}p_{\eta_M} &= \frac{p_{\eta_{M-1}}^2}{Q_{M-1}} - k_BT \;, \\
\frac{d}{dt}\eta_i &= \frac{p_{\eta_i}}{Q_i} \;,
\end{aligned} \tag{9}$$

in which the pseudofriction coefficients p_{η_j} are recursively thermalized, forming a *chain thermostat* of length M. For the harmonic oscillator, chains of length 4 reliably produce ergodic motion.

A different generalization of the original scheme (3) is due to Kusnezov, Bulgac, and Bauer. Two additional degrees of freedom, so-called *demons*, are used to replicate the interaction of the original system with a heat bath. The first one, ξ, is coupled to the equations of motion of the positions, the second one, ζ, to the momenta of the particles:

$$\frac{d}{dt}q_i = \frac{\partial H}{\partial p_i} - g_2'(\xi)\, F_i(q,p) \,, \qquad \frac{d}{dt}p_i = -\frac{\partial H}{\partial q_i} - g_1'(\zeta)\, G_i(q,p) \,. \tag{10}$$

The phase space distribution function that is to be sampled in the extended phase space is chosen to be

$$f(q,p,\zeta,\xi) = \exp\left(-\beta\Big(H(q,p) + \frac{1}{\kappa_1}g_1(\zeta) + \frac{1}{\kappa_2}g_2(\xi)\Big)\right) \,. \tag{11}$$

An analysis of the Liouville equation (4) in the extended phase space $x = (q,p,\xi,\zeta)$ imposing the constraints

$$\frac{\partial\dot{\zeta}}{\partial\zeta} = 0 \,, \qquad \frac{\partial\dot{\xi}}{\partial\xi} = 0 \,, \tag{12}$$

leads to the following equations of motion for the demons,

$$\frac{d}{dt}\zeta = \kappa_1 \sum_{i=1}^{N}\left(\frac{\partial H}{\partial p_i}G_i - \frac{1}{\beta}\frac{\partial G_i}{\partial p_i}\right) \,, \tag{13}$$

$$\frac{d}{dt}\xi = \kappa_2 \sum_{i=1}^{N}\left(\frac{\partial H}{\partial q_i}F_i - \frac{1}{\beta}\frac{\partial F_i}{\partial q_i}\right) \,.$$

In principle, since the choice of the functions F, G, g_1, g_2 is arbitrary, this method offers a lot of freedom. The most prominent choice of functions is the following so-called cubic coupling scheme:

$$g_1 = \frac{1}{4}\zeta^4 \,, \qquad g_2 = \frac{1}{2}\xi^2 \,, \qquad F_i = q_i^3 \,, \qquad G_i = p_i \,. \tag{14}$$

These equations provide ergodic behavior in all examples given in [10].

Classical MD is nowadays a widely used theoretical tool that is applied in physics, chemistry, and biology to model properties of liquids, solids, molecular clusters etc. The thermostat methods of the present section are employed to allow simulations in the canonical ensemble. A review of the huge number of applications of MD is beyond our scope, and we refer the reader to [4].

3 Path Integral Applications

The thermostat methods of the preceding section are applicable to systems of classical point particles. It is noteworthy that the same techniques can also be used in order to solve quantum statistical problems. The *path integral method* is the basis for this and will be sketched briefly in this section.

All static properties of a quantum system in thermal equilibrium at a temperature T can be determined from the thermal density matrix,

$$\underset{\sim}{\rho}(\beta) = \mathrm{e}^{-\beta \underset{\sim}{H}} , \tag{15}$$

where $\underset{\sim}{H}$ is the Hamiltonian of the system and $\beta = 1/k_B T$. The thermal expectation value of a quantum observable $\underset{\sim}{B}$ is given by

$$\langle\langle\, \underset{\sim}{B} \,\rangle\rangle = \frac{1}{Z} \mathrm{Tr} \underset{\sim}{\rho} \underset{\sim}{B} , \tag{16}$$

with

$$Z = \mathrm{Tr} \underset{\sim}{\rho} . \tag{17}$$

The density matrix $\rho(\beta)$ has the following property,

$$\underset{\sim}{\rho}(\beta_1 + \beta_2) = \underset{\sim}{\rho}(\beta_1)\, \underset{\sim}{\rho}(\beta_2) , \tag{18}$$

which in position representation leads to the so-called convolution equation,

$$\langle R | \underset{\sim}{\rho}(\beta_1 + \beta_2) | R' \rangle = \int \mathrm{d}R'' \langle R | \mathrm{e}^{-\beta_1 \underset{\sim}{H}} | R'' \rangle \langle R'' | \mathrm{e}^{-\beta_2} \underset{\sim}{H} | R' \rangle , \tag{19}$$

or, in a modified notation for the matrix elements,

$$\rho(R, R'; \beta_1 + \beta_2) = \int \mathrm{d}R'' \rho(R, R''; \beta_1)\, \rho(R'', R'; \beta_2) . \tag{20}$$

We repeat this process M times,

$$\rho(R_0, R_M; \beta) = \int \cdots \int \mathrm{d}R_1 \mathrm{d}R_2 \ldots \mathrm{d}R_{M-1} \\ \rho(R_0, R_1; \tau) \rho(R_1, R_2; \tau) \ldots \rho(R_{M-1}, R_M; \tau) , \tag{21}$$

where $\tau = \beta/M$ is called the time step, and the succession of points $(R_0, R_1, \ldots R_M)$ is called a path. Equation (21) is exact for any $M > 0$.

If the Hamiltonian is composed of a kinetic and a potential operator, $\underset{\sim}{T}$ and $\underset{\sim}{V}$, we can use the Trotter formula to find an approximation of the density matrix at small τ,

$$\mathrm{e}^{-\beta \underset{\sim}{H}} = \lim_{M \to \infty} \left(\mathrm{e}^{-\frac{\beta}{M} \underset{\sim}{T}} \mathrm{e}^{-\frac{\beta}{M} \underset{\sim}{V}} \right)^M . \tag{22}$$

The most elementary approximation neglects all commutators between $\underset{\sim}{T}$ and $\underset{\sim}{V}$,

$$\rho(R, R'; \tau) \approx \int \mathrm{d}R''\, \langle R | \mathrm{e}^{-\tau \underset{\sim}{T}} | R'' \rangle \langle R'' | \mathrm{e}^{-\tau \underset{\sim}{V}} | R' \rangle . \tag{23}$$

In addition, we assume that $\underset{\sim}{V}$ is diagonal in position representation,

$$\langle R''|\mathrm{e}^{-\tau \underset{\sim}{V}}|R'\rangle = \mathrm{e}^{-\tau V(R')}\,\delta(R''-R')\ . \tag{24}$$

The kinetic part of the density matrix can be evaluated using the eigenfunctions of $\underset{\sim}{T}$. In a cube of side length L with periodic boundary conditions, the exact eigenfunctions of $\underset{\sim}{T}$ are given by plane waves, $L^{-3N/2}\exp(iK_{\mathbf{n}}R)$, with eigenvalues $\lambda K_{\mathbf{n}}^2$, with $K_{\mathbf{n}} = 2\pi\mathbf{n}/L$ and $\mathbf{n}$ a $3N$-dimensional integer vector. For distinguishable particles, it follows that

$$\begin{aligned}\rho_f(R,R';\tau) &= \langle R|\mathrm{e}^{-\tau \underset{\sim}{T}}|R'\rangle \\ &= \frac{1}{L^{3N}}\sum_{\mathbf{n}}\mathrm{e}^{-\tau\lambda K_{\mathbf{n}}^2 - iK_{\mathbf{n}}(R-R')} \\ &= (4\pi\lambda\tau)^{-3N/2}\exp\left(-\frac{(R-R')^2}{4\lambda\tau}\right)\ ,\end{aligned} \tag{25}$$

if the thermal wavelength of one step τ is small compared to the size of the box,

$$\tau\lambda \ll L^2\ , \tag{26}$$

which permits to approximate the sum by an integral. Altogether, using (24) and (25) in (23), and inserting the result in (21), we get the so-called primitive approximation for (21),

$$\begin{aligned}\rho(R_0,R_M;\beta) = \int\cdots\int &\mathrm{d}R_1\mathrm{d}R_2\ldots\mathrm{d}R_{M-1} \\ &(4\pi\lambda\tau)^{-3N/2}\exp\left(-\sum_{m=1}^{M}\left(\frac{R_{m-1}-R_m}{4\lambda\tau}+\tau V(R_m)\right)\right)\ .\end{aligned} \tag{27}$$

As a result, this representation relates the quantum density matrix at any temperature to integrals over the path $R_1\ldots R_{M-1}$ of an expression that has the form of a classical distribution function. In particular, the integrand is always non-negative and can be interpreted as a probability. This is an important property to allow for Monte Carlo simulations.

For MD calculations, it is useful to introduce a set of M momentum integrations,

$$\begin{aligned}Z = \mathcal{N}\int\cdots\int &\mathrm{d}R_1\mathrm{d}R_2\ldots\mathrm{d}R_{M-1}\mathrm{d}P_1\ldots\mathrm{d}P_{M-1} \\ &\exp\left(-\beta\sum_{m=1}^{M}\left(\frac{P_m^2}{2M_m}+\frac{R_{m-1}-R_m}{4\lambda\tau}+\tau V(R_m)\right)\right)\ .\end{aligned} \tag{28}$$

The additional momentum integrations lead to a different numerical prefactor included in $\mathcal{N}$. (28) is in the form of a classical phase space integral of a system with harmonic next-neighbor-interactions in an external potential [2]. This permits the application of classical isothermal MD techniques to sample the distribution function specified by (28). Therefore, the argument of the exponential is used as a Hamiltonian and a set of equations of motion is derived such that the thermal weight function is sampled in time, a problem completely analogous to the one addressed in Sect. 2.

Beyond, the generalization of the path integral MD approach to N quantum particles is highly nontrivial if the particles obey Bose or Fermi statistics [11], the latter case still being unsolved due to the sign problem.

4 Coherent States

In an effort to apply the classical techniques of isothermal MD more directly to quantum mechanics, we are led to the study of *coherent states* [12] that feature a quasi-classical time evolution in a harmonic oscillator potential. Coherent states are defined as eigenstates of the annihilation operator $\underset{\sim}{a}$,

$$\underset{\sim}{a}\,|\alpha\rangle = \alpha\,|\alpha\rangle \;, \tag{29}$$

the Hamiltonian of the harmonic oscillator being $\underset{\sim}{H}^{(1)} = \hbar\omega(\underset{\sim}{a}^\dagger\underset{\sim}{a} + \frac{1}{2})$ for a single particle. The eigenvalue α of $\underset{\sim}{a}$ is complex, and we write

$$\alpha = \sqrt{\frac{m\omega}{2\hbar}}\,r + \frac{i}{\sqrt{2m\hbar\omega}}\,p \;. \tag{30}$$

It turns out that in position representation, the wave function $\langle\, x \,|\, \alpha \,\rangle \equiv \langle\, x \,|\, r, p \,\rangle$ is given by a shifted Gaussian with mean position r and mean momentum p [12,13].

The elementary calculation

$$\begin{aligned} \mathrm{e}^{-i\omega t\underset{\sim}{a}^\dagger\underset{\sim}{a}}|\alpha\rangle &= \mathrm{e}^{-\frac{1}{2}|\alpha|^2}\sum_{n=0}^{\infty}\frac{\alpha^n}{\sqrt{n!}}\mathrm{e}^{-in\omega t}|n\rangle \\ &= |\mathrm{e}^{-i\omega t}\alpha\rangle \end{aligned} \tag{31}$$

implies that the time evolution of a coherent state in an external harmonic oscillator potential is given by a uniform rotation of its label, α, in a clockwise manner in the complex plane. This can be cast into a form that is familiar from the classical oscillator,

$$\frac{d}{dt}r = \frac{p}{m} \;, \qquad \frac{d}{dt}p = -m\omega^2 r \;. \tag{32}$$

This fact is of decisive importance for the direct translation of the Nosé–Hoover method to quantum dynamics.

In analogy to (31), it is also possible to determine matrix elements of the canonical density operator,

$$\begin{aligned} \mathrm{e}^{-\beta\hbar\omega \underset{\sim}{a}^\dagger \underset{\sim}{a}}|\alpha\rangle &= \mathrm{e}^{-\beta\hbar\omega \underset{\sim}{a}^\dagger \underset{\sim}{a}} \mathrm{e}^{-\frac{1}{2}|\alpha|^2} \sum_{n=0}^{\infty} \frac{\alpha^n}{\sqrt{n!}}|n\rangle \\ &= \mathrm{e}^{-\frac{1}{2}|\alpha|^2} \sum_{n=0}^{\infty} \frac{(\mathrm{e}^{-\beta\hbar\omega}\alpha)^n}{\sqrt{n!}}|n\rangle \\ &= \mathrm{e}^{-\frac{1}{2}|\alpha|^2(1-\mathrm{e}^{-2\beta\hbar\omega})}|\mathrm{e}^{-\beta\hbar\omega}\alpha\rangle \ . \end{aligned} \tag{33}$$

Due to the following representation of the unity operator in terms of coherent states,

$$\underset{\sim}{1} = \int \frac{\mathrm{d}^2\alpha}{\pi} \ |\alpha\rangle\langle\alpha| \ , \quad \mathrm{d}^2\alpha = \mathrm{d}\,(\mathrm{Re}\alpha)\,\mathrm{d}\,(\mathrm{Im}\alpha) \ , \tag{34}$$

the trace involved in a quantum statistical average may be transformed as follows,

$$\begin{aligned} \langle\langle\, \underset{\sim}{B} \,\rangle\rangle &= \frac{1}{Z^{(1)}(\beta)} \int \frac{\mathrm{d}^2\alpha}{\pi} \ \langle\alpha|\underset{\sim}{B}\mathrm{e}^{-\beta \underset{\sim}{H}^{(1)}}|\alpha\rangle \\ &= \frac{1}{Z^{(1)}(\beta)} \int \frac{\mathrm{d}^2\alpha}{\pi} \ \langle\alpha|\mathrm{e}^{-\frac{1}{2}\beta \underset{\sim}{H}^{(1)}} \underset{\sim}{B} \mathrm{e}^{-\frac{1}{2}\beta \underset{\sim}{H}^{(1)}}|\alpha\rangle \\ &= \frac{1}{Z^{(1)}(\beta)} \int \frac{\mathrm{d}^2\alpha}{\pi} \ \mathrm{e}^{-\frac{1}{2}\beta\hbar\omega} \ \mathrm{e}^{-|\alpha|^2\left(1-\mathrm{e}^{-\beta\hbar\omega}\right)} \ \langle\mathrm{e}^{-\frac{1}{2}\beta\hbar\omega}\alpha|\underset{\sim}{B}|\mathrm{e}^{-\frac{1}{2}\beta\hbar\omega}\alpha\rangle \\ &= \frac{1}{Z^{(1)}(\beta)} \int \frac{\mathrm{d}^2\alpha}{\pi} \ \mathrm{e}^{\frac{1}{2}\beta\hbar\omega} \ \mathrm{e}^{-|\alpha|^2\left(\mathrm{e}^{\beta\hbar\omega}-1\right)} \ \langle\alpha|\underset{\sim}{B}|\alpha\rangle \ . \end{aligned} \tag{35}$$

Defining

$$\tilde{Z}^{(1)}(\beta) = \mathrm{e}^{-\frac{1}{2}\beta\hbar\omega} \ Z^{(1)}(\beta) = \frac{1}{\mathrm{e}^{\beta\hbar\omega}-1} \ , \tag{36}$$

$$w^{(1)}(\alpha) = \mathrm{e}^{-|\alpha|^2\left(\mathrm{e}^{\beta\hbar\omega}-1\right)} = \mathrm{e}^{-\left(\frac{p^2}{2m}+\frac{1}{2}m\omega^2 r^2\right)\left(\mathrm{e}^{\beta\hbar\omega}-1\right)/(\hbar\omega)} \ , \tag{37}$$

$$\mathcal{B}(r,p) = \langle r,p|\underset{\sim}{B}|r,p\rangle \ , \tag{38}$$

we may write

$$\langle\langle\, \underset{\sim}{B} \,\rangle\rangle = \frac{1}{\tilde{Z}^{(1)}(\beta)} \int \frac{\mathrm{d}r\,\mathrm{d}p}{2\pi\hbar} \ w^{(1)}(r,p)\ \mathcal{B}(r,p) \ , \tag{39}$$

which has the form of a classical canonical phase space average, (1).

In the case of two indistinguishable quantum particles ($\underset{\sim}{H}^{(2)} = \underset{\sim}{H}^{(1)} \otimes \underset{\sim}{1} + \underset{\sim}{1} \otimes \underset{\sim}{H}^{(1)}$), we need to consider fermions and bosons, characterized by the following (anti-)symmetric two-particle states,

$$|A_\varepsilon\rangle = \begin{cases} \underset{\sim}{S}_-|\alpha_1, \alpha_2\rangle & \text{if } \varepsilon = - \\ \underset{\sim}{S}_+|\alpha_1, \alpha_2\rangle & \text{if } \varepsilon = + \,. \end{cases} \tag{40}$$

$\underset{\sim}{S}_+$ ($\underset{\sim}{S}_-$) is the projector onto the (anti-)symmetric two-particle subspace, and we have $\langle\, A_\varepsilon \,|\, A_\varepsilon \,\rangle = (1 + \varepsilon \mathrm{e}^{-|\alpha_1 - \alpha_2|^2})/2$. An analysis analogous to the one-particle case yields

$$\langle\langle\, \underset{\sim}{B} \,\rangle\rangle = \frac{1}{\tilde{Z}_\varepsilon^{(2)}(\beta)} \iint \frac{\mathrm{d}^2\alpha_1}{\pi} \frac{\mathrm{d}^2\alpha_2}{\pi}\, w_\varepsilon^{(2)}(\alpha_1, \alpha_2) \frac{\langle A_\varepsilon|\underset{\sim}{B}|A_\varepsilon\rangle}{\langle\, A_\varepsilon \,|\, A_\varepsilon \,\rangle} \,, \tag{41}$$

with

$$w_\varepsilon^{(2)}(\alpha_1, \alpha_2) = \mathrm{e}^{-|\alpha_1|^2(\mathrm{e}^{\beta\hbar\omega}-1)} \mathrm{e}^{-|\alpha_2|^2(\mathrm{e}^{\beta\hbar\omega}-1)} \langle\, A_\varepsilon \,|\, A_\varepsilon \,\rangle \,, \tag{42}$$

and

$$\begin{aligned} \tilde{Z}_\varepsilon^{(2)}(\beta) &= \iint \frac{\mathrm{d}^2\alpha_1}{\pi} \frac{\mathrm{d}^2\alpha_2}{\pi}\, w_\varepsilon^{(2)}(\alpha_1, \alpha_2) \\ &= \frac{1}{2}(\tilde{Z}^{(1)}(\beta)^2 + \varepsilon \tilde{Z}^{(1)}(2\beta)) \,. \end{aligned} \tag{43}$$

5 Quantum Thermostat Methods

From (37), it can be inferred that in the case of a single quantum particle in a harmonic oscillator, the function $w^{(1)}(r,p)/\tilde{Z}^{(1)}$ needs to be sampled in time on the parameter space of coherent states in order that time averages coincide with canonical ensemble averages. $w^{(1)}(r,p)$ is a two-dimensional Gaussian distribution function, and if we insert the following set of equations of motion,

$$\begin{aligned} \frac{d}{dt} r &= \frac{p}{m} \,, \qquad \frac{d}{dt} p = -m\omega^2 r - p\frac{p_\eta}{Q} \,, \\ \frac{d}{dt} p_\eta &= \frac{1}{\beta}\left(\frac{p^2}{m}\frac{\mathrm{e}^{\beta\hbar\omega} - 1}{\hbar\omega} - 1\right) \,, \end{aligned} \tag{44}$$

into the generalized Liouville equation (4) with $f = w^{(1)}(r,p)\, \exp(-\beta p_\eta^2/2Q)$, we find that this distribution function is a stationary solution of (4). Of course, (44) can be derived precisely from this condition [7].

The set of (44) is the analogue of the classical Nosé–Hoover scheme (3) for the quantum harmonic oscillator. Since in both cases, Gaussian distributions are involved, one can also expect ergodicity problems for the scheme (44). However, the idea of the ergodic chain thermostat, (9), can be transferred directly to the present case and resolves the problem, see Fig. 1. This applies also to the quantum demon method [7].

The case of two identical quantum particles is more subtle, since the distribution function $w_\varepsilon^{(2)}(\alpha_1, \alpha_2)$ is *entangled*, by which we mean that it is not separable into a product of two functions each of which depends only on

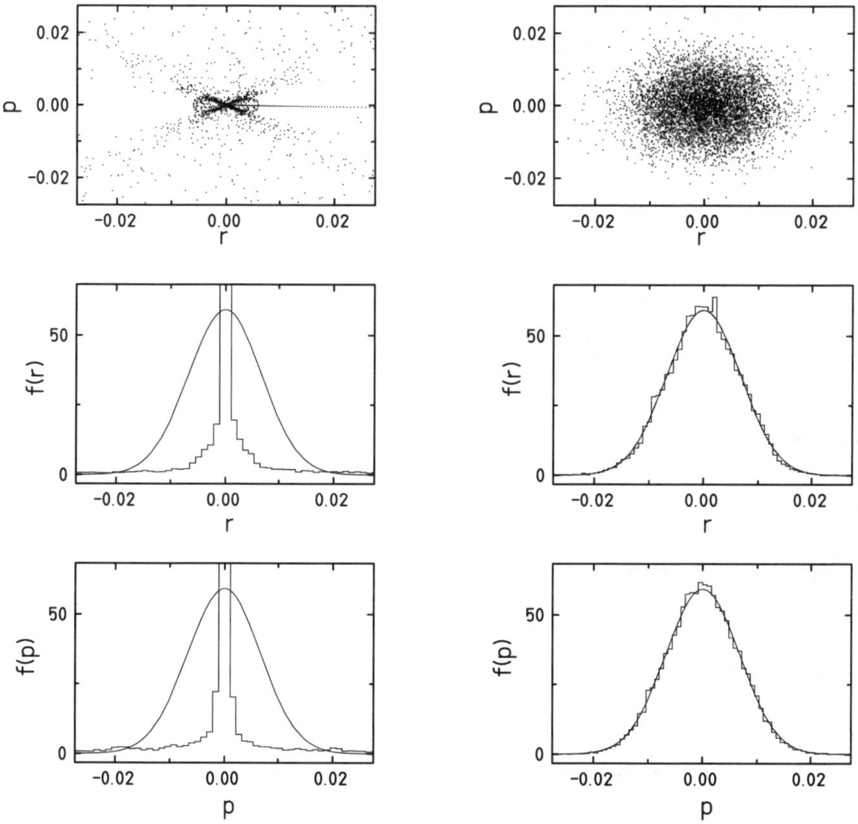

Fig. 1. *Left panel*: Results of time averaging with a simple Nosé–Hoover scheme, $T = 0.1$. From above: (r,p)–density plot, position distribution, momentum distribution. The *solid line* shows the respective normalized marginal of $w^{(1)}$, e.g. $f(r) = (1/\tilde{Z}^{(1)}) \int (\mathrm{d}p/(2\pi\hbar)) w^{(1)}(r,p)$. The distributions sampled by time averaging are presented as histograms. *Right panel*: Same description, but a thermostat chain of length $M = 4$ was employed

α_1 or α_2, respectively. However, the ansatz

$$\frac{d}{dt}r_1 = \frac{p_1}{m}\ , \qquad \frac{d}{dt}r_2 = \frac{p_2}{m}\ ,$$
$$\frac{d}{dt}p_1 = -m\omega^2 r_1 - p_1\frac{p_{\eta_1}}{Q_1}\ ,$$
$$\frac{d}{dt}p_2 = -m\omega^2 r_2 - p_2\frac{p_{\eta_2}}{Q_2}\ , \tag{45}$$

for the equations of motion of the coherent state parameters can be inserted successfully into the Liouville equation (4) with

$$f = w_\varepsilon^{(2)} \exp\left(-\beta\left(\frac{p_{\eta_1}^2}{2Q_1} + \frac{p_{\eta_2}^2}{2Q_2}\right)\right) \tag{46}$$

and leads to the equations

$$\dot{p}_{\eta_1} = \frac{1}{\beta}\left(\frac{p_1^2}{m}\frac{e^{\beta\hbar\omega}-1}{\hbar\omega} - 1 + \varepsilon p_1 \frac{p_1 - p_2}{m\hbar\omega}\frac{1}{e^V + \varepsilon 1}\right)\ ,$$
$$\dot{p}_{\eta_2} = \frac{1}{\beta}\left(\frac{p_2^2}{m}\frac{e^{\beta\hbar\omega}-1}{\hbar\omega} - 1 - \varepsilon p_2 \frac{p_1 - p_2}{m\hbar\omega}\frac{1}{e^V + \varepsilon 1}\right)\ , \tag{47}$$

for the two pseudofriction coefficients. Beyond the terms familiar from the one-particle case, we find additional terms that account for Bose-attraction and Pauli-blocking, see Fig. 2. These effects substantially improve the ergodicity of the method as compared to the simple Nosé–Hoover method for a single particle which is not ergodic even at very high temperatures. Figure 3 illustrates that above $T \gtrsim 1.2$, the agreement of the sampled histograms and

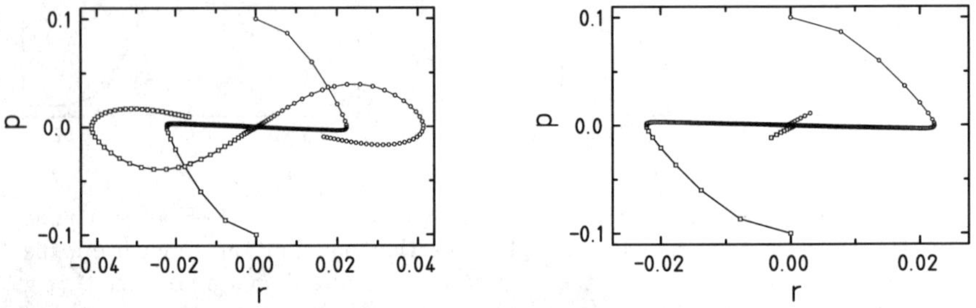

Fig. 2. *Left figure*: Nosé–Hoover dynamics of two identical fermions. The initial values of $\{r_1, p_1, r_2, p_2\}$ are $\{0, -0.1, 0, 0.1\}$. Initially, both fermions are "cooled" to the origin of the harmonic potential, where they undergo a sudden acceleration away from each other due to the Pauli blocking. The total integration time is 6.5 periods. *Right figure*: Same description, but for two bosons. However, the total integration time is 8.1 periods, since the bosons stick close to one another and do not separate earlier

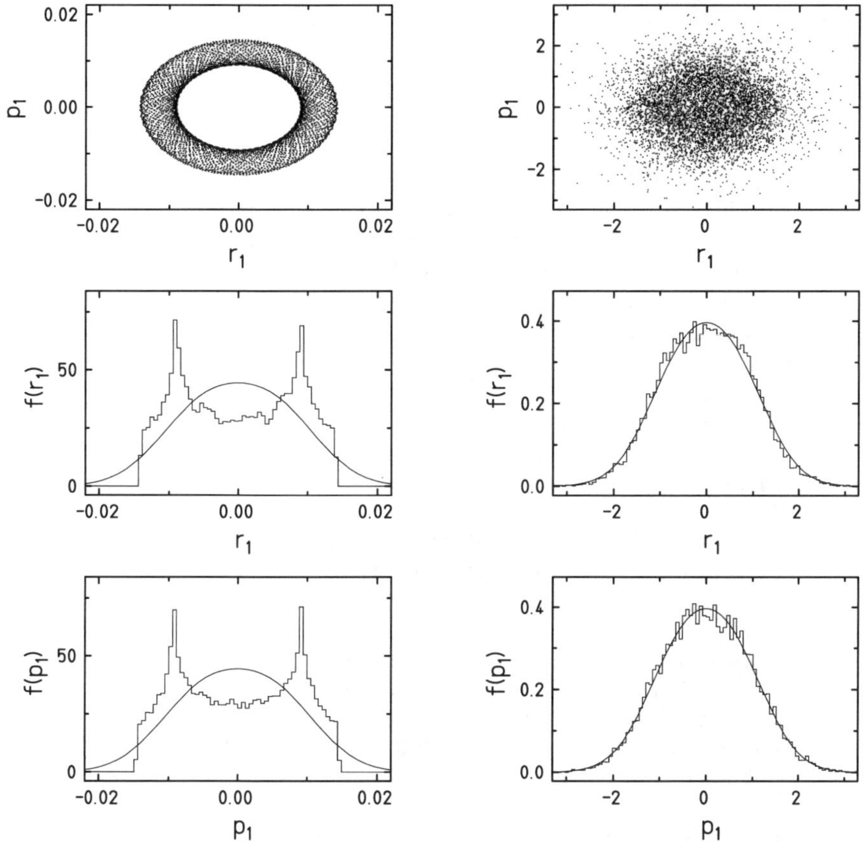

Fig. 3. Nosé–Hoover dynamics of two identical fermions. *Left panel*: $T = 0.1$, *right panel*: $T = 1.2$. The initial conditions are identical in both cases. The more complex dynamics resulting from Fermi-blocking leads to ergodic motion in the temperature range $T \gtrsim 1.2$, whereas in the one-particle case, the simple Nosé–Hoover scheme is non-ergodic at any temperature

the theoretical marginals is good. Remaining ergodicity problems encountered at low temperature values are again resolved by employing a chain thermostat.

6 Summary and Outlook

In this contribution, we have outlined thermostat methods that are commonly used in classical MD simulations for the determination of canonical ensemble averages. Due to the path integral formulation of quantum statistical mechanics, they may also be applied to quantum systems. We have also presented a different approach to the determination of quantum canonical averages using an isothermal quantum molecular dynamics scheme that is closely related to the classical methods, but at its present stage limited to the case of noninteracting particles in a harmonic confining potential.

Given the rich dynamical behavior of ultra-cold trapped quantum gases depending on the value of the s-wave scattering length [14,15], it is timely to investigate how this thermostating method can be combined with powerful approximate quantum dynamics schemes to deal with interacting quantum systems at finite temperature. One possible approach is to cool the system of interest "sympathetically", i.e. via an interaction between particles that are coupled to a Nosé–Hoover thermostat and the physical system under investigation. This corresponds precisely to the currently employed experimental technique of *sympathetic cooling* [16].

References

1. S. Nosé: Prog. Theor. Phys. Suppl. **103**, 1 (1991)
2. D. Ceperley: Rev. Mod. Phys. **67**, 279 (1995)
3. R.P. Feynman: *Statistical Mechanics*, (W.A. Benjamin, Inc., Reading, MA, 1972)
4. M.E. Tuckerman, G.J. Martyna: J. Phys. Chem. B **104**, 159 (2000)
5. W. Kob: J. Phys. Cond. Matter **11**, R85 (1999)
6. H. Feldmeier, J. Schnack: Rev. Mod. Phys. **72**, 655 (2000)
7. D. Mentrup, J. Schnack: Physica A **297**, 337 (2001)
8. M.E. Tuckerman, C.J. Mundy, G.J. Martyna: Europhys. Lett. **45**, 149 (1999)
9. G.J. Martyna, M.L. Klein, M. Tuckerman: J. Chem. Phys. **97**, 2635 (1992)
10. D. Kusnezov, A. Bulgac, W. Bauer: Ann. of Phys. **204**, 155 (1990)
11. M.E. Tuckerman, A. Hughes: Proc. CECAM conference on "Computer simulation of rare events and quantum dynamics in condensed phase systems", 311 (1997)
12. J.R. Klauder, B.-S. Skagerstam: *Coherent states*, (World Scientific Publishing, Singapore, 1985)
13. W.P. Schleich: *Quantum Optics in Phase Space*, (Wiley-VCH, Berlin, 2001)
14. F. Dalfovo, S. Giorgini, L.P. Pitaevskii, S. Stringari: Rev. Mod. Phys. **71**, 463 (1999)
15. H. Saito, M. Ueda: Phys. Rev. Lett. **86**, 1406 (2001)
16. C.J. Myatt, E.A. Burt, R.W. Ghrist, E.A. Cornell, C.E. Wieman: Phys. Rev. Lett. **78**, 586 (1997)

Non-local Kinetic Theory

Klaus Morawetz and Pavel Lipavský

Abstract. The spectral concept as provided by the Green's function technique is able to cover the Landau variational concept of quasiparticles as well as the Beth–Uhlenbeck equation of state. To this end the extended quasiparticle picture is presented. The corresponding nonequilibrium kinetic theory shows that the collisions have to be considered as non-local and non-instantaneous. One consequence of this is that the duration time or collision delay leads to correlated density, momentum and energy in agreement with the Beth–Uhlenbeck equation of state but exceeding the Landau quasiparticle concept. In the instantaneous approximation the connection between Landau's quasiparticles and the spectral quasiparticles are given by the rearrangement energy which is a consequence of the energy gain during a collision.

1 Two Concepts of Quasiparticles

The understanding of the time dependence of dense correlated systems is a long standing task. It would allow one to describe the formation of complex structures which appear in many physical systems. One approach to such correlated many body systems is the statistical kinetic theory trying to find the time evolution of a distribution function from a kinetic equation. Since the formulation of the Boltzmann kinetic equation [1] 133 years ago, there has been a continuous activity to derive and generalize it in a number of ways from classical and/or quantum statistics. All these alternative approaches agree on the simplest level of the non-degenerate system in which the free-particle drift is in balance with the dissipation described by local and instant collisions. Beyond this basic level, complete agreement of the resulting kinetic theories is met rather accidentally.

Before we will present a consistent extension of the original Boltzmann equation towards dense interacting Fermi systems, we should give a short overview about the underlying problems and their known solution.

1.1 Landau Theory versus the Beth–Uhlenbeck Equation of State

One of the most successful and applied phenomenological concepts is the Landau theory of quasiparticles [2,3]. It postulates that the density of quasiparticles equals the density of constituent particles,

$$n = \int \frac{dk}{(2\pi)^3} \tilde{f}_k, \tag{1}$$

where $\tilde{f}_k$ is the quasiparticle distribution in momentum space. Since quasiparticles cover all degrees of freedom of the system, the quasiparticle energy ϵ can be defined as the variation of the total energy $\mathcal{E}$,

$$\tilde{\epsilon}_k = \frac{\delta \mathcal{E}}{\delta \tilde{f}_k}, \tag{2}$$

and from the entropy follows that the equilibrium distribution is of the Fermi–Dirac form, $\tilde{f}_k = f_{FD}(\tilde{\epsilon}_k - \tilde{\mu})$.

The variational approach works only if binary collisions are treated within the instantaneous and elastic approximations. To show why, assume a phenomenological kinetic equation

$$\frac{\partial \tilde{f}_k}{\partial t} = \tilde{I}_k, \tag{3}$$

where $\tilde{I}_k$ is an unspecified collision integral. To conserve the number of particles,

$$\frac{dn}{dt} = \int \frac{dk}{(2\pi)^3} \frac{\partial \tilde{f}_k}{\partial t} = \int \frac{dk}{(2\pi)^3} \tilde{I}_k, \tag{4}$$

the collision integral has to satisfy $\int dk \tilde{I}_k = 0$ which is possible only for instantaneous collisions. In addition, from the energy balance,

$$\frac{d\mathcal{E}}{dt} = \int \frac{dk}{(2\pi)^3} \frac{\delta \mathcal{E}}{\delta \tilde{f}_k} \frac{\partial \tilde{f}_k}{\partial t} = \int \frac{dk}{(2\pi)^3} \tilde{\epsilon}_k \tilde{I}_k, \tag{5}$$

it follows that the total energy is conserved, $\frac{d\mathcal{E}}{dt} = 0$, only if the sum of variational quasiparticle energies are conserved in collisions, $\int dk \tilde{\epsilon}_k \tilde{I}_k = 0$.

In contrast to this concept which is applicable to dense Fermi systems, there exists the well known equation of state for dense classical gases. The correlation beyond the ideal gas are described by the temperature dependent second virial coefficient $B(T)$ in the pressure $p = T(n_f + n_f^2 B(T))$ or for the density

$$n = n_f + 2n_f^2 B(T). \tag{6}$$

This equation of state decomposes the total number of particles n into the number of free particles n_f and the number of correlated pairs $n_f^2 B(T)$.

As long as classical gases are concerned, the equation of state (6) is a small density expansion and not suited for dense quantum systems like the Fermi liquid described by the Landau theory (1). However it was already recognized in 1936 by Beth and Uhlenbeck [4,5] that the equation of state (6) can be obtained also for quantum systems where now the second virial coefficient becomes density dependent itself, and can be written for nondegenerate systems in terms of the two-particle scattering phase shift ϕ of the scattering channel α [6]

$$B(T,n) = \frac{2^{3/2}}{n_f^2 \lambda^3} \mathrm{e}^{2\mu/T} \int\limits_{-\infty}^{\infty} dE \mathrm{e}^{-E/T} \sum_{\alpha} c_\alpha \left[\delta(E - E_\alpha) + \frac{1}{\pi} \frac{d}{dE} \phi_\alpha(E) \right] . \quad (7)$$

Here μ is the chemical potential and $\lambda^2 = 2\pi\hbar^2/mT$ the thermal de Broglie wave length. This Beth–Uhlenbeck equation of state separates between bound states sharply peaked around the energies E_α and the scattering contributions. If the scattering contributions are small and if only one bound state is present with a binding energy E_b, (7) reduces to the mass action law,

$$n_p = n_f^2 B = n_f^2 \frac{3\sqrt{2}}{8} \lambda^3 \mathrm{e}^{-E_b/T} , \quad (8)$$

giving the number of bound (correlated) pairs for a given temperature and density of free particles.

The Beth–Uhlenbeck equation of state (6) is in clear contradiction to the Landau equation of state (1). We will clarify this puzzle within a unified non-local kinetic theory.

1.2 Spectral Concept

To shed some light on the aforementioned problem, we recall that theoretically the Beth–Uhlenbeck equation of state follows from the spectral approach to quasiparticles which will be introduced later in detail. Let us assume for the moment that the interaction is described by a selfenergy dependent on momentum, space, frequency and time with a real part $\sigma(p, r, \omega, t)$ linked to the imaginary part $\frac{1}{2}\gamma(p, r, \omega, t)$ by the Kramers–Kronig relation

$$\sigma_\omega = \int \frac{d\omega'}{2\pi} \frac{\gamma'_\omega}{\omega - \omega'} . \quad (9)$$

Then the spectral function as the momentum resolved density of states reads

$$a = \frac{\gamma}{(\omega - \frac{k^2}{2m} - \sigma)^2 + \frac{1}{4}\gamma^2} . \quad (10)$$

For simplicity we will consider only nonrelativistic systems with the kinetic energy of free particles $k^2/2m$. The total density is then given by

$$n = \int \frac{d\omega}{2\pi} a \, f_{FD} \quad (11)$$

with the Fermi–Dirac equilibrium distribution function f_{FD} in the equilibrium case. As long as the quasiparticle damping γ is sufficiently small the spectral function has a pronounced peak or pole around the quasiparticle energy ϵ

$$\epsilon = \frac{k^2}{2m} + \sigma(k, \epsilon). \tag{12}$$

The spectral quasiparticle picture appears in the zero order expansion in γ

$$a^{\mathrm{QP}} = 2\pi\delta(\omega - \epsilon). \tag{13}$$

Within the quasiparticle picture (13), the number of quasiparticles equals the number of constituent particles. This follows from the definition (11). The spectral function (13) fulfills the sum rule of number of state conservation,

$$\int \frac{d\omega}{2\pi} a = 1, \tag{14}$$

however, the quasiparticle picture (13) does not satisfy the next order sum rule

$$\int \frac{d\omega}{2\pi} \omega\, a = \frac{k^2}{2m} + \sigma^{\mathrm{HF}}. \tag{15}$$

The approximation (13) yields instead $\int \frac{d\omega}{2\pi} \omega a^{\mathrm{QP}} = \frac{k^2}{2m} + \sigma$. The sum rule (15) is violated beyond the mean field approximation, σ^{HF}, since the quasiparticle energy (12) contains further correlations.

Therefore three questions arise: (i) What approximation satisfies both sum rules (14) and (15)? (ii) How does the correlated density enter the quasiparticle concept? (iii) How relates the spectral quasiparticle energy ϵ of (12) to the variational quasiparticle energy $\tilde{\epsilon}$ of (2)?

The answers can be found if we expand (10) up to first order in damping or small scattering rates [6–11]

$$a^{\mathrm{EQP}} = \left(1 + \frac{\partial \sigma}{\partial \omega}\right) 2\pi\delta(\omega - \epsilon) + \frac{\wp'}{\omega - \epsilon}\gamma, \tag{16}$$

where $\wp'$ denotes the derivative of the principal value. This spectral function establishes the extended quasiparticle picture. It is superior to the quasiparticle picture since it completes the spectral sum rules (14) as proved in [12] and (15) as proved in [13]. This answers already the first question. In the following we will show that the Beth–Uhlenbeck equation of state (6) follows from (11) for the self energy in T-matrix approximation, $T^R = |T^R| \mathrm{e}^{i\phi}$ and in particular the answer to question (iii) how the spectral and variational pictures are linked.

1.3 Non-Local Collisions

We now turn to the question of the relation between spectral and variational quasiparticle concepts. In this article we will show that the kinetic equation corresponding to the extended quasiparticle picture is of non-local form.

Moreover, we will show that both results, the variational Landau theory as well as the Beth–Uhlenbeck equation of state, are limiting cases of the non-local kinetic theory based on the spectral concept. Therefore the spectral approach is considered as the more fundamental one.

According to conservation laws, within the spectral concept collisions have to be non-local and non-instantaneous events. For simplicity let us neglect the spatial dependence for the moment. As shown below in Sect. 3.1, the collision duration is given by the energy derivative of the scattering phase shift $\Delta_t = \frac{\partial \phi}{\partial \omega}$. During the collision particles can gain energy from the background, $\Delta_E = -\frac{1}{2}\frac{\partial \phi}{\partial t}$, given by the time derivative of the phase shift. In contrast to the instantaneous collision integral (3) appropriate for the kinetic equation in the Landau theory, a collision of two quasiparticles with momentum k and p into momenta $k-q$ and $p+q$ is now characterized by this time delay and energy gain (75)

$$\frac{\partial f_k}{\partial t} = I_k[f(t-\Delta_t), \delta(\epsilon_k + \epsilon_p - \epsilon_{k-q} - \epsilon_{p+q} - \Delta_E)]. \tag{17}$$

The answer to the third question is found from the balance equations for the density and energy given in Sect. 3.3. The density conservation now appears from (17) as the time derivative of the Beth–Uhlenbeck formula

$$\frac{dn}{dt} = \frac{d}{dt}\int \frac{dk}{(2\pi)^3} f_k + \frac{d}{dt}\int d\mathcal{P}\Delta_t, \tag{18}$$

where the probability per time to form a correlated pair is expressed in Sect. 3.3 in terms of the scattering T-matrix as

$$d\mathcal{P} = \frac{dkdpdq}{(2\pi)^8}|T|^2 \delta(\epsilon_k + \epsilon_p - \epsilon_{k-q} - \epsilon_{p+q}) f_k f_p (1 - f_{k-q} - f_{p+q}). \tag{19}$$

This is the correlated density, $n^{\text{cor}} = \int d\mathcal{P}\Delta_t$, known from the equilibrium Green's functions [6]. From (18) it follows that the number of spectral quasiparticles does not equal the number of constituent particles.

The energy balance found from (17),

$$\frac{d\mathcal{E}}{dt} = \int \frac{dk}{(2\pi)^3}\epsilon_k \frac{\partial f_k}{\partial t} + \frac{d}{dt}\int d\mathcal{P}\frac{\epsilon_k + \epsilon_p}{2}\Delta_t - \int d\mathcal{P}\Delta_E, \tag{20}$$

includes the energy of correlated particles $\propto \Delta_t$ and the mean energy gain $\propto \Delta_E$. The right hand side of balance equation (20) does not have a transparent form of the total time derivative, however, it can be shown that (20) conserves the energy [13,14]

$$\mathcal{E} = \int \frac{dk}{(2\pi)^3} f_k \frac{k^2}{2m} + \frac{1}{2}\int \frac{dkdp}{(2\pi)^6} f_k f_p \text{Re}T^R(\epsilon_k + \epsilon_p, k, p, 0) + \int d\mathcal{P}\frac{\epsilon_k + \epsilon_p}{2}\Delta_t. \tag{21}$$

One can check that (20) results from the time derivative of (21). The energy of correlated particles corresponds directly to the second term of (20). The time derivative of the second term of (21) splits into the selfenergy part of the quasiparticle energy and into the mean energy gain [13]. The mean energy gain follows exclusively from the time derivative of $\mathrm{Re}T^R$ which confirms that the effect of medium on the binary collision is necessary for this mechanism of the energy transfer.

The energy (21) follows also from the nonequilibrium expression, in equilibrium [15],

$$\mathcal{E} = \int \frac{dk d\omega}{(2\pi)^4} \frac{1}{2} \left(\omega + \frac{k^2}{2m} \right) f_{FD}(\omega) a(k,\omega), \tag{22}$$

evaluated within the extended quasiparticle picture (16) as is shown in [13]. Approximations included in the kinetic equation and in the spectral function are thus consistent and one can evaluate observables and conserving quantities either from the kinetic equation or from their definitions.

1.4 Limit of Landau Theory

Now we are able to see what approximations are necessary to obtain the variational concept of the Landau theory. First of all the scatterings have to be in a local approximation $\Delta_t \to 0$. Then the number of quasiparticles equals the number of particles and we can identify the Landau quasiparticle distribution $\tilde{f}$ with the quasiparticle distribution of the spectral concept f.

If the collisions are nearly instantaneous, one might think that the energy gain is also absent. This is not the case. The energy gain gives the missing link between the quasiparticle energies of the variational approach $\tilde{\epsilon}$ of (2) and the ϵ of (12) [14]. The variational selfenergy indeed mimics the last two terms of (20) by the rearrangement contribution to the quasiparticle energy as we will see now.

In the instantaneous approximation we have from (20)

$$\frac{d\mathcal{E}}{dt} = \int \frac{dk}{(2\pi)^3} \epsilon_k \frac{\partial \tilde{f}_k}{\partial t} - \int d\mathcal{P} \Delta_E . \tag{23}$$

The last term can be rewritten using $\Delta_E = -\frac{1}{2}\frac{\partial \phi}{\partial t}$

$$-\int d\mathcal{P} \Delta_E = \frac{1}{2} \int d\mathcal{P} \frac{\partial \phi}{\partial t} = \int \frac{dk}{(2\pi)^3} \int d\mathcal{P} \frac{\delta \phi}{\delta \tilde{f}_k} \frac{\partial \tilde{f}_k}{\partial t} \equiv \int \frac{dk}{(2\pi)^3} \epsilon^{\Delta} \frac{\partial \tilde{f}_k}{\partial t}, \tag{24}$$

where we have introduced the rearrangement energy

$$\epsilon_k^{\Delta} = \int d\mathcal{P} \frac{\delta \phi}{\delta \tilde{f}_k} . \tag{25}$$

Using (24) in (23) we end up with the variational expression of the Landau theory (5)

$$\frac{d\mathcal{E}}{dt} = \int \frac{dk}{(2\pi)^3}(\epsilon_k + \epsilon_k^{\Delta})\frac{\partial \tilde{f}_k}{\partial t}. \tag{26}$$

Therefore we conclude that the relation between the quasiparticle energy of Landau's variational concept (2) and the spectral quasiparticle energy (12) is just given by the rearrangement energy (25), $\tilde{\epsilon} = \epsilon + \epsilon^{\Delta}$.

According to (24), the elastic collision integral, $\propto \delta(\tilde{\epsilon}_k + \tilde{\epsilon}_p - \tilde{\epsilon}_{k-q} - \tilde{\epsilon}_{p+q})$ with the variational quasiparticle energy $\tilde{\epsilon} = \epsilon + \epsilon^{\Delta}$, yields the same energy conservation as the non-elastic one, $\propto \delta(\epsilon_k + \epsilon_p - \epsilon_{k-q} - \epsilon_{p+q} - 2\Delta_E)$ with the spectral quasiparticle energy. Without a sacrifice of the energy conservation, one can thus circumvent an inconvenience of non-elastic collision integrals by an incorporation of the rearrangement energy provided the number of correlated pairs can be neglected.

In summary, in highly excited systems, the phenomenological quasiparticles introduced via variational concepts are not identical to elementary excitations observed in the single-particle spectrum. For collisions of a finite duration, these two pictures result in different densities of quasiparticles and one can hardly establish any connection between them. In the limit of instantaneous collisions the densities become identical, a difference, however, remains in the quasiparticle energy. The variational quasiparticle energy includes the rearrangement energy which mimics the non-elasticity of binary collisions caused by the time-dependence of the Pauli blocking during collisions.

2 Kinetic Theory

2.1 Historical Remark

Now we turn to the formulation of the kinetic theory. The firmly established concept of the equilibrium virial expansion has been extended to nonequilibrium systems only recently [13] although a number of attempts have been made to modify the Boltzmann equation so that its equilibrium limit would cover at least the second virial coefficient [16–18]. The corrections to the Boltzmann equation have a form of gradient or non-local contributions to the scattering integral.

For a hard-sphere gas, the non-local correction is trivial, therefore the main theoretical focus was on the statistical correlations [19–30]. It turned out that the treatment of higher order contributions is far from trivial as the dynamical statistical correlations result in divergences that are cured only after a re-summation of an infinite set of contributions. Naturally, this re-summation leads to non-analytic density corrections to the scattering integral [26–30]. The moral of these hard-sphere studies is that beyond the non-local corrections one must sum up an infinite set of contributions; the plain expansion leads to incorrect results. From this point of view, the approximate statistical virial corrections implemented by [31], although possibly reasonable, are not sufficiently justified by the theory.

Real particles do not interact like hard spheres, in particular when their de Broglie wave lengths are comparable with the potential range. An effort to describe the virial corrections for more realistic systems resulted in various generalizations of Enskog's equation [16,18,32–55]. By closer inspection one finds that all tractable quantum theories deal exclusively with the non-local corrections. The statistical correlations in quantum systems would require an adequate solution of three-particle collisions (say from Fadeev equations) which are now intensively studied. A systematic incorporation of three-particle collisions into the kinetic equation, however, is not yet fully understood, therefore we discuss only binary processes.

An alternative way to describe the correlation has been developed for Fermi systems at very low temperatures. Since the Pauli exclusion principle excludes all but the zero-angle scattering channels, the correlation can be re-cast into a renormalization of the single-particle energies known as the quasiparticle energies. While quasiparticles behave in many aspects like free particles, the density dependence of the quasiparticle energies results in non-trivial thermodynamic properties. In accordance with the focus either on zero-angle or finite-angle scattering, in the literature one can find two distinct classes of quantum kinetic equations.

The zero-angle class are equations determining the time evolution of the momentum and space dependent quasiparticle distribution function $\tilde{f}(k,r,t)$. This leads to Landau–Silin-type kinetic equations

$$\frac{\partial \tilde{f}}{\partial t}+\frac{\partial \tilde{\varepsilon}}{\partial k}\frac{\partial \tilde{f}}{\partial r}-\frac{\partial \tilde{\varepsilon}}{\partial r}\frac{\partial \tilde{f}}{\partial k}=I_{\rm in}-I_{\rm out}. \tag{27}$$

Here the drift term is characterized by the variational quasiparticle energies $\tilde{\varepsilon}(k,r,t)$ and the collision integral consists of scattering-in and -out terms. This kinetic equation is local in time and space, i.e., it describes the collisions as instantaneous point-like events. Similarly to the Boltzmann equation, the local collisions yield no virial corrections. The virial corrections are thus covered exclusively by the quasiparticle energy and no correlated density appears.

The finite-angle class are equations which include the finite duration of collisions, i.e., they have non-Markovian scattering integrals. These kinetic equations are usually developed for the reduced density matrix or the Wigner function $\rho(k,r,t)$. The Wigner function obeys a Levinson-type of kinetic equation [56,57]

$$\frac{\partial \rho_t}{\partial t}+\frac{\partial \epsilon^{\rm HF}}{\partial k}\frac{\partial \rho_t}{\partial r}-\frac{\partial \epsilon^{\rm HF}}{\partial r}\frac{\partial \rho_t}{\partial k}=2\,{\rm Im}\int\limits_{-\infty}^{t} d\bar{t}\, G^R_{t,\bar{t}}\left[(1-\rho_{\bar{t}})\Sigma^<_{\bar{t},t}-\rho_{\bar{t}}\Sigma^>_{\bar{t},t}\right]. \tag{28}$$

Here the drift term is essentially determined by the Hartree–Fock mean field, $\epsilon^{\rm HF} = k^2/2m + \sigma^{\rm HF}$, and the double-time selfenergies $\Sigma^{\gtrless}_{\bar t,t}$ are given as functionals of the Wigner function. Explicit time arguments of the collision integral, $\bar t < t$, show the retardation which requires to include the propagation, $G^R_{t,\bar t}$, from the retarded time $\bar t$ to the time t of balancing changes. Having the non-Markovian form, this kinetic equation is capable of describing finite duration effects and corresponding virial corrections.

There is a remarkable difference in the actual physical contents of these common ingredients, namely $\rho \neq f$, $\varepsilon \neq \epsilon^{\rm HF}$, and the right hand sides of (27) and (28) are rather different. It was shown that both approaches are equivalent [58], the Levinson equation (28) contains terms on the collisional side which should be possible to rearrange into gradients such that the quasiparticle energies of the drift side in the Landau–Silin equation (27) are obtained. The mutual relation between the Wigner function ρ and the quasiparticle distribution f is of essential importance for such a rearrangement. In [58] we argued that the differences between the quasiparticle picture, (27), and the equations derived from the Born–Bogoliubov–Green–Kirkwood–Yvon (BBGKY) hierarchy of reduced density matrices (28) can be characterized in terms of the retardation of the collision integral of (28). There are various contributions to the total retardation, each responsible for different features seen in the quasiparticle picture extended by non-local corrections.

A part of the retardation of the Levinson equation (28) describes the off-shell motion of particles. This off-shell motion can be eliminated from the kinetic equation which requires to introduce an effective distribution (the quasiparticle distribution f) from which the Wigner function ρ can be constructed (70)

$$\rho = f + \int \frac{d\omega}{2\pi} \frac{\wp}{\omega - \varepsilon} \frac{\partial}{\partial \omega} \left((1-f)\sigma^<_\omega - f\sigma^>_\omega \right) . \tag{29}$$

This relation is the extended quasiparticle picture derived for small scattering rates (12). The limit of small scattering rates has been first introduced by Craig [7]. An inverse functional $f[\rho]$ has been constructed in [59] which had not lead to a closed theory, however. For equilibrium non-ideal plasmas this approximation has been employed in [12,60] and under the name of the generalized Beth–Uhlenbeck approach it has been used in [61] for studies of the correlated density in nuclear matter. The authors in [11] have used this approximation with the name extended quasiparticle picture for the study of the mean removal energy and high-momenta tails of the Wigner function. The non-equilibrium form has been derived finally as the modified Kadanoff and Baym ansatz [58,62]. We call it the extended quasiparticle picture.

The retardation is responsible for gradient terms by which the mean-field drift of the Levinson equation (28) differs from the quasiparticle drift of the Landau–Silin equation (71). Remaining parts of the retardation are a piece which is compensated by the decay of internal propagators, and a piece which describes the collision delay. Due to the importance of the ansatz (29) and validity in nonequilibrium, we will give in the following the derivation from Green's function technique. This will give insight into the underlying physics and approximations of kinetic equations which is usually non-transparent and uncontrolled in direct numerical methods.

2.2 Green's Function Technique

First we show that the Levinson equation can be transformed into the kinetic equation of the Landau–Silin type. To this end we derive the kinetic equation for the Wigner function from the real-time Green's functions.

We consider the two independent correlation functions for fermionic operators $G^>(1,2) = \langle a_1 a_2^+ \rangle$ and $G^<(1,2) = \langle a_2^+ a_1 \rangle$, where cumulative variables mean time and space, $1 \equiv t, x$. The average is performed with the unknown nonequilibrium statistical operator. The time diagonal part of $G^<$ yields the Wigner function, $\rho(x_1, x_2, t) = G^<(1,2)_{t_{1,2}=t}$. Employing the Heisenberg equation of motion $i\partial a/\partial t = [a, H]$ one can derive the Martin–Schwinger hierarchy for the time ordered, T, causal Green's function $iG = \langle T a_1 a_2^+ \rangle = \Theta(t_1 - t_2) G^> \pm \Theta(t_2 - t_1) G^<$ which reads

$$G_0^{-1} G(1,1') = \delta_{1,1'} + \int d3 V(1,3) G_2(1,3,1',3^+) \tag{30}$$

with the Hartree–Fock drift term

$$G_0^{-1}(1,2) = \left(i\frac{\partial}{\partial t_1} + \frac{\nabla_1^2}{2m} \right) \delta(1-2) - \Sigma^{\mathrm{HF}}(x_1, x_2, t_1) \delta(t_1 - t_2). \tag{31}$$

This hierarchy couples the one-particle Green's function to the two-particle Green's function $i^2 G_2(1,2,3,4) = \langle T a_1 a_2 a_4^+ a_3^+ \rangle$ and so on. The plus sign in the coordinate 3^+ denotes an infinitesimal larger time than 3. Formally this hierarchy can be enclosed by introducing the selfenergy

$$\int d3 V(1,3) G_2(1,3,1',3^+) \equiv \int\limits_C d3 \Sigma(1,3) G(3,1'). \tag{32}$$

The integration path C is specified such that physical boundary conditions are completed. Since for the system correlations count only if they have happened within the memory time, in the far past history the correlations beyond Hartree–Fock are assumed to be absent $\lim\limits_{t,t' \to -\infty} \Sigma G = 0$. This asymptotic weakening of initial correlations leads to a special contour integration selected to start at the infinite past, reach the times of observation and return to the past as we will see. The asymptotic weakening of initial correlations used explicitly breaks the time symmetry such that the resultant equation can be irreversible, while the basic Heisenberg equation of motions are of course reversible. This assumption can be considered as the weakest form of chaotic assumption analogous to the Boltzmann chaos ansatz.

The integration along the complex path can be decomposed into parts along the real time axis. The compact and transparent notation of this step provides the generalized Kadanoff and Baym formalism (GKB) [63,64]. Let us shortly outline the formal developments and concentrate on the time integration. If the integration path $\mathcal{C}$ would run along the real time axes we would obtain in the limit of infinite past [without loss of generality $t < t'$]

$$\begin{aligned}\int_{-\infty}^{\infty} \Sigma(t,\tilde{t})G(\tilde{t},t') &= \int_{-\infty}^{t} \Sigma^{>}(t,\tilde{t})G^{<}(\tilde{t},t') - \int_{t}^{t'} \Sigma^{<}(t,\tilde{t})G^{<}(\tilde{t},t') \\ &\quad + \int_{t'}^{\infty} \Sigma^{<}(t,\tilde{t})G^{>}(\tilde{t},t') \\ &\to \int_{-\infty}^{\infty} \Sigma^{<}(t,\tilde{t})G^{>}(\tilde{t},t')|_{t=t'=-\infty} \end{aligned} \tag{33}$$

where we applied the above definition of the causal functions. Therefore in order to complete the asymptotic weakening of initial correlations, the integration path is specified by subtracting the remaining part of (33) as

$$\int_{\mathcal{C}} d\bar{1}\Sigma(1,\bar{1})G(\bar{1},1') = \int_{-\infty}^{+\infty} d\bar{1}\left\{\Sigma(1,\bar{1})G(\bar{1},1') - \Sigma^{<}(1,\bar{1})G^{>}(\bar{1},1')\right\}. \tag{34}$$

This form we will abbreviate as $\Sigma G - \Sigma^{<}G^{>}$ and it can be written formally in matrix notation which would correspond to the Keldysh technique. More conveniently, one can establish simple algebraic rules for obtaining the correlated parts out of causal product which is the GKB formalism. Provided a causal form is given by $C = AB$, where the product is understood as integrations over intermediate variables (34), then the time pieces or correlation functions can be obtained following the same steps as in (33) by

$$\begin{aligned} C^{\gtrless} &= A^R B^{\gtrless} + A^{\gtrless} B^A \\ C^{R/A} &= A^{R/A} B^{R/A}, \end{aligned} \tag{35}$$

where the internal time integration runs along the real time axis. The retarded and advanced functions are introduced as $B^R(1,2) = -i\theta(t_1-t_2)(B^{>}+B^{<})$ and $B^A(1,2) = i\theta(t_2-t_1)(B^{>}+B^{<})$. These Lengreth and Wilkins rules (35) have the great advantage of simplifying complicated matrix notations towards algebraic expressions of the relevant physical quantities. With the help of (35) from (30) and (32) the Kadanoff and Baym equation of motion for the correlation function $G^{<}$ follows [15]

$$\left(G_0^{-1}G^{<} - G^{<}G_0^{-1}\right) = \left(\Sigma^R G^{<} - G^{<}\Sigma^A\right) - \left(G^R\Sigma^{<} - \Sigma^{<}G^A\right). \tag{36}$$

The Kadanoff and Baym equation (36) is more general than any of the above kinetic equations (28) or (27).

As we will see, when the gradient expansion is applied [62,65,66], the kinetic equation for the quasiparticle distribution function is of Landau–Silin type (71). Such derivations do not include the rearrangement energy, however. The Levinson equation (28) for the Wigner function is the time diagonal element of (36) [64,67,68]. In the right hand side of (36) one needs an additional approximation by which the double-time correlation functions are constructed from the Wigner function.

2.3 Precursor of the Levinson Equation

We have demonstrated how the Levinson equation can be converted into the Landau–Silin equation [58]. Here we repeat these steps. On the time diagonal, $t_{1,2} = t$, the Kadanoff and Baym equation (36) yields an identity

$$\begin{aligned} &- i\big(G_0^{-1}G^< - G^<G_0^{-1}\big)(t,x_1,t,x_2) \\ &= \int\limits_{-\infty}^{t} dt' \int dx' \Big(G^>(t,x_1,t',x')\Sigma^<(t',x',t,x_2) \\ &\quad +\Sigma^<(t,x_1,t',x')G^>(t',x',t,x_2)\Big) \\ &- \int\limits_{-\infty}^{t} dt' \int dx' \Big(G^<(t,x_1,t',x')\Sigma^>(t',x',t,x_2) \\ &\quad +\Sigma^>(t,x_1,t',x')G^<(t',x',t,x_2)\Big). \end{aligned} \tag{37}$$

On the left hand side there is the Hartree–Fock drift and the right hand side contains a non-Markovian collision integral. Note that correlations beyond the Hartree–Fock field are exclusively in the collision integral.

The Wigner representation. Now we concentrate on the gradient expansion of the collision integral. The quasi-classical approximation is achieved using the Wigner mixed representation for space arguments

$$\hat\sigma\left(k,\frac{x_1+x_2}{2},\frac{t_1+t_2}{2},t_1-t_2\right) = \int d(x_1-x_2)\mathrm{e}^{-ik(x_1-x_2)}\Sigma(t_1,x_1,t_2,x_2). \tag{38}$$

As a rule, we use the lower case to denote operators in the full Wigner representation (momentum, space, time, frequency) and the hat to denote the Wigner representation in space and double-time representation. Note that time arguments of $\hat\sigma$ now denote the center-of-mass time, $\frac{t_1+t_2}{2}$, and the difference time, $t_1 - t_2$. These two Wigner representations are identical for ρ and $\sigma^{\rm HF}$ which do not depend on the difference time.

In the representation (38) the identity (37) reads

$$\left(\frac{\partial}{\partial t}+\frac{k}{m}\frac{\partial}{\partial r}\right)\rho(k,r,t)+2\sin\left(\frac{1}{2}(\partial_r^{1}\partial_k^{2}-\partial_k^{1}\partial_r^{2})\right)\sigma^{\rm HF}(k,r,t)\rho(k,r,t)$$
$$=\exp\left(\frac{i}{2}\left(\partial_r^{1}\partial_k^{2}-\partial_k^{1}\partial_r^{2}\right)\right)\int\limits_0^\infty d\tau\left(\hat{g}^{>}\left(k,r,t-\frac{\tau}{2},\tau\right)\hat{\sigma}^{<}\left(k,r,t-\frac{\tau}{2},-\tau\right)\right.$$
$$+\hat{\sigma}^{<}\left(k,r,t-\frac{\tau}{2},\tau\right)\hat{g}^{>}\left(k,r,t-\frac{\tau}{2},-\tau\right)$$
$$-\hat{g}^{<}\left(k,r,t-\frac{\tau}{2},\tau\right)\hat{\sigma}^{>}\left(k,r,t-\frac{\tau}{2},-\tau\right)$$
$$\left.-\hat{\sigma}^{>}\left(k,r,t-\frac{\tau}{2},\tau\right)\hat{g}^{<}\left(k,r,t-\frac{\tau}{2},-\tau\right)\right). \tag{39}$$

Here, the superscripts *1*, *2* denote that the partial derivatives apply only to the first and second function in the product, e.g., $(\partial_r^{1}\partial_k^{2})^3 ab=\frac{\partial^3 a}{\partial^3 r}\frac{\partial^3 b}{\partial^3 k}$.

Gradient approximation. The power expansion of the goniometric functions in (39) defines the expansion in space gradients. This expansion goes to infinite order, but in the following we will restrict our attention to the linear expansion. It is customary to abbreviate the linear gradient terms with the help of the Poisson brackets,

$$2\sin\left(\frac{1}{2}(\partial_r^{1}\partial_k^{2}-\partial_k^{1}\partial_r^{2})\right)ab\approx\{a,b\}=\frac{\partial a}{\partial k}\frac{\partial b}{\partial r}-\frac{\partial a}{\partial r}\frac{\partial b}{\partial k}. \tag{40}$$

The linear approximation in space gradients is straightforward, however, one has to be careful about time arguments of functions in the collision integral of (39). For example, the first and second product seem to form an anticommutator in which the linear gradients in space cancel. This is not true because of time arguments.

The gradient approximation in time requires a special treatment because of the lower integration limit. Due to this limit the integral does not define a standard matrix product with respect to the time variables. We will give this expansion an explicit notation.

In equilibrium, the functions $\hat{g}^{\gtrless}$ and $\hat{\sigma}^{\gtrless}$ do not depend on the center-of-mass time. For slowly evolving systems the center-of-mass dependence is smooth and weak. Accordingly, in the collision integral of (39), we can expand the center-of-mass time dependence around the time t in powers of τ. Since all functions in (39) have the same center-of-mass time, $t-\frac{\tau}{2}$, it is possible to write the linear expansion of (39) with respect to time gradients in a compact form

$$\frac{\partial}{\partial t}\rho+\{\epsilon^{\rm HF},\rho\}=I+\frac{\partial}{\partial t}R\,, \tag{41}$$

with $\epsilon^{\rm HF} = \frac{k^2}{2m} + \sigma^{\rm HF}$. The zero order gradient term, $I = I^> - I^<$, reads

$$I^< = \left(1 + \frac{i}{2}\left(\partial_r^1 \partial_k^2 - \partial_k^1 \partial_r^2\right)\right) \int_0^\infty d\tau \left(\hat\sigma^>(t,\tau)\hat g^<(t,-\tau) + \hat g^<(t,\tau)\hat\sigma^>(t,-\tau)\right). \tag{42}$$

$I^>$ is obtained from $I^<$ via the interchange of particles and holes, $>\leftrightarrow<$. In the first order gradient term, $R = R^> - R^<$, is given by

$$R^< = -\frac{1}{2}\left(1+\frac{i}{2}\left(\partial_r^1 \partial_k^2 - \partial_k^1 \partial_r^2\right)\right) \int_0^\infty d\tau\, \tau \left(\hat\sigma^>(t,\tau)\hat g^<(t,-\tau) + \hat g^<(t,\tau)\hat\sigma^>(t,-\tau)\right). \tag{43}$$

Now we are ready to turn all functions into the Wigner representation

$$\sigma_\omega(k,r,t) = \int d\tau\, \mathrm{e}^{i\omega\tau}\, \hat\sigma(k,r,t,\tau). \tag{44}$$

We will write the energy argument as the subscript and keep other arguments implicit in most of the formulas.

After a substitution of σ's and g's of the Wigner representation (44) into (42), one can integrate out the difference time τ. The integration over time results in the δ-function which describes processes on the energy shell, and the principle value terms which represent off-shell processes,

$$\int_0^\infty d\tau\, \mathrm{e}^{i(\omega-\omega')\tau} = \pi\delta(\omega-\omega') + i\frac{\wp}{\omega-\omega'}. \tag{45}$$

The δ-function can be readily integrated out leaving both functions with identical energy arguments. The principle value part, $\wp$, is an odd function of energies ω and ω', therefore its non-gradient contributions cancel but the linear gradients survive. Formula (42) thus turns into

$$I^< = \int \frac{d\omega}{2\pi} \sigma_\omega^> g_\omega^< + \int \frac{d\omega d\omega'}{(2\pi)^2} \frac{\wp}{\omega'-\omega} \left\{\sigma_{\omega'}^>, g_\omega^<\right\}. \tag{46}$$

In the correlated part we express the time integral as

$$\int_0^\infty d\tau\, \tau\, \mathrm{e}^{i(\omega-\omega')\tau} = -i\frac{\partial}{\partial\omega}\int_0^\infty d\tau\, \mathrm{e}^{i(\omega-\omega')\tau} = -i\pi\delta'(\omega-\omega') + \frac{\wp'}{\omega-\omega'}. \tag{47}$$

The meaning of δ' and $\wp'$ is seen from comparison with (45). The δ' and $\wp'$ are odd and even functions in $\omega-\omega'$, respectively. The gradient expansion of (43) thus reads

$$R^{<} = -\int \frac{d\omega d\omega'}{(2\pi)^2} \frac{\wp'}{\omega' - \omega} \sigma^{>}_{\omega'} g^{<}_{\omega} + \int \frac{d\omega}{2\pi} \left\{ \sigma^{>}_{\omega}, \partial_\omega g^{<}_{\omega} \right\} . \tag{48}$$

The term R contributes to the kinetic equation (41) as a gradient correction linear in time. Since the second term of $R^{<}$ in (48) is proportional to space gradients, we neglect this term leading to the second order contribution in gradients.

2.4 Connection Between the Wigner and Quasiparticle Distributions

At very low temperatures, dissipative processes are known to vanish due to the Pauli exclusion principle. It is desirable to reorganize the kinetic equation so that the scattering integral will vanish in this limit too. As can be seen from (48), the term $\partial_t R$ remains finite even for very low temperatures. It is possible to formally remove $\partial_t R$ if we shift this term on the drift side of (41) and introduce a new distribution function,

$$f = \rho - R. \tag{49}$$

We have denoted the new function as f to anticipate that it is indeed the quasiparticle distribution as will be shown now.

Since we want to arrive at the kinetic equation for f, it is advantageous to write relation (49) in the opposite way so that we obtain ρ as a functional of f, $\rho = f + R$. Using (48) and the particle-hole conjugated term, $R^{>}$, one finds

$$\rho = f - \int \frac{d\omega' d\omega}{(2\pi)^2} \frac{\wp'}{\omega' - \omega} \left(\sigma^{>}_{\omega'} g^{<}_{\omega} - \sigma^{<}_{\omega'} g^{>}_{\omega} \right) . \tag{50}$$

This relation provides the so called extended quasiparticle picture of the Wigner function.

Wave-function renormalization. To separate the on-shell and off-shell contributions of the correlation term, we use the imaginary part of the selfenergy, $\gamma = \sigma^{>} + \sigma^{<}$ and the spectral function, $a = g^{>} + g^{<}$ to rewrite (50) as

$$\rho = f - \int \frac{d\omega' d\omega}{(2\pi)^2} \frac{\wp'}{\omega' - \omega} \left(\gamma_{\omega'} g^{<}_{\omega} - \sigma^{<}_{\omega'} a_{\omega} \right) . \tag{51}$$

With the help of the Kramers–Kronig relation

$$\int \frac{d\omega'}{2\pi} \frac{\wp'}{\omega - \omega'} \gamma_{\omega'} = \frac{\partial \sigma_\omega}{\partial \omega}, \tag{52}$$

where σ is the real part of the selfenergy, we obtain

$$\rho = f + \int \frac{d\omega}{2\pi} \frac{\partial \sigma_\omega}{\partial \omega} g^{<}_{\omega} + \int \frac{d\omega' d\omega}{(2\pi)^2} \frac{\wp'}{\omega' - \omega} \sigma^{<}_{\omega'} a_{\omega}. \tag{53}$$

Since we develop a theory up to first order in γ, we can use $a_\omega = 2\pi\delta(\omega-\varepsilon)$ and $g^<_\omega = f\,2\pi\delta(\omega-\varepsilon)$, in the correlation terms (53) which yields

$$\rho = \left(1+\left.\frac{\partial\sigma}{\partial\omega}\right|_{\omega=\varepsilon}\right) f + \int\frac{d\omega'}{2\pi}\frac{\wp'}{\omega'-\varepsilon}\sigma^<_{\omega'} . \tag{54}$$

In (54) the on-shell (quasiparticle) contribution is reduced by the wave-function renormalization,

$$z = 1+\left.\frac{\partial\sigma}{\partial\omega}\right|_{\omega=\varepsilon} , \tag{55}$$

and the term with $\sigma^<$ provides the off-shell contributions.

The exact renormalization is $z^{-1} = 1-\partial_\omega\sigma$, therefore (55) is only the linear approximation in $\partial_\omega\sigma$. From the Kramers–Kronig relation (52) one can see that this corresponds to the linear approximation in the scattering rate γ. We will keep all terms only up to this order in γ.

Extended quasiparticle picture. From the relation between the Wigner and the quasiparticle distributions (54) one can deduce an approximative construction of the correlation function $g^<$ as a functional of the quasiparticle distribution f. Since $\rho = \int\frac{d\omega}{2\pi}g^<$, we write (54) as

$$\int\frac{d\omega}{2\pi}g^<_\omega = \int\frac{d\omega}{2\pi}\left(f\,z\,2\pi\delta(\omega-\varepsilon)+\frac{\wp'}{\omega-\varepsilon}\sigma^<_\omega\right). \tag{56}$$

One can see that the correlation function in the extended quasiparticle picture [62,66],

$$g^<_\omega = f\,z\,2\pi\delta(\omega-\varepsilon)+\frac{\wp'}{\omega-\varepsilon}\sigma^<_\omega , \tag{57}$$

satisfies (56). Perhaps it is not necessary to recall that the function f has been introduced here merely to suppress the non-dissipative part of the collision integral, $\partial_t R$. In [62,66], the function f has been defined as the factor of the singularity of $g^<$ which justifies its interpretation in terms of the quasiparticle distribution. As relation (56) shows, both approaches lead to identical results.

In parallel with (57), for the hole correlation function one finds

$$g^>_\omega = (1-f)\,z\,2\pi\delta(\omega-\varepsilon)+\frac{\wp'}{\omega-\varepsilon}\sigma^>_\omega . \tag{58}$$

From $a = g^> + g^<$, it follows that (57) and (58) are consistent with the extended quasiparticle picture of the spectral function (16).

2.5 Landau–Silin Equation

So far we have treated only the time gradients finding that the identity (41) can be expressed as

$$\frac{\partial f}{\partial t} + \{\epsilon^{\mathrm{HF}}, \rho\} = I. \tag{59}$$

Now we rearrange this identity into the Landau–Silin equation for f.

The time-local remainder I of the collision integral still includes the space gradients, $I = I_B - I_\nabla$. The non-gradient part [the first term of (46)] is the Boltzmann-type scattering integral of the Landau–Silin equation

$$I_B = \int \frac{d\omega}{2\pi} (g_\omega^> \sigma_\omega^< - g_\omega^< \sigma_\omega^>) = z\left((1-f)\sigma_\varepsilon^< - f\sigma_\varepsilon^>\right). \tag{60}$$

In the second step we have used (57). The off-shell contributions to (60) have not been neglected but cancel exactly.

The gradient part which is the second term of (46) reads

$$\begin{aligned} I_\nabla &= -\int \frac{d\omega d\omega'}{(2\pi)^2} \frac{\wp}{\omega' - \omega} \left(\{\sigma_{\omega'}^>, g_\omega^<\} - \{\sigma_{\omega'}^<, g_\omega^>\}\right) \\ &= \int \frac{d\omega}{(2\pi)} \left(\{\sigma_\omega, g_\omega^<\} - \{\sigma_\omega^<, g_\omega\}\right). \end{aligned} \tag{61}$$

Here (52) has been used to remove the ω'-integration for $\sigma^> + \sigma^<$ and a similar Kramers–Kronig relation for the ω-integration over the spectral function $g^> + g^<$ has been applied. The function $g = \mathrm{Re} g^{R,A} = \frac{1}{2}(g^R + g^A)$ is the off-shell part of the propagators.

The mean-field drift $\{\epsilon^{\mathrm{HF}}, \rho\}$ in (59) is not compatible with the quasiparticle distribution in the time derivative and the scattering integral. Moreover, the space gradients from the collision integral, I_∇, have to be accounted for. Our aim is now to show that all gradients can be collected into the quasiparticle drift,

$$\{\epsilon^{\mathrm{HF}}, \rho\} + I_\nabla = \{\varepsilon, f\}. \tag{62}$$

The quasiparticle energy differs from the mean-field energy by the pole value of the selfenergy, $\varepsilon = \epsilon^{\mathrm{HF}} + \sigma_\varepsilon$.

We will proceed in two steps. First we observe that the right hand side of (62) includes only the on-shell contribution, therefore we show that the on-shell part of the left hand side equals the right hand side. Second, we show that the off-shell part of the left hand side of (62) vanishes. The on-shell parts of (57) used in (61) result in

$$\begin{aligned} I_\nabla^{\mathrm{on}} &= \int d\omega \{\sigma_\omega, zf\delta(\omega - \varepsilon)\} \\ &= \{\sigma_\varepsilon, zf\} - \sigma_\varepsilon' \{\varepsilon, zf\} + zf\{\sigma_\varepsilon', \varepsilon\} \\ &= \{\sigma_\varepsilon, zf\} + (1-z)\{\varepsilon, zf\} + zf\{z-1, \varepsilon\}. \end{aligned} \tag{63}$$

Now we add the on-shell part of the commutator $\{\epsilon^{\rm HF},\rho\}^{\rm on}=\{\epsilon^{\rm HF},zf\}$. Using a rearrangement

$$\{\epsilon^{\rm HF},zf\}=\{\varepsilon-\sigma,zf\}=\{\varepsilon,f\}+(z-1)\{\varepsilon,f\}+f\{\varepsilon,z\}-\{\sigma,zf\} \tag{64}$$

we obtain

$$\{\epsilon^{\rm HF},\rho\}^{\rm on}+I_\nabla^{\rm on}=\{\varepsilon,f\}-\{\varepsilon,(1-z)^2f\}. \tag{65}$$

The last term is of higher order than linear in the damping and can be neglected which confirms the relation (62) already from the on-shell parts.

It remains to prove that all off-shell contributions on the left hand side of (62), $\mathcal{O}=\{\epsilon^{\rm HF},\rho\}^{\rm off}+I_\nabla^{\rm off}$ cancel, i.e., $\mathcal{O}=0$. From the off-shell term of (57) we find

$$\mathcal{O}=\int\frac{d\omega}{2\pi}\left(\left\{\epsilon^{\rm HF}+\sigma_\omega,\frac{\wp'}{\omega-\varepsilon}\sigma_\omega^<\right\}-\{g_\omega,\sigma_\omega^<\}\right). \tag{66}$$

The term with $\epsilon^{\rm HF}$ results from $\{\epsilon^{\rm HF},\rho\}$, while the others from I_∇. In the last term we use the extended quasiparticle picture of the real part of the propagator

$$g_\omega=z\frac{\wp}{\omega-\varepsilon}+\frac{\wp'}{\omega-\varepsilon}(\sigma_\omega-\sigma_\varepsilon), \tag{67}$$

which directly results from (16) and the Kramers–Kronig relation. Using $\left\{\frac{\wp}{\omega-\varepsilon},\sigma_\omega^<\right\}=-\left\{\varepsilon,\frac{\wp'}{\omega-\varepsilon}\sigma_\omega^<\right\}$ we can rewrite $\mathcal{O}$ as

$$\mathcal{O}=\int\frac{d\omega}{2\pi}\left(\left\{\sigma_\omega-\sigma_\varepsilon,\frac{\wp'}{\omega-\varepsilon}\sigma_\omega^<\right\}-\left\{\frac{\wp'}{\omega-\varepsilon}(\sigma_\omega-\sigma_\varepsilon),\sigma_\omega^<\right\}\right). \tag{68}$$

The linear expansion in the vicinity of the pole, $\sigma_\omega-\sigma_\varepsilon=(\omega-\varepsilon)(z-1)+o(\gamma^2)$, yields

$$\mathcal{O}=\int\frac{d\omega}{2\pi}\left\{\frac{\wp''}{\omega-\varepsilon},(z-1)\sigma_\omega^<\right\}. \tag{69}$$

The product $(z-1)\sigma^<$ is of the second order in γ, i.e., the off-shell term $\mathcal{O}$ is negligible. This completes the proof of relation (62).

In summary, the requirement that the scattering integral vanishes at very low temperatures directly leads to the concept of quasiparticles represented by relation (50) between the Wigner and the quasiparticle distributions

$$\rho=f+\int\frac{d\omega}{2\pi}\frac{\wp}{\omega-\varepsilon}\frac{\partial}{\partial\omega}\left((1-f)\sigma_\omega^<-f\sigma_\omega^>\right). \tag{70}$$

The space gradients of the collision integral renormalize the mean-field drift into the familiar quasiparticle drift (62). From (59) collecting (60) and (62) the resulting kinetic equation for the quasiparticle distribution is of the Landau–Silin type,

$$\frac{\partial f}{\partial t} + \frac{\partial \varepsilon}{\partial k}\frac{\partial f}{\partial r} - \frac{\partial \varepsilon}{\partial r}\frac{\partial f}{\partial k} = z\left((1-f)\sigma_\varepsilon^< - f\sigma_\varepsilon^>\right), \tag{71}$$

however, there is no rearrangement energy in the drift. Moreover, the scattering integral is not local and instantaneous.

3 Non-local Scattering Integral

The link between the Levinson and Landau–Silin equations shows that the collision integral of the Levinson equation cannot be fully interpreted in terms of the scattering processes but it includes various renormalization features. In the above treatment we have constructed a closed equation for the quasiparticle distribution function while the reduced density matrix or Wigner function is given by an additional functional of this quasiparticle distribution. There arises an important question: whether it is not possible to construct the correlation functions, $g^{>,<}$, directly from the Wigner distribution so that the identity (39) turns into a closed equation for the Wigner function. The problems and drawbacks which appears in this way are discussed in [58]. To illustrate the problems commonly overseen in the literature, let us discuss the simple expansion of the Levinson equation (28) up to the first order in memory [68]

$$\frac{\partial}{\partial t}\rho = \mathcal{I} + \frac{\partial}{\partial t}\mathcal{R}. \tag{72}$$

To obtain balance equations for the density including virial corrections, one has to integrate (72) over momentum k, see [68,69]. Following the arguments developed for the Boltzmann-type equations, the wrong assumption could be used so that the correlated part of the collision integral, $\tilde{n}_c = -\int dk\mathcal{R}$, combines with the left hand side of (72), $\tilde{n}_f = \int dk\rho$, to establish the density conservation, $\frac{\partial}{\partial t}(\tilde{n}_f + \tilde{n}_c) = 0$. The correlated density derived in this way resembles the result known from equilibrium [5,6,60,61,70] but has a wrong sign and is two times too large. Note that in this picture, the Wigner function in the left hand side of the Levinson equation is wrongly treated as the quasiparticle distribution, done usually in density operator studies [68,69,71]. The wrong sign follows from this last misinterpretation. By definition the momentum integral over the Wigner distribution yields instead the full density, $\int dk\rho = n$. The conservation law thus says that either $\int dk\mathcal{R} = 0$ or the Levinson equation does not conserve the number of particles in these treatments.

In contrast, as shown in [58] the scattering integral $\mathcal{I}$ contains additional gradient terms which exactly compensate the explicit time gradients (used to derive $\tilde{n}_f$). Therefore, the correct method is that the correlated observables and virial corrections found from the off-shell contribution to the Wigner function appear as inner gradients. The collision integral of the Levinson equation then conserves the number of particles, as it should. It turns out that the consistent way to derive solvable kinetic equations consists in the systematic use of the extended quasiparticle picture (70). This circumvents internal double counts which are unavoidable in Levinson-type equations and which are a serious drawback. For details and further discussion see [58].

3.1 Quasi-Classical Limit

The consistent treatment of gradient contributions is provided by the quasi-classical limit, i.e., the linear expansion in gradients. The quasi-classical limit and the limit of small scattering rates explicitly determine how to evaluate the scattering integral from the selfenergy $\sigma^<$. For non-degenerate systems, a very similar scheme was carried out by Bärwinkel [16,72]. One can see in Bärwinkel's papers, that the scattering integral is troubled by a large set of gradient corrections. This formal complexity seems to be the main reason why most authors either neglect gradient corrections at all [65,73] or provide them buried in multi-dimensional integrals [45,46,74]. For a degenerate system, the set of gradient corrections to the scattering integral is even larger than for rare gases studied by Bärwinkel, see [75]. To avoid manipulations with long and obscure formulas, the gradient corrections have to be sorted and expressed in a comprehensive form. To this end, we will express all gradient corrections in the form of shifted arguments, as we have done above for the collision delay.

The quasi-classical limit of the self energy with all linear gradients retained is a tedious but straightforward algebraic exercise, see [58] for practical rules. Though any many-body approximation for the selfenergy can be treated with these rules, we use here only the ladder approximation given by the T-matrix $t^{R,A} = |t|e^{\mp i\phi}$, and which reads [13,58]

$$
\begin{aligned}
\sigma^<(\omega,k,r,t) = & \int \frac{dp}{(2\pi)^3}\frac{dE}{2\pi}\frac{dq}{(2\pi)^3}\frac{d\Omega}{2\pi}\left(1-\frac{1}{2}\frac{\partial\Delta_2}{\partial r}\right) \\
& \times\left|t\left(\omega+E-\Delta_E,k-\frac{1}{2}\Delta_K,p-\frac{1}{2}\Delta_K,q,r-\Delta_r,t-\frac{1}{2}\Delta_t\right)\right|^2 \\
& \times g^>(E,p,r-\Delta_2,t) \\
& \times g^<(\omega-\Omega-\Delta_E,k-q-\Delta_K,r-\Delta_3,t-\Delta_t) \\
& \times g^<(E+\Omega-\Delta_E,p+q-\Delta_K,r-\Delta_4,t-\Delta_t).
\end{aligned} \tag{73}
$$

The gradient terms are expressed via Δ's which are derivatives of the scattering phase shift ϕ, according to the following list [13,75]

$$\begin{aligned}
\Delta_t &= \left.\frac{\partial\phi}{\partial\Omega}\right|_{\varepsilon_1+\varepsilon_2} & \Delta_2 &= \left(\frac{\partial\phi}{\partial p}-\frac{\partial\phi}{\partial q}-\frac{\partial\phi}{\partial k}\right)_{\varepsilon_1+\varepsilon_2} \\
\Delta_E &= -\frac{1}{2}\left.\frac{\partial\phi}{\partial t}\right|_{\varepsilon_1+\varepsilon_2} & \Delta_3 &= -\left.\frac{\partial\phi}{\partial k}\right|_{\varepsilon_1+\varepsilon_2} \\
\Delta_K &= \frac{1}{2}\left.\frac{\partial\phi}{\partial r}\right|_{\varepsilon_1+\varepsilon_2} & \Delta_4 &= -\left(\frac{\partial\phi}{\partial k}+\frac{\partial\phi}{\partial q}\right)_{\varepsilon_1+\varepsilon_2}
\end{aligned} \tag{74}$$

and $\Delta_r = \frac{1}{4}(\Delta_2+\Delta_3+\Delta_4)$. Numerical values of the delay Δ_t and the displacements Δ_{2-4} for the scattering of isolated particles are presented in [13,76].

Sending all Δ's to zero one recovers the instantaneous and local approximation of the selfenergy, (73). The non-instantaneousness and non-locality given by Δ's appear in three ways: in arguments of the correlation functions, in arguments of the T-matrix, and as a fore-factor. In correlation functions, the Δ's describe displacements of the initial and final positions of colliding particles, the final state of one of the particles is fixed to the balanced phase-space point (k,r,t). The arguments of the T-matrix correspond to the center of arguments of all initial and final states.

The pre-factor has no special physical meaning as it depends on the actual choice of energy and momentum variables. For instance, using in (73) an integral over $E' = E - \Delta_E$ instead of the integral over E, a corresponding factor, $dE/dE' = 1 + \partial_\omega \Delta_E = 1 - \frac{1}{2}\partial_t \Delta_t$, appears.

3.2 Extended Quasiparticle Picture

Now we can complete the derivation of the kinetic equation. Since the scattering integral is already proportional to the 'small' scattering rate, we can use from the extended quasiparticle picture (57) the pole part, $g^<(\omega,k,r,t) = f(k,r,t)2\pi\delta(\omega-\varepsilon(k,r,t))$, to convert selfenergies $\sigma^{>,<}$ into functionals of the quasiparticle distribution. The selfenergy $\sigma^<$ is given by (73) and $\sigma^>$ is obtained by the interchange $>\longleftrightarrow<$. The resulting kinetic equation from (71) reads

$$\begin{aligned}
&\frac{\partial f_1}{\partial t}+\frac{\partial\varepsilon_1}{\partial k}\frac{\partial f_1}{\partial r}-\frac{\partial\varepsilon_1}{\partial r}\frac{\partial f_1}{\partial k} = \int\frac{dpdq}{(2\pi)^6}2\pi\delta\left(\varepsilon_1+\bar\varepsilon_2-\bar\varepsilon_3-\bar\varepsilon_4-2\Delta_E\right) \\
&\times\left(1-\frac{1}{2}\frac{\partial\Delta_2}{\partial r}-\frac{\partial\bar\varepsilon_2}{\partial r}\frac{\partial\Delta_2}{\partial\omega}\right) \\
&\times\left|t\left(\varepsilon_1+\bar\varepsilon_2-\Delta_E,k-\frac{1}{2}\Delta_K,p-\frac{1}{2}\Delta_K,q,r-\Delta_r,t-\frac{1}{2}\Delta_t\right)\right|^2 \\
&\times\left((1-f_1)(1-\bar f_2)\bar f_3\bar f_4 - f_1\bar f_2(1-\bar f_3)(1-\bar f_4)\right).
\end{aligned} \tag{75}$$

The abbreviated notation of f's and ε's means

$$\begin{aligned}\varepsilon_1 &\equiv \varepsilon(k,r,t), & \bar{\varepsilon}_3 &\equiv \varepsilon(k-q-\Delta_K, r-\Delta_3, t-\Delta_t),\\ \bar{\varepsilon}_2 &\equiv \varepsilon(p,r-\Delta_2,t), & \bar{\varepsilon}_4 &\equiv \varepsilon(p+q-\Delta_K, r-\Delta_4, t-\Delta_t).\end{aligned} \tag{76}$$

The bar indicates that all Δ's appear in arguments with negative signs. In the Δ's and their derivatives the energy $\omega + E = \varepsilon_1 + \bar{\varepsilon}_2$ is substituted after all derivatives are taken. Note that the pre-factor has changed compared to (73). In (73), $\bar{\varepsilon}_2$ depends on Δ_2 which depends on the energy E. This E-dependence has resulted in a norm of the singularity.

The non-local kinetic equation (75) represents the main result of the papers [13,58,75]. In the following we provide conservation laws following from the non-local and non-instantaneous scattering integral.

3.3 Conservation Laws

If we neglect all shifts except the energy gain, we can rewrite the kinetic equation in a form of Landau theory as has been discussed in the introduction. The time invariant observables are then the quasiparticle density, momentum, energy and stress tensor in the form

$$\begin{aligned} n^{\rm qp} &= \sum_k f & \mathcal{Q}^{\rm qp} &= \sum_k k\, f & j^{\rm qp} &= \sum_k \frac{\partial \varepsilon}{\partial k} f \\ \mathcal{E}^{\rm qp} &= \sum_k (\frac{k^2}{2m} + \frac{1}{2}\sigma_{\rm mf}) f_k & \mathcal{J}_{ij}^{\rm qp} &= \sum_k \left(k_j \frac{\partial \varepsilon}{\partial k_i} + \delta_{ij}\varepsilon \right) f - \delta_{ij}\mathcal{E}^{\rm qp}, \end{aligned} \tag{77}$$

where we abbreviated the meanfield part $\sigma_{\rm mf} = \sigma^{\rm HF} + \int \frac{dp}{(2\pi)^3} t^R(\omega + \varepsilon_p, k, p, 0) f_p$ of the selfenergy.

Taking now into account the binary correlations which we have reformulated into the shifts, we obtain from the non-local kinetic equation the modified balance equations [13]

$$\frac{\partial (n^{\rm free} + n^{\rm mol})}{\partial t} + \frac{\partial (j^{\rm qp} + j^{\rm mol})}{\partial r} = 0 \tag{78}$$

$$\frac{\partial (\mathcal{Q}_j^{\rm qp} + \mathcal{Q}_j^{\rm mol})}{\partial t} + \sum_i \frac{\partial (\mathcal{J}_{ij}^{\rm qp} + \mathcal{J}_{ij}^{\rm mol})}{\partial r_i} = 0 \tag{79}$$

$$\frac{\partial \mathcal{E}}{\partial t} = \frac{\partial (\mathcal{E}^{\rm qp} + \mathcal{E}^{\rm mol})}{\partial t} = 0. \tag{80}$$

This shows that the conserving observables now consists of the sum of the quasiparticle parts (77) and the correlated parts which we can interprete as the parts coming from correlated pairs or molecules

$$\begin{aligned}
n^{\rm mol} &= \int d\mathcal{P} \Delta_t & j^{\rm mol} &= \int d\mathcal{P} \Delta_3 \\
\mathcal{Q}^{\rm mol} &= \int d\mathcal{P} \frac{k+p}{2} \Delta_t & \mathcal{E}^{\rm mol} &= \int d\mathcal{P} \frac{\epsilon_k + \epsilon_p}{2} \Delta_t \\
\mathcal{J}_{ij}^{\rm mol} &= \frac{1}{2} \int d\mathcal{P} \Big\{ k_j \Delta_{3i} + p_j (\Delta_{4i} - \Delta_{2i}) + q_j (\Delta_{4i} - \Delta_{3i}) \Big\}, & & \qquad (81)
\end{aligned}$$

where $d\mathcal{P}$ is the probability to form a pair per time (19). This immediate interpretation of $d\mathcal{P}$ is obvious from the expressions (81). The density of pairs is given if $d\mathcal{P}$ is multiplied with the life time of the molecule Δ_t. The energy, the momentum and mass current which is carried by the pairs results analogously. As one can see, the correlated part of the stress tensor (81) takes a form of a virial which is well known for the collision flux in dense gases. The corresponding momenta are connected to the corresponding offsets. The diagonal part of this stress tensor is of course the correlated part of the pressure.

The final proof that these conservation laws indeed describe the time invariance of the correct observables and the check of the inner consistency of the theory is to show that the correlated parts (81) can be obtained directly from the correlated part of the reduced density matrix (70). The first part of (70) gives immediately the Landau quasiparticle observables (77). The correlated parts of the observables (81) indeed follow from (77) which is a lengthy calculation. All these proofs can be found in detail in the appendix G of [13].

The correlated density is proportional to the energy-derivative of the phase shift expressed here via the collision delay and is indeed the generalized Beth–Uhlenbeck equation [5,6,10,60,61,68,70]. This correlated density is consistent with the equation of continuity (78) found from the kinetic equation (75). It should be noted that the number of quasiparticles is not conserved, $\partial_t n_f \neq 0$, because during the time interval Δ_t the colliding particles are excluded from the single-particle statistics being in two-particle scattering states. One can see that instantaneous approximations of the kinetic equation cannot capture the correlated density. At the same time, the correct description of the quasiparticle density makes the choice (74) of the collision delay preferable to other characteristic times one can define in the spirit of Wigner from the asymptotic behavior of final states of binary collisions.

4 Summary and Conclusions

It was presented that the asymptotic behavior of nonequilibrium Green's functions differs from the Landau theory of quasiparticle transport in two points. First, the drift does not include the rearrangement energy. Second, the scattering integral is non-local and non-instantaneous. By these features, the asymptotic kinetic equation is capable to cover the Beth–Uhlenbeck equation of states resembling the virial expansions of dense gases.

The theory includes two families of corrections beyond the standard Landau picture, the off-shell motion is treated within the extended quasiparticle picture, and non-local extensions of the collisions are treated within the first order gradient corrections. These two approximations are mutually consistent.

Due to the finite duration of collisions, the kinetic equation is non-Markovian as the initial and final states of the collision are at distinct times. This non-Markovian behavior has to be distinguished from the non-Markovian character of the Levinson equation. From the derivation we have seen that most gradient contributions and non-Markovian effects of the Levinson equation can be reformulated into wavefunction renormalizations and non-local and non-instantaneous shifts in the collision integral. For slowly varying systems, we have shown that one half of the Levinson-type retardation describes the off-shell motion and corresponding renormalizations of single-particle functions. The second half compensates the decay of propagators during the integration and can be eliminated leaving the pole contribution to the scattering integral identical to the one obtained from the quasiparticle picture.

The resulting balance equations from the non-local kinetic equation show, in addition to the quasiparticle observables, contributions from correlated pairs where the sum of both are conserved in time. These correlated parts exceed the Landau theory and are expressed in terms of the different shifts. The collision delay resulting from the energy dependence of the scattering phase shift is a non-Markovian correction and determines density of correlated particles. It reproduces the correlated density obtained within the generalized Beth–Uhlenbeck approach. The virial corrections to the balance equations appear from intrinsic gradients instead of correlation parts in the equation of the reduced density matrix. In [13] it is shown that the non-local kinetic equation leads to complete balance equations for the density, energy and stress tensor which establish conservation laws including correlated parts.

The energy gain from the background or medium plays a special role. It gives rise to the rearrangement energy which links the Landau quasiparticle energy as a variational concept to the quasiparticle energy as a pole of the spectral approach. The non-instantaneous and non-local corrections given by the Δ's do not change the structure and the overall interpretation of the scattering integral but only slightly renormalize its ingredients. The exclusive dependence of the non-local and non-instantaneous corrections on the scattering phase shift confirms results from the theory of gases [43,44,47,52] obtained by very different technical tools. The non-instantaneous and non-local scattering integral in the form (75) parallels the classical Enskog's equation, therefore it can be treated with numerical tools developed for the theory of classical gases, see e.g. [77] at least as long as $1 - f_1 - f_2$ remains positive.

The non-local kinetic equation requires no more numerical efforts beyond solving the Boltzmann equation. The numerical solution and application has been demonstrated in [78–81]. The non-local corrections with parameters found from experimental data on collisions of two isolated nucleons [76] have been implemented into numerical simulations of heavy ion reactions in the non-relativistic regime [78,81]. The experimentally observed high temperature of mono-nucleon products of central reactions has been recovered while the temperature of more complex cluster remains unchanged [78]. For the midrapidity distribution in very peripheral reactions we have achieved agreement between theoretical predictions and experimental data [81].

References

1. L. Boltzmann, Wien. Ber. **66**, 275 (1872).
2. L.D. Landau, Soviet Phys. JETP **3**, 920 (1957).
3. A.A. Abrikosov, L.P. Gorkov, and I.E. Dzyaloshinski, *Methods of Quantum Field Theory in Statistical Physics* (Prentice Hall, New York, 1963).
4. G.E. Beth and E. Uhlenbeck, Physica **3**, 729 (1936).
5. G.E. Beth and E. Uhlenbeck, Physica **4**, 915 (1937).
6. M. Schmidt, G. Röpke, and H. Schulz, Ann. Phys. (NY) **202**, 57 (1990).
7. R.A. Craig, Ann. Phys. **40**, 416 (1966).
8. R.A. Craig, Ann. Phys. **40**, 434 (1966).
9. V. Špička, K. Morawetz, and P. Lipavský, Phys. Rev. E **64**, 046107 (2001).
10. T. Bornath, D. Kremp, W.D. Kraeft, and M. Schlanges, Phys. Rev. E **54**, 3274 (1996).
11. H.S. Köhler and R. Malfliet, Phys. Rev. C **48**, 1034 (1993).
12. D. Kremp, W.D. Kraeft, and A.D.J. Lambert, Physica A **127**, 72 (1984).
13. P. Lipavský, K. Morawetz, and V. Špička, *Kinetic equation for strongly interacting dense Fermi systems*, Vol. 26, 1 of *Annales de Physique* (EDP Sciences, Paris, 2001), ISBN: 2-86883-541-4.
14. P. Lipavský, V. Špička, and K. Morawetz, Phys. Rev. E **59**, R1291 (1999).
15. L.P. Kadanoff and G. Baym, *Quantum Statistical Mechanics* (Benjamin, New York, 1962).
16. K. Bärwinkel, Z. Naturforsch. A **24**, 38 (1969).
17. K. Bärwinkel, in *Proceedings of the 14th International Symposium on Rarified Gas Dynamics* (University of Tokyo Press, Tokyo, 1984).
18. R.F. Snider, J. Stat. Phys. **63**, 707 (1991).
19. D. Enskog, in *Kinetic theory*, edited by S. Brush (Pergamon Press, New York, 1972), Vol. 3, orig.: *K. Svenska Vet. Akad. Handl.* **63**(4) (1921).
20. S. Chapman and T.G. Cowling, *The Mathematical Theory of Non-uniform Gases* (Cambrigde University Press, Cambridge, 1990), third edition Chap. 16.
21. J.O. Hirschfelder, C.F. Curtiss, and R.B. Bird, *Molecular Theory of Gases and Liquids* (Wiley, New York, 1964), chapts. 6.4a and 9.3.
22. P.P.J.M. Schram, *Kinetic Theory of Gases and Plasmas* (Kluwer Academic Publishers, Dordrecht, 1991).
23. E. Cohen, *Fundamental Problems in Statistical Mechanics* (North-Holland, Amsterdam, 1962).
24. J. Weinstock, Phys. Rev. **132**, 454 (1963).
25. J. Weinstock, Phys. Rev. **132**, 470 (1963).

26. J. Weinstock, Phys. Rev. A **140**, 460 (1965).
27. K. Kawasaki and I. Oppenheim, Phys. Rev. A **139**, 1763 (1965).
28. J.R. Dorfman and E.G. Cohen, J. Math. Phys. **8**, 282 (1967).
29. R. Goldman and E.A. Frieman, J. Math. Phys. **8**, 1410 (1967).
30. H. van Beijeren and M.H. Ernst, J. Stat. Phys. **21**, 125 (1979).
31. A. Bonasera, F. Gulminelli, and J. Molitoris, Phys. Rep. **243**, 1 (1994).
32. L. Waldmann, Z. Naturforsch. A **12**, 660 (1957).
33. L. Waldmann, Z. Naturforsch. A **13**, 609 (1958).
34. L. Waldmann, Z. Naturforsch. A **15**, 19 (1960).
35. R.F. Snider, J. Chem. Phys. **32**, 1051 (1960).
36. R.F. Snider, J. Math. Phys. **5**, 1580 (1964).
37. M.W. Thomas and R.F. Snider, J. Stat. Phys. **2**, 61 (1970).
38. R.F. Snider and B.C. Sanctuary, J. Chem. Phys. **55**, 1555 (1971).
39. J.C. Rainwater and R.F. Snider, J. Chem. Phys. **65**, 4958 (1976).
40. R. Balescu, *Equilibrium and Nonequilibrium Statistically Mechanics* (Wiley, New York, 1975).
41. J.A. McLennan, *Introduction to Nonequilibrium Statistical Mechanics* (Prentice-Hall, Englewood Cliffs, 1989).
42. F. Laloë, J. Phys. (Paris) **50**, 1851 (1989).
43. G. Tastevin, P. Nacher, and F. Laloë, J. Phys. (Paris) **50**, 1879 (1989).
44. P. Nacher, G. Tastevin, and F. Laloë, J. Phys. (Paris) **50**, 1907 (1989).
45. D. Loos, J. Stat. Phys. **59**, 691 (1990).
46. D. Loos, J. Stat. Phys. **61**, 467 (1990).
47. M. de Haan, Physica A **164**, 373 (1990).
48. H. de Haan, Physica A **165**, 224 (1990).
49. H. de Haan, Physica A **170**, 571 (1991).
50. F. Laloë and W.J. Mullin, J. Stat. Phys. **59**, 725 (1990).
51. R.F. Snider, J. Stat. Phys. **61**, 443 (1990).
52. P.J. Nacher, G. Tastevin, and F. Laloë, Ann. Phys. (Leipzig) **48**, 149 (1991).
53. P.J. Nacher, G. Tastevin, and F. Laloë, Journal de Physique I **1**, 181 (1991).
54. R.F. Snider, J. Stat. Phys. **80**, 1085 (1995).
55. R.F. Snider, W.J. Mullin, and F. Laloë, Physica A **218**, 155 (1995).
56. I.B. Levinson, Fiz. Tverd. Tela Leningrad **6**, 2113 (1965).
57. I.B. Levinson, Zh. Eksp. Teor. Fiz. **57**, 660 (1969), [Sov. Phys.–JETP **30**, 362 (1970)].
58. K. Morawetz, P. Lipavský, and V. Špička, Ann. of Phys. **294**, 134 (2001).
59. B. Bezzerides and D.F. DuBois, Phys. Rev. **168**, 233 (1968).
60. H. Stolz and R. Zimmermann, phys. stat. sol. (b) **94**, 135 (1979).
61. M. Schmidt and G. Röpke, phys. stat. sol. (b) **139**, 441 (1987).
62. V. Špička and P. Lipavský, Phys. Rev. B **52**, 14615 (1995).
63. D.C. Langreth and J.W. Wilkins, Phys. Rev. B **6**, 3189 (1972).
64. P. Lipavský, V. Špička, and B. Velický, Phys. Rev. B **34**, 6933 (1986).
65. W. Botermans and R. Malfliet, Phys. Rep. **198**, 115 (1990).
66. V. Špička and P. Lipavský, Phys. Rev. Lett **73**, 3439 (1994).
67. A.P. Jauho and J.W. Wilkins, Phys. Rev. B **29**, 1919 (1984).
68. K. Morawetz and G. Röpke, Phys. Rev. E **51**, 4246 (1995).
69. Y.L. Klimontovich, *Kinetic Theory of Nonideal Gases and Nonideal Plasmas* (Pergamon Press, Oxford, 1982).
70. R. Zimmermann and H. Stolz, phys. stat. sol. (b) **131**, 151 (1985).

71. V.V. Belyi, Y.A. Kukharenko, and J. Wallenborn, Phys. Rev. Lett. **76**, 3554 (1996).
72. K. Bärwinkel, Z. Naturforsch. A **24**, 22 (1969).
73. P. Danielewicz, Ann. Phys. (NY) **152**, 239 (1984).
74. V.G. Morozov and G. Röpke, Physica A **221**, 511 (1995).
75. V. Špička, P. Lipavský, and K. Morawetz, Phys. Lett. A **240**, 160 (1998).
76. K. Morawetz, P. Lipavský, V. Špička, and N.H. Kwong, Phys. Rev. C **59**, 3052 (1999).
77. F.J. Alexander, A.L. Garcia, and B.J. Alder, Phys. Rev. Lett. **74**, 5212 (1995).
78. K. Morawetz *et al.*, Phys. Rev. Lett. **82**, 3767 (1999).
79. K. Morawetz, Phys. Rev. C **62**, 44606 (2000).
80. K. Morawetz, S. Toneev, and M. Ploszajczak, Phys. Rev. C **62**, 64602 (2000).
81. K. Morawetz *et al.*, Phys. Rev. C **63**, 034619 (2001).

Collisional Absorption of Laser Radiation in a Strongly Correlated Plasma

Thomas Bornath, Manfred Schlanges, Paul Hilse, and Dietrich Kremp

Abstract. Starting from quantum kinetic theory, collisional absorption of laser radiation is investigated for dense plasmas. The electrical current and the energy balance equations are formulated within the framework of nonequilibrium Green's function methods. Quantum statistical expressions are derived for two-temperature plasmas in which the coupling between electrons and ions is weak due to the influence of the strong high-frequency laser field, however, the electron and the ion components may be strongly coupled. Consequently, the expressions for, e.g., the electrical current, the cycle-averaged energy absorption rate, and energy transfer rates between electrons and ions contain the dynamical structure factors and the dielectric function of the strongly correlated electron and ion subsystems. The expressions are valid for arbitrary field strength assuming the nonrelativistic case.

Numerical results are presented to discuss these quantities as a function of the applied laser field and for different plasma parameters. In particular, nonlinear phenomena such as higher harmonics generation and multiphoton absorption in the inverse Bremsstrahlung processes are considered. The influence of correlation effects incorporated in structure factor and in the dielectric function will be discussed. Furthermore, the significance to include quantum effects is demonstrated comparing our results with previous ones obtained from classical theories.

1 Introduction

Due to the impressive progress in laser technology, that makes femtosecond lasers pulses of very high intensity available in laboratory experiments [1], the laser–matter interaction has become a field of current interest. If the solid target is irradiated by such a laser pulse, dense plasmas can be created relevant for astrophysics and inertial confinement fusion. Especially, at high intensities the quiver velocity can be large compared to the thermal velocity and interesting nonlinear effects have to be expected.

One of the important mechanisms of energy deposition is inverse bremsstrahlung (IB), i.e., laser light absorption via collisional processes between the plasma particles. In strongly ionized plasmas, this absorption process is essentially governed by the electron–ion interaction usually described in terms of the electron–ion collision frequency.

In several papers, various approaches were used to calculate the electron–ion collision frequency and the dynamic conductivity, respectively, for classical plasmas under different conditions [2–8]. They cannot be applied to situations of laser–plasma interaction where quantum effects become important. Quantum effects in dense plasmas can be expected (i) if the Landau length $l = e^2/k_B T_e$ is comparable to the thermal wave length $\lambda = (2\pi\hbar^2/m_e k_B T_e)^{1/2}$, i.e. $l/\lambda \leq 1$, (ii) for $\hbar\omega/k_B T_e > 1$ with ω being the laser frequency, and (iii) if the electrons with number density n_e have to be described by Fermi statistics in degenerate plasmas, i.e. $n_e\lambda^3 > 1$. Quantum mechanical treatments were given by several authors, e.g. [9–13]. Rigorous quantum kinetic approaches, however, to the inverse bremsstrahlung absorption in dense plasmas were missing until recently. Kremp et al. [14] derived a quantum kinetic equation for dense plasmas in strong laser fields using nonequilibrium Green's function techniques. In this approach, the different interaction processes can be taken into account by appropriate approximations of the generalized field–dependent scattering rates including nonlinear field effects such as multiphoton processes and higher harmonics generation. Subsequently, quantum statistical expressions for the electron–ion collision frequency were derived, and time–dependent phenomena were studied by numerical solution of this equation [15,16]. Quantum expressions for the collision term and the electron–ion collision frequency including dynamic screening were given first in references [17,19–21]. The main focus of the present paper is on generalizations of the approach to study effects of strong electron–electron and ion–ion correlations on the collisional absorption rate [22,23].

2 Kinetic Equation for Plasmas in Electromagnetic Fields

We consider a plasma under the influence of intense laser radiation. The plasma is assumed to be fully ionized consisting of electrons with number densities n_e and ions of charge $e_i = Ze$ with number density n_i. Equilibrium and nonequilibrium properties of strongly correlated plasmas are successfully described using the method of real–time Green's functions. In this framework, the nonequilibrium plasma state is given by the two–time correlation functions which are averages over creation and annihilation operators $\psi^\dagger$ and ψ

$$g_a^>(1,1') = \frac{1}{i\hbar}\langle\psi_a(1)\psi_a^\dagger(1')\rangle\,, \qquad g_a^<(1,1') = -\frac{1}{i\hbar}\langle\psi_a^\dagger(1')\psi_a(1)\rangle, \qquad (1)$$

where $1 \equiv (\mathbf{r}_1, t_1, s_1^3)$, and a labels the particle species. For equal times $t_1 = t_1'$, the density matrix or Wigner function, respectively, follows from the correlation function $g_a^<$. The time evolution of $g_a^{\gtrless}$, which contain the complete dynamical and statistical information, is determined by the Kadanoff and Baym equations [14]

$$\left[i\hbar\frac{\partial}{\partial t_1} - \frac{1}{2m_a}\left(\frac{\hbar}{i}\nabla_1 - \frac{e_a}{c}\mathbf{A}(1)\right)^2 - e_a\phi(1)\right] g_a^{\gtrless}(1,1')$$
$$= \sum_b \int \mathrm{d}2\, V_{ab}(1-2)\, g_{ab}(12, 1'2^+)$$
$$= \int \mathrm{d}\bar{1}\, \left[\Sigma_a^R(1,\bar{1})\, g_a^{\gtrless}(\bar{1},1') + \Sigma_a^{\gtrless}(1,\bar{1})\, g_a^A(\bar{1},1')\right] \tag{2}$$

with g_{ab} being the respective two-particle Green's function and $V_{ab}(1-2) = V_{ab}(\mathbf{r}_1 - \mathbf{r}_2)\delta(t_1 - t_2)$ being the Coulomb potential. In the last line, introduction of the self-energy functions $\Sigma_a^{\gtrless}$ and Σ_a^R decouples the hierarchy of equations.

We consider in the following a spatially homogeneous electric field using vector potential gauge ($\mathbf{A}$ and ϕ denote the vector and scalar potential, respectively)

$$\mathbf{A}(t) = -\int_{-\infty}^{t} \mathrm{d}\bar{t}\, \mathbf{E}(\bar{t}); \quad \phi = 0\,. \tag{3}$$

From the Kadanoff–Baym equations there follows for equal times $t_1 = t_1'$ an equation for the Wigner function that reads in the spatially homogeneous case

$$\frac{\partial}{\partial t} f_a(\mathbf{p},t) = 2\,\mathrm{Re}\int_{t_0}^{t} \mathrm{d}\bar{t}\Big\{g_a^{>}(\mathbf{p};t,\bar{t})\ \Sigma_a^{<}(\mathbf{p};\bar{t},t) - g_a^{<}(\mathbf{p};t,\bar{t})\ \Sigma_a^{>}(\mathbf{p};\bar{t},t)\Big\}$$
$$= I_a(\mathbf{p},t)\,. \tag{4}$$

2.1 Gauge Invariant Green's Functions

It is advantageous to have a gauge invariant description (the definition of macroscopic variables, e.g., should not depend on the gauge) which is achieved by the following transform [24,25] (see also [26] and references cited there for the case of a strong static electric field)

$$\tilde{g}_a(\mathbf{k};t,t') = \int \mathrm{d}^3 r\, \mathrm{e}^{-\frac{i}{\hbar}\mathbf{r}\cdot\left(\mathbf{k}+e_a\int_{t'}^{t}\mathrm{d}\bar{t}\,\frac{\mathbf{A}(\bar{t})}{t-t'}\right)} g_a(\mathbf{r};t,t')\,. \tag{5}$$

The connection between the Fourier transform $g_a(\mathbf{p},tt')$ and the gauge invariant function $\tilde{g}_a(\mathbf{k},tt')$ is given by

$$g_a(\mathbf{p},tt') = \int \mathrm{d}^3 r\, \mathrm{e}^{-\frac{i}{\hbar}\mathbf{p}\cdot\mathbf{r}} g_a(\mathbf{r},tt')$$
$$= \int \mathrm{d}^3 r\, \mathrm{e}^{-\frac{i}{\hbar}\mathbf{p}\cdot\mathbf{r}} \int \frac{\mathrm{d}^3 k}{(2\pi\hbar)^3}\, \mathrm{e}^{\frac{i}{\hbar}\mathbf{r}\cdot\left(\mathbf{k}+e_a\int_{t'}^{t}\mathrm{d}\bar{t}\,\frac{\mathbf{A}(\bar{t})}{t-t'}\right)} \tilde{g}_a(\mathbf{k},tt')$$
$$= \tilde{g}_a\left(\mathbf{p} - e_a\int_{t'}^{t}\mathrm{d}\bar{t}\,\frac{\mathbf{A}(\bar{t})}{t-t'}, tt'\right), \tag{6}$$

whereas the respective single-time distribution functions are connected by

$$f_a(\mathbf{p},t) = \tilde{f}_a\left(\mathbf{p} - e_a\mathbf{A}(t), t\right) . \tag{7}$$

Using these relations in the above time–diagonal Kadanoff–Baym equation we get [14]

$$\begin{aligned}\left\{\frac{\partial}{\partial t} + e_a\mathbf{E}(t)\cdot\nabla_{k_a}\right\}\tilde{f}_a(\mathbf{k}_a,t) &= -2\mathrm{Re}\int_{t_0}^{t}\mathrm{d}\bar{t}\{\tilde{\Sigma}_a^{>}\tilde{g}_a^{<} - \tilde{\Sigma}_a^{<}\tilde{g}_a^{>}\} \\ &= \tilde{I}_a(\mathbf{k}_a,t)\,,\end{aligned} \tag{8}$$

where a and b label the particle species, and with the arguments on the rhs written explicitly as

$$\begin{aligned}\tilde{g}_a\tilde{\Sigma}_a = \tilde{g}_a&\left(\mathbf{k} + e_a\mathbf{A}(t) - e_a\int_{\bar{t}}^{t}\mathrm{d}t'\,\frac{\mathbf{A}(t')}{t-\bar{t}};t,\bar{t}\right) \\ &\times\tilde{\Sigma}_a\left(\mathbf{k} + e_a\mathbf{A}(t) - e_a\int_{\bar{t}}^{t}\mathrm{d}t'\,\frac{\mathbf{A}(t')}{t-\bar{t}};\bar{t},t\right).\end{aligned} \tag{9}$$

The functions carry a field induced momentum shift.

The kinetic equation (8) is still very general. Two steps are necessary to find a closed form with an explicit expression for the collision term: (i) The self-energy functions or, equivalently, the two-particle Green's function, have to be specified in a certain approximation, and (ii) the two-time correlation functions $g_a^{\gtrless}$ have to be expressed in terms of the Wigner distribution function. The latter task is known as the reconstruction problem [27]. Approximately, the *generalized Kadanoff–Baym ansatz* (GKBA) can be used

$$g_a^{\gtrless}(\mathbf{p};t,t') = i\hbar g_a^{R}(\mathbf{p};t,t')g_a^{\gtrless}(\mathbf{p};t',t') - i\hbar g_a^{\gtrless}(\mathbf{p};t,t)g_a^{A}(\mathbf{p};t,t'). \tag{10}$$

with $-i\hbar g_a^{<}(\mathbf{p};t',t') = f_a(\mathbf{p};t')$, and $i\hbar g_a^{>}(\mathbf{p};t',t') = 1 - f_a(\mathbf{p};t')$. With help of (6) and (7), it is easy to transform this ansatz for the gauge invariant quantities.

Powerful schemes are available to determine appropriate approximations for the self-energy function taking into account nonlinear field dependence as well as many-body and quantum effects relevant for high-density plasmas.

2.2 Kinetic Equation in the Statically Screened Born Approximation

In order to keep the expression as simple as possible, we discuss in this subsection a rather simple approximation for the self-energy which determines the collision integral in the kinetic equation: the second Born approximation with statically screened potential

$$\Sigma_a^{\gtrless}(\mathbf{k}_a;t_1,t_1') = \times \sum_b \int \frac{\mathrm{d}^3k_b \mathrm{d}^3\bar{k}_a \mathrm{d}^3\bar{k}_b}{(2\pi\hbar)^9} \left|V_{ab}(\mathbf{k}_a - \bar{\mathbf{k}}_a)\right|^2 (2\pi\hbar)^3 \delta(\mathbf{k}_a + \mathbf{k}_b - \bar{\mathbf{k}}_a - \bar{\mathbf{k}}_b) \times \hbar^2 g_a^{\gtrless}(\bar{\mathbf{k}}_a;t_1,t_1') g_b^{\gtrless}(\bar{\mathbf{k}}_b;t_1,t_1') g_b^{\lessgtr}(\mathbf{k}_b;t_1',t_1). \tag{11}$$

The correlation functions $g_a^{\gtrless}$ are expressed by the Wigner functions with the help of the GKBA (10). Using for the retarded and advanced Green's functions the expressions of free particles,

$$g_a^{R/A}(\mathbf{p};t,t') = \mp\frac{i}{\hbar}\Theta(\pm[t-t'])\exp\left[-\frac{i}{\hbar}\int_{t'}^{t} \mathrm{d}\bar{t}\,[\mathbf{p} - e_a\mathbf{A}(\bar{t})]^2/2m_a\right], \tag{12}$$

one is lead to the following kinetic equation [14]

$$\left\{\frac{\partial}{\partial t} + e_a\mathbf{E}(t)\cdot\nabla_{k_a}\right\}\tilde{f}_a(\mathbf{k}_a,t) = \sum_b \tilde{I}_{ab}(\mathbf{k}_a,t) \tag{13}$$

with the collision integral

$$\tilde{I}_{ab}(\mathbf{k}_a,t) = 2\int \frac{\mathrm{d}^3k_b \mathrm{d}^3\bar{k}_a \mathrm{d}^3\bar{k}_b}{(2\pi\hbar)^6}\frac{1}{\hbar^2}\left|V_{ab}(\mathbf{k}_a - \bar{\mathbf{k}}_a)\right|^2 \delta(\mathbf{k}_a + \mathbf{k}_b - \bar{\mathbf{k}}_a - \bar{\mathbf{k}}_b) \times \int_{t_0}^{t} \mathrm{d}\bar{t}\,\mathrm{Re}\,\mathrm{e}^{\left\{\frac{i}{\hbar}\left[(\epsilon_{ab}-\bar{\epsilon}_{ab})(t-\bar{t})-(\mathbf{k}_a-\bar{\mathbf{k}}_a)\cdot\mathbf{R}_{ab}(t,\bar{t})\right]\right\}} \times \left\{\bar{\tilde{f}}_a\bar{\tilde{f}}_b\left[1-\tilde{f}_a\right]\left[1-\tilde{f}_b\right] - \tilde{f}_a\tilde{f}_b\left[1-\bar{\tilde{f}}_a\right]\left[1-\bar{\tilde{f}}_b\right]\right\}\Big|_{\bar{t}}, \tag{14}$$

where we denoted $\epsilon_{ab} = \epsilon_a + \epsilon_b$, $\epsilon_a = k_a^2/2m_a$ and $\tilde{f}_a = \tilde{f}_a[\mathbf{k}_a + \mathbf{Q}_a(t,\bar{t}),\,\bar{t}]$, the quantities $\mathbf{Q}_a$ and $\mathbf{R}_{ab}$ are defined below.

This kinetic equation is more general than Boltzmann-like kinetic equations. It is a non-Markovian equation which conserves the full energy (in second Born approximation). The time dependent field modifies the collision integral in several ways. (i) There are shifts of the momenta, $\mathbf{Q}_a(t,t') = -e_a\int_{t'}^{t}\mathrm{d}t_1\,\mathbf{E}(t_1)$, produced by the field during the collision time (intra-collisional field contribution). (ii) In addition to the usual collisional energy broadening there appears a field dependent broadening given by

$$R_{ab}(t,\bar{t}) = \left(\frac{e_a}{m_a} - \frac{e_b}{m_b}\right)\int_{\bar{t}}^{t}\mathrm{d}t'\int_{t'}^{t}\mathrm{d}t''E(t''). \tag{15}$$

(iii) An important feature is the nonlinear dependence of the collision integral on the field strength which leads to interesting effects. We find by Fourier expansion of the collision integral that Coulomb collisions in a strong electric field (a) are accompanied by emission and absorption of multiple photons, and (b) give rise to generation of higher harmonics in the time dependence of the distribution function. A numerical solution of the above kinetic equation was presented in [16]. Analytical results for collisional absorption basing on this equation were given in [18].

In subsequent papers [17,20], we have given a generalization using the so-called V^s-approximation. It reads for the gauge-invariant Fourier transform

$$\tilde{\Sigma}_a^{\gtrless}(\mathbf{k};t,t') = i\hbar \int \frac{\mathrm{d}^3 q}{(2\pi\hbar)^3} \tilde{g}_a^{\gtrless}(\mathbf{k}-\mathbf{q};t,t')\, \tilde{V}_{aa}^{s\gtrless}(\mathbf{q};t,t')\,, \tag{16}$$

and after inserting this expression into (8) we have the starting point to calculate the properties of interest. The details of the calculation are shown in [20]. The V^s-approximation corresponds to a dynamically screened Born approximation and is therefore applicable to weakly coupled laser plasmas. Below we will give a generalization to the case of strong coupling [22].

3 Balance Equations

In the following we want to consider collisional absorption by the dense plasma. It is therefore obvious to start from the balance equation for the energy and the electrical current resulting from of the kinetic equation. Mean velocity and mean kinetic energy are defined in terms of the gauge invariant function $\tilde{f}_a$ by

$$n_a \mathbf{v}_a(t) = \int \frac{\mathrm{d}^3 k}{(2\pi\hbar)^3} \frac{\mathbf{k}}{m_a} \tilde{f}_a(\mathbf{k},t) \tag{17}$$

$$W_a^{\mathrm{kin}}(t) = \int \frac{\mathrm{d}^3 k}{(2\pi\hbar)^3} \frac{\mathbf{k}^2}{2m_a} \tilde{f}_a(\mathbf{k},t)\,. \tag{18}$$

Using the kinetic equation (4) and relation (7), we get

$$n_a \frac{\partial}{\partial t}\mathbf{v}_a(t) - n_a \frac{e_a}{m_a}\mathbf{E}(t) = \int \frac{\mathrm{d}^3 p}{(2\pi\hbar)^3} \frac{\mathbf{p}}{m_a} I_a(\mathbf{p},t)\,, \tag{19}$$

and

$$\frac{\mathrm{d}}{\mathrm{d}t} W_a^{\mathrm{kin}} - \mathbf{j}_a(t)\cdot\mathbf{E} = \int \frac{\mathrm{d}^3 p}{(2\pi\hbar)^3} \frac{(\mathbf{p}-e_a\mathbf{A}(t))^2}{2m_a} I_a(\mathbf{p},t)\,, \tag{20}$$

where $\mathbf{j}_a = e_e n_a \mathbf{v}_a$ is the electrical current density of species a.

For our purposes it is convenient to express the collision integral not in terms of the self energies but in terms of the four-point function L_{ab} which is connected to the two-particle Green's function by $L_{ab} = g_{ab} - g_a g_b$. We get

$$I_a(\mathbf{p},t) = \sum_b \frac{1}{V} \int \frac{\mathrm{d}^3 q\, \mathrm{d}^3 p_1\, \mathrm{d}^3 p_2}{(2\pi\hbar)^6} [\delta(\mathbf{p}-\mathbf{p}_1-\mathbf{q}) - \delta(\mathbf{p}-\mathbf{p}_1)]$$
$$\times V_{ab}(\mathbf{q}) L_{ab}^{<}(\mathbf{p}_1,\mathbf{p}_1+\mathbf{q},t;\mathbf{p}_2,\mathbf{p}_2-\mathbf{q},t)_{|c}\,, \tag{21}$$

where we have $L^{<}_{ab}(11't;22't')_{|c} = L^{<}_{ab}(11't;22't') - \delta_{ab}L^{0<}_{aa}(11't,22't')$ with $L^{<}_{ab}$ being the correlation function of density fluctuations

$$(i\hbar)L^{<}_{ab}(11't,22't') = \langle \delta\hat{\rho}_b(22't')\,\delta\hat{\rho}_a(11't)\rangle \ , \tag{22}$$

$\delta\hat{\rho}_a(11't) = \Psi^{+}_a(1',t)\Psi_a(1,t) - \langle \Psi^{+}_a(1',t)\Psi_a(1,t)\rangle$, and with $L^{0<}_{aa}(11't,22't') = -i\hbar\, g^{<}_a(1t,2't')\, g^{>}_a(2t',1't)$.

Now the general collision integral I_a is inserted into the balance equations (19) and (20). The right hand side of the energy balance equation (20) can be shown to consist of one part describing the change of potential energy of species a and another describing the energy transfer with the other species. Therefore we have

$$\frac{\mathrm{d}}{\mathrm{d}t}W^{\mathrm{kin}}_a + \frac{\mathrm{d}}{\mathrm{d}t}W^{\mathrm{pot}}_a = \mathbf{j}_a(t)\cdot\mathbf{E} + \sum_b Z_{ab}\,. \tag{23}$$

The total energy balance in the system is [20]

$$\frac{\mathrm{d}}{\mathrm{d}t}\sum_a \left(W^{\mathrm{kin}}_a + W^{\mathrm{pot}}_a\right) = \mathbf{j}\cdot\mathbf{E} \tag{24}$$

with $\mathbf{j} = \sum_a \mathbf{j}_a$. Thus the change of total energy of the plasma equals the energy loss of the electromagnetic field due to Poynting's theorem.

For the calculation of this collisional absorption, one has to calculate the electric current in the system which follows from the balance equation for the mean velocities (19). We find

$$\frac{\mathrm{d}}{\mathrm{d}t}\mathbf{j}_a(t) - n_a\frac{e_a^2}{m_a}\,\mathbf{E}(t) = \sum_{b\neq a}\int \frac{\mathrm{d}^3q}{(2\pi\hbar)^3}\,\frac{e_a\mathbf{q}}{m_a}\,V_{ab}(q)\,L^{<}_{ba}(\mathbf{q};t,t) \tag{25}$$

with

$$L^{<}_{ab}(\mathbf{q};tt') = \frac{1}{V}\int \frac{\mathrm{d}^3p_1\mathrm{d}^3p_2}{(2\pi\hbar)^6}\,L^{<}_{ab}(\mathbf{p}_1,\mathbf{p}_1-\mathbf{q},t;\mathbf{p}_2,\mathbf{p}_2+\mathbf{q},t')\,. \tag{26}$$

The two-time correlation function L_{ab} of the density fluctuations is determined by the following general equations of motion (on the so-called Keldysh contour [28,29], for brevity all arguments are suppressed)

$$L_{ab} = \Pi_{ab} + \sum_{c,d}\Pi_{ac}\,V_{cd}\,L_{db}\,. \tag{27}$$

In a plasma in a strong laser field, the coupling between species with different charges can be considered to be weak, whereas the coupling between particles with equal charges in the subsystem is not affected by the field. Then the so-called polarization functions Π_{ab} can adopted to be diagonal, $\Pi_{ab} = \delta_{ab}\Pi_a$, and an approximation in lowest order of V_{ie} is appropriate. We find

$$L_{ei}^{\gtrless}(\mathbf{q};t,t') = \int \mathrm{d}\bar{t}\,\Big[\mathcal{L}_{ee}^{\gtrless}(\mathbf{q};t,\bar{t})\,V_{ei}(q)\,\mathcal{L}_{ii}^{A}(\mathbf{q};\bar{t},t') + \mathcal{L}_{ee}^{R}(\mathbf{q};t,\bar{t})\,V_{ei}(q)\,\mathcal{L}_{ii}^{\gtrless}(\mathbf{q};\bar{t},t')\Big]\,. \tag{28}$$

Here the two auxiliary functions $\mathcal{L}_{ee}$ and $\mathcal{L}_{ii}$ are defined by

$$\mathcal{L}_{ee} = \Pi_e + \Pi_e\,V_{ee}\,\mathcal{L}_{ee}\,, \quad \mathcal{L}_{ii} = \Pi_i + \Pi_i\,V_{ii}\,\mathcal{L}_{ii}\,. \tag{29}$$

Consequently, the functions $\mathcal{L}_{aa}^{R/A}$ and $\mathcal{L}_{aa}^{\gtrless}$ are density response functions and correlation functions of density fluctuations, respectively, of the two subsystems.

For the electron current there follows that

$$\frac{\mathrm{d}}{\mathrm{d}t}\mathbf{j}_e(t) - n_e\frac{e_e^2}{m_e}\,\mathbf{E}(t) = \mathrm{Re}\int \frac{\mathrm{d}^3q}{(2\pi\hbar)^3}\,\frac{e_e\mathbf{q}}{m_e\hbar}\,V_{ei}(q)\,2\pi i\int_{t_0}^{t}\mathrm{d}\bar{t}\Big[\mathcal{S}_{ee}(\mathbf{q};t,\bar{t}) \times V_{ei}(q)\mathcal{L}_{ii}^{A}(\mathbf{q};\bar{t},t) + \mathcal{L}_{ee}^{R}(\mathbf{q};t,\bar{t})\,V_{ei}(q)\,\mathcal{S}_{ii}(\mathbf{q};\bar{t},t)\Big]\,, \tag{30}$$

where we introduced the dynamical structure factor

$$2\pi\mathcal{S}_{aa}(\mathbf{q};t,\bar{t}) = \frac{i\hbar}{2}\Big[\mathcal{L}_{aa}^{>}(\mathbf{q};t,\bar{t}) + \mathcal{L}_{aa}^{<}(\mathbf{q};t,\bar{t})\Big]\,. \tag{31}$$

The equation for the ion's current is similar.

4 Collisional Absorption in a Strong High-Frequency Electrical Field

The field dependence can be made explicite. One has to take into account that the functions $\mathcal{L}_{aa}$ are functionals of the electron correlation functions $g_a^{\gtrless}$ alone. Then one finds [20,22]

$$\mathcal{L}_{aa}(\mathbf{q};tt') = \exp\left\{\frac{i}{\hbar}\mathbf{q}\cdot\frac{e_a}{m_a}\int_{t'}^{t}\mathrm{d}\bar{t}\,\mathbf{A}(\bar{t})\right\}\,\mathcal{L}_{aa}(\mathbf{q};t-t')\,, \tag{32}$$

where the function $\mathcal{L}_{aa}(\mathbf{q};t-t')$ is a functional of local equilibrium functions $G_a^{\gtrless}$. The Fourier transforms of $G_a^{\gtrless}$ are given by

$$\begin{aligned} -i\hbar G_a^{<}(\mathbf{p};\omega) &= A_a(\mathbf{p};\omega)f_a(\omega,T_a) \\ i\hbar G_a^{>}(\mathbf{p};\omega) &= A_a(\mathbf{p};\omega)\left[1-f_a(\omega,T_a)\right]\,, \end{aligned} \tag{33}$$

with $f_a(\omega,T_a) \equiv 1/[\exp\{(\hbar\omega-\mu_a)/(k_BT_a)\}+1]$, and A_a being the spectral function of the strongly correlated subsystem. Chemical potential and temperatures are still time dependent, however on a slow time scale.

The electric field occurs in (32) in an exponential factor and causes thus nonlinear effects like multi-photon absorption and the occurence of higher harmonics in the current. In the following, a harmonic electric field $\mathbf{E} = \mathbf{E}_0 \cos \omega t$ will be considered. The exponential pre-factor in (32) can be expanded into a Fourier series. The current balance is given then by

$$\frac{\mathrm{d}}{\mathrm{d}t}\mathbf{j}_e(t) - n_e \frac{e_e^2}{m_e} \mathbf{E}(t) = \mathrm{Re} \int \frac{\mathrm{d}^3 q}{(2\pi\hbar)^3} \frac{2\pi e_e \mathbf{q}}{m_e \hbar} V_{ei}^2(q)$$
$$\times \sum_{m=-\infty}^{\infty} \sum_{n=-\infty}^{\infty} (-i)^{m+1} J_n\left(\frac{\mathbf{q}\cdot\mathbf{w}_{ei}^0}{\hbar\omega}\right) J_{n-m}\left(\frac{\mathbf{q}\cdot\mathbf{w}_{ei}^0}{\hbar\omega}\right) \mathrm{e}^{im\omega t}$$
$$\times \int_{-\infty}^{\infty} \frac{\mathrm{d}\bar{\omega}}{2\pi} \left[\mathcal{S}_{ee}(\mathbf{q};\bar{\omega}-n\omega)\,\mathcal{L}_{ii}^A(\mathbf{q};\bar{\omega}) + \mathcal{L}_{ee}^R(\mathbf{q};\bar{\omega}-n\omega)\,\mathcal{S}_{ii}(\mathbf{q};\bar{\omega})\right], \quad (34)$$

where J_l is the Bessel function of lth order and with $\mathbf{w}_{ei}^0 = (e_e/m_e - e_i/m_i)\mathbf{E}_0/\omega$.

The ion dynamic structure factor $\mathcal{S}_{ii}$ and the response function $\mathcal{L}_{ii}$ are localized in the low-frequency region, i.e., for a high-frequency electric field, $\bar{\omega}$ can be neglected in comparison with $n\omega$. In this case, the first term in the brackets vanishes because $\int \mathrm{d}\bar{\omega}\mathcal{L}_{ii}^A(\mathbf{q};\bar{\omega}) = 0$, and for the current follows

$$\mathbf{j}_e(t) - \int_{-\infty}^{t} \mathrm{d}\bar{t}\, \frac{n_e e_e^2}{m_e} \mathbf{E}(\bar{t}) = \mathrm{Re} \int \frac{\mathrm{d}^3 q}{(2\pi\hbar)^3} \sum_m \sum_n \frac{e_e}{m_e} \frac{\mathbf{q}}{m\hbar\omega} V_{ei}^2(q)$$
$$\times (-i)^{m+2} J_n\left(\frac{\mathbf{q}\cdot\mathbf{w}_{ei}^0}{\hbar\omega}\right) J_{n-m}\left(\frac{\mathbf{q}\cdot\mathbf{w}_{ei}^0}{\hbar\omega}\right) \mathcal{L}_{ee}^R(\mathbf{q};-n\omega)\, n_i\, \mathcal{S}_{ii}(\mathbf{q})\, \mathrm{e}^{im\omega t} \quad (35)$$

with the static structure factor $\mathcal{S}_{ii}(\mathbf{q})$ [30,33] defined by

$$\mathcal{S}_{ii}(\mathbf{q}) \equiv \frac{1}{n_i} \int_{-\infty}^{\infty} \mathrm{d}\bar{\omega}\, \mathcal{S}_{ii}(\mathbf{q},\bar{\omega})\,. \quad (36)$$

The Fourier coefficients of the current, $\mathbf{j}(t) = \sum_{m=-\infty}^{\infty} \mathbf{j}_m(\omega)\,\mathrm{e}^{-im\omega t}$ with $j_m = j_{-m}^*$, can be identified easily from (35). Only the odd harmonics are allowed due to the symmetry of the interaction, which is characterized by $\mathrm{Re}\,\mathcal{L}_{ee}^R(\mathbf{q},\omega) = \mathrm{Re}\,\mathcal{L}_{ee}^R(-\mathbf{q},-\omega)$ and $\mathrm{Im}\,\mathcal{L}_{ee}^R(\mathbf{q},\omega) = -\mathrm{Im}\,\mathcal{L}_{ee}^R(-\mathbf{q},-\omega)$. In particular, we get for the real parts ($l = 0, 1, 2...$)

$$\mathrm{Re}\,\mathbf{j}_{2l+1}(\omega) = \frac{(-1)^l}{(2l+1)} \int \frac{\mathrm{d}^3 q}{(2\pi\hbar)^3} \frac{e_e}{m_e\hbar\omega} \mathbf{q}\, V_{ei}^2(q)\, n_i\, \mathcal{S}_{ii}(\mathbf{q}) \sum_{n=0}^{\infty} J_n\left(\tfrac{\mathbf{q}\cdot\mathbf{v}_e^0}{\hbar\omega}\right)$$
$$\times \left[J_{n-(2l+1)}\left(\frac{\mathbf{q}\cdot\mathbf{w}_{ei}^0}{\hbar\omega}\right) + J_{n+(2l+1)}\left(\frac{\mathbf{q}\cdot\mathbf{w}_{ei}^0}{\hbar\omega}\right)\right] \mathrm{Im}\,\mathcal{L}_{ee}^R(\mathbf{q};-n\omega)\,, \quad (37)$$

whereas the imaginary parts are given by

$$\mathrm{Im}\,\mathbf{j}_{2l+1}(\omega) = \delta_{l,0}\frac{n_e e_e^2}{m_e\omega}\frac{\mathbf{E}_0}{2} + \frac{(-1)^l}{(2l+1)}\int \frac{\mathrm{d}^3q}{(2\pi\hbar)^3}\frac{e_e}{m_e\hbar\omega}\,\mathbf{q}\,V_{ei}^2(q)\,n_i\,\mathcal{S}_{ii}(\mathbf{q})$$
$$\times\left\{J_0\left(\frac{\mathbf{q}\cdot\mathbf{w}_{ei}^0}{\hbar\omega}\right)J_{-(2l+1)}\left(\frac{\mathbf{q}\cdot\mathbf{w}_{ei}^0}{\hbar\omega}\right)\mathrm{Re}\,\mathcal{L}_{ee}^R(\mathbf{q};0)\right.$$
$$+\sum_{n=1}^{\infty}J_n\left(\frac{\mathbf{q}\cdot\mathbf{w}_{ei}^0}{\hbar\omega}\right)$$
$$\times\left[J_{n-(2l+1)}\left(\frac{\mathbf{q}\cdot\mathbf{w}_{ei}^0}{\hbar\omega}\right) - J_{n+(2l+1)}\left(\frac{\mathbf{q}\cdot\mathbf{w}_{ei}^0}{\hbar\omega}\right)\right]$$
$$\left.\times\mathrm{Re}\,\mathcal{L}_{ee}^R(\mathbf{q};-n\omega)\right\}\,. \tag{38}$$

The energy dissipation in an electrical field $\mathbf{E} = \mathbf{E}_0\cos\omega t$ is then given by

$$\mathbf{j}(t)\cdot\mathbf{E}(t) = \mathbf{E}_0\cdot\left\{\mathrm{Re}\,\mathbf{j}_1(\omega) + \sum_{l=1}^{\infty}\left[\mathrm{Re}\Big(\mathbf{j}_{2l+1}(\omega)+\mathbf{j}_{2l-1}(\omega)\Big)\cos(2l\omega t)\right.\right.$$
$$\left.\left.+\mathrm{Im}\Big(\mathbf{j}_{2l+1}(\omega)+\mathbf{j}_{2l-1}(\omega)\Big)\sin(2l\omega t)\right]\right\}, \tag{39}$$

containing beside the constant term even harmonics only.

The dissipation of energy averaged over one oscillation cycle is given by

$$\langle\mathbf{j}\cdot\mathbf{E}\rangle \equiv \frac{1}{T}\int_{t-T}^{t}\mathrm{d}t'\,\mathbf{j}(t')\cdot\mathbf{E}(t') = E_0\cdot\mathrm{Re}\,\mathbf{j}_1(\omega)\,.$$

This averaged quantity will be considered below in different approximations. Often also the electron–ion collision frequency ν_{ei} (more precisely it is $\mathrm{Re}\nu_{ei}(\omega)$) is discussed which is defined for the high-frequency case by (ω_p – plasma frequency)

$$\nu_\mathrm{ei} = \frac{\omega^2}{\omega_\mathrm{p}^2}\frac{\langle\,\mathbf{j}\cdot\mathbf{E}\,\rangle}{\langle\,\epsilon_0\mathbf{E}^2\,\rangle}. \tag{40}$$

5 Weakly Coupled Plasma

Equation (35) is a generalization of the theory developed in [20]. Approximating $\mathcal{S}_{ii}(\mathbf{q}) \approx 1$ and using $\mathcal{L}_{ee}^R$ in random phase approximation (RPA), one gets the results of Sect. IV in that former paper. The dissipated energy is given in this approximation by

$$\langle\mathbf{j}\cdot\mathbf{E}\rangle = n_i e_i^2\int\frac{\mathrm{d}^3q}{(2\pi\hbar)^3}\,V(q)\sum_{n=1} n\omega\,J_n^2\left(\frac{\mathbf{q}\cdot\mathbf{v}_e^0}{\hbar\omega}\right)2\mathrm{Im}\,\frac{1}{\varepsilon_\mathrm{RPA}(\mathbf{q};-n\omega)}\,. \tag{41}$$

This result has a similar form as that of the nonlinear Dawson–Oberman model [3]. We want to stress, however, that in the above formula, the dielectric function is given by the quantum Lindhard form, whereas the dielectric theory of Decker et al. leads to the classical Vlasov dielectric function.

Finally, using $\mathrm{Im}\,\varepsilon^{-1} = -\mathrm{Im}\,\varepsilon/|\varepsilon|^2$, we get

$$\langle \mathbf{j}\cdot\mathbf{E}\rangle = n_i e_i^2 \int \frac{\mathrm{d}^3 q}{(2\pi\hbar)^3}\, V(q) \sum_{n=1} n\omega\, J_n^2\left(\frac{\mathbf{q}\cdot\mathbf{v}_e^0}{\hbar\omega}\right) \frac{2\mathrm{Im}\,\varepsilon_{\mathrm{RPA}}(\mathbf{q};n\omega)}{|\varepsilon_{\mathrm{RPA}}(\mathbf{q};n\omega)|^2}\,. \quad (42)$$

The Lindhard dielectric function has to be calculated numerically, what can be done for arbitrary degeneracy. For a transparent discussion of the different quantum effects, however, it is advantageous to consider especially the nondegenerate case, in which some necessary integrations can be done analytically. For this case of a Maxwellian electron distribution function, we get [19]

$$\langle \mathbf{j}\cdot\mathbf{E}\rangle = \frac{8\sqrt{2\pi}Z^2 e^4 n_e n_i \sqrt{m_e}}{(4\pi\varepsilon_0)^2 (k_B T)^{3/2}}\,\omega^2 \sum_{n=1}^{\infty} n^2 \int_0^{\infty} \frac{\mathrm{d}k}{k^3}\, \frac{1}{|\varepsilon_{\mathrm{RPA}}(k,n\omega)|^2}$$

$$\times \mathrm{e}^{-\frac{n^2 m_e \omega^2}{2k_B T\, k^2}}\, \mathrm{e}^{-\frac{\hbar^2 k^2}{8 m_e k_B T}}\, \frac{\sinh \frac{n\hbar\omega}{2k_B T}}{\frac{n\hbar\omega}{2k_B T}} \int_0^1 \mathrm{d}z\, J_n^2\left(\frac{eE_0 k}{m_e\omega^2}\, z\right)\,. \quad (43)$$

In the classical limit, $\hbar \to 0$, this expression is well-known. Within the classical kinetic theory it was derived first by Klimontovich [5]. Later Decker et al. [3] got such an expression in the framework of the nonlinear Dawson–Oberman model. The classical formulae have the well-known problem of a divergency at large k which is solved by some cut-off procedures. In contrast, in our quantum approach no divergencies exist.

Quantum effects, indicated by $\hbar$, occur here at different places. The first place is one of the exponential functions in (43) describing the quantum diffraction effect at large momenta k. This exponential function ensures the convergence of the integral. The second place is the term with the sinh function which is connected with the Bose statistics of multiple photon emission and absorption. Finally, quantum effects enter also the calculation of $|\varepsilon(q,n\omega)|^2$ itself.

In the following results are shown for a hydrogen plasma which is assumed to be fully ionized.

In Fig. 1 the collision frequency in such a plasma is shown as a function of the quiver velocity. For comparison there are given curves (dashed line) following from the asymptotic formulae of Silin for the cases of small and of big quiver velocities, respectively. These formulae read [4]

$$\nu_{ei} = \frac{4}{3} \frac{\sqrt{2\pi} e^4 Z^2 n_i}{(4\pi\varepsilon_0)^2 m_e^2 v_{\rm th}^3} \ln \frac{k_{\max}}{k_{\min}}, \quad v_0 \ll v_{\rm th}, \tag{44}$$

$$\frac{\nu_{ei}}{\omega_p} = \frac{Z}{\pi^2} \frac{1}{n\lambda_D^3} \left(\frac{v_{\rm th}}{v_0}\right)^3 \left[\ln\left(\frac{v_0}{2v_{\rm th}}\right) + 1\right] \ln \frac{k_{\max}}{k_{\min}}, \quad v_0 \gg v_{\rm th}, \tag{45}$$

with λ_D being the Debye length. In the high-frequency field under consideration, $k_{\min}$ is given as usual by $k_{\min} = \omega_0/v_{\rm th}$. For the upper cut-off we take here for the comparison the classical one, $k_{\max} = 4\pi\varepsilon_0 k_B T/(Ze^2)$.

Further, the classical expression of Decker et al. is evaluated (dash-dotted line). Our quantum expression, (40) with (42), was evaluated with the quantum Lindhard dielectric function fully dynamical and, for comparison, also with static screening in the denominator in (42) leading to lower values of ν_{ei}. These results are given in Fig. 1 with solid lines. In the given logarithmic scale, however, there is almost no difference between these two cases to be observed. The qualitative behaviour of the results from the classical dielectric theory of Decker et al. and from our quantum approach is very similar. The collision frequency is nearly constant for small field strengths up to $v_0/v_{\rm th} = 1$ and decreases then rapidly for higher fields. This is in agreement with the asymptotic formulae of Silin, too. There are, however, quantitative differences. These can be attributed to the use of Coulomb logarithms in the classical approaches which correspond to cutting procedures in the integral over momentum. The difference to the classical results of Decker et

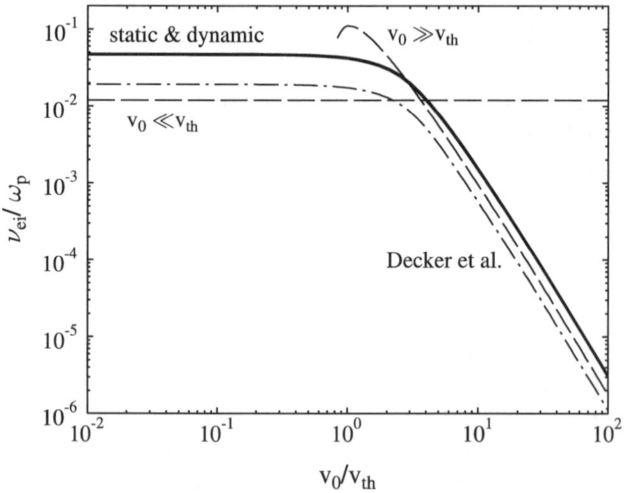

Fig. 1. Electron–ion collision frequency as a function of the quiver velocity $v_0 = eE/\omega m_e$ for a hydrogen plasma in a laser field ($Z = 1; n_e = 10^{22}\text{cm}^{-3}; T = 3 \cdot 10^5$ K; $\omega/\omega_p = 5$). For comparison, results of Decker et al. (*dash-dotted line*) and of the asymptotic formulae (44,45) by Silin (*dashed line*) are given

al. is smaller for the case of a higher temperature (lower coupling parameter $\Gamma = (e^2/4\pi\varepsilon_0)/dk_BT$ with $d = (4\pi n_i/3)^{-1/3}$) [20].

The dependence of the collision frequency on the coupling parameter will be considered now more in detail. In Fig. 2, the collision frequency is shown as a function of coupling parameter Γ for a small quiver velocity, i.e. a small field strength. Results of the evaluation of (40,41) are given by the upper solid line. The static screening results are the lower solid curve. Again the asymptotic formula (44) of Silin is given, and the classical expression of Decker et al. was evaluated. Furthermore, numerical results of Cauble and Rozmus [12] are plotted. They considered small field strengths and used a memory function kinetic approach which allows to consider plasmas up to strong coupling. The points in Fig. 2 correspond to their so-called Debye–Hückel mean field approximation, cf. [12]. Finally, we compare in Fig. 2 also with numerical simulation results of Pfalzner and Gibbon [8]. They applied a tree code method to classical molecular dynamics simulations using a soft Coulomb potential.

According to Fig. 2 the collision frequency increases with increasing coupling Γ. For small Γ, the dielectric theory and our theory give almost the same results. The values of the asymptotic formula of Silin are slightly bigger. With increasing Γ this asymptotic formula as well as the dielectric approach reach a maximum around $\Gamma \sim 0.2$ and sharply drop down afterwards. This behaviour is governed by the Coulomb logarithm used in these approaches. It results from a cut-off procedure at large momenta k. Such a cut-off, inherent in many classical approaches, is avoided in our approach because the k integration is automatically convergent, cf. the second exponential function

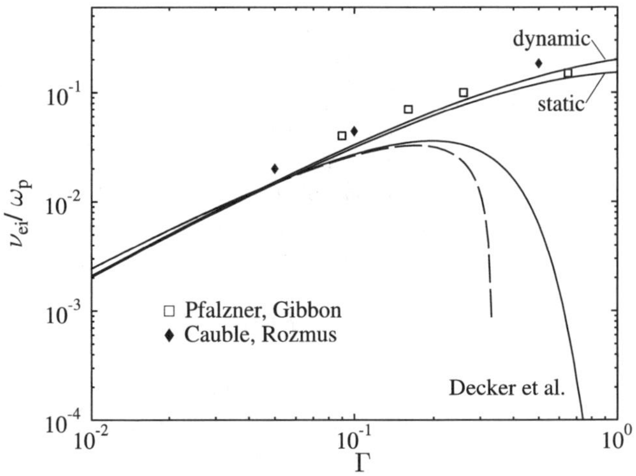

Fig. 2. Electron–ion collision frequency as a function of the coupling parameter Γ for a hydrogen plasma in a laser field ($Z = 1; v_0/v_{\rm th} = 0.2$; $n_e = 10^{22}\,{\rm cm}^{-3}$; $\omega/\omega_p = 3$). Comparison is given with the theory of Decker et al. and with the asymptotic formula (44) of Silin (*dashed line*)

in (43). Therefore the range of applicability of our approach is extended to higher values of Γ.

The agreement with the results of Cauble et al. and of Pfalzner et al. is rather good in the considered range of Γ with the values of the present theory being slightly smaller. One has to take into account, however, that our approximation is a weak coupling theory whereas the approaches we compare with include the correlations in higher approximations. A generalization will be considered in the next section. We want to mention that our results do not depent solely on Γ, but depend on temperature and on density (as well as the results of Cauble et al. do because of the usage of a modified potential taking into account short-range quantum effects).

Now the consequences of the quantum approach in contrast to the classical dielectric theory will be investigated. We consider such parameters that the plasma can be assumed to be nondegenerate, i.e. (43) can be used. In this case a direct comparison with the classical dielectric theory of Decker et al. is possible. Quantum effects indicated by $\hbar$ occur in (43) at two places. One is the quantum diffraction effect ensuring the convergence of the integral at big k (cf. the second exponential function). The other is the factor $\sinh x/x$ with $x = n\hbar\omega/(2k_B T)$. The classical theory uses, instead, a cut-off $k_{\mathrm{max}} = m v_{\mathrm{th}}^2 4\pi\varepsilon_0/Ze^2$, and the sinh factor is missing.

An important feature of the expressions (41)–(43) is the sum over n which can be interpreted as a sum over the different multiphoton processes, i.e. the emission and absorption of energies $n\hbar\omega$. The different contributions ν_n in the sum $\nu_{ei} = \sum_n \nu_n$ are dependent on the field strength. It is obvious that with increasing field the number of terms contributing essentially to the sum is also increasing. The following two figures, Fig. 3 and Fig. 4, showing ν_n vs. n (full solution of (43) – solid line) for two different field strengths illustrate this issue. Moreover, we compare with the classical dielectric theory (dotted line) and the case in which the factor $\sinh x/x$ in (43) is set to unity (dashed line). One observes that the differences between these three cases grow with increasing photon number n. The reason for the faster decreasing contributions in the classical approach is the hard cut-off in the k integration while the maximum of the integrand is shifted to higher k due to the exponential factor $\exp[-(n^2\omega^2 m_e)/(2k_B T k^2)]$. Thus the relativ error of the ν_n increases with n.

The other quantum effect is connected with the $\sinh x$ factor which behaves for large $x(n)$ as $\mathrm{e}^x/2x$. Therefore this factor becomes more important for large n, the processes involving large numbers of photons are enhanced. This can be seen in Fig. 4 which considers the case $v_0/v_{\mathrm{th}} = 10$. The solid curve corresponding to the full solution extends to much higher n values than that curve which results from a neglection of the sinh term. An interesting feature is the plateau-like behavior up to $n \sim 350$ with the subsequent sharp drop down. We can conclude at this point, that especially in the strong field case where multiphoton processes play an increasing role, it is important to

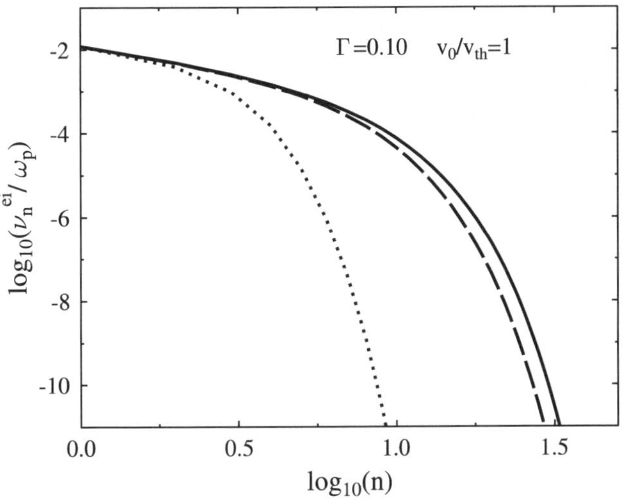

Fig. 3. Contributions ν_n vs. photon number n in a hydrogen plasma ($n_e = 10^{22}\,\mathrm{cm}^{-3}$; $\omega/\omega_p = 5$; $\Gamma = 0.1$) for $v_0/v_{\mathrm{th}} = 1.0$. Present approach (*solid line*), sinh term neglected (*dashed line*), classical dielectric theory (*dotted line*)

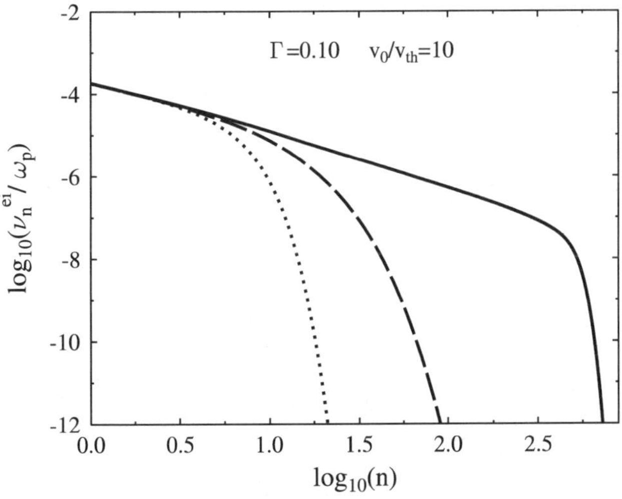

Fig. 4. Contributions ν_n vs. photon number n in a hydrogen plasma ($n_e = 10^{22}\,\mathrm{cm}^{-3}$; $\omega/\omega_p = 5$; $\Gamma = 0.1$) for $v_0/v_{\mathrm{th}} = 10$. Present approach (*solid line*), sinh term neglected (*dashed line*); classical dielectric theory (*dotted line*)

treat the problem on the basis of quantum mechanics. This problem was also discussed in [21] basing on an asymptotic solution of expression (43) for strong fields.

6 Strongly Coupled Plasma

Now we give a generalization of the above results for the case of a stronger coupling. Starting now from (37) we get

$$\langle \mathbf{j}\cdot\mathbf{E}\rangle = 2\int \frac{\mathrm{d}^3q}{(2\pi\hbar)^3}\sum_{m=1}^{\infty} m\omega\, J_m^2(z)\, V_{ei}^2(q) n_i\, \mathcal{S}_{ii}(\mathbf{q})\, \mathrm{Im}\,\mathcal{L}_{ee}^R(\mathbf{q}; -m\omega)\,. \tag{46}$$

With the definition of a dielectric function of the electron subsystem,

$$\mathrm{Im}\,\varepsilon_{ee}^{-1}(\mathbf{q}; -m\omega) = e^2 V(q) \mathrm{Im}\,\mathcal{L}_{ee}(\mathbf{q}; -m\omega)\,, \tag{47}$$

we arrive at

$$\langle \mathbf{j}\cdot\mathbf{E}\rangle = 2n_i e_i^2\int \frac{\mathrm{d}^3q}{(2\pi\hbar)^3}\sum_{m=1}^{\infty} m\omega\, J_m^2(z)\, V(q)\, \mathcal{S}_{ii}(\mathbf{q})\, \mathrm{Im}\,\varepsilon_{ee}^{-1}(\mathbf{q}; -m\omega)\,. \tag{48}$$

There are two generalizations. The first one is the occurence of the static ion–ion structure factor

$$\mathcal{S}_{ii}(\mathbf{q}) = 1 + n_i \int \mathrm{d}^3r\, [g_{ii}(\mathbf{r}) - 1]\, \mathrm{e}^{-\frac{i}{\hbar}\mathbf{q}\cdot\mathbf{r}}\,, \tag{49}$$

where $g_{ii}(r)$ is the pair correlation function. This enables us to consider a correlated ion subsystem.

Further, the function $\mathcal{L}_{ee}^R$ is the exact density response function of the electron subsystem, not only the RPA-function as in [20]. The electron-electron interactions can be included therefore on a higher level. Appropriate approximations can be expressed via local field corrections (LFC), see e.g. [31,32],

$$\mathcal{L}_{ee}^R(\mathbf{q},\omega) = \frac{\chi_e^0(\mathbf{q},\omega)}{1 - V_{ee}(q) G(q) \chi_e^0(\mathbf{q},\omega)} \tag{50}$$

with χ_e^0 being the usual free-electron Lindhard polarizability, and the local field factor G given in our notation by

$$G(\mathbf{q}) = -n_e^{-1}\int \frac{\mathrm{d}^3q'}{(2\pi\hbar)^3}\,\frac{\mathbf{q}\cdot\mathbf{q}'}{q'^2}\,[S_{ee}(\mathbf{q}-\mathbf{q}') - 1]\,.$$

Figure 5 shows the influence of the LFC (the structur factor is calculated in hypernetted chain (HNC) approximation). The collision frequency is given as a function of the coupling parameter Γ. For weak and moderate electric fields, ($v_0/v_{\mathrm{th}} = 0.2$ and 3), there occur deviations in the region $\Gamma > 1$ which increase with increasing coupling. For rather strong fields, the LFC has no influence up to a coupling of about $\Gamma = 10$. Furthermore, one can see that for strong coupling the influence of the field strength becomes smaller.

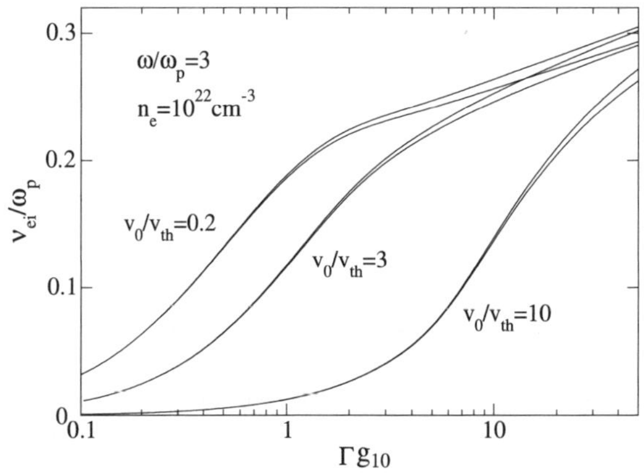

Fig. 5. Electron–ion collision frequency as a function of the coupling parameter Γ for different values of the quiver velocity. LFC in accordance with Ichimaru and Utsumi [32] (*solid*), without LFC (*dashed*). S_{ii} is calculated in HNC approximation. The quiver velocity is defined as $v_0 = eE_0/m_e\omega$, $v_{\rm th} = (k_{\rm B}T_e/m_e)^{1/2}$

Let us consider now the influence of the ion–ion correlations. Effects of such correlations via static structure factors were discussed already by Dawson and Oberman [2] starting from the linearized Vlasov equation and recently by Hazak et al. [23] using the quantum BBGKY hierarchy.

In Fig. 6 the collision frequency is shown using different approximations for the static ion–ion structur factor S_{ii} (here the LFC is calculated as in [32]). The structur factor was calculated numerically using a HNC code. For comparison we used a semi-analytical formula of Baus and Hansen [33] which

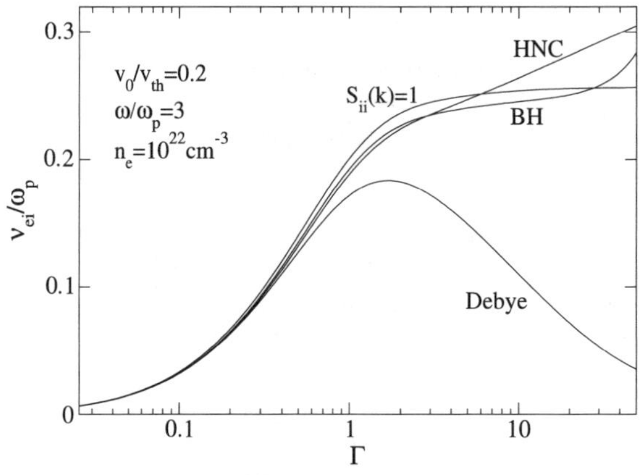

Fig. 6. Electron–ion collision frequency as a function of the coupling parameter Γ. Ion structur factor in different approaches: HNC, Debye, BH (Baus–Hansen formula)

reads

$$S(k) = \frac{1}{1 - c(k)} = \frac{1}{1 + (k_D^2/k^2)h_p(kr_0)}, \tag{51}$$

where $k_D = (4\pi e^2 n/k_B T)^{1/2}$ is the Debye wave number and

$$h_p(y) = d_p(y)(2p - 1)!!/y^p ,$$

with $d_p(y)$ being the spherical Bessel function of order p. Here r_0 is an effective radius which can serve as a fit parameter, in a rough approximation it is the mean ion-spere radius. For the details we refer to [33]. In our calculations we used the formula for $p = 1$. For $r_0 \to 0$, we get the Debye–Hückel result.

Inclusion of the structur factor decreases the collision frequency for small and moderate coupling up to a value of $\Gamma \approx 5$. For high values of the coupling parameter there is a strong increase of the collision frequency. This increase is in the HNC calculation even stronger than in the semi-analytical formula of Baus and Hansen. As expected, the Debye approximation can be applied only for weak coupling.

Acknowledgements

This work was supported by the Deutsche Forschungsgemeinschaft (Schwerpunkt "Wechselwirkung intensiver Laserfelder mit Materie" and Sonderforschungsbereich 198). The authors would like to thank V. Bezkrovniy for providing his HNC code.

References

1. M.D. Perry and G. Mourou: Science **264**, 917 (1994)
2. C. Oberman, A. Ron, and J. Dawson: Phys. Fluids **5**, 1514 (1962); J.M. Dawson and C. Oberman: Phys. Fluids **6**, 394 (1963)
3. C.D. Decker, W.B. Mori, J.M. Dawson, and T. Katsouleas: Phys. Plasmas **1**, 4043 (1994)
4. V.P. Silin: Zh. Eksp. Teor. Fiz. **47**, 2254 (1964), [JETP **20**, 1510 (1965)]
5. Yu.L. Klimontovich: *Kinetic Theory of Nonideal Gases and Nonideal Plasmas* (russ.), (Nauka, Moscow 1975) [Engl. transl.: Pergamon Press, Oxford 1982]
6. P. Mulser and A. Saemann: Contrib. Plasma Phys. **37**, 211 (1997); P. Mulser, F. Cornolti, E. Bèsuelle and R. Schneider: Phys. Rev. E **63**, 63 (2000)
7. A.B. Langdon: Phys. Rev. Lett. **44**, 575 (1980)
8. S. Pfalzner and P. Gibbon: Phys. Rev. **E 57**, 4698 (1998)
9. S. Rand: Phys. Rev. **136**, B 231 (1964)
10. L. Schlessinger and J. Wright: Phys. Rev. **A20**, 1934 (1979)
11. V.P. Silin and S.A. Uryupin: Zh. Eksp. Teor. Fiz. **81**, 910 (1981), [JETP **54**, 485 (1981)]

12. R. Cauble and W. Rozmus: Phys. Fluids **28**, 3387 (1985)
13. H. Reinholz, R. Redmer, G. Röpke, and A. Wierling: Phys. Rev. E **62**, 5648 (2000)
14. D. Kremp, Th. Bornath, M. Bonitz and M. Schlanges: Phys. Rev. E **60**, 4725 (1999)
15. D. Kremp, Th. Bornath, P. Hilse, H. Haberland, M. Schlanges, and M. Bonitz: Contrib. Plasma Phys. **41**, 259 (2001)
16. H. Haberland, M. Bonitz, and D. Kremp: Phys. Rev. E **64**, 26405 (2001)
17. M. Bonitz, Th. Bornath, D. Kremp, M. Schlanges and W.D. Kraeft: Contrib. Plasma Phys. **39**, 329 (1999)
18. Th. Bornath, M. Schlanges, P. Hilse, D. Kremp and M. Bonitz: Laser and Particle Beams **18**, 535 (2000)
19. M. Schlanges, Th. Bornath, D. Kremp, M. Bonitz, P. Hilse: J. Phys. IV France **10**, Pr5–323 (2000)
20. Th. Bornath, M. Schlanges, P. Hilse, and D. Kremp: Phys. Rev. E **64**, 26414 (2001)
21. H.-J. Kull and L. Plagne: Phys. of Plasmas **8**, 5244 (2001)
22. Th. Bornath, M. Schlanges, P. Hilse, and D. Kremp: J. Phys. A: Math. Gen. **36**, 5941 (2003);
M. Schlanges, P. Hilse, Th. Bornath, and D. Kremp in: *Progress in Nonequilibrium Green's Functions II* ed. by M. Bonitz and D. Semkat, (World Scientific, Singapore 2003), pp. 50–65
23. G. Hazak, N. Metzler, M. Klapisch, and J. Gardner: Physics of Plasmas **9**, 345 (2002)
24. S. Fujita: *Introduction to Nonequilibrium Quantum Statistical Mechanics* (W.B. Saunders Comp., Philadelphia, London, 1966)
25. H. Haug and A.P. Jauho, *Quantum Kinetics in Transport and Optics of Semiconductors* (Springer-Verlag, Heidelberg, New York, 1996)
26. K. Morawetz: Phys. Rev. E **50**, 4625 (1994)
27. P. Lipavský, V. Špička, and B. Velický: Phys. Rev. B **34**, 6933 (1986)
28. L.V. Keldysh: Zh. Eksp. Teor. Fiz. **47**, 1515 (1964) [Sov. Phys.–JETP **20**, 1018 (1965)]
29. D.F. DuBois in: *Lectures in Theoretical Physics*, edited by W.E. Brittin (Gordon and Breach, N.Y. 1967), pp. 469–619
30. J.-P. Hansen and I.R. McDonald: *Theory of Simple Liquids* (Academic Press, San Diego 1990)
31. J. Hubbbard: Proc. R. Soc. London Ser. A **243**, 336 (1957)
32. S. Ichimaru and K. Utsumi: Phys. Rev. B **24**, 7385 (1981)
33. M. Baus and J.-P. Hansen: J. Phys. C: Solid State Phys. **12**, L55 (1979)

Correlations and Equilibration in Relativistic Quantum Systems*

Wolfgang Cassing and Sascha Juchem

Abstract. In this article we study the time evolution of an interacting field theoretical system, i.e. ϕ^4-field theory in 2+1 space-time dimensions, on the basis of the Kadanoff–Baym equations for a spatially homogeneous system including the self-consistent tadpole and sunset self-energies. We find that equilibration is achieved only by inclusion of the sunset self-energy. Simultaneously, the time evolution of the scalar particle spectral function is studied for various initial states. We also compare associated solutions of the corresponding Boltzmann equation to the full Kadanoff–Baym theory. This comparison shows that a consistent inclusion of the spectral function has a significant impact on the equilibration rates only if the width of the spectral function becomes larger than 1/3 of the particle mass. Furthermore, based on these findings, the conventional transport of particles in the on-shell quasiparticle limit is extended to particles of finite life time by means of a dynamical spectral function $A(X, \mathbf{p}, M^2)$. The off-shell propagation is implemented in the Hadron-String-Dynamics (HSD) transport code and applied to the dynamics of nucleus-nucleus collisions.

1 Introduction

The many-body theory of strongly interacting particles out of equilibrium is a challenging problem since a couple of decades. Many approaches based on the Martin–Schwinger hierarchy of Green's functions [1] have been formulated [2–6] and applied to model cases. Nowadays, the dynamical description of strongly interacting systems out of equilibrium is dominantly based on on-shell transport theories and efficient numerical recipies have been set up for the solution of the coupled channel transport equations [7–13] (and Refs. therein). These transport approaches have been derived either from the Kadanoff–Baym equations [14] in [15–19] or from the hierarchy of connected equal-time Green's functions [2,20] in [3,10,21] by applying a Wigner transformation and restricting to first order in the derivatives of the phase-space variables (X, p).

However, as recognized early in these derivations [3,17], the on-shell quasiparticle limit, that invoked additionally a reduction of the $8N$-dimensional phase-space to $7N$ independent degrees of freedom, where N denotes the number of particles in the system, should not be adequate for particles of short life time and/or high collision rates. Therefore, transport formulations

* supported by GSI Darmstadt

for quasiparticles with dynamical spectral functions have been presented in the past [18] providing a formal basis for an extension of the presently applied transport models. Apart from the transport extensions mentioned above there is a further branch discussing the effects from non-local collisions terms (cf. [22–27]).

The basic questions in this field are:
i) What are the microscopic mechanisms that lead to thermalization though the underlying quantum theory is time reversible?
ii) What are the effects of off-shell transitions in case of strong coupling?
iii) Is there an adequate semiclassical limit to the full quantum theory?

In this article we will attempt to shed some light on these questions for fully relativistic self-interacting systems and review some recent progress achieved in the last years. Our work is organized as follows: In Sect. 2 we will study numerically the self-interacting scalar ϕ^4-theory in 2+1 dimensions on the basis of the Kadanoff–Baym equations for homogeneous systems and concentrate on the time evolution of correlations and the single-particle spectral function. A comparison to the respective Boltzmann limit is performed in Sect. 3. Furthermore, the derivation of a semiclassical off-shell transport theory for inhomogeneous systems is presented in Sect. 4 with special emphasis on the time evolution of spectral functions.

2 The Kadanoff–Baym Equations

We briefly recall the basic equations for Green's functions and particle self-energies as well as their symmetry properties that will be exploited throughout this work.

2.1 Preliminaries

Within the framework of the *closed-time-path* formalism [1,14] Green's functions G and self-energies Σ are given as path ordered quantities. They are defined on the time contour consisting of two branches from $(+)$ $-\infty$ to ∞ and $(-)$ from ∞ to $-\infty$. For convenience these propagators and self-energies are transformed into a 2×2 matrix representation according to their path structure [4], i.e. according to the chronological $(+)$ or antichronological $(-)$ branch for the time coordinates x_0 and y_0. Explicitly the Green's functions are given by

$$
\begin{aligned}
i\,G^{c}_{xy} = i\,G^{++}_{xy} &= \langle\, T^{c}\,\{\,\phi(x)\,\phi^{\dagger}(y)\,\}\,\rangle\,,\\
i\,G^{<}_{xy} = i\,G^{+-}_{xy} &= \eta\;\langle\,\{\,\phi^{\dagger}(y)\,\phi(x)\,\}\,\rangle\,,\\
i\,G^{>}_{xy} = i\,G^{-+}_{xy} &= \langle\,\{\,\phi(x)\,\phi^{\dagger}(y)\,\}\,\rangle\,,\\
i\,G^{a}_{xy} = i\,G^{--}_{xy} &= \langle\, T^{a}\,\{\,\phi(x)\,\phi^{\dagger}(y)\,\}\,\rangle.
\end{aligned}
\qquad (1)
$$

where the subscript $\cdot_{xy}$ denotes the dependence on the coordinate space variables x and y and $T^c\,(T^a)$ represent the (anti-)time-ordering operators. In the definition of $G^<$ the factor $\eta = +1$ stands for bosons and $\eta = -1$ for fermions. In the following we will consider a theory for scalar bosons. The full Green's functions are determined via the Dyson–Schwinger equations for path-ordered quantities, here given in 2×2 matrix representation as

$$\begin{pmatrix} G^c & G^< \\ G^> & G^a \end{pmatrix}_{xy} = \begin{pmatrix} G^c_o & G^<_o \\ G^>_o & G^a_o \end{pmatrix}_{xy} + \begin{pmatrix} G^c_o & G^<_o \\ G^>_o & G^a_o \end{pmatrix}_{xz} \odot \begin{pmatrix} \Sigma^c & -\Sigma^< \\ -\Sigma^> & \Sigma^a \end{pmatrix}_{zz'} \odot \begin{pmatrix} G^c & G^< \\ G^> & G^a \end{pmatrix}_{z'y} . \tag{2}$$

The self-energies $\Sigma^{\cdot}$ are also defined according to their time structure while the symbol '$\odot$' implies an integration over the intermediate space-time coordinates from $-\infty$ to ∞. Linear combinations of diagonal and off-diagonal matrix elements give the retarded and advanced Green's functions $G^{\mathrm{ret/adv}}$ and self-energies $\Sigma^{\mathrm{ret/adv}}$

$$\begin{aligned} G^{\mathrm{ret}}_{xy} &= G^c_{xy} - G^<_{xy} = G^>_{xy} - G^a_{xy}\,, \qquad & G^{\mathrm{adv}}_{xy} &= G^c_{xy} - G^>_{xy} = G^<_{xy} - G^a_{xy}\,, \\ \Sigma^{\mathrm{ret}}_{xy} &= \Sigma^c_{xy} - \Sigma^<_{xy} = \Sigma^>_{xy} - \Sigma^a_{xy}\,, \qquad & \Sigma^{\mathrm{adv}}_{xy} &= \Sigma^c_{xy} - \Sigma^>_{xy} = \Sigma^<_{xy} - \Sigma^a_{xy}\,. \end{aligned} \tag{3}$$

Resorting equations (2) one obtains Dyson–Schwinger equations for the retarded (advanced) Green's functions (where only the respective self-energies are involved)

$$\hat{G}^{-1}_{o,x}\; G^{\mathrm{ret/adv}}_{xy} = \delta_{xy} + \Sigma^{\mathrm{ret/adv}}_{xz} \odot G^{\mathrm{ret/adv}}_{zy}\,, \tag{4}$$

and the wellknown Kadanoff–Baym equation for the Wightman function $G^<$,

$$\hat{G}^{-1}_{o,x}\; G^<_{xy} = \Sigma^{\mathrm{ret}}_{xz} \odot G^<_{zy} + \Sigma^<_{xz} \odot G^{\mathrm{adv}}_{zy}\,. \tag{5}$$

In these equations $\hat{G}^{-1}_{o,x}$ denotes the (negative) Klein–Gordon differential operator which for bosonic field quanta of mass m is given by $\hat{G}^{-1}_{o,x} = -(\partial^\mu_x \partial^x_\mu + m^2)$. The Klein–Gordon equation is solved by the free propagators G_o as

$$\hat{G}^{-1}_{o,x} \begin{pmatrix} G^c_o & G^<_o \\ G^>_o & G^a_o \end{pmatrix}_{xy} = \delta_{xy} \begin{pmatrix} 1 & 0 \\ 0 & -1 \end{pmatrix}, \qquad \hat{G}^{-1}_{o,x}\; G^{\mathrm{ret/adv}}_{o,xy} = \delta_{xy} \tag{6}$$

with the δ-distribution $\delta_{xy} \equiv \delta^{(d+1)}(x-y)$ for $d+1$ space-time dimensions.

2.2 Equilibration within the Scalar ϕ^4-Theory

The scalar ϕ^4-theory is an example for a fully relativistic field theory of interacting scalar particles that allows to test theoretical approximations without coming to the problems of gauge invariant truncation schemes [28–31]. Its Lagrangian density is given by ($x = (t, \mathbf{x})$)

$$\mathcal{L}(x) = \frac{1}{2}\, \partial_\mu \phi(x)\, \partial^\mu \phi(x) - \frac{1}{2}\, m^2\, \phi^2(x) - \frac{\lambda}{4!}\, \phi^4(x)\,, \tag{7}$$

where m denotes the mass and λ is the coupling strength determining the interaction strength of the scalar fields. Its quantization is performed in the canonical form (in d spatial dimensions)

$$[\,\phi(x)\,,\,\partial_{y_0}\phi(y)\,]_{x_0=y_0} \;=\; i\,\delta^{(d)}(\mathbf{x}-\mathbf{y})\,. \tag{8}$$

Since for the real boson theory (7) the relation $G^>(x,y) = G^<(y,x)$ holds (1) the knowledge of the Green's functions $G^<(x,y)$ for all x, y characterizes the system completely. Nevertheless, we will give the equations for $G^<$ and $G^>$ explicitly since this is the familiar representation for general field theories [5]. Self-consistent equations of motion are obtained by a loop expansion of the two-particle irreducible (2PI) effective action Γ on the closed time path. It is given by the Legendre transform of the generating functional of the connected Green's functions W as

$$\Gamma[G] \;=\; \Gamma^o \,+\, \frac{i}{2}\,[\,ln(1-\odot_p\,G_o\odot_p\,\Sigma) \,+\, \odot_p\,G\,\odot_p\,\Sigma\,] \,+\, \Phi[G]\,. \tag{9}$$

in case of vanishing vacuum expectation value $\langle 0|\phi(x)|0\rangle = 0$ [32]. In (9) Γ^o depends only on free Green's functions and is treated as a constant while the symbols $\odot_p$ represent convolution integrals over the closed time path. The functional Φ is the sum of all closed $2PI$ diagrams built up by full propagators G; it determines the self-energies by functional variation as

$$\Sigma(x,y) \;=\; 2i\,\frac{\delta\Phi}{\delta G(y,x)}\,. \tag{10}$$

Taking into account contributions up to the 3-loop order the Φ functional reads explicitly for the problem (7)

$$i\Phi \;=\; \frac{i\lambda}{8}\int_C d^{d+1}x\;G(x,x)^2 \;-\; \frac{\lambda^2}{48}\int_C d^{d+1}x\int_C d^{d+1}y\;G(x,y)^4. \tag{11}$$

By using the stationarity condition of the action $\delta\Gamma/\delta G = 0$ and resolving the time structure of the path ordered quantities we obtain the Kadanoff–Baym equations for the time evolution of the Wightman function [5,29]:

$$\begin{aligned}
-\left[\partial_\mu^x\partial_x^\mu + m^2\right]\,G^<(x,y) &= \Sigma_{\rm tad}(x)\,G^<(x,y)\\
&\quad+\int_{t_o}^{x_o}\!dz_o\int\!d^dz\;\left[\Sigma^>(x,z)-\Sigma^<(x,z)\right]\,G^<(z,y)\\
&\quad-\int_{t_o}^{y_o}\!dz_o\int\!d^dz\;\Sigma^<(x,z)\,\left[G^>(z,y)-G^<(z,y)\right],
\end{aligned}$$

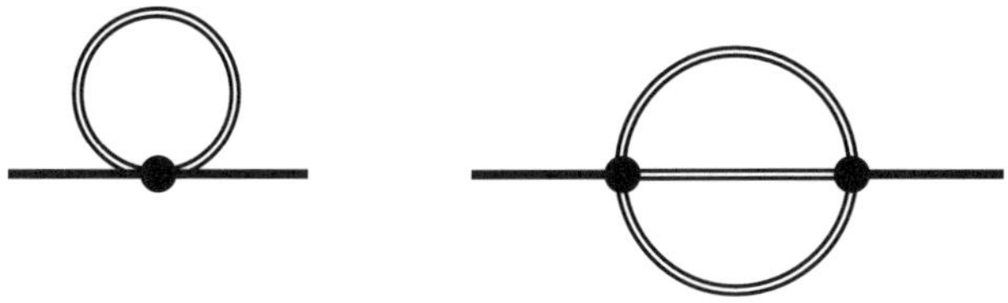

Fig. 1. Self-energies of the Kadanoff–Baym equation: tadpole self-energy (l.h.s.) and sunset self-energy (r.h.s.). Since the lines represent full Green's functions the self-energies are self-consistent (see text)

$$
\begin{aligned}
- \left[\partial_\mu^y \partial_y^\mu + m^2\right] G^<(x,y) = \Sigma_{\text{tad}}(y)\, G^<(x,y) \\
+ \int_{t_o}^{x_o} dz_o \int d^dz \; \left[G^>(x,z) - G^<(x,z)\right] \, \Sigma^<(z,y) \\
- \int_{t_o}^{y_o} dz_o \int d^dz \; G^<(x,z) \, \left[\Sigma^>(z,y) - \Sigma^<(z,y)\right],
\end{aligned}
\tag{12}
$$

where d denotes the spatial dimension of the problem ($d = 2$ in the case considered below). Within the 3-loop approximation for the effective action (i.e. the Φ functional (11)) we get two different self-energies: In leading order of the coupling constant only the tadpole diagram (l.h.s. of Fig. 1) contributes and leads to the generation of an effective mass for the field quanta. This self-energy (in coordinate space) is given by

$$
\Sigma_{\text{tad}}(x) \;=\; \frac{\lambda}{2}\, i\, G^<(x,x)\,, \tag{13}
$$

and is local in space and time. In next order in the coupling constant (i.e. λ^2) the non-local sunset self-energy (r.h.s. of Fig. 1) enters the time evolution as

$$
\Sigma^{</>}(x,y) \;=\; -\frac{\lambda^2}{6}\, G^{</>}(x,y)\, G^{</>}(x,y)\, G^{>/<}(y,x) \tag{14}
$$

$$
\longrightarrow \quad \Sigma^{</>}(x,y) \;=\; -\frac{\lambda^2}{6} \left[G^{</>}(x,y)\right]^3 . \tag{15}
$$

Thus the Kadanoff–Baym equation (16) in our case includes the influence of a mean-field on the particle propagation – generated by the tadpole diagram – as well as of scattering processes as inherent in the sunset diagram.

The Kadanoff–Baym equation describes the full quantum nonequilibrium time evolution on the two-point level for a system prepared at an initial time t_0, i.e. when higher order correlations are discarded. The causal structure of this initial value problem is obvious since the time integrations are performed over the past up to the actual time x_0 (or y_0, respectively) and do not extend to the future.

In the following we will restrict to homogeneous systems in space. To obtain a numerical solution of the Kadanoff–Baym equation, Eq. (12) is transformed to momentum space:

$$
\begin{aligned}
\partial^2_{t_1} G^<(\mathbf{p},t_1,t_2) &= -[\,\mathbf{p}^2 + m^2 + \bar{\Sigma}_{\rm tad}(t_1)\,]\, G^<(\mathbf{p},t_1,t_2) \\
&\quad - \int_{t_0}^{t_1} dt' \, \left[\Sigma^>(\mathbf{p},t_1,t') - \Sigma^<(\mathbf{p},t_1,t')\right] \, G^<(\mathbf{p},t',t_2) \\
&\quad + \int_{t_0}^{t_2} dt' \, \Sigma^<(\mathbf{p},t_1,t') \, \left[G^>(\mathbf{p},t',t_2) - G^<(\mathbf{p},t',t_2)\right] \\
&= -[\,\mathbf{p}^2 + m^2 + \bar{\Sigma}_{\rm tad}(t_1)\,]\, G^<(\mathbf{p},t_1,t_2) + I_1^<(\mathbf{p},t_1,t_2)\,,
\end{aligned}
\tag{16}
$$

where we have summarized both memory integrals into the function $I_1^<$. The equation of motion in the second time direction t_2 is given analogously. In two-time and momentum space $(\mathbf{p},t,t')$ representation the self-energies read

$$
\begin{aligned}
\bar{\Sigma}_{\rm tad}(t) &= \frac{\lambda}{2} \int \frac{d^d p}{(2\pi)^d} \; i\, G^<(\mathbf{p},t,t)\,, \\
\Sigma^<(\mathbf{p},t,t') &= -\frac{\lambda^2}{6} \int \frac{d^d q}{(2\pi)^d} \int \frac{d^d r}{(2\pi)^d} \; G^<(\mathbf{q},t,t')\, G^<(\mathbf{r},t,t')\, G^>(\mathbf{q}+\mathbf{r}-\mathbf{p},t',t) \\
&= -\frac{\lambda^2}{6} \int \frac{d^d q}{(2\pi)^d} \int \frac{d^d r}{(2\pi)^d} \; G^<(\mathbf{q},t,t')\, G^<(\mathbf{r},t,t')\, G^<(\mathbf{p}-\mathbf{q}-\mathbf{r},t,t')\,.
\end{aligned}
\tag{17}
$$

2.3 Renormalization

In 2+1 space-time dimensions both self-energies incorporated (1) are ultraviolet divergent. Since we consider particles with a finite mass no problems arise from the infrared momentum regime. For the renormalization of the divergences we only suppose that the time-dependent nonequilibrium distribution functions are decreasing for large momenta comparable to the equilibrium ones, i.e exponentially. Thus we can apply the usual finite temperature renormalization scheme. By separating the real time (equilibrium) Green's functions into vacuum ($T = 0$) and thermal parts it becomes apparent that only the pure vacuum contributions of the self-energies are divergent. For the linear divergent tadpole diagram we introduce a mass counterterm (at renormalized mass m) as

$$\delta m^2_{\rm tad} = \int \frac{d^2p}{(2\pi)^2} \frac{1}{2\omega_{\mathbf{p}}} , \qquad \omega_{\mathbf{p}} = \sqrt{\mathbf{p}^2 + m^2} , \tag{18}$$

that cancels the contribution from the momentum integration of the vacuum part of the Green's function. In case of the sunset diagram only the logarithmically divergent pure vacuum part requires renormalization, while it remains finite as long as at least one temperature line is involved. Contrary to the case of 3+1 dimensions it is not necessary to employ the involved techniques developed for the renormalization of self-consistent theories (in equilibrium) in [33]. Since the divergence only appears (in energy-momentum space) in the real part of the Feynman self-energy Σ^c at $T = 0$ (and equivalently in the real part of the retarded/advanced self-energies $\Sigma^{\rm ret/adv}$), it can be absorbed by another mass counterterm

$$\begin{aligned} \delta m^2_{\rm sun} &= -{\rm Re}\, \Sigma^c_{T=0}(p^2) \;=\; -{\rm Re}\, \Sigma^{\rm ret/adv}_{T=0}(p^2) \\ &= \frac{\lambda^2}{6} \int \frac{d^2q}{(2\pi)^2} \int \frac{d^2r}{(2\pi)^2} \frac{1}{4\,\omega_{\mathbf{q}}\,\omega_{\mathbf{r}}\,\omega_{\mathbf{q+r-p}}} \frac{\omega_{\mathbf{q}}+\omega_{\mathbf{r}}+\omega_{\mathbf{q+r-p}}}{[\,\omega_{\mathbf{q}}+\omega_{\mathbf{r}}+\omega_{\mathbf{q+r-p}}\,]^2 - p_0^2} \end{aligned} \tag{19}$$

at given 4-momentum $p = (p_0, \mathbf{p})$ and renormalized mass m.

2.4 Numerical Implementation

For the solution of the Kadanoff–Baym equations we generate a closed set of first order differential equations in time for the Green's functions $i\,G^<_{\phi\phi}(x,y) = \langle \phi(y)\,\phi(x)\rangle$, $i\,G^<_{\pi\phi}(x,y) = \langle \phi(y)\,\pi(x)\rangle$ and $i\,G^<_{\pi\pi}(x,y) = \langle \pi(y)\,\pi(x)\,\rangle$ with the canonical field momentum $\pi(x) = \partial_{x_0}\phi(x)$. The disadvantage of having more Green's functions in this scheme is compensated by its good accuracy. We especially take into account the propagation along the time diagonal which leads to an improved numerical precision. The set of differential equations is solved by means of a Runge–Kutta algorithm of 4th order. For the calculation of the self-energies we apply a Fourier method similar to the one used in [15,34]. The self-energies (17) are calculated in coordinate space, where they are products of coordinate-space Green's functions – that are available by Fourier transformation – and are finally transformed to momentum space again. We note that the momentum space is discretized by a finite number of momentum states fixed by periodic boundary conditions in a finite volume $V = a^2$.

2.5 Equilibration

As already observed in the 1+1 dimensional case [29–31] the inclusion of the mean field does not lead to an equilibration of arbitrary initialized momentum distributions, since it only modifies the propagation of the particles by the generation of an effective mass. Thermalization in 2+1 dimensions requires the inclusion of the collisional self-energies from the sunset diagram. To demonstrate the phenomenon of equilibration we consider the time evolution of several initial conditions characterized by the same energy density. The initial equal-time Green's functions $iG^<(\mathbf{p}, t=0, t=0)$ adopted are displayed in Fig. 2 (l.h.s.) as a function of the momentum p_x (for $p_y = 0$). We concentrate on polar symmetric configurations due to the large numerical expense for this first investigation[1]. Since the equal-time Green's functions $G^<(\mathbf{p}, t, t)$ are purely imaginary we display (the real part of) $i\,G^<$. Furthermore, the corresponding initial distribution functions $n_o(\mathbf{p}, t=0)$ are shown in Fig. 2 (r.h.s.). While the initial distributions D1, D2 have the shape of (polar symmetric) 'tsunami' waves with maxima at different momenta, the initial distribution DT corresponds to a free Bose gas at a given initial temperature. The difference between the Green's functions and the distribution functions is given by the vacuum contribution which has its maximum at small momenta, i.e. $2\omega_{\mathbf{p}}\, i\,G^<_{\phi\phi}(\mathbf{p}, t=0, t=0) = 2n_o(\mathbf{p}, t=0) + 1$. So even for the distributions D1, D2 the corresponding Green's functions are nonvanishing at low momentum. The time evolution of various momentum modes of the equal-time Green's function for these three different initial states is shown in Fig. 3. Starting from different initial conditions (as shown in Fig. 2) the single momentum modes finally approach the equilibrium values. For large times all modes reach a static limit as characteristic for an equilibrated system. On the other hand, for small times one observes a damped oscillating behaviour. This can be identified as a typical 'switching-on' effect where the system – being in a static (free) situation – is excited by a sudden increase of the coupling constant. In the present calculation the initial state is given by a free state with mass $m = 1$ and the coupling $\lambda/m = 16$ is switched on for times

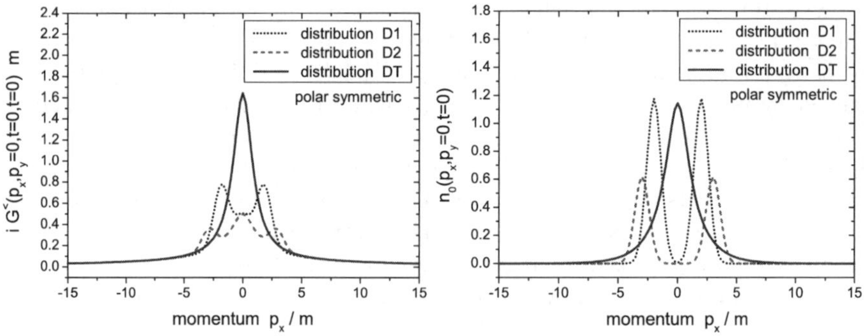

Fig. 2. Initial Green's functions $iG^<(|\mathbf{p}|, t=0, t=0)$ (*left*) and corresponding initial distribution functions $n_o(|\mathbf{p}|, t{=}0)$ (*right*) for the distributions D1, D2 and DT in momentum space (cut of the polar symmetric distribution in p_x for $p_y = 0$)

[1] In Sect. 3 we will present calculations for non-symmetric systems.

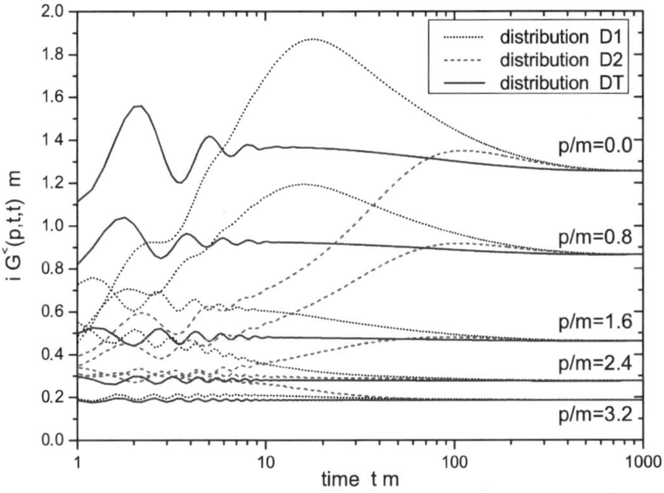

Fig. 3. Time evolution of various momentum modes of the equal-time Green's function for $|\mathbf{p}|/m = 0.0, 0.8, 1.6, 2.4, 3.2$ (from *top* to *bottom*) for three different initial configurations D1, D2, DT (characterized by the different line type) with the same energy density. For the rather strong coupling constant $\lambda/m = 16$ the initial oscillations are damped rapidly and have already disappeared at $t \cdot m = 20$. Finally all momentum modes assume the same respective equilibrium value independent of the initial state

larger than 0. One might also start with an effective initial mass due to the self-consistent tadpole contribution [31]. For the study of the equilibration process, however, this does not make any significant difference. Due to the character of the self-energies the 'switching-on' oscillations are damped in the course of time. The damping depends on the coupling strength and is much larger for strongly coupled systems. For very strong interactions the initial oscillations are even hard to recognize. The final part of the time evolution is characterized by an approximately exponential approach to the equilibrium value. In contrast to the calculations performed for ϕ^4-theory in 1+1 space-time dimensions [31] there is no intermediate region, where the momentum modes show a power-law behaviour. Furthermore, we observe that – depending on the initial conditions and the coupling strength – the momentum modes can temporarily exceed their respective equilibrium value. This can be seen for the lowest momentum modes of distribution D1 in Fig. 3 which possesses initially a maximum at relatively small momentum. Especially the momentum mode $|\mathbf{p}| = 0$ of the equal-time Green's function $G^<$, which starts at around 0.5, is rising to a value of $\sim$1.8 before it is decreasing again to its equilibrium value of about 1.26. Thus the evolution towards the final equilibrium value is – after the initial phase with damped oscillations – not necessarily monotonic. On the other hand, initial configurations like the distribution DT, where the system initially is given by a free gas of particles at a temperature T_0, do not show this property. Although they are – of course – not the equilibrium state of the interacting theory, they are much closer to it than the distributions D1 and D2. Therefore, the evolution towards the equilibrium distribution runs less violently.

2.6 Time Evolution of the Spectral Function

Within the Kadanoff–Baym calculations the full quantum character of the two-point functions is taken into account. Consequently one fully incorporates the spectral properties of the nonequilibrium system during its time evolution. The spectral function $A(x, y)$ in our case is given by

$$A(x,y) = \langle [\phi(x), \phi(y)]_- \rangle = i \left[G^>(x,y) - G^<(x,y) \right] . \tag{20}$$

For each system time $T = (t_1 + t_2)/2$ the spectral function in Wigner-space is obtained via Fourier transformation with respect to the relative time coordinate $\Delta t = t_1 - t_2$:

$$A(\mathbf{p}, p_0, T) = \int_{-\infty}^{\infty} d\Delta t \, e^{i\Delta t \, p_0} \;\; A(\mathbf{p}, t_1 = T + \Delta t/2, t_2 = T - \Delta t/2). \tag{21}$$

A damping of the function $A(\mathbf{p}, t_1, t_2)$ in relative time Δt corresponds to the generation of a finite width of the spectral function in Wigner-space. This width in turn can be interpreted as the inverse life time of the scalar particle. We recall that the spectral function obeys – for every mean time T and for all momenta $\mathbf{p}$ – the normalization

$$\int_{-\infty}^{\infty} \frac{dp_0}{2\pi} \, p_0 \, A(\mathbf{p}, p_0, T) \;=\; 1 \qquad \forall \, \mathbf{p}, \, T, \tag{22}$$

which is a reformulation of the equal-time commutation relation (8).

In Fig. 4 we present the time evolution of the spectral function for the initial distributions D1, D2 and DT for two different momentum modes $|\mathbf{p}|/m = 0.0$ and $|\mathbf{p}|/m = 1.6$. Since the spectral functions are antisymmetric in energy $A(\mathbf{p}, -p_0, T) = -A(\mathbf{p}, p_0, T)$ we only show the positive energy part. For our initial value problem in two-time space the Fourier transformation (21) is restricted for system time T to an interval $\Delta t \in [-2T, 2T]$. Thus in the very early phase the spectral function assumes a finite width already due to the limited support of the Fourier transformation and a Wigner representation is not very meaningful. We therefore present the spectral functions for various system times starting from $t \cdot m = 20$ up to $t \cdot m = 400$.

For the free thermal initialization DT the evolution of the spectral function is very smooth. It is comparable to the smooth evolution of the equal-time Green's function for this initialization as discussed in the last Subsection. The spectral function is already close to the equilibrium shape at small times being initially only slightly broader than for late times. The maximum of the

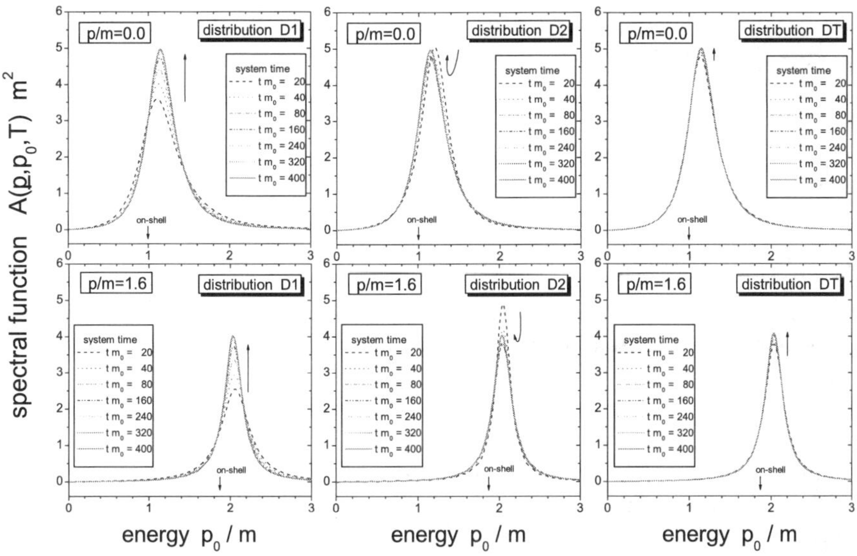

Fig. 4. Time evolution of the spectral function $A(\mathbf{p}, p_0, T)$ for the initial distributions D1, D2 and DT (*from left to right*) for the two momenta $|\,\mathbf{p}\,|/m = 0.0$ (*upper row*) and $|\,\mathbf{p}\,|/m = 1.6$ (*lower row*). The spectral function is shown for several times $t \cdot m =$ 20, 40, 80, 160, 240, 320, 400 as indicated by the different line types

spectral function lies (for all momenta) higher than the (bare) on-shell value and nearly keeps its position over the whole time evolution. It results from a positive tadpole mass shift which is only partly compensated by a downward shift from the sunset diagram.

The time evolution of the initial distributions D1 and D2 has a richer structure. For the initial distribution D1 the spectral function is broad for small system times (see the line for $t \cdot m = 20$) and is getting a little sharper in the course of the system evolution (as presented for the momentum mode $|\,\mathbf{p}\,|/m = 0.0$ as well as for $|\,\mathbf{p}\,|/m = 1.6$). At the same time the height of the spectral function is increasing (as demanded by the normalization property (22)) with time. This is indicated by the small arrow close to the peak position. Furthermore, the maximum of the spectral function (which is approximately the on-shell energy) is shifted slightly upwards for the zero mode and downwards for the mode with higher momentum. Although the real part of the (retarded) sunset self-energy leads (in general) to a lowering of the effective mass, the on-shell energy of the momentum modes is still higher than the one for initial mass m (indicated by the 'on-shell' arrow) due to the positive mass shift from the tadpole contribution. For the initial distribution D2 we find an opposite behaviour. Here the spectral function develops at intermediate times a slightly higher width than in the beginning before it is approaching the narrower static shape. The corresponding evolution of the maximum is again indicated by the (bent) arrow. Finally all spectral functions assume the (same) equilibrium form.

As already observed above for the equal-time Green's functions, we emphazise that there is no unique type of evolution for the nonequilibrium system. In fact, the evolution of the system during the equilibration process is sensitive to the initial conditions.

2.7 Generation of Correlations

The time evolution within the Kadanoff–Baym equation is characterized by the generation of correlations. This can be seen from Fig. 5, where all energy density contributions are displayed as a function of time. The kinetic energy density $\varepsilon_{\rm kin}$ is given by all parts of the total energy density that do not depend on the coupling constant ($\propto \lambda^0$). All terms proportional to λ^1 are summarized by the tadpole energy density $\varepsilon_{\rm tad}$ including the actual tadpole term as well as the corresponding tadpole mass counterterm. The contributions from the sunset diagram ($\propto \lambda^2$) – given by the correlation integral $I_1^<$ as well as by the sunset mass counterterm – are represented by the sunset energy density $\varepsilon_{\rm sun}$,[2]

$$\varepsilon_{\rm tot}(t) = \varepsilon_{\rm kin}(t) + \varepsilon_{\rm tad}(t) + \varepsilon_{\rm sun}(t),$$

$$\varepsilon_{\rm kin}(t) = \frac{1}{2}\int \frac{d^dp}{(2\pi)^d}\,(\mathbf{p}^2+m^2)\, i\,G^<_{\phi\phi}(\mathbf{p},t,t) + \frac{1}{2}\int \frac{d^dp}{(2\pi)^d}\, i\,G^<_{\pi\pi}(\mathbf{p},t,t),$$

$$\varepsilon_{\rm tad}(t) = \frac{1}{4}\int \frac{d^dp}{(2\pi)^d}\,\bar{\Sigma}_{\rm tad}(t)\, i\,G^<_{\phi\phi}(\mathbf{p},t,t) + \frac{1}{2}\int \frac{d^dp}{(2\pi)^d}\,\delta m^2_{\rm tad}\, i\,G^<_{\phi\phi}(\mathbf{p},t,t),$$

$$\varepsilon_{\rm sun}(t) = -\frac{1}{4}\int \frac{d^dp}{(2\pi)^d}\, i\,I_1^<(\mathbf{p},t,t) + \frac{1}{2}\int \frac{d^dp}{(2\pi)^d}\,\delta m^2_{\rm sun}\, i\,G^<_{\phi\phi}(\mathbf{p},t,t)\,. \qquad (23)$$

The calculation in Fig. 5 has been performed for the initial distribution DT (which corresponds to a free gas of Bose particles at temperature $T_0 = 1.59\,m$) with a coupling constant $\lambda/m = 16$. This state is a stationary solution for the well-known Boltzmann equation, but not for the Kadanoff–Baym equation. In the latter treatment the system evolves from the uncorrelated state and the correlation energy density $\varepsilon_{\rm sun}$ decreases with time. The decrease of the correlation energy $\varepsilon_{\rm sun}$, which is – with exception of the sunset mass counterterm contribution – initially zero, is mainly compensated by an increase of the kinetic energy density $\varepsilon_{\rm kin}$. The remaining difference is equal

[2] Speaking of powers of the coupling constant we mean the 'superficial' order of the corresponding diagrams. Since our self-energies Σ are built up self-consistently by full Green's functions G always higher orders of the coupling constant are resummed.

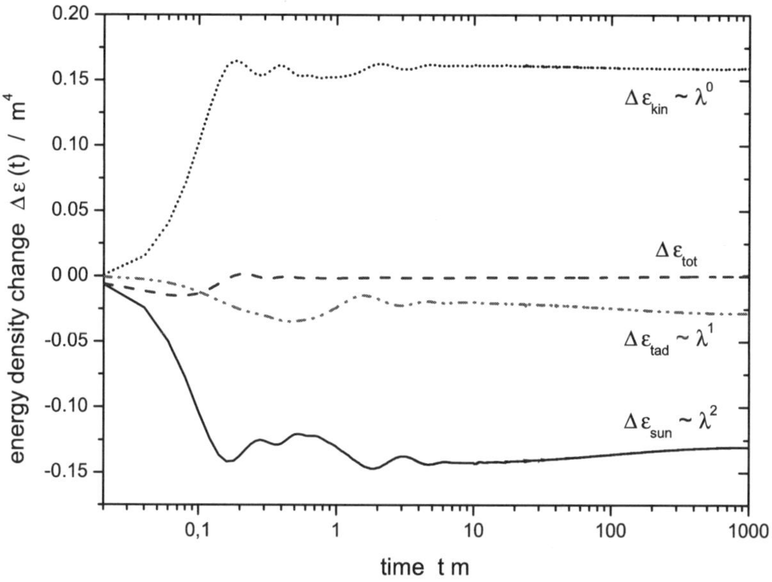

Fig. 5. Change of the different contributions to the total energy density in time. The sunset energy density $\varepsilon_{\rm sun}$ decreases as the system correlates. This is mostly compensated by an increase of the kinetic energy density $\varepsilon_{\rm kin}$. Together with the smaller tadpole contribution $\varepsilon_{\rm tad}$ the total energy $\varepsilon_{\rm tot}$ is conserved

to the change of the tadpole energy density $\varepsilon_{\rm tad}$ such that the total energy density is conserved. While the sunset and the kinetic term have always the behaviour displayed in the figure, the change of the tadpole energy density depends on the initial configuration and can be positive or negative. Since the self-energies are obtained within a Φ derivable scheme (9), the fundamental conservation laws, as e.g. energy conservation, are respected to all orders in the coupling constant [32].

3 Comparison to the Boltzmann Limit

As mentioned above repeatedly the Kadanoff–Baym equation represents the full quantum field theoretical dynamics on the one-particle level. However, its numerical solution is quite involved and it is of strong interest to investigate, in how far approximate schemes deviate from the full calculation. Nowadays, transport models are widely used in the description of quantum system out of equilibrium (cf. Introduction). Most of these models work in the 'quasiparticle' picture, where all particles obey a fixed energy-momentum relation and the energy is no independent degree of freedom anymore; it is determined by the momentum and the (effective) mass of the particle. Accordingly, these particles are treated with their δ-function spectral shape as infinitely long living, i.e. stable objects. This assumption is rather questionable e.g. for high-energy heavy ion reactions, where the particles achieve a large width due to the frequent collisions with other particles in the high density and/or high energy regime. Furthermore, this is doubtful for particles that are unstable even in the vacuum. The question, in how far the quasiparticle approximation influences the dynamics in comparison to the full Kadanoff–Baym calculation, is of general interest [15,34].

In order to investigate this question we formulate the Boltzmann limit in analogy to [35]. For the detailed steps and assumptions in the actual derivation we refer the reader to [36]. The Boltzmann equation describes the time evolution of the momentum distribution function

$$N(\mathbf{p},t) \;=\; \frac{\omega_{\mathbf{p}}}{2}\, i\, G^{<}_{\phi\phi}(\mathbf{p},t,t) \;+\; \frac{1}{2\omega_{\mathbf{p}}}\, i\, G^{<}_{\pi\pi}(\mathbf{p},t,t) \;-\; \mathrm{Re}\, G^{<}_{\pi\phi}(\mathbf{p},t,t) \qquad (24)$$

by $2 \leftrightarrow 2$ on-shell scattering processes, i.e.

$$\partial_t N(\mathbf{p},t) = \frac{\lambda^2}{2\omega_{\mathbf{p}}} \int \frac{d^dq}{(2\pi)^d} \int \frac{d^dr}{(2\pi)^d} \int \frac{d^ds}{(2\pi)^d}\, \frac{1}{8\omega_{\mathbf{q}}\omega_{\mathbf{r}}\omega_{\mathbf{s}}}\, (2\pi)^d\, \delta^{(d)}(\mathbf{p}+\mathbf{q}-\mathbf{r}-\mathbf{s})$$
$$\left\{\bar{N}_{\mathbf{p}}\, \bar{N}_{\mathbf{q}}\, N_{\mathbf{r}}\, N_{\mathbf{s}} \;-\; N_{\mathbf{p}}\, N_{\mathbf{q}}\, \bar{N}_{\mathbf{r}}\, \bar{N}_{\mathbf{s}}\right\}\, \pi\, \delta(\omega_{\mathbf{p}}+\omega_{\mathbf{q}}-\omega_{\mathbf{r}}-\omega_{\mathbf{s}})\,, \qquad (25)$$

using $N_{\mathbf{p}} = N(\mathbf{p},t)$ and $\bar{N}_{\mathbf{p}} = N(\mathbf{p},t)+1$ for the corresponding Bose factors.

The numerical procedure for the solution of (25) is basically the same as the one developed for the solution of the Kadanoff–Baym equation. Moreover, we calculate the actual momentum dependent on-shell energy $\omega_{\mathbf{p}}$ for every momentum mode by a solution of the dispersion relation including contributions from the tadpole and the real part of the (retarded) sunset self-energy. Thus it is guaranteed, that the particles are treated as quasiparticles with the correct energy-momentum relation at every time.

For the comparison between the full Kadanoff–Baym dynamics and the Boltzmann approximation we concentrate on equilibration times. As a measure for the equilibration time we consider the time scale on which the initially non-isotropic distribution proceeds to the polar symmetric equilibrium value. We define a moment $Q(t)$ for a given momentum distribution $n(\mathbf{p},t)$ at time t by

$$Q(t) \;=\; \frac{\int \frac{d^2p}{(2\pi)^2}\, (p_x^2 - p_y^2)\, n(\mathbf{p},t)}{\int \frac{d^2p}{(2\pi)^2}\, n(\mathbf{p},t)}\,, \qquad (26)$$

which vanishes for symmetric systems, e.g. for the equilibrium state. For the Kadanoff–Baym case we calculate the actual distribution function by

$$n(\mathbf{p},t) \;=\; \sqrt{G^{<}_{\phi\phi}(\mathbf{p},t,t)\; G^{<}_{\pi\pi}(\mathbf{p},t,t)} \;-\; \frac{1}{2}\,. \qquad (27)$$

The moment $Q(t)$ shows an approximately exponential decrease in time such that we can define a relaxation rate Γ_Q via the relation $Q(T) \propto e^{-\Gamma_Q t}$.

In Fig. 6 the relaxation rate Γ_Q (scaled by the coupling strength squared) is displayed for the Kadanoff–Baym and the Boltzmann calculation for two different initial configurations. The initial distribution 2 corresponds to the initial state of two (on the p_x-axis) separated particle accummulations (cf. Fig. 7). The peak at low momenta again comes from the vacuum contribution of the Green's function. The time evolution within the full Kadanoff–Baym theory is presented in Fig. 7 by several snapshots at times $t \cdot m = 0$, 10.5, 21, 35, 70, 105 showing the propagation towards an isotropic (static) final state. For distribution 1 the position and the width of the two bumps is modified.

Figure 6 shows for both initializations that the relaxation in the full quantum calculation happens faster for large coupling constants than in the quasi-classical approximation, whereas for small couplings the equilibration times of the full and the approximate evolutions are comparable. While the scaled relaxation rate Γ_Q/λ^2 is nearly constant in the Boltzmann case, it increases with the coupling strength in the Kadanoff–Baym calculation (especially for initial distribution 2).

Since the free Green's function – as used in the Boltzmann calculation – has only support on the mass shell, only $(2 \longleftrightarrow 2)$ scattering processes are described in the Boltzmann limit. All other processes with a different number of incoming and outgoing particles vanish. Within the full Kadanoff–Baym calculation this is very much different since here the spectral function – deter-

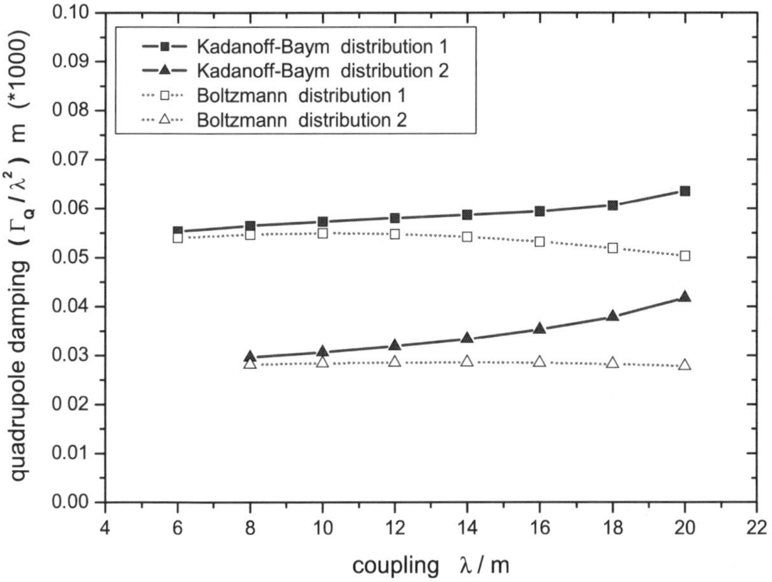

Fig. 6. Relaxation rate (divided by the coupling squared) for Kadanoff–Baym and Boltzmann calculations as a function of the interaction strength. For the two different initial configurations the full Kadanoff–Baym evolution leads to a faster equilibration

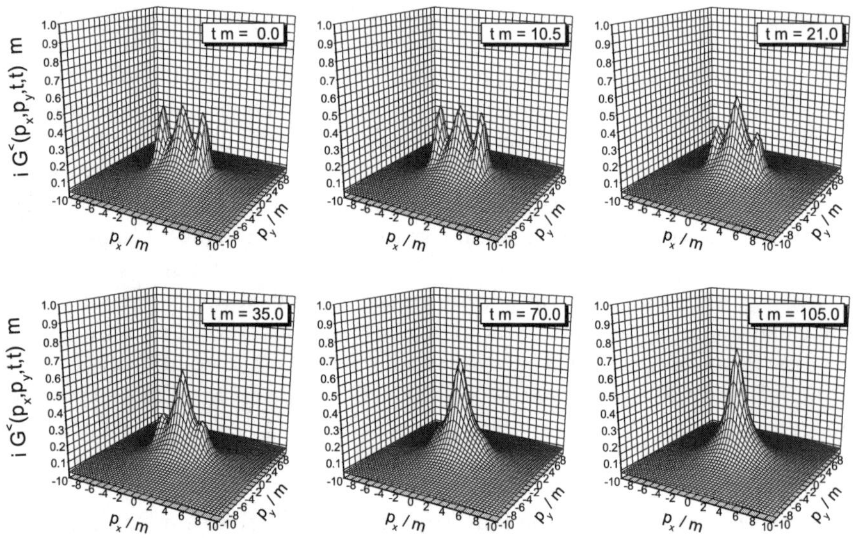

Fig. 7. Evolution of the Green's function in momentum space. The equal time Green's function is displayed for various times $t \cdot m = 0,\ 10.5,\ 21,\ 35,\ 70,\ 105$. Starting from an initially non-isotropic shape it develops towards a symmetric final distribution

mined from the self-consistent Green's function – aquires a finite width. The Green's function has support at all energies – although it drops far off the mass shell. Especially for large coupling constants, where the spectral function is sufficiently broad, the three particle production process gives a significant contribution to the collision integral. Since the width of the spectral function increases with the interaction strength, such processes become more important in the high coupling regime. As a consequence the difference between both approaches is larger for stronger interactions as observed in Fig. 6. For small couplings λ/m in both approaches basically the usual $2 \longleftrightarrow 2$ scattering contributes and the results for the thermalization rate Γ_Q are quite similar.

In summarizing this section we point out that the full solution of the Kadanoff–Baym equations does include $0 \longleftrightarrow 4$, $1 \longleftrightarrow 3$ and $2 \longleftrightarrow 2$ off-shell collision processes which – in comparison to the Bolzmann on-shell $2 \longleftrightarrow 2$ collision limit – become important when the spectral width of the particles reaches about 1/3 of the particle mass. On the other hand, the simple Boltzmann limit works surprisingly well for smaller couplings and those cases, where the spectral function is sufficiently narrow.

4 Derivation of Semiclassical Transport Equations for Particles with Dynamical Life Times

The Kadanoff–Baym equations (5) presently cannot be solved for general inhomogeneous problems due to the high complexity of these equations and the huge 'computer storage' required. It is thus of interest to obtain a semiclassical limit which is much easier to solve e.g. by current testparticle methods.

We start again with the Kadanoff–Baym equation (5), however, change to the Wigner representation via Fourier transformation of the rapidly oscillating relative coordinate $(x-y)$. The theory is then formulated in terms of the coordinates $X=(x+y)/2$ and the momentum p,

$$F_{Xp} \;=\; \int d^4(x-y)\; e^{ip_\mu(x^\mu - y^\mu)}\; F_{xy}\,. \tag{28}$$

Since convolution integrals convert under Wigner transformations as

$$\int d^4(x-y)\; e^{ip_\mu(x^\mu - y^\mu)}\; F_{1,xz} \odot F_{2,zy} \;=\; e^{-i\diamond}\; F_{1,Xp}\, F_{2,Xp}\,, \tag{29}$$

one has to deal with an infinite series in the differential operator $\diamond$ which is a four-dimensional generalization of the Poisson-bracket,

$$\diamond\{\,F_1\,\}\{\,F_2\,\} \;=\; \frac{1}{2}\left(\frac{\partial F_1}{\partial X_\mu}\,\frac{\partial F_2}{\partial p^\mu} \;-\; \frac{\partial F_1}{\partial p_\mu}\,\frac{\partial F_2}{\partial X^\mu}\right)\,. \tag{30}$$

As a standard approximation of kinetic theory only contributions up to first order in the gradients are considered. This is justified if the gradients in the mean spatial coordinate X and momentum p are small.
Applying this approximation scheme (Wigner transformation and neglecting all gradient terms of order $n \geq 2$) to the Dyson–Schwinger equations of the retarded and advanced Green's functions one ends up with

$$\begin{aligned}
(\,p^2 - M_0^2 - \mathrm{Re}\Sigma^{\mathrm{ret}}_{Xp}\,)\,\mathrm{Re}\,G^{\mathrm{ret}}_{Xp} \;&=\; 1 \;-\; \frac{1}{4}\,\Gamma_{Xp}\,A_{Xp}\,,\\
(\,p^2 - M_0^2 - \mathrm{Re}\Sigma^{\mathrm{ret}}_{Xp}\,)\,A_{Xp} \;&=\; \Gamma_{Xp}\,\mathrm{Re}\,G^{\mathrm{ret}}_{Xp}\,,
\end{aligned} \tag{31}$$

where we have separated the retarded and advanced Green's functions as well as the self-energies into real and imaginary contributions

$$G^{\mathrm{ret,adv}}_{Xp} \;=\; \mathrm{Re}\,G^{\mathrm{ret}}_{Xp} \;\mp\; \frac{i}{2}\,A_{Xp}\,, \qquad \Sigma^{\mathrm{ret,adv}}_{Xp} \;=\; \mathrm{Re}\Sigma^{\mathrm{ret}}_{Xp} \;\mp\; \frac{i}{2}\,\Gamma_{Xp}\,. \tag{32}$$

The imaginary part of the retarded propagator is given (up to a factor) by the normalized spectral function A_{Xp} (20) while the (negative) imaginary part of the self-energy is half the width Γ_{Xp}. Above (31) are given for a (renormalized) vacuum mass M_0 and a (renormalized) retarded self-energy

$\Sigma^{\rm ret}$. In the presence of an additional space-time local self-energy (e.g. tadpole self-energy) the mass term is shifted accordingly $M_0^2 \to M_0^2 + \Sigma_{\rm tad}$. From the algebraic equations (31) we obtain a direct relation between the real and the imaginary part of the propagator (provided $\Gamma_{Xp} \neq 0$):

$$\mathrm{Re}\, G^{\rm ret}_{Xp} \; = \; \frac{p^2 - M_0^2 - \mathrm{Re}\Sigma^{\rm ret}_{Xp}}{\Gamma_{Xp}} \, A_{Xp} \, . \tag{33}$$

The algebraic solution for the spectral function shows a Lorentzian shape with space-time and four-momentum dependent width Γ_{Xp}. This result is valid for bosons to first order in the gradient expansion,

$$A_{Xp} \; = \; \frac{\Gamma_{Xp}}{(p^2 - M_0^2 - \mathrm{Re}\Sigma^{\rm ret}_{Xp})^2 \, + \, \Gamma^2_{Xp}/4} \, . \tag{34}$$

For the real part of the retarded Green's function we get also algebraically

$$\mathrm{Re}\, G^{\rm ret}_{Xp} \; = \; \frac{p^2 - M_0^2 - \mathrm{Re}\Sigma^{\rm ret}_{Xp}}{(p^2 - M_0^2 - \mathrm{Re}\Sigma^{\rm ret}_{Xp})^2 \, + \, \Gamma^2_{Xp}/4} \, . \tag{35}$$

4.1 Transport Equations

The Kadanoff–Baym equation (5) gives in the same semiclassical approximation scheme a generalized transport equation,

$$\begin{aligned} &\diamond\{\, p^2 - M_0^2 - \mathrm{Re}\Sigma^{\rm ret}_{Xp} \,\}\, \{\, G^<_{Xp} \,\} \; - \; \diamond\{\, \Sigma^<_{Xp} \,\}\, \{\mathrm{Re}\, G^{\rm ret}_{Xp} \,\} \\ &= \frac{i}{2} \left[\, \Sigma^>_{Xp}\, G^<_{Xp} \; - \; \Sigma^<_{Xp}\, G^>_{Xp} \,\right] , \end{aligned} \tag{36}$$

and a generalized mass-shell equation,

$$\begin{aligned} &[\, p^2 - M_0^2 - \mathrm{Re}\Sigma^{\rm ret}_{Xp} \,]\, G^<_{Xp} \; - \; \Sigma^<_{Xp}\, \mathrm{Re}\, G^{\rm ret}_{Xp} \\ &= \frac{1}{2} \diamond \{\, \Sigma^<_{Xp} \,\}\, \{\, A_{Xp} \,\} \; - \; \frac{1}{2} \diamond \{\, \Gamma_{Xp} \,\}\, \{\, G^<_{Xp} \,\} \, . \end{aligned} \tag{37}$$

In the transport equation (36) one recognizes on the l.h.s. the drift term $p^\mu \partial_\mu \bullet$ as generated by the contribution $\diamond\, \{p^2 - M_0^2\}\{\bullet\}$, as well as the Vlasov term determined by the real part of the retarded self-energy. On the other hand the r.h.s. represents the collision term with its 'gain and loss' structure. To evaluate the $\diamond\{\Sigma^<\}\{\mathrm{Re}\, G^{\rm ret}\}$-term in (36), which does not contribute in the quasiparticle limit, it is useful to introduce distribution functions for the Green's functions and self-energies as

$$\begin{aligned} i\, G^<_{Xp} \; &= \; N_{Xp}\, A_{Xp} \, , & i\, G^>_{Xp} \; &= \; (\, 1 \, + \, N_{Xp} \,)\, A_{Xp} \, , \\ i\, \Sigma^<_{Xp} \; &= \; N^\Sigma_{Xp}\, \Gamma_{Xp} \, , & i\, \Sigma^>_{Xp} \; &= \; (\, 1 \, + \, N^\Sigma_{Xp} \,)\, \Gamma_{Xp} \, . \end{aligned} \tag{38}$$

Following the argumentation of Botermans and Malfliet [16] the distribution functions N and N^{Σ} in (38) should be equal in the second term of the l.h.s. of (36) within a consistent first order gradient expansion. As a consequence the self-energy $\Sigma^<$ can be replaced by $G^< \cdot \Gamma/A$ in the term $\diamond\{\Sigma^<\}\{\mathrm{Re}\, G^{\mathrm{ret}}\}$. The general transport equation (36) then can be written as

$$\left[\diamond\{p^2 - M_0^2 - \mathrm{Re}\Sigma^{\mathrm{ret}}_{Xp}\}\{G^<_{Xp}\} - \frac{1}{\Gamma_{Xp}} \diamond\{\Gamma_{Xp}\}\{(p^2 - M_0^2 - \mathrm{Re}\Sigma^{\mathrm{ret}}_{Xp})\, G^<_{Xp}\}\right]$$
$$\times\; A_{Xp}\,\Gamma_{Xp} \;=\; i\,\left[\Sigma^>_{Xp}\, G^<_{Xp} \;-\; \Sigma^<_{Xp}\, G^>_{Xp}\right]. \qquad (39)$$

4.2 Test Particle Representation

In order to obtain an approximate solution to the transport equation (39) we use a test particle ansatz for the Green's function $G^<$, more specifically for the real and positive semidefinite quantity

$$\begin{aligned} F_{Xp} &= A_{Xp}\, N_{Xp} \\ &= i\, G^<_{Xp} \sim \sum_{i=1}^{N} \delta^{(3)}(\mathbf{X}-\mathbf{X}_i(t))\, \delta^{(3)}(\mathbf{p}-\mathbf{p}_i(t))\, \delta(p_0-\epsilon_i(t))\,. \end{aligned} \qquad (40)$$

In the most general case (where the self-energies depend on four-momentum p, time t and the spatial coordinates $\mathbf{X}$) the equations of motion for the test particles read

$$\frac{d\mathbf{X}_i}{dt} = \frac{1}{1-C_{(i)}}\,\frac{1}{2\epsilon_i}\left[2\mathbf{p}_i + \nabla_{p_i}\mathrm{Re}\Sigma^{\mathrm{ret}}_{(i)} + \frac{\epsilon_i^2 - \mathbf{p}_i^2 - M_0^2 - \mathrm{Re}\Sigma^{\mathrm{ret}}_{(i)}}{\Gamma_{(i)}}\,\nabla_{p_i}\Gamma_{(i)}\right], \qquad (41)$$

$$\frac{d\mathbf{p}_i}{dt} = \frac{-1}{1-C_{(i)}}\,\frac{1}{2\epsilon_i}\left[\nabla_{X_i}\mathrm{Re}\Sigma^{\mathrm{ret}}_i + \frac{\epsilon_i^2 - \mathbf{p}_i^2 - M_0^2 - \mathrm{Re}\Sigma^{\mathrm{ret}}_{(i)}}{\Gamma_{(i)}}\,\nabla_{X_i}\Gamma_{(i)}\right], \qquad (42)$$

$$\frac{d\epsilon_i}{dt} = \frac{1}{1-C_{(i)}}\,\frac{1}{2\epsilon_i}\left[\frac{\partial \mathrm{Re}\Sigma^{\mathrm{ret}}_{(i)}}{\partial t} + \frac{\epsilon_i^2 - \mathbf{p}_i^2 - M_0^2 - \mathrm{Re}\Sigma^{\mathrm{ret}}_{(i)}}{\Gamma_{(i)}}\,\frac{\partial \Gamma_{(i)}}{\partial t}\right], \qquad (43)$$

where the notation $F_{(i)}$ implies that the function is taken at the coordinates of the test particle at time t, i.e. $F_{(i)} \equiv F(t, \mathbf{X}_i(t), \mathbf{p}_i(t), \epsilon_i(t))$.

In (41–43) a common multiplication factor $(1 - C_{(i)})^{-1}$ appears, which contains the energy derivatives of the retarded self-energy

$$C_{(i)} = \frac{1}{2\epsilon_i}\left[\frac{\partial}{\partial \epsilon_i}\mathrm{Re}\Sigma^{\mathrm{ret}}_{(i)} + \frac{\epsilon_i^2 - \mathbf{P}_i^2 - M_0^2 - \mathrm{Re}\Sigma^{\mathrm{ret}}_{(i)}}{\Gamma_{(i)}}\,\frac{\partial}{\partial \epsilon_i}\Gamma_{(i)}\right]. \qquad (44)$$

It yields a shift of the system time t to the 'eigentime' of particle i defined by $\tilde{t}_i = t/(1 - C_{(i)})$. As the reader immediately verifies, the derivatives with respect to the 'eigentime', i.e. $d\mathbf{X}_i/d\tilde{t}_i$, $d\mathbf{p}_i/d\tilde{t}_i$ and $d\epsilon_i/d\tilde{t}_i$ then emerge without this renormalization factor for each test particle i when neglecting higher order time derivatives in line with the semiclassical approximation scheme. In the limiting case of particles with vanishing gradients of the width Γ_{Xp} these equations of motion reduce to the well-known transport equations of the quasiparticle picture.

Following [37] we take $M^2 = p^2 - \mathrm{Re}\,\Sigma^{\mathrm{ret}}$ as an independent variable instead of p_0, which then fixes the energy (for given $\mathbf{p}$ and M^2) to

$$p_0^2 = \mathbf{p}^2 + M^2 + \mathrm{Re}\,\Sigma^{\mathrm{ret}}_{X\mathbf{p}M^2} . \tag{45}$$

Equation (43) then turns to

$$\frac{d(M_i^2 - M_0^2)}{dt} = \frac{M_i^2 - M_0^2}{\Gamma_{(i)}} \frac{d\Gamma_{(i)}}{dt} \tag{46}$$

for the time evolution of the test particle i in the invariant mass squared as derived in [37]. For applications of the semiclassical off-shell transport approach we refer the reader to [37].

5 Summary

In this article we have studied the time evolution of an interacting field theoretical system, i.e. ϕ^4-field theory in $2+1$ space-time dimensions, on the basis of the Kadanoff–Baym equations for a spatially homogeneous system including the self-consistent tadpole and sunset self-energies. We find that equilibration is achieved only by inclusion of the sunset self-energy. Simultaneously, the time evolution of the single-particle spectral function has been calculated for various initial conditions. A comparison of the full solution of the Kadanoff–Baym equations with the solution for the corresponding Boltzmann equation shows that a consistent inclusion of the spectral function has a sizeable impact on the equilibration rates if the width of the spectral function becomes larger than $\sim 1/3$ of the particle mass.

Furthermore, the conventional transport of particles in the on-shell quasiparticle limit has been extended to particles of finite life time by means of a dynamical spectral function $A(X, \mathbf{p}, M^2)$. Starting again from the Kadanoff–Baym equations we have derived in consistent first order gradient expansion equations of motion for test particles with respect to their time evolution in $\mathbf{X}, \mathbf{p}$ and M^2. This off-shell propagation has been examined for a couple of model cases in [37] as well as for nucleus-nucleus collisions showing that – at subthreshold energies – the off-shell dynamics play an important role for the production of energetic particles.

Acknowledgements

The authors acknowledge valuable discussions with C. Greiner and S. Leupold throughout these studies.

References

1. J. Schwinger, J. Math. Phys. **2**, 407 (1961).
2. S.J. Wang and W. Cassing, Ann. Phys. (N.Y.) **159**, 328 (1985).
3. W. Cassing and S.J. Wang, Z. Phys. A **337**, 1 (1990).
4. K. Chou, Z. Su, B. Hao, and L. Yu, Phys. Rept. **118**, 1 (1985).
5. S. Mrówczyński and P. Danielewicz, Nucl. Phys. B **342**, 345 (1990).
6. S. Mrówczyński and U. Heinz, Ann. Phys. (N.Y.) **229**, 1 (1994).
7. H. Stöcker and W. Greiner, Phys. Rept. **137**, 277 (1986).
8. G.F. Bertsch and S. Das Gupta, Phys. Rept. **160**, 189 (1988).
9. W. Cassing, V. Metag, U. Mosel, and K. Niita, Phys. Rept. **188**, 363 (1990).
10. W. Cassing and U. Mosel, Prog. Part. Nucl. Phys. **25**, 235 (1990).
11. C. Fuchs and T. Gaitanos, nucl-th/0211091; this volume.
12. S. Bass, M. Belkacem, M. Bleicher et al., Prog. Part. Nucl. Phys. **41**, 255 (1998).
13. W. Cassing and E.L. Bratkovskaya, Phys. Rept. **308**, 65 (1999).
14. L.P. Kadanoff and G. Baym, *Quantum statistical mechanics*, Benjamin, New York, 1962.
15. P. Danielewicz, Ann. Phys. (N.Y.) **152**, 239 (1984); *ibid.* 305.
16. W. Botermans and R. Malfliet, Phys. Rept. **198**, 115 (1990).
17. R. Malfliet, Prog. Part. Nucl. Phys. **21**, 207 (1988).
18. P.A. Henning, Nucl. Phys. A **582**, 633 (1995); Phys. Rept. **253**, 235 (1995); this book page 67.
19. C. Greiner and S. Leupold, Ann. Phys. (N.Y.) **270**, 328 (1998).
20. S.J. Wang, W. Zuo, and W. Cassing, Nucl. Phys. A **573**, 245 (1994).
21. W. Cassing, K. Niita, and S.J. Wang, Z. Phys. A **331**, 439 (1988).
22. R. Malfliet, Nucl. Phys. A **545**, 3 (1992).
23. R. Malfliet, Phys. Rev. B **57**, R11027 (1998).
24. P. Danielewicz and S. Pratt, Phys. Rev. C **53**, 249 (1996).
25. V. Špička, P. Lipavský, and K. Morawetz, Phys. Rev. B **55**, 5095 (1997); Phys. Lett. A **240**, 160 (1998); this book page 114.
26. P. Lipavský, V. Špička, and K. Morawetz, Phys. Rev. E **59**, 1291 (1999).
27. P. Lipavský, K. Morawetz, and V. Špička, Annales de Physique **26**, 1 (2001).
28. A. Peter et al., Z. Phys. A **358**, 91 (1997); Z. Phys. C **71**, 515 (1997).
29. J. Berges and J. Cox, Phys. Lett. B **517**, 369 (2001).
30. J. Berges and G. Aarts, Phys. Rev. D **64**, 105010 (2001).
31. J. Berges, Nucl. Phys. A **699**, 847 (2002).
32. Y.B. Ivanov, J. Knoll, and D.N. Voskresensky, Nucl. Phys. A **657**, 413 (1999).
33. H. van Hees and J. Knoll, Phys. Rev. D **65**, 025010 (2002); Phys. Rev. D **65**, 105005 (2002).
34. H.S. Köhler, Phys. Rev. C **51**, 3232 (1995).
35. D. Boyanovsky, I.D. Lawrie, and D.-S. Lee, Phys. Rev. D **54**, 4013 (1996).
36. S. Juchem, W. Cassing and C. Greiner, Phys. Rev. D in press, hep-ph/0307353
37. W. Cassing and S. Juchem, Nucl. Phys. A **665**, 377 (2000); Nucl. Phys. A **672**, 417 (2000); Nucl. Phys. A **677**, 445 (2000).

Consequences of Kinetic Nonequilibrium for the Nuclear Equation-of-state in Heavy Ion Collisions

Christian Fuchs

Abstract. The equation of state (EOS) in ground state nuclear matter is discussed in terms of relativistic many-body theory. At supra-normal densities the EOS can only be tested in heavy ion collisions. However, highly compressed nuclear matter created in relativistic heavy collisions is to large extent governed by local non-equilibrium. We discuss the influence of such kinetic non-equilibrium on the corresponding EOS. As an idealized scenario for typical situation which occur in heavy ion reactions colliding nuclear matter configurations are studied. An effective EOS constructed for anisotropic momentum configurations shows a significant net softening compared to ground state nuclear matter.

1 Introduction

One major goal of relativistic heavy ion physics is to explore the behavior of the nuclear equation-of-state (EOS) far away from saturation, i.e. at high densities and non-zero temperature. Over the last three decades a large variety of observables has been investigated both from the experimental and theoretical side motivated by the search for the nuclear EOS. The collective particle flow is thereby intimately connected to the dynamics during the compressed high density phase of such reactions [1,2]. E.g., the elliptic flow which develops in the early compression phase is thought to be a suitable observable to extract information on the EOS [3]. But also the production of strange particles is a good probe to study dense matter [4]. Recent precision measurements of the K^+ production at SIS energies strongly support the scenario of a soft EOS [5,6].

The temporal evolution of the collision from a highly anisotropic initial configuration in phase space to an – at least partially – equilibrated final configuration is successfully described by microscopic transport models like BUU or QMD. In these type of models the nuclear mean field is usually based on phenomenological parameterizations [7–10]. Such parameterizations allow different extrapolations to high densities, summarized by referring to a 'hard' or a 'soft' equation-of-state [1], which can be tested in heavy ion collisions. A more microscopic approach is to start from free NN interactions, to determine the correlated two-body interaction in nuclear matter and to use this type of interaction in heavy ion reactions.

For reaction in the SIS energy range the relaxation time needed by the system to equilibrate coincides more or less with the high density phase of the reaction. Hence, anisotropy effects are present over the compression phase where one essentially intends to study the EOS at high densities. Experimental evidence for incomplete equilibration even in central collisions at SIS energies has recently been reported in [11]. Transport simulations [12,13] show thereby that the local phase space in the overlapping zone of two inter-penetrating nuclei can to large extent be idealized by colliding nuclear matter, i.e. two fluids of counter-streaming matter. The initial phase space distribution in the participant zone is that of two currents of nuclear matter colliding with beam velocity. The local momentum distributions of colliding nuclear matter configurations, called CNM in the following, are given by two Fermi ellipsoids, i.e. two boosted Fermi spheres separated by a relative velocity $v_{\rm rel}$. Such configurations describe the early stages of the reaction and have been a subject of a variety of theoretical investigations [14–17]. In the course of the reaction the mid-rapidity region is increasingly populated due to binary collisions and the system is heated up ($T \simeq 30$–$80\,\mathrm{MeV}$). The originally cold and sharp momentum ellipsoids become diffuse and merge together. The energy density depends in the course of the reaction therefore on both, the *local* relative velocity and the *local* temperature: $\epsilon(\varrho, v_{\rm rel}, T = 0) \longrightarrow \epsilon(\varrho, v_{\rm rel}, T) \longrightarrow \epsilon(\varrho, v_{\rm rel} = 0, T)$. The time evolution from highly anisotropic momentum space to almost equilibrated matter can be approximated by sequences of CNM configurations varying the densities and the relative streaming velocities of the currents. Thereby we neglect the temperature dependence and discuss only the case $\epsilon(\varrho, v_{\rm rel}, T = 0)$. Here we will connect configurations which contain a high momentum space anisotropy to the ground state EOS $\epsilon(\varrho, v_{\rm rel} = 0, T = 0)$.

2 The EOS of Ground State Nuclear Matter

A perturbative approach to the strongly interacting nuclear systems is not possible. However, a systematic summation of diagrams up to infinite order in terms of the Brueckner hole-line expansion turned out to be the appropriate treatment. Already in lowest order, which corresponds to standard Brueckner theory the saturation of nuclear matter, can be described at least qualitatively. The formalism is based on an effective quantum field theory for mesons and nucleons and thus ignores the underlying quark nature of the nucleon. In the Brueckner model the T-matrix (or Brueckner G-matrix) serves as an effective in-medium two-body interaction. It is determined by a self-consistent summation of the ladder diagrams in a quasi-potential approximation (Thompson equation) to the Bethe–Salpeter equation. The character of the bare nucleon-nucleon interaction, in particular the repulsive short range part (hard-core) requires an account of two-body correlations in a self-consistent way. The effect of the correlations on the two-nucleon wave function in the medium is schematically shown in Fig. 1.

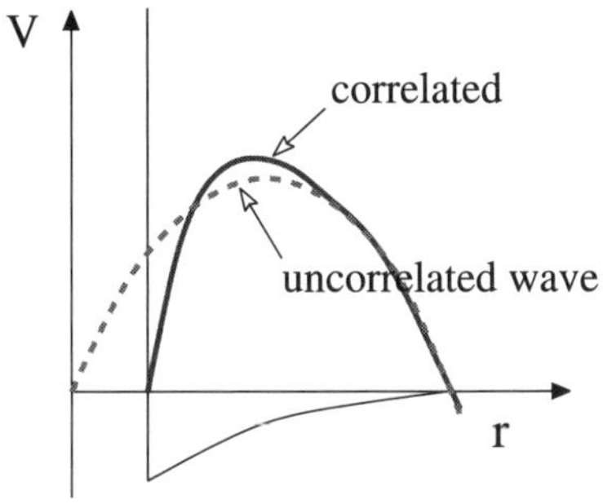

Fig. 1. Effect of the two-body correlations on the two-nucleon wave function in nuclear matter. The figure is taken from [18]

However, non-relativistic Brueckner calculations are not able to meet the empirical saturation point of nuclear matter ($\rho_{\text{sat}} = 0.17\ \text{fm}^{-3}$, $E_{\text{bind}} = -16\,\text{MeV}$). In contrast, the saturation points obtained for various types of NN-potentials were all located on the so-called Coester line (for a review see [18]) in the ϵ–ϱ plane. A breakthrough was achieved when first relativistic (Dirac-) Brueckner (DB) calculations were performed in the late eighties [19,20]. Now the Coester line was shifted much closer towards the empirical area of saturation. The reason for the success of the relativistic approach is due to the fact that the dressing of the in-medium spinors introduces a density dependence to the interaction which is missing in the non-relativistic treatment. In the latter case the inclusion of three-body forces leads to similar effects [21]. Presently it is, however, not completely clear up to which degree such three-body effects are already effectively included in the relativistic approach on the two-body level which is then due to the nonlocal structure of the relativistic one-boson-exchange (OBE) potentials [22].

However, relativistic Brueckner calculations are not straightforward and the approaches of various groups [19,20,23,24] are similar but differ in detail, depending on solution techniques and the particular approximations made. E.g. the T-matrix has to be decomposed into Lorentz components, i.e. scalar, vector, tensor, etc. contributions. This procedure is not free from ambiguities. Due to identical matrix elements for positive energy states pseudo-scalar and pseudo-vector components cannot uniquely be disentangled for on-shell scattering. However, with a pseudo-scalar vertex the pion couples maximally to negative energy states which are not included in the standard Brueckner approach. This is inconsistent with the potentials used since the OBE potentials are based on the no-sea approximation. Hence, pseudo-scalar contributions due to the one-π exchange lead to large and spurious contributions from negative energy states. In [23] it was shown that such spurious contributions dominate the momentum dependence of the nuclear self-energy, and, in particular, lead to an artificially strong momentum dependence inside the Fermi sea. It was further demonstrated in [23] that the method used in [19] fails to cure this problem and in [24] a new and reliable method was proposed to remove those spurious contributions from the T-matrix. The correspond-

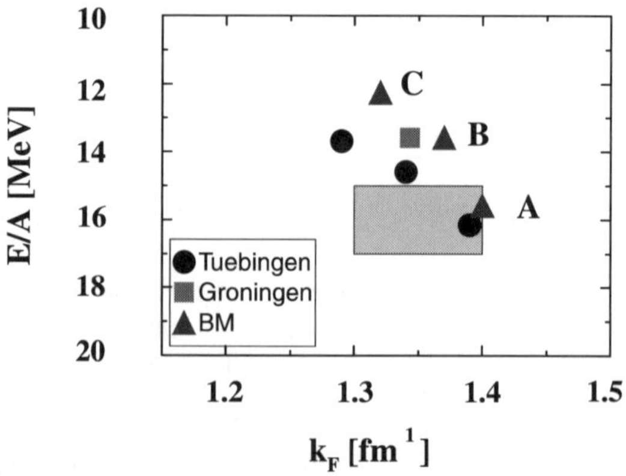

Fig. 2. Nuclear matter saturation points (binding energy over Fermi momentum k_F) obtained in [24] (*circles*) using the Bonn A,B, and C potentials. The hatched area shows the empirical region of saturation. Results are compared to calculations from [19] (*triangle*) and [20] (*square*)

ing saturation properties for nuclear matter obtained in [24] using the Bonn A, B, and C potentials (different parameter sets differ mainly in the tensor strength) are shown in Fig. 2 and compared to DB results from [19,20].

3 Colliding Nuclear Matter

The phase space distribution of colliding nuclear matter (CNM)

$$\Theta_{12} = \Theta_1 + \Theta_2 - \Theta_1 \cdot \Theta_2 \tag{1}$$

is composed by the momentum distributions of two counter-streaming matters $\Theta_i = \Theta(\mu^*(k_{F_i}) - k^*_\nu u^\nu_i)$, i.e. two boosted Fermi ellipsoids. Θ is the step function, k_{F_i} are the Fermi momenta and $u^\nu_i = (\gamma_i, \gamma_i \mathbf{u}_i)$ are the streaming velocities of the two subsystem currents. The last term in (1) ensures that the Pauli principle is fulfilled for small velocities where the two ellipsoids might overlap. The total baryon density has thereby to be restored. Details

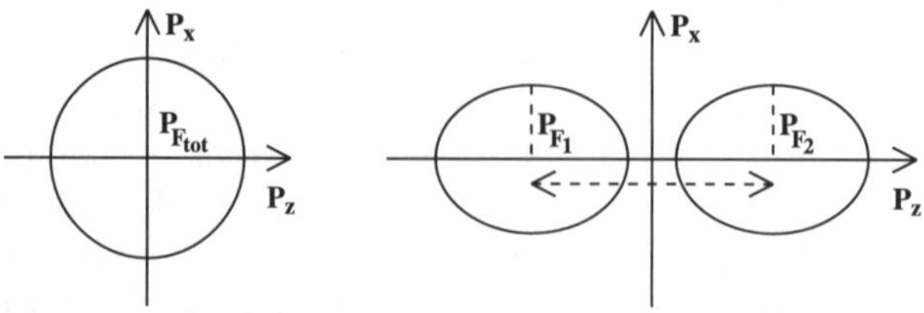

Fig. 3. The momentum space configurations of ground state (*left*) and colliding nuclear matter (*right*) for the temperature zero case

of this procedure can be found in [17]. The momentum space configurations in ground state and colliding nuclear matter are displayed in Fig. 3.

3.1 Energy-Momentum Tensor

The interaction of nucleons with the surrounding medium is expressed in terms of a self energy which dresses the particles, respectively the in-medium propagator. Analogous to spin-isospin saturated nuclear matter at rest, the self energy in colliding matter can be decomposed into scalar and vector components expressed covariantly by

$$\Sigma(k) = \Sigma_S(k) - \gamma_\mu \Sigma^\mu(k) \ , \ \Sigma_\mu(k) = \Sigma_0(k) u_{\mu 12} + \Sigma_v(k) \Delta_{12}^{\mu\nu} k_\nu \ . \quad (2)$$

Here, the Σ_i with $i = S, 0, v$ are scalar functions. The four–velocity u_{12}^μ is the streaming velocity of the combined system derived from the total current $j_{12}^\mu = \varrho_{12} u_{12}^\mu$. Since ϱ_{12} is the invariant baryon density in the c.m. frame of the two currents, $\varrho_{12} = \sqrt{j_{12}^2} = \langle 1 \rangle_{12}|_{\text{c.m.}}$, it is natural to work in that frame, i.e. the frame where the sum of the spatial components of the two counter-streaming currents vanish $\mathbf{j}_{12} = 0$. With the help of the projector $\Delta_{12}^{\mu\nu} = g^{\mu\nu} - u_{12}^\mu u_{12}^\nu$ perpendicular to u_{12}^μ, components (2) are manifestly covariant. The effective mass and the kinetic momenta of the quasi-particles in the medium are given by

$$M^* = M + \Sigma_S(k), \ k_\mu^* = k_\mu + \Sigma_\mu(k), \ E^* = \sqrt{\mathbf{k}^{*2} + M^{*2}(k)} \ . \quad (3)$$

The self-energy has now to be derived from some model, e.g. from an effective two-body interaction like the in-medium T-matrix. In Hartree-Fock approximation the nucleon self-energy is given by the integral over the anti-symmetrized scalar and vector amplitudes [24]

$$\Sigma(k) = \frac{\kappa}{(2\pi)^3} \int \frac{d^3\mathbf{q}}{\tilde{E}^*(\mathbf{q})} \left[\tilde{M}^* T_{\mathrm{S}}(k,q;\chi) + \tilde{\not{q}}^* T_{\mathrm{V}}(k,q;\chi) \right] \Theta_{12}(q;\chi) \ . \quad (4)$$

The scalar T_{S} and the vector T_{V} amplitude are Lorentz invariant functions which depend (besides the parametric dependence on the configuration expressed by χ) on the c.m. momentum, the c.m. relative momentum and the c.m. scattering angle between k and q. In this general case the self-energy components (2–3) are explicitely momentum dependent. A full solution of the Bethe–Salpeter equation in non-equilibrium, i.e. in the framework of the relativistic DB model as proposed in [25], has presently not yet been achieved. Thus we approximate the amplitudes $T_{S,V}$ in (4) by ground states amplitudes [17,27].

The energy momentum tensor in CNM reads

$$T^{\mu\nu} = \langle \tilde{k}^{*\mu} k^{*\nu} / \tilde{E}^* \rangle_{12} - V^{\mu\nu} - \frac{1}{2} g^{\mu\nu} \left\{ \overline{\Sigma}_S \, \varrho_{S_{12}} - V_\lambda^\lambda \right\} \ . \quad (5)$$

The configuration averaged scalar field and the terms with the vector field read

$$\overline{\Sigma}_S = \langle \Sigma_S(k)\tilde{M}^*/\tilde{E}^*\rangle_{12}/\varrho_{S_{12}} \ , \quad V^{\mu\nu} = \langle \tilde{k}^{*\mu}\Sigma^\nu(k)/\tilde{E}^*\rangle_{12} \ . \tag{6}$$

The total scalar density and the total baryon current are given by

$$\varrho_{S_{12}} = \langle \tilde{M}^*/\tilde{E}^*\rangle_{12} \ , \ j_{\mu\,12} = \langle \tilde{k}^*_\mu/\tilde{E}^*\rangle_{12} \ , \tag{7}$$

where

$$\langle X\rangle_{12} = \frac{\kappa}{(2\pi)^3}\int d^3\mathbf{k}X(k)\Theta_{12}(k;\chi) \tag{8}$$

denotes the summation over all occupied states. In spin-isospin saturated nuclear matter the phase space occupancy factor in (8) equals $\kappa = 4$. The set of parameters which determines the colliding configuration is denoted by $\chi = \{k_{F_1}, k_{F_2}, u_1, u_2, \tilde{M}^*\}$. Thus the effective mass $\tilde{M}^*$, the scalar density $\varrho_{S_{12}} = \langle \tilde{M}^*/\tilde{E}^*\rangle_{12}$ and the configuration (1) itself are coupled by non-linear equations.

3.2 Mean Field Approximation

In order to examine the structure of the energy-momentum tensor it is instructive to consider the mean field approximation. Here the self-energies (2) are not explicitely momentum dependent which has the advantage that all integrals in (5) can be solved analytically. To do so we use the Walecka model (QHD-I) [26]. In QHD-I the T-matrix amplitudes are replaced by corresponding coupling constants for a scalar σ and a vector ω meson field, $T_S \longmapsto \Gamma_S = g_\sigma^2/m_\sigma^2$ and $T_V \longmapsto \Gamma_V = g_\omega^2/m_\omega^2$, and thus the double integrals decouple. The energy-momentum tensor has now the simpler form:

$$T^{\mu\nu} = \langle k^{*\mu}k^{*\nu}/E^*\rangle_{12} - \Gamma_V j_{12}^\mu j_{12}^\nu - \frac{1}{2}g^{\mu\nu}\left\{\Gamma_S\varrho_{S_{12}}\varrho_{S_{12}} - \Gamma_V j_{12}^\lambda j_{\lambda\,12}\right\} \ . \tag{9}$$

In nuclear matter one obtains the kinetic energy and pressure densities as

$$\epsilon_{\text{kin}} = \langle E^*(\mathbf{k})\rangle = \frac{3}{4}E_F\varrho(k_F) + \frac{1}{4}M^*\varrho_S \tag{10}$$

$$p_{\text{kin}} = \frac{1}{3}\langle \mathbf{k}^{*2}/E^*\rangle = \frac{1}{4}E_F\varrho(k_F) - \frac{1}{4}M^*\varrho_S \ . \tag{11}$$

The corresponding expressions in CNM are similar and transparent as long as the momentum distributions of the two currents do not overlap. Otherwise the contributions arising from the Pauli correction in (1) complicate the expressions considerably. To be transparent we discuss the formulas only for the case where the relative velocity of the two currents is large enough that their boosted Fermi ellipsoids do not overlap, i.e. $\Theta_1(k,u_1)\Theta_2(k,u_2) = 0$. To further simplify (9) we consider symmetric configurations ($k_{F_{1,2}} =$

$k_F, u_{1,2} = \pm u$). The total (c.m.) current is then given by $j_{\mu\,12} = j_{\mu\,1} + j_{\mu\,2} = (2\gamma(u)\varrho_0(k_F), \mathbf{0})$ where $\varrho_0(k_F)$ is the density of a subsystem current in its rest frame. Analogously the total scalar density reads

$$
\begin{aligned}
\varrho_{S_{12}} &= \varrho_{S_1} + \varrho_{S_2} \\
&= 2\varrho_S(M^*) = \frac{2\kappa}{4\pi^2} M^* \left[k_F E_F - M^{*2} \ln \left(\frac{k_F + E_F}{M^*} \right) \right]
\end{aligned} \tag{12}
$$

with $M^* = M - \Gamma_S \varrho_{S_{12}}$. Energy density $\epsilon = T^{00}$, transverse pressure $p_\perp = T^{11} = T^{22}$ and longitudinal pressure $p_\| = T^{33}$ are given by

$$
\begin{aligned}
T^{00} &= 2\left[\frac{3}{4} E_F \varrho_0(k_F) + \frac{1}{4} M^* \, \varrho_S(M^*) + E_F \varrho_0(k_F) \, (\gamma^2 - 1) \right] \\
&\quad + \frac{1}{2} \left\{ \Gamma_V \, (2\gamma\varrho_0(k_F))^2 + \Gamma_S \, (2\varrho_S(M^*))^2 \right\} && (13) \\
p_\perp &= 2\left[\frac{1}{4} E_F \varrho_0(k_F) - \frac{1}{4} M^* \, \varrho_S(M^*) \right] \\
&\quad + \frac{1}{2} \left\{ \Gamma_V \, (2\gamma\varrho_0(k_F))^2 - \Gamma_S \, (2\varrho_S(M^*))^2 \right\} && (14) \\
p_\| &= 2\left[\frac{1}{4} E_F \varrho_0(k_F) - \frac{1}{4} M^* \, \varrho_S(M^*) + E_F \varrho_0(k_F) \, (\gamma^2 - 1) \right] \\
&\quad + \frac{1}{2} \left\{ \Gamma_V \, (2\gamma\varrho_0(k_F))^2 - \Gamma_S \, (2\varrho_S(M^*))^2 \right\} && (15)
\end{aligned}
$$

To compare CNM with NM it is instructive to rewrite (13–15). The separation of projectile and target nucleons in momentum space increases the phase space volume in CNM. The boosted ellipsoids are elongated in longitudinal direction but in transverse direction the Fermi momenta k_F of the subsystem currents (which are fixed by the current rest densities $\varrho_0(k_F)$) are not enhanced. This feature can be accounted for if we integrate over one Fermi sphere at rest $\Theta(k; k_F, u = 0)$, however, with a doubled phase space occupancy factor 2κ

$$
\langle X \rangle_{2\kappa} = \frac{2\kappa}{(2\pi)^3} \int d^3\mathbf{k} X(k) \Theta(k; k_F, u = 0) \; . \tag{16}
$$

This description yields directly the CNM scalar density (12)

$$
\varrho_{S_{12}} = \left\langle \frac{M^*}{E^*} \right\rangle_{2\kappa} \tag{17}
$$

and $2\varrho_0(k_F) = \langle 1 \rangle_{2\kappa}$. One can now rewrite (13–15) separating thereby the static parts from those contributions which depend on the streaming velocity:

$$
\begin{aligned}
T^{00} &= T^{00}|_{2\kappa} + 2\gamma^2\mathbf{u}^2 \left[E_F \varrho_0(k_F) + \Gamma_V \, \varrho_0^2(k_F) \right] && (18) \\
p_\perp &= p_\perp|_{2\kappa} + 2\gamma^2\mathbf{u}^2 \, \Gamma_V \, \varrho_0^2(k_F) && (19) \\
p_\| &= p_\||_{2\kappa} + 2\gamma^2\mathbf{u}^2 \left[E_F \varrho_0(k_F) + \Gamma_V \, \varrho_0^2(k_F) \right] \; . && (20)
\end{aligned}
$$

Equations (18–20) are exact as long as the ellipsoids do not overlap. It can be directly seen from there that CNM is characterized by two effects which act in opposite direction: The first one is the separation of the nucleons belonging to the different currents in phase space. This effect acts like an additional degree of freedom. Evidently it reduces the pressure of the Fermi motion. Since the scalar part of the potential energy is not affected by the spread in momentum space, besides the fact that the contributions of projectile and target are superposed, it further deepens the scalar potential. However, the vector repulsion gives rise to an additional velocity dependent pressure in both transverse and longitudinal directions (terms proportional to Γ_V). The elongation of the boosted Fermi ellipsoids in the longitudinal direction enhances the baryon vector density which leads to an additional kinetic pressure in the longitudinal direction ($2\gamma^2\mathbf{u}^2 E_F \varrho_0(k_F)$). Both these contributions are related to the kinetic energy of the relative motion in the system and enhance the energy density.

If we want now to compare the non-equilibrium configuration to equilibrated NM this should be done at comparable rest densities, i.e. the two currents with their rest densities $\varrho_0(k_F)$ should be compared to NM at the same total density

$$\varrho_{NM}(k_{F_{\rm tot}}) = 2\varrho_0(k_F) \;\; ; \;\; k_{F_{\rm tot}} = 2^{\frac{1}{3}} k_F \; . \tag{21}$$

By this procedure one prevents a distortion of the comparison by the purely kinematical enhancement of the baryon density due to the boosts of the subsystem currents which do not affect the transverse degrees of freedom. A comparison of CNM and NM at identical total baryon densities would mean to interpret the Lorentz contraction in the anisotropic system as a compression when related to the isotropic configuration. Condition (21) is also necessary for a meaningful comparison of CNM configurations with different streaming velocities. Doing so, one can estimate the phase space effects comparing the contribution $T^{00}|_{2\kappa}(k_F)$ in (18) with $T^{00}|_{NM}(k_{F_{\rm tot}})$ in NM.

3.3 The Effective EOS

In Fig. 4 the equation-of-state's in symmetric colliding nuclear matter are shown for streaming velocities of the subsystem currents $u = |\mathbf{u}_i| = 0.2/0.4/0.6/0.8$ (in units of c) which correspond to incident energies $E_{\rm lab} = 0.08/0.36/1.05/3.34$ GeV/nucleon. $u = 0$ is the isotropic case (ground state NM). The energy per particle is defined in the usual way as

$$E_{12}(\varrho_{12}, u) = \mathrm{T}^{00}/\varrho_{12} - M \tag{22}$$

with $\varrho_{12} = \sqrt{j_{12}^2}$ the invariant baryon density in the c.m. frame of the two currents. At high streaming velocities the "EOS" is significantly stiffer than in ground state nuclear matter since the energy per particle E_{12} includes the

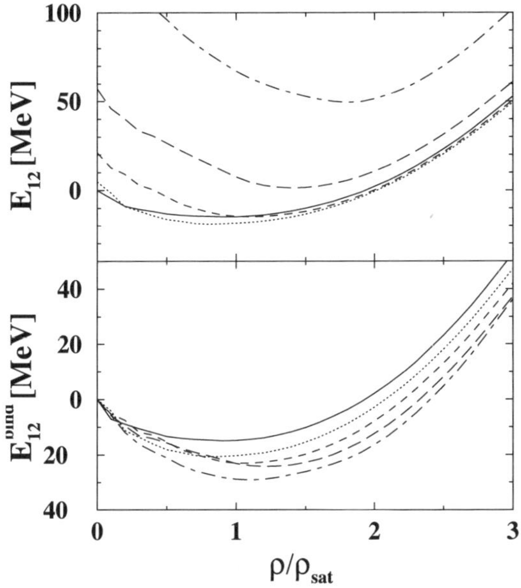

Fig. 4. EOS in nuclear matter (*solid*) and colliding nuclear matter determined in the DB model. The *upper part* shows the total energy per particle E_{12} as a function of the subsystem rest densities $2\varrho_0$. The streaming velocities are u=0.2 (*dotted*), 0.4 (*dashed*), 0.6 (*long-dashed*), 0.8 (*dot-dashed*). The *lower part* shows the effective EOS, i.e. the binding energy per particle E_{12}^{bind} where the kinetic energy of the relative motion in CNM has been subtracted

contribution form the relative motion of the two currents. However, a meaningful discussion of non-equilibrium effects with respect to the ground state EOS should be based on the *free energy* and thus the contribution from the relative motion of the two currents has to be subtracted. From there one obtains the binding energy per particle. This leads to an effective EOS in colliding matter which is directly linked to the hydrodynamical picture [1]. The kinetic energy of a nucleon inside the medium is given by $E^*_{\mathrm{kin}} = E^*(k) - M^*(k)$. Analogously, we identify the kinetic energy density of the relative motion in (5) with

$$\mathcal{E}_{\mathrm{kin}} = \langle \tilde{k}^{*0}(k^{*0} - M^*)/\tilde{E}^* \rangle_{12} = \langle E^*(k) - M^*(k) \rangle_{12} \ . \tag{23}$$

The subtraction of the effective mass M^* instead of the bare mass in (23) accounts for the fact that the two currents are interacting by momentum dependent forces. Such a momentum dependence occurs in the relativistic treatment already on the mean field level where the self-energies do not explicitely depend on momentum. The magnitude of M^* characterizes thereby the strength of the momentum dependence. In the non-relativistic limit M^* plays the role of the Landau mass which is used to parameterize the momentum dependence of the mean field.

The energy of the relative motion of two counter-streaming and interacting currents is more subtle to define. The present description is inspired by the fact that in the non-relativistic case this should be the energy which is necessary to shift the two Fermi spheres in their c.m. frame against each other. In the relativistic framework this corresponds to the difference of the kinetic energy of the combined system and the corresponding NM configuration at twice the subsystem rest density $2\varrho(k_F)$

$$\mathcal{E}_{\rm rel}(\varrho_{12}, u) = \langle E^* - M^* \rangle_{12} - \langle E^* - M^* \rangle_{u=0} \ . \qquad (24)$$

By definition $\mathcal{E}_{\rm rel}$ contains the kinetic energy arising from the separation of the two distributions, respecting thereby the Pauli principle, and accounts for the interaction between the two currents by the presence of the effective mass in (24). Now one obtains the binding energy per particle

$$E_{12}^{\rm bind}(\varrho_{12}, u) = \frac{T^{00} - \mathcal{E}_{\rm rel}}{\varrho_{12}} - M \qquad (25)$$

as a function of the total c.m. density ϱ_{12} and the c.m. streaming velocity $\pm u$.

The binding energy, i.e. the effective EOS in colliding nuclear matter, is shown in the lower part of Fig. 4. The effective EOS appears softer and even more attractive compared to ground state nuclear matter. The two-Fermi-ellipsoid geometry leads to a softening of the effective EOS at moderate streaming velocities which is due to the reduced Fermi pressure in transverse direction and the enlarged scalar attraction as compared to NM at rest. The vector potential is not affected by the subtraction scheme for the kinetic energy of the relative motion and fully contributes to $E_{12}^{\rm bind}$. Thus, the enhancement of the vector repulsion acts in opposite direction and makes the EOS compared to ground state NM harder at large densities and/or high streaming velocities.

4 Conclusions

To summarize, the equation of state probed by the compression phase in energetic heavy ion reactions is to large extent governed by local non-equilibrium. As an idealized scenario (cold) colliding nuclear matter was considered. For such configurations the separation of phase space acts in lowest order like an effectively new degree of freedom which counteracts the additional repulsion between the currents due to their relative motion. This is a model independent feature, i.e. the same effect occurs when different interactions like QHD-I or non-linear versions of the Walecka model are used [27]. To compare to ground state matter an effective EOS was constructed which excludes the kinetic energy from the relative motion. The net softening of the ' "effective EOS" ' by about 10 MeV due to the enlarged phase space in CNM is hardly seen on the scale of the total energy per particle but on the scale of an effective binding energy. We conclude that this effect should be taken into account when conclusions on the EOS are drawn from heavy ion collisions.

References

1. H. Stöcker, W. Greiner, Phys. Rep. **137**, 277 (1986)
2. W. Reisdorf, H.G. Ritter, Annu. Rev. Nucl. Part. Sci. **47**, 663 (1997); N. Herrmann, J.P. Wessels, T. Wienold, Annu. Rev. Nucl. Part. Sci. **49**, 581 (1999)
3. P. Danielewicz, Roy A. Lacey et al., Phys. Rev. Lett. **81**, 2438 (1998); C. Pinkenburg et al., Phys. Rev. Lett. **83**, 1295 (1999)
4. J. Aichelin, C.M. Ko, Phys. Rev. Lett. **55**, 2661 (1985)
5. C. Sturm et al. (KaoS Collaboration), Phys. Rev. Lett. **86**, 39 (2001)
6. C. Fuchs, A. Faessler, E. Zabrodin, Y.-M. Zheng, Phys. Rev. Lett. **86**, 1974 (2001)
7. B. Blättel, V. Koch, W. Cassing, U. Mosel, Phys. Rev. C **38**, 1767 (1988)
8. J. Aichelin, Phys. Rep. **202**, 233 (1991)
9. G.M. Welke et al., Phys. Rev. C **38**, 2101 (1988); C. Gale et al., Phys. Rev. C **41**, 1545 (1990)
10. P. Danielewicz, Nucl. Phys. A **673**, 275 (2000)
11. F. Rami et al. (FOPI Collaboration), Phys. Rev. Lett. **84**, 1120 (2000)
12. C. Fuchs, P. Essler, T. Gaitanos, H.H. Wolter, Nucl. Phys. A **626**, 987 (1997)
13. A. Lang, B. Blättel, W. Cassing, V. Koch, U. Mosel, K. Weber, Z. Phys. A **340**, 287 (1991)
14. L.W. Neise, H. Stöcker, W. Greiner, J. Phys. G **13**, L181 (1987)
15. Y.B. Ivanov, V.N. Russkihk, M. Schönhofen, M. Cubero, B.L. Friman, W. Nörenberg, Z. Phys. A **340**, 385 (1991)
16. H. Elsenhans, L. Sehn, A. Faessler, H. Müther, N. Ohtzuka, H.H. Wolter, Nucl. Phys. A **536**, 750 (1992)
17. L. Sehn, H.H. Wolter, Nucl. Phys. A **601**, 473 (1996); C. Fuchs, L. Sehn, H.H. Wolter, Nucl. Phys. A **A601**, 505 (1996)
18. H. Müther, A. Polls, Prog. Part. Nucl. Phys. **45**, 243 (2000)
19. B. ter Haar, R. Malfliet, Phys. Rep. **149**, 207 (1987)
20. R. Borckmann, R. Machleidt, Phys. Rev. C **42**, 1965 (1990)
21. W. Zuo, A. Lejeune, U. Lombardo, J.F. Mathiot, Nucl. Phys. A **706**, 418 (2002)
22. R. Machleidt, *The Meson Theories of Nuclear Forces and Nuclear Structure*, Advances in Nuclear Physics **19**, 189 (1989)
23. C. Fuchs, T. Waindzoch, A. Faessler, D.S. Kosov, Phys. Rev. C **58**, 2022 (1998)
24. T. Gross-Boelting, C. Fuchs, A. Faessler, Nucl. Phys. A **648**, 105 (1999)
25. W. Botermans, R. Malfliet, Phys. Rep. **198**, 115 (1990)
26. B.D. Serot, J.D. Walecka, *Advances in Nuclear Physics* **16**, 1 (1986)
27. C. Fuchs, T. Gaitanos, Nucl. Phys. Nucl. Phys. A **714**, 643 (2003)

Part III

Response to Short Laser Pulses in Semiconductors

Ultrafast Spectroscopy in the Quantum Hall Regime

Christian Schüller, Neil A. Fromer, Ilias E. Perakis and Daniel S. Chemla

Abstract. We review recent experimental and theoretical investigations of the ultrafast nonlinear optical response of the two-dimensional electron gas (2DEG) in the quantum Hall effect regime. We have investigated the dynamics of the strong 2DEG correlations using ultrafast time-resolved four-wave mixing experiments in GaAs-AlGaAs modulation-doped quantum wells. We discuss how the correlations lead to memory effects and to a strong resonant coupling between the two lowest Landau level magnetoexcitons. We compare our experimental results with simulations based on a recent theory of the nonlinear optical response of systems with a strongly correlated ground state and find a good qualitative agreement.

1 Introduction

Semiconductor nanostructures are a subject of great and continuously growing interest since many years. In particular, the two-dimensional electron gas (2DEG) in modulation-doped semiconductor structures allows the study of fundamental properties of strongly correlated electrons and collective effects in reduced dimensions. Maybe the most prominent example is the formation of strongly correlated phases of the 2D electrons in a strong magnetic field that exhibit such unique electronic transport properties as the integer and fractional quantum Hall effects (see, e.g., [1–3]). Even though correlation effects of *photoexcited* electron–hole (e–h) pairs in *undoped* semiconductors were extensively studied during the past decade by time-resolved nonlinear spectroscopy (see, for example, [4–7], and references therein), until recently very little was known about the ultrafast dynamics of strongly correlated electron systems. During the past three years, degenerate four-wave mixing (FWM) experiments have been successfully employed to highlight this fascinating regime of an interacting electron system [8–13].

Most theoretical treatments of the ultrafast dynamics of *photoexcited* semiconductor structures assume that the ground state is rigid and just provides the band structure and the high frequency dielectric screening [14]. This is a good approximation in undoped semiconductors, where the conduction band is empty and the valence band is full in the ground state, while the Coulomb-induced band coupling is negligible. In this case, the only Coulomb correlations that need to be considered are dynamically generated by the optical excitation [15]. For this situation, a commonly used approach is the dynamically controlled truncation scheme (DCTS) [16–18]. This model however breaks down if strongly correlated carriers exist in the sample prior to optical excitation and interact with the photoexcited electrons. The situation becomes particularly dramatic if the 2DEG enters the quantum Hall effect regime and supports low-energy, intra- and inter–Landau level collective excitations (see, e.g., [2,3]). The coupling of such slow degrees of freedom to the *e*–*h* pairs leads to a failure of the Boltzmann semiclassical picture of collisions localized in time and space. Here, quantum mechanical interference effects dominate the interaction process. It has been shown that in the presence of the low-energy collective excitations of a 2DEG in a strong magnetic field, one needs to account for the time evolution of the complete *coupled* system, i.e., the photoexcited *e*–*h* pairs plus the 2DEG [8,10,11,13].

The chapter is organized as follows. In the next two sections, an introduction into the linear-optical properties of a 2DEG in a strong external magnetic field, and into the experimental optical technique will be given. After setting these basics, in the fourth section, we will discuss FWM experiments, where electrons are excited into the Landau level which hosts the 2DEG [8]. With increasing magnetic field, a crossover from Markovian to non-Markovian behavior, as well as large jumps in the decay time of the FWM signal at even Landau level filling factors were observed. The main observations can be qualitatively reproduced by a model which takes into account scattering by the collective excitations of the two-dimensional electron gas and Pauli blocking.

In the fifth section, an investigation of the dynamics of the 2DEG inter–Landau level excitations will be presented [9–13]. These results are compared directly with measurements of an undoped quantum-well structure. We have observed strong time-dependent Coulomb coupling between the Landau levels induced by the presence of the 2DEG. The time dependence of the nonlinear response reveals non-Markovian and memory effects of the photoexcited system which cannot be understood in terms of the Random Phase Approximation (RPA). We introduce a new theoretical approach in the sixth section that treats the interactions of the magnetoexcitons with the 2DEG excitations and qualitatively accounts for the most salient experimental results in terms of shake-up of the 2DEG [10–13].

2 Magneto–Optics of Quantum Hall Systems

In photoexcited semiconductors, the time Δt after the creation of an *e*–*h* pair by an optical photon can be roughly divided into four temporal regimes [7]: (i) the *coherent* regime ($\Delta t < 500\,\mathrm{fs}$), where the *e*–*h* pair is in phase coherence with the exciting light field. (ii) the *non-thermal* regime ($\Delta t < 5\,\mathrm{ps}$): here, the *e*–*h* pairs have already lost their phase coherence due to, e.g., scattering with phonons or ground state electrons, but the distribution of electrons and holes cannot yet be described by a thermal distribution function with a single temperature. (iii) the *hot-excitation* regime ($\Delta t \approx 5$–$100\,\mathrm{ps}$); electrons are still not in an equilibrium distribution but can be described by a single-temperature distribution function. Finally, (iv) the *isothermal* regime ($\Delta t >$ 100 ps), where recombinations of electrons and holes (photoluminescence) takes place.

Our main focus here is on the initial coherent regime. The direct exciton–exciton interactions that dominate the nonlinear response in undoped semiconductors are screened in modulation–doped quantum wells, and the coherent nonlinear response is determined by the Fermi sea pair excitations. For zero magnetic field and resonant photoexcitation, the optical dynamics is dominated by inelastic electron-electron (e–e) scattering processes [19,20]. For *below–resonance* photoexcitation, the dissipation processes are suppressed and coherent effects dominate. A novel dynamics of the Fermi Edge Singularity is then observed [21,22], due to many-body correlations of the photoexcited holes with the Fermi sea [22–24]. The application of a magnetic field, which leads to strongly correlated ground states, extends the duration of the coherent regime. The decoherence times are critically dependent on the filling factor of the Landau levels. Importantly, the non–instantaneous Coulomb correlations lead to novel dynamics in the coherent time regime, which are discussed in this chapter.

The systems which we are considering here are 2DEG's formed in modulation-doped GaAs-AlGaAs quantum-well (MDQW) structures grown by molecular-beam epitaxy. Figure 1a shows a schematic picture of the conduction and valence-band structure of a typical sample which we used for the ultrafast experiments. The sample consists of 12 nm-wide GaAs quantum wells which were embedded in 42 nm $Al_{0.3}Ga_{0.7}As$ barriers. The 12 nm-wide central region of each barrier layer was n-doped with Si. This modulation doping leads to the formation of a 2DEG in each of the quantum wells. Due to the spatially separated positive and negative charges in the barriers and the quantum wells, respectively, a bending of the conduction and valence-band edges results (schematically shown in Fig. 1a). The electrons in the conduction band (and photoexcited holes in the valence band) can move freely perpendicular to the growth direction, inside each layer, which leads to the formation of subbands with nearly parabolic dispersion relation for the electrons (see Fig. 1b). The dispersion of the holes is more complex. We will come back to this point later when we discuss absorption experiments. In the experiments, different samples, containing one to thirty GaAs quantum wells, were used. Unless otherwise is stated, the widths of the quantum wells was 12 nm. All optical experiments were performed at low temperature, $T = 1.7$ K, where the sample was immersed in liquid superfluid helium in an optical split-coil magnet cryostat.

We will start our discussion of the optical properties of these systems by looking at experiments, which can be described essentially within the last temporal regime mentioned above, the *isothermal* regime, where electrons and holes have relaxed into quasi-thermal distributions. In this regime, *incoherent*

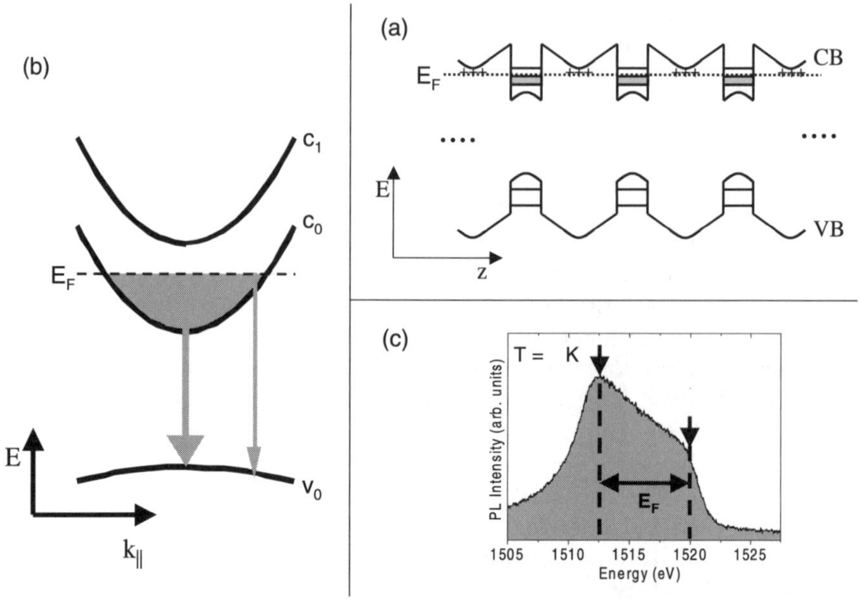

Fig. 1. (**a**) Schematical drawing of the valence- and conduction-band structure in growth direction of a modulation-doped GaAs-AlGaAs multiple quantum-well structure as used in the ultrafast experiments. (**b**) Sketch of conduction- and valence-band subbands in momentum space. Possible recombinations of electrons of the 2D Fermi sea with photoexcited holes are indicated by *arrows*. (**c**) Measured PL spectrum of a 2DEG sample with electron concentration $n = 2.3 \times 10^{11}\,\mathrm{cm}^{-2}$. For excitation, the 632 nm line of a He-Ne-laser was used

recombinations of electrons and holes, i.e., photoluminescence (PL), takes place. In Fig. 1b, in a single-particle picture, the highest and lowest energy recombinations of electrons out of the 2D Fermi sea with photoexcited holes in the valence band are indicated by vertical arrows. Figure 1c shows an experimental PL spectrum of a one-sided doped, 25 nm-wide single quantum-well structure. The signal shows an onset at the recombination energies of electrons and holes sitting at the band edges, and a cutoff at the Fermi energy (cf. Fig. 1b). Neglecting the small curvature of the hole subband, one can directly determine the Fermi energy out of this measurement as indicated in Fig. 1c. The shape of the PL signal is determined, mainly, by the density of occupied states of electrons and holes, which is constant for a 2DEG, and by the distribution of the relaxed photoexcited carriers.

The situation changes if a perpendicular magnetic field is applied. Then, the constant density of states condenses into discrete Landau levels (see Fig. 2a). Each Landau level exhibits a Zeeman splitting (omitted in Fig. 2a), $g\mu_B B$. The g-factor of electrons in GaAs, $g = -0.44$, arises from band-structure effects (μ_B is the Bohr magneton), while additional splittings arise

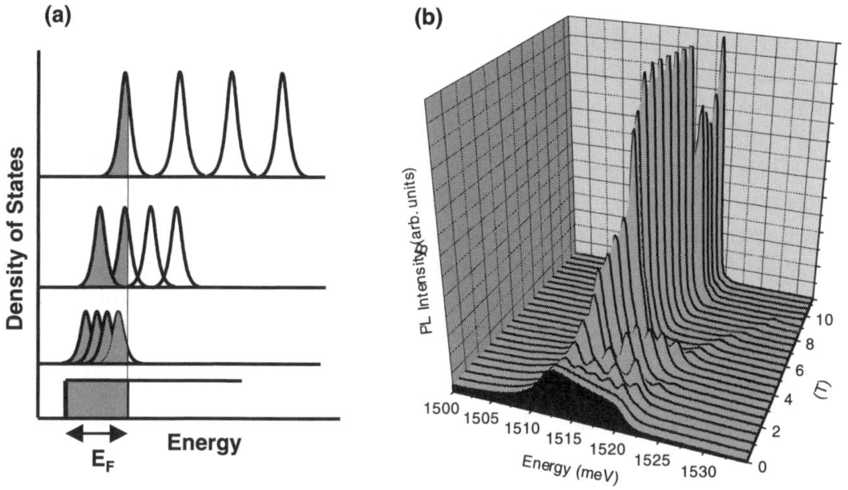

Fig. 2. (**a**) Sketch of the density of states of a 2DEG for increasing magnetic fields (from *bottom* to *top*). Levels which are occupied by electrons of the 2DEG are shaded. Due to residual disorder, the Landau levels in real systems have a finite widths. (**b**) Series of experimental PL spectra for different magnetic fields. One can see the emission lines which originate of electrons of the 2DEG which recombine with photoexcited holes

from the strong Coulomb correlations. Each spin-split Landau level has a degeneracy, which is determined by the number of magnetic flux quanta per area, eB/h. The filling factor ν of the system is defined as the number of electrons in the 2DEG devided by the number of flux quanta, $\nu = nh/(eB)$, where n is the density of electrons in the 2DEG. This means that if ν is an integer, we have ν completely filled spin-split Landau levels. At these integer values, plateaus are observed in a Hall measurement. Figure 2b displays a series of experimental PL spectra for different magnetic fields of the same sample as shown in Fig. 1c. One can see the evolution of occupied Landau levels with magnetic field. At even ν (spin-splitting is not resolved here) the Fermi energy jumps from one Level to the next lower level, and the PL from the higher level, which is empty now, decreases abruptly (e.g., at $B = 4.5\,\mathrm{T}$ in Fig. 2b where $\nu = 2$). The weak signal from the second Landau level (LL1[1]) for magnetic fields larger than 4.5 T in Fig. 2b stems from photoexcited e–h pairs. At the same time, at integer ν, there are pronounced anomalies in the intensity and energetic position of the recombination from the lowest Landau

[1] Since the spin splitting is small compared to the cyclotron energy, $\hbar eB/m^*$, for simplicity, in the labeling of the Landau levels, LLN, we do not consider the spin-splitting, i.e., for $0 < \nu < 2$ the Fermi energy is in LL0, for $2 < \nu < 4$ the Fermi energy is in LL1, etc.

level (LL0). These can be especially seen at $\nu = 2$ ($B = 4.5$ T in Fig. 2b) and $\nu = 1$ ($B = 9$ T in Fig. 2b) where the 2DEG is completely spin polarized. There is a variety of literature on these many-particle interaction induced effects in PL (see, e.g., [25–29]), which shall not be the focus of this work.

We turn now to the discussion of linear absorption spectra, which give access to the density of states of the *unoccupied* levels. Figure 3 displays a series of spectra of a typical multiple quantum-well sample as it was used for FWM experiments. The spectra were measured using a broadband incoherent light source. At $B = 0$ T, the spectrum reflects the constant density of states of the unoccupied levels, with two maxima at the onset, which are due to the Fermi edge singularity effect for the heavy-hole and light-hole transitions [24,23,21]. Around $B = 4$ T, the Fermi energy jumps into the lowest spin-down Landau level (filling factor $\nu = 2$), and, for magnetic fields $B > 4$ T, we observe absorption into this level (LL0 in Fig. 3) due to the increasing number of empty states. A closer look onto the spectra shown in Fig. 3 already exposes some of the interaction effects caused by the presence of the 2DEG, which will be studied in detail later in this chapter: While the linewidth of LL0 is approximately constant, the linewidth of LL1 increases significantly once LL0 starts to appear in the spectrum. The increase in the LL1 linewidth for smaller filling factors implies an increase in the dephasing rate of this state

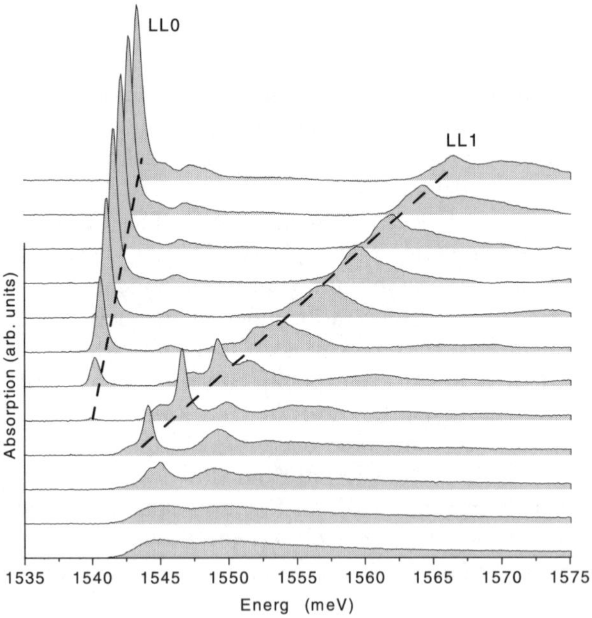

Fig. 3. Linear absorption spectra of a multiple quantum-well sample with ten 12 nm-wide GaAs quantum wells for different magnetic fields. The spectra are taken with right-circularly polarized light ($\sigma+$)

due to scattering with the 2DEG, which for these fields is entirely in LL0. The increased dephasing rate is a result of additional scattering mechanisms available for LL1 excitations due to the presence of the 2DEG. We should note here that the additional structures, which are present in the spectra in Fig. 3, especially close to the LL1 transition, are caused by transitions from different hole Landau levels: in fact, the structure of the hole levels is more complicated than the simple electron Landau levels. This is caused by the mixing of heavy- and light-hole states at finite magnetic field (see, e.g., [30,31]) which leads to finite oscillator strengths for transitions from more than only one hole Landau level to the second electron Landau level (LL1 in Fig. 3). However, for right-circularly polarized light (σ^+) the situation for LL0 is quite simple [31]. Here, only one heavy-hole level with $m_j = -3/2$ contributes to the transition LL0 to the $m_j = -1/2$ electron state, which gives rise to the sharp LL0 peak in Fig. 3 (m_j is the z-component of the angular-momentum quantum number). In σ^- polarization there are several transitions to both of the lowest spin-split electron levels ($m_j = \pm 1/2$) [11,31], which complicates the situation. Therefore, for simplicity, in the remainder of this chapter we restrict the discussion to experiments with σ^+ light.

3 The Ultrafast Optical Technique

Nowadays there are pulsed solid state lasers available with pulse durations down to a few femtoseconds. There is a broad variety of ultrafast nonlinear optical techniques (for an overview see, e.g., [7]), which, in the past decade, revealed a wealth of very interesting Coulomb correlation effects in – in most cases undoped – semiconductors. In our investigations of quantum Hall systems we used degenerate FWM experiments to explore the dynamics of e–h pairs interacting with a quantum Hall liquid. Figure 4 shows a schematic picture of these ultrafast time-resolved optical experiments. The experiments were performed with two laser beams ($\mathbf{k_1}$ and $\mathbf{k_2}$ in Fig. 4) of equal intensity which were in resonance with transitions from valence-band to conduction-band Landau levels. The beams were circularly polarized and separated by a time delay Δt. The first pulse ($\mathbf{k_2}$) creates a polarization in the sample. If Δt is within the *coherent* regime, the interactions of the polarization waves propagating along $\mathbf{k_1}$ and $\mathbf{k_2}$ create an interference pattern inside the sample. Then, photons from $\mathbf{k_2}$ are self-diffracted into the nonlinear direction $2\mathbf{k_2} - \mathbf{k_1}$, forming the FWM signal. In time-integrated (TI-FWM) experiments, the total signal intensity was measured by a photomultiplier tube, while for spectrally resolved experiments (SR-FWM) a spectrometer with 0.75 m focal length and a charge coupled-device camera were used to record the spectrum of the emitted signal. In Figs. 4b–d, experimental situations as used for the FWM experiments are sketched. Either the wavelength of the femtosecond laser pulse was chosen to resonantly excite electrons into the electron Landau level which hosts the Fermi energy

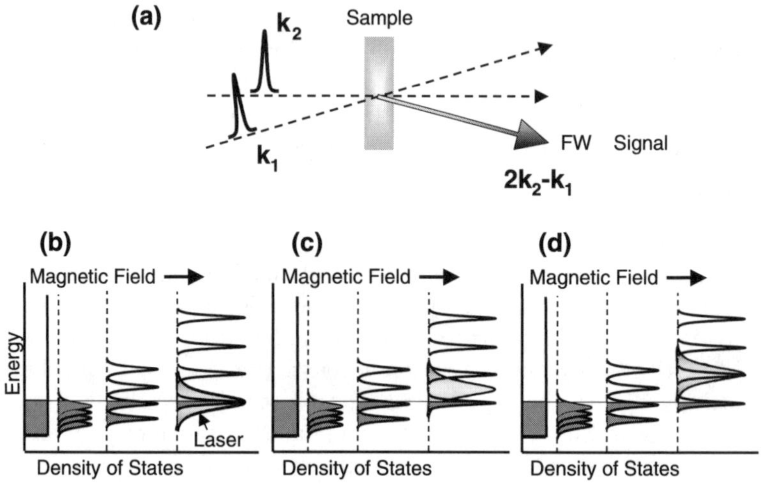

Fig. 4. (**a**) Schematic picture of the experimental arrangement for FWM experiments. (**b**)–(**d**) Sketches of the densities of states of a 2DEG for different magnetic fields. Occupied levels are *gray-shaded*. The spectrum of a laser pulse is drawn in (**b**)–(**d**) for different experimental situations, as explained in the text

(Fig. 4b), or, to excite both, the occupied and the next higher unoccupied level (Fig. 4c), or, electrons were excited into the next higher unoccupied level (Fig. 4d).

4 Intra–Landau Level Excitations

In this section we discuss experiments where we have investigated the interband polarization in the presence of intra–Landau level excitations. This was done by tuning a spectrally narrow laser pulse of $\tau = 300\,\text{fs}$ duration into resonance with the Landau level which hosts the Fermi energy, i.e., which is partially filled with electrons (cf. Fig. 4b). The excitation intensity was kept low enough for the density of photo-created e–h pairs, n_{eh}, to remain small compared to the density n of the 2DEG, typically $n_{eh} < 0.1 \times n$. We have observed very strong variation of the interband dephasing time, T_2, as a function of the filling factor, as well as direct evidence of memory effects in the optical dynamics [8]. In a strong magnetic field, such that the 2DEG occupies only LL0, there are no interactions between the photoexcited e–h pairs, unless there is an asymmetry between electron and hole wavefunctions [32]. When LL0 is partially filled, the dephasing originates mainly from the scattering of the photoexcited carriers with the intra–Landau level collective excitations of the strongly correlated 2DEG [33–35].

In these series of experiments we have investigated two multiple quantum-well samples (30 periods each), which differ in the carrier densities of the

2DEG. Sample A has a carrier density of $n_A = 2.6 \times 10^{11}\,\text{cm}^{-2}$, and sample B has $n_B = 4.9 \times 10^{11}\,\text{cm}^{-2}$. Figure 5 displays a typical SR-FWM spectrum of sample A for a magnetic field of $B = 11\,\text{T}$. At this field, the Fermi energy is in LL0 ($\nu \approx 0.94$), and the splitting of the Landau levels, $\sim \hbar\omega_c \approx 38\,\text{meV}$, is large enough so that a laser pulse with $\approx 7\,\text{meV}$ energetic width, tuned in resonance with LL0, excites solely e–h pairs within this level. In the semi-logarithmic plot in Fig. 5 we can immediately see two important experimental facts: (i) the decay of the SR-FWM signal (S_{SR}) is not a single exponential. It shows a kink around $\Delta t = 3\,\text{ps}$. (ii) The spectral width and shape of S_{SR} depend strongly on Δt. The shape narrows for longer delay times Δt. In the following we will first discuss this experimental behavior in more detail and then give an interpretation.

A detailed investigation of the magnetic-field dependence of the decay of S_{SR} is shown in Fig. 6. It displays traces through the energetic maxima of S_{SR} of sample A for different magnetic fields, i.e., different filling factors of the 2DEG. One can see that for low fields the decay is almost a single exponential with an unusually long decay time, while for larger fields it becomes more and more nonexponential, developing the above mentioned kink around $\Delta t =$ 3 ps. A strong variation of the decay of S_{SR} is observed when ν is tuned from values larger than 2 to values below 2, i.e., when the Fermi energy jumps from LL1 to LL0. By extracting an overall decay time we can get a direct measure of the interband dephasing time T_2. Figure 7 displays the experimentally determined decay times of samples A and B versus magnetic field. It is striking to note the very large jump of T_2 each time the system passes through even filling factors. Since these features are reproducible as

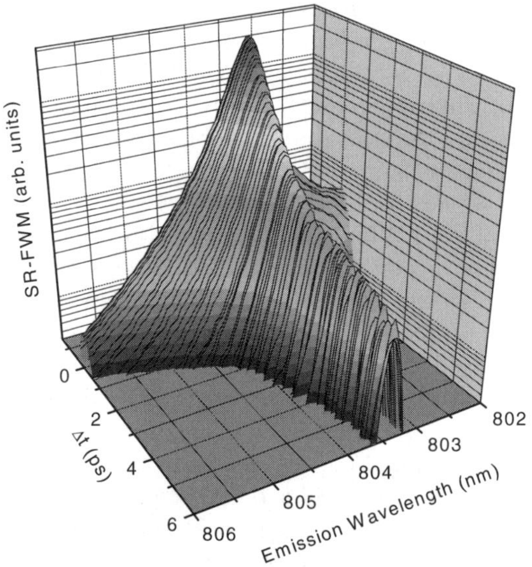

Fig. 5. Spectrally-resolved FWM spectra of sample A on a logarithmic scale for a magnetic field of $B = 11\,\text{T}$, where the Fermi energy is in LL0 ($\nu \approx 0.94$). The laser is tuned to resonantly excite LL0 only

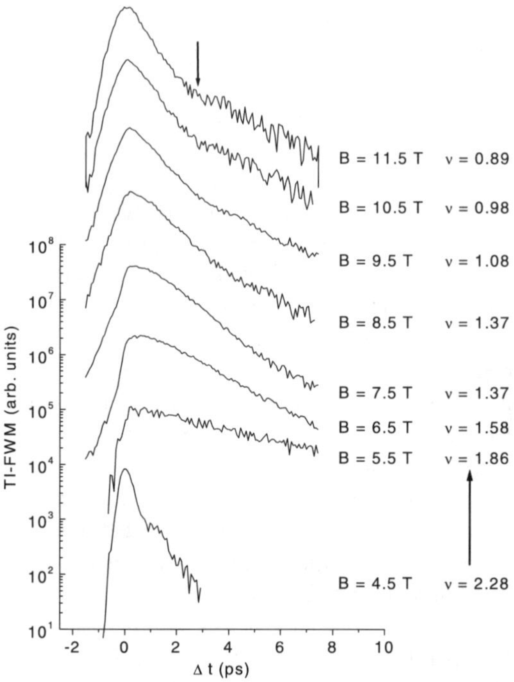

Fig. 6. TI-FWM spectra of sample A for different magnetic fields

a function of ν for samples with different densities, we can assert that this is an effect of the cold 2DEG. Notice that as we approach filling factor $\nu = 2$ from below (higher field), the signal becomes too weak to measure, since the oscillator strength of the LL0 transition goes to zero. Because of this, we can not measure the maximum value of T_2 in this case. Similarly, for $\nu > 2$ (lower field), T_2 becomes very short, but we are unable to accurately measure this since the decay time becomes shorter than the 300 fs pulse duration. In fact, the jump in dephasing time may be larger than what we have presented here.

The nonexponential behavior of the FWM signal at high fields is characterized by a change of the slope that occurs in sample A at $\Delta t \approx 4.2\,\text{ps} \rightarrow 2.5\,\text{ps}$ as $B \approx 7.5\,\text{T} \rightarrow 11.5\,\text{T}$. This change in slope indicates memory effects in the polarization dynamics, which are also seen in the frequency domain. Figure 8a displays S_{SR} at fixed $\Delta t = 0$ for different magnetic field, and Fig. 8b shows S_{SR} at fixed field $B = 11\,\text{T}$ for different $\Delta t = 0...6$ ps. Clearly, the $S_{SR}(\omega)$ profile changes from a Lorentzian with a constant width, $\Gamma \propto T_2^{-1}$, to an asymmetric one that would correspond to a frequency dependent linewidth, $\Gamma(\omega)$. Such a profile indicates a polarization relaxation term $\propto \Gamma(\omega)P(\omega)$. A frequency dependent scattering rate $\Gamma(\omega)$ is a result of non-Markovian dynamics, or dephasing with a memory kernel, which gives

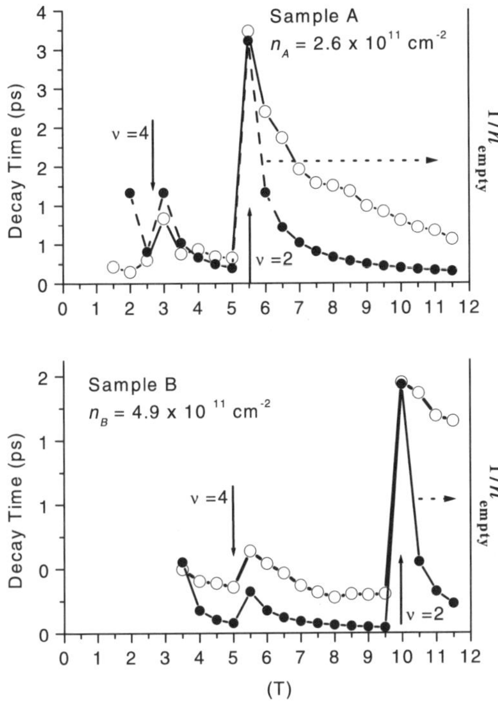

Fig. 7. Experimental decay times (*open circles*) of the TI-FWM signals of samples A and B versus magnetic field. The *full circles* are results of a simple model as described in the text

a scattering term

$$\frac{\partial P}{\partial t} = \int_{-\infty}^{t} \Gamma(t-t')P(t')dt' \tag{1}$$

in the time domain. We also note that if the $S_{SR}(\omega)$ spectra are asymmetric, they are redshifted from the absorption spectra, while if they are Lorentzian they almost coincide with the peaks in the absorption (see Fig. 8a).

In the following we want to give an interpretation of the observed experimental results. One can expect, on general physical grounds, that the dephasing of a photo-created electron is proportional to the number of available states into which the electron can be scattered (Pauli blocking). If the Fermi energy is in the *Nth* Landau level (neglecting spin), then the number of empty states n_{empty} inside this level is proportional to

$$n_{\text{empty}} \propto (2(N+1) - \nu)/\nu. \tag{2}$$

The factor of 2 is for the spin. In the case that the Fermi energy is in LL0, Eq. (2) reduces to $2\nu^{-1} - 1$. Then, the memory kernel of Eq. (1) can be presented as $\Gamma(t-t') = (2\nu^{-1} - 1)\kappa(t-t')$, where $\kappa(t-t')$ represents all other

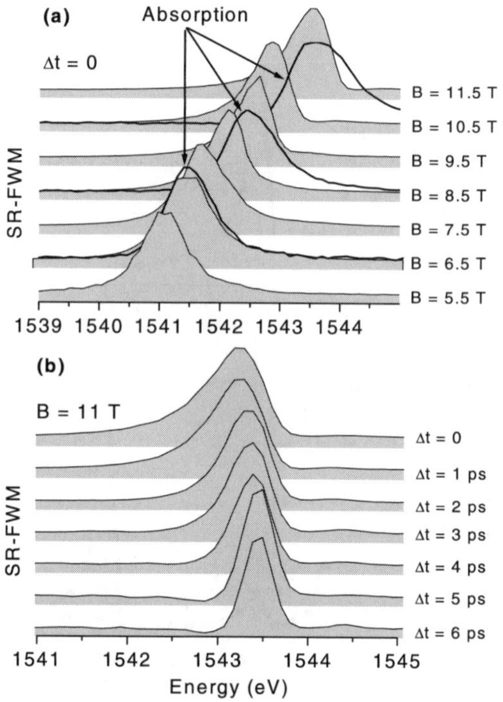

Fig. 8. (**a**) SR-FWM spectra of sample A at fixed $\Delta t = 0$ for different magnetic field. For three magnetic-field values the linear absorption spectra are shown for comparison. (**b**) SR-FWM spectra of sample A for fixed $B = 11\,\mathrm{T}$ for different Δt

dephasing processes, besides Pauli blocking. In addition to scattering with intra–Landau level collective excitations, there are several other relaxation processes which contribute to the dephasing at weaker fields, e.g., phonon and impurity scattering, Auger-like processes, etc. [36]. These background processes, however, lead to Markovian dephasing, i.e., $\kappa(t) \to \delta(t)$, which would lead to symmetric lineshapes of $S_{SR}(\omega)$. In Fig. 7 we have plotted n_{empty}^{-1} as a function of magnetic fields (solid circles in Fig. 7), which describes the above Pauli-blocking contribution to the dephasing. We have fitted these curves to the experimental data by setting the maximum points equal to one another. As mentioned above, we are unsure exactly how high the maximum T_2 is, or how low the minimum value is. Nevertheless, the agreement is striking in terms of the location and magnitude of the steps. However, there are significant differences in the B-dependence of T_2 for strong field. In particular, the change in behavior occurs for sample A at $B > 7.5\,\mathrm{T}$, where we begin to see the nonexponential behavior in Fig. 6, or the asymmetry in Fig. 8. We attribute this observed transition from Markovian to non-Markovian behavior to the non–instantaneous scattering of the photoexcited magnetoexcitons with the intra–LL0 collective excitations of the 2DEG, which, for LL0 photoexcitation, dominates as the cyclotron energy, $\hbar\omega_c$, exceeds the other characteristic energies of the system. Because of the breakdown of the perturbation and standard RPA theories [37] due to the Landau level degeneracy, one must account for the true excitations of the strongly correlated 2DEG, which are collective in nature (e.g. the magnetorotons [1]). In [8] we sketch how the magnetoexciton–collective 2DEG excitation interactions lead to a frequency–dependent dephasing width. The latter is determined by the dynamical screening of the Coulomb interaction by the intra–Landau level collective excitations.

5 Inter–Landau Level Dynamics

In this section, we present an investigation of the nonlinear dynamics of the 2DEG inter–Landau level excitations. The nonlinear measurements were performed on a 10-period MDQW sample with carrier density $n = 2.1 \times 10^{11}\,\mathrm{cm}^{-2}$. The linear absorption spectra of this sample was shown in Fig. 3. For most of the measurements in this study, the total number of carriers excited by the laser was kept below $2 \times 10^{10}\,\mathrm{cm}^{-2}$, or $0.1 \times n$. Comparison measurements were made on an undoped quantum-well (QW) sample with similar well and barrier sizes to clearly identify the effects related to the presence of the 2DEG in the MDQW. We performed SR-FWM experiments with a laser pulse duration of $100\,\mathrm{fs} < \tau < 200\,\mathrm{fs}$. The laser was tuned to excite varying proportions of LL0 and LL1, and the beams were σ^+ circularly polarized. We will first describe the experimental results, and then, in the next section, give the theoretical interpretation.

First, we want to discuss experiments were the laser energy was tuned to excite almost equal proportions of LL0 and LL1. This corresponds to an experimental situation as sketched in Fig. 4c. Typical SR-FWM signals, $S_{SR}(\Delta t, \omega)$, for both the doped and the undoped samples are shown in Fig. 9. The spectra of the laser and the linear absorption are projected on the back panels, for comparison. Several unusual features are immediately apparent in the signal from the doped sample, $S_{SR}^{\mathrm{doped}}(\Delta t, \omega)$, Fig. 9a. The most striking is that despite an equal excitation of both Landau levels, the MDQW shows a LL0 signal which is 35 times larger than the LL1 signal. On the other hand, the measurement of the undoped QW, $S_{SR}^{\mathrm{undoped}}(\Delta t, \omega)$, Fig. 9b, shows almost equal emission from both Landau levels, in proportion to the excitation. We find that the spectral distribution of the signal from the undoped sample approximately follows the mean–field Hartree–Fock (HF) theory [39], but the signal from the MDQW sample is drastically different. Furthermore, although we see emission almost entirely from LL0, the signal has very pronounced beats as a function of Δt, with a period given by the inverse of the energy difference between LL0 and LL1. Such strong beating in Δt from only a single emission energy is a clear signal of non–Markovian dynamics. Comparing this

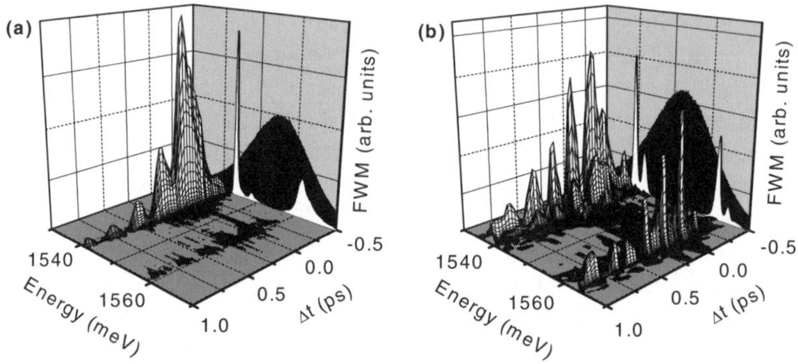

Fig. 9. (**a**) SR-FWM spectra of a MDQW sample for equal excitation of LL0 and LL1. (**b**) SR-FWM spectra of an undoped quantum well for the same excitation conditions

to the signal from the undoped QW, we see that $S_{SR}^{\text{undoped}}(\Delta t, \omega)$ also shows beats, but from both emission peaks, as expected from the HF theory.

The picture is just as unusual when we tune the laser frequency to excite almost entirely into LL1, exciting 60 times the carriers into LL1 than into LL0. Figure 10a shows $S_{SR}(\Delta t = 0, \omega)$, the FWM spectra for $\Delta t = 0$, for both samples under these excitation conditions. It is clear that the signal from LL0 is greatly enhanced relative to LL1 in the MDQW. In the undoped sample, there is almost no signal from LL0, as expected from the excitation (60:1 for LL1:LL0), while in the doped sample the LL0 signal is comparable to the LL1 signal. In addition to the transfer of oscillator strength to LL0, $S_{SR}^{\text{doped}}(\Delta t, \omega)$ also shows a very unique dependence on Δt when we preferentially excite LL1. According to the HF theory for FWM in semiconductors, the rise time of the $\Delta t < 0$ signal should be 1/2 the decay time for $\Delta t > 0$, and this is the

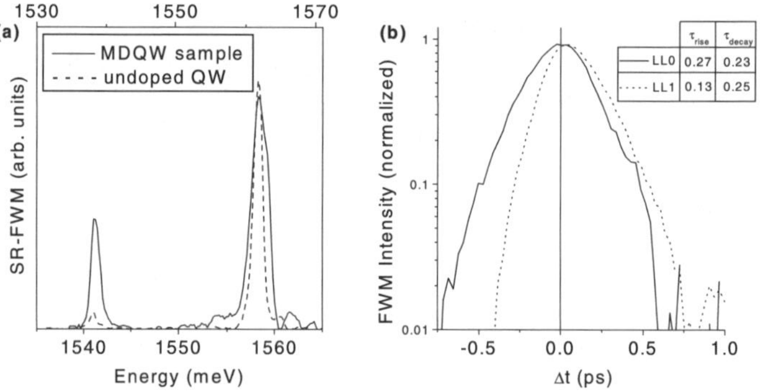

Fig. 10. (**a**) Comparison of SR-FWM spectra of a MDQW sample and an undoped QW for resonant excitation of LL1. (**b**) Time dependence of FWM intensities of the MDQW at the energies of LL0 and LL1

measured result for the undoped QW sample. This is also the measured result for the signal from LL1 in the MDQW, but surprisingly the signal from LL0 is almost symmetric as a function of Δt, with comparable signals for $\Delta t < 0$ and $\Delta t > 0$. Figure 10b shows the dependence of S_{SR} on Δt at the emission maxima of LL0 and LL1 for the MDQW sample. Such a large signal for $\Delta t < 0$ can only be a result of correlation effects beyond the HF theory [40]. However, the effect is only seen in the signal from LL0, and only in the doped sample, which implies that in this case the correlations are induced by the presence of the 2DEG in the doped sample.

Since we are exciting the first two Landau levels in these experiments, we expect that the inter–Landau level excitations of the 2DEG, the magnetoplasmons (MP), are important for understanding these results. At long wavelength, the MP energy is close to the inter–Landau level spacing, so we must account for the almost resonant creation and destruction of the MP excitations non-perturbatively. In particular, it is possible for a photoexcited LL1 electron to scatter into LL0 while exciting the 2DEG. The scattering to this new state provides additional dephasing for the LL1 photoexcited carriers, which will affect the FWM signal.

By changing the magnetic field, we confirmed that the beat frequency is seen in the signal from LL0, when we excite both levels, and is very close to the Landau-level spacing. This is shown in Fig. 11, which shows $S_{SR}^{\text{doped}}(\Delta t)$ at the LL0 energy for $B = 6, 8$, and $10\,\text{T}$ when we excite both, LL0 and LL1, equally. Further detailed experiments on exciting only into LL1 revealed that the enhanced LL0 signal is only present for filling factors $\nu < 2$, i.e., if LL0 is

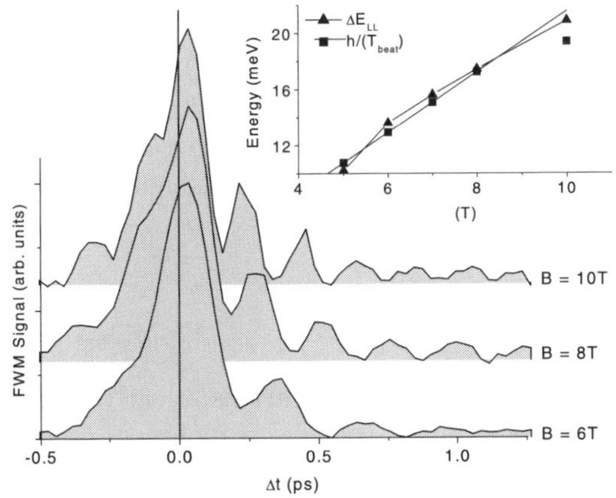

Fig. 11. FWM emission from LL0 versus time delay for the MDQW sample as a function of magnetic field, when the laser is tuned to excite LL0 and LL1 equally. The *inset* shows the comparison between the Landau level spacing measured in the absorption spectrum (*triangles*) and the inverse of the beat period T_{beat} seen in the LL0 FWM signal (*squares*)

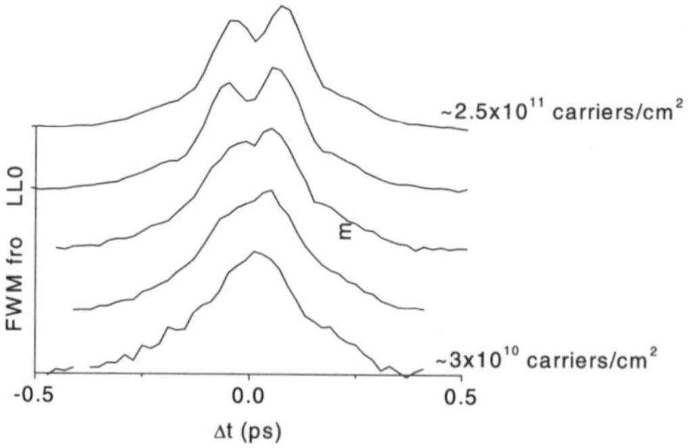

Fig. 12. FWM emission from LL0 versus time delay for the MDQW sample as a function of excitation density, at $B = 10$ T, with only LL1 directly excited by the laser

partly empty. We can conclude that while the 2DEG strongly enhances the signal from LL0 relative to LL1, this enhancement only exists at magnetic fields for which there is available space in LL0 before the excitation.

Finally, we have also measured $S_{SR}^{\text{doped}}(\Delta t, \omega)$ and $S_{SR}^{\text{undoped}}(\Delta t, \omega)$ as a function of the incident power, varying the photo-carrier density in the range $0.1 \times n \to n$, where n is the density of the 2DEG in the MDQW ($n = 2.1 \times 10^{11}\,\text{cm}^{-2}$). The evolution of the FWM signal as the excitation power is increased is shown in Fig. 12. The LL0 emission begins to develop weak beats as a function of Δt, with a very pronounced minimum at $\Delta t = 0$. In detailed measurements [12] we found that the enhancement of the LL0 signal in the MDQW decreases with increasing excitation power, and the doped and undoped samples begin to look more similar in their overall nonlinear optical response. This can be understood qualitatively, since as the density of photoexcited carriers approaches that of the electron gas, the exciton–exciton interactions begin to dominate over the signal due to the cold 2DEG correlations.

6 Theoretical Interpretation

To understand the experimental observations, clearly a detailed theory for describing the ultrafast nonlinear optical response of a system with strong ground state correlations is required. Such a theory was developed in [11]. To connect with the experiments and clarify the main physical ideas involved, we derived a simple model from the full theory. Such "Average Polarization Models" have greatly aided in understanding the many-body processes responsible for the FWM signal in undoped semiconductors [40,42]. We note that the qualitative features of the dynamics are robust and do not depend sensitively on our assumptions about the different interaction parameters. For a detailed description, we refer the reader to [11,13].

We only consider the two Landau levels excited in the experiments, LL0 and LL1. Using the conventions and approximations discussed in [11], we extract the following model equations from the full theory:

$$\begin{aligned} &i\partial_t P_0(t) = (\Omega_0 - i\Gamma_0)P_0(t) - V_{01}P_1(t) - W\bar{P}(t) \\ &+2\mu\mathcal{E}(t)P_0^L(t)P_0^{L*}(t) + 2V_{01}\left[P_1^{L*}(t) - P_0^{L*}(t)\right]P_1^L(t)P_0^L(t) \\ &+W_M\mathcal{M}^*(t)\left[P_1^L(t) - P_0^L(t)\right]. \end{aligned} \quad (3)$$

Here, P_n and P_n^L describe the nonlinear and linear polarization, respectively, of the n-th Landau level. The first line of Eq. (3) contains the energy (Ω_0) and damping (Γ_0) of the LL0 magnetoexciton, as well as the coupling between the Landau levels, $V_{01}P_1(t)$.

The correlation function $\bar{P}(t)$ describes the time propagation of the X+MP states (denoted as Y in the following) that arise from the scattering of the photoexcited exciton (X) with the 2DEG, with strength described by the parameter W defined in [11]. Neglecting the frequency–dependence of the self–energy due to the X–MP scattering and the excitation–induced nonlinear contributions to such dephasing, the equation of motion for $\bar{P}(t)$ reads [11]

$$i\partial_t \bar{P}(t) = (\bar{\Omega} - i\gamma)\bar{P}(t) + [P_1(t) - P_0(t)]. \quad (4)$$

The above treatment of the X–MP scattering, which occurs via the exchange of center of mass momentum between the X and the MP, is similar to the average polarization model treatment of the X–X scattering of magnetoexcitons [43], which is valid for strong X–X interactions and magnetic fields. In the above equation, $\bar{\Omega}$ is the average energy of the X+MP states, and exceeds the zero–momentum magnetoexciton energy Ω_1 due to the strong contribution of finite center of mass momentum MPs close to the magnetoroton minimum [1,44].

The second line of Eq. (3) gives the driving terms for $P_0(t)$, which are similar to the HF theory for undoped semiconductors of [39]. The first term of the second line is the familiar Pauli blocking nonlinearity, which exists even in atomic systems, and comes from the fact that the excitations obey the Pauli exclusion principle. It is proportional to the coherent density $|P_n^L(t)|^2$, and can be thought of as the scattering of a laser photon with the coherent density of photoexcited carriers. The second term is the nonlinearity due to the HF X–X interactions. Similar to the Pauli blocking term above, we can describe this term as the scattering of the photoexcited polarization with the coherent density of photoexcited carriers. We refer to this nonlinearity as the bare Coulomb interaction. While these effects are also found in the HF model of FWM in undoped semiconductors, the additional dephasing and screening from the 2DEG lead to a qualitative difference in the FWM spectrum.

The third line of Eq. (3) describes excitation–induced dephasing effects entirely absent in the undoped case. This source term, which describes the

MP correlations, comes from the photoexcitation and time evolution of the long–lived MP state, described by the MP correlation function $\mathcal{M}(t)$, that obeys the equation of motion [11]

$$i\partial_t \mathcal{M} = (\Omega^M - i\gamma_M)\mathcal{M} - P_1^L P_0^{L*} + P_0^L P_0^{L*} + P_1^L P_1^{L*} - P_0^L P_1^{L*}. \qquad (5)$$

The weak MP damping, described by the dephasing rate γ_M, enhances the non-Markovian dephasing effects. As already seen in inelastic light scattering spectra, the disorder leads to the strong photoexcitation of finite momentum MP's close to the magnetoroton energy [44]. The energy Ω_M is the average energy of the coupled MP states, and exceeds the cyclotron energy that gives the zero momentum MP energy. The above term describes FWM processes such as the one shown schematically in Fig. 13. A photoexcited X decays into a Y (X+MP) excitation, with a rate determined by W_M as defined in [11]. The e–h pair in the four–particle excitation Y recombines leading to the coherent emission, which leaves the 2DEG in an excited state. This MP propagates in time and then interacts with the second photoexcited X into a new X state, which is subsequently annihilated by the optical field. It is interesting to note the similarity of this process to the familiar one of coherent anti-Stokes Raman scattering [41] that, however, involves phonons.

The equation of motion for $P_1(t)$ contains similar source terms,

$$\begin{aligned}
&i\partial_t P_1(t) = (\Omega_1 - i\Gamma_1)P_1(t) - V_{10}(1-\nu_0)P_0(t) + W\bar{P}(t) \\
&+2\mu\mathcal{E}(t)P_1^L(t)P_1^{L*}(t) - 2V_{01}(1-\nu_0)\left[P_1^{L*}(t) - P_0^{L*}(t)\right]P_1^L(t)P_0^L(t) \\
&-W_M(1-\nu_0)\mathcal{M}^*(t)\left[P_1^L(t) - P_0^L(t)\right]. \qquad (6)
\end{aligned}$$

Here, ν_0 is the filling factor of the LL0 states occupied by the 2DEG. By fitting the calculated linear polarization spectrum to the linear absorption measurements taken to characterize our sample, we can fix the parameters entering in the above equations to within ±50%.

For the case where we excite LL0 and LL1 equally, the simulated FWM signal $S_{SR}^{\mathrm{model}}(\Delta t, \omega)$ is presented in Fig. 14b, along with the experimental results $S_{SR}^{\mathrm{doped}}(\Delta t, \omega)$ (Fig. 14a). As the figure clearly shows, the simulations

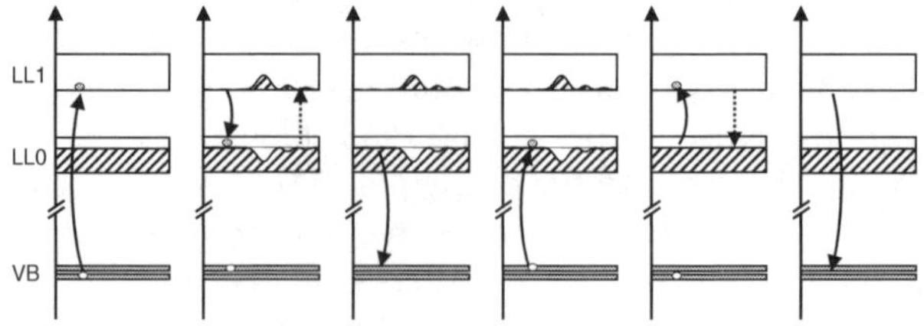

Fig. 13. Stokes (*first three panels*) and anti-Stokes (*last three panels*) magnetoplasmon correlation contributions to the FWM signal

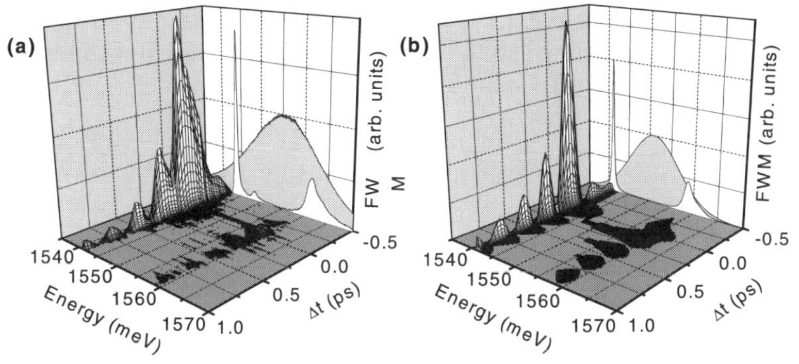

Fig. 14. (**a**) SR-FWM signal from the MDQW sample at $B = 8\,\mathrm{T}$ (same as Fig. 9a). (**b**) The simulated signal for the same conditions as (**a**). The laser pulse and absorption spectra are projected on the *back panel*

are able to recreate both the transfer of signal strength to LL0 and the pronounced beats coming from only the single level. The beat period in the simulation is given by the inverse of the Landau-level energy difference, as in the experiment.

We note that when we move the laser to excite only into LL1, we can also recreate the transfer of signal strength from LL1 to LL0 observed in the experiment. For the simulations we find a value of this enhancement which is within 15% of the experimental one [12]. The model is able to describe the time dependence of this signal as well. Recall that for excitation of LL1, the LL0 signal had a very large $\Delta t < 0$ signal, so that the signal was almost symmetric as a function of time delay. In Fig. 15 we show $S_{SR}^{\mathrm{model}}(\Delta t)$ at the LL0 and LL1 emission energies, compared to the experimental results. The

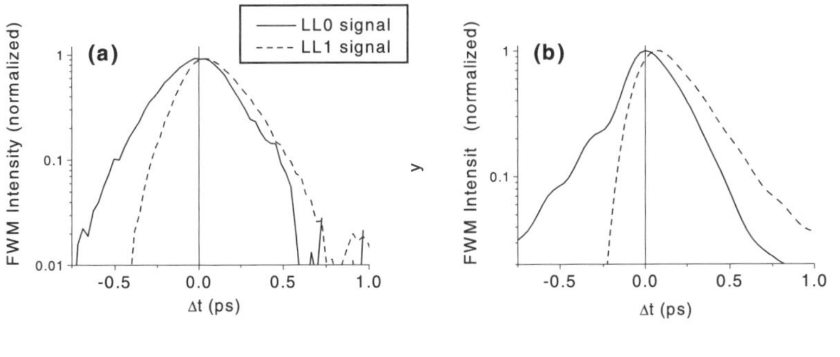

Fig. 15. (**a**) Experimental FWM signal from the MDQW sample at $B = 8\,\mathrm{T}$ when we excite LL1 preferentially (60:1 excitation of LL1:LL0), at the L00 emission energy (*solid curve*) and the LL1 emission energy (*dashed curve*). (**b**) The simulated signal for the same conditions as (**a**)

$\Delta t < 0$ signal is much larger from LL0 than from LL1, as in the experiment. This behavior can be traced to the non-Markovian dephasing of the polarization and the enhanced LL1 dephasing due to the non-instantaneous X $\rightarrow$ Y scattering.

7 Conclusion

In conclusion, we have investigated the quantum coherence of electron–hole pairs in modulation-doped GaAs-AlGaAs quantum wells in the quantum Hall regime. We have shown that the presence of the 2DEG leads to a time-dependent Coulomb coupling of the photoexcited carriers which cannot be treated using the conventional theories for understanding correlations in undoped semiconductors. At high magnetic fields, the FWM signal shows strong evidence of memory effects. We proposed a model based on scattering of the photoexcited electrons with magnetorotons in the lowest Landau level that accounts for the main features of experiments, where the lowest Landau level is excited resonantly. Experiments also on the next higher Landau level exhibit a strong inter–Landau level coupling with an unusual time dependence due to the propagation in time of long-lived magnetoplasmon excitations and their interactions with the photoexcited carriers.

Acknowledgements

We would like to thank many colleagues who contributed to this work as listed in the references. The work was supported by the Alexander von Humboldt Foundation, and the Deutsche Forschungsgemeinschaft via project SCHU1171/1 and a Heisenberg grant SCHU1171/2 (C. S.), the U.S. DOE, under contract No. DE-AC03-76SF00098 (Berkeley), and by U.S. DOE, Grant No. DE-FG02-01ER45916 and DARPA/SPINS (I. E. P.).

References

1. T. Chakraborty and P. Pietiläinen, *The Quantum Hall Effects* (Springer, Berlin, Heidelberg, New York, 1995).
2. *Perspectives in Quantum Hall Effects*, edited by Das Sarma and A. Pinczuk (Wiley, New York, 1997).
3. H.L. Stormer, D.C. Tsui, and A.C. Gossard, Rev. Mod. Phys. **71**, S298 (1999).
4. S. Mukamel, *Principles of nonlinear optical spectroscopy* (Oxford University Press, New York, 1995).
5. D.S. Chemla and J. Shah, Nature (London) **411**, 549 (2001), and references therein.
6. D.S. Chemla, in *Nonlinear Optics in Semiconductors*, edited by E. Garmire and A. Kost (Academic Press, San Diego, 1999), Vol. 58, p. 175, and references therein.

7. J. Shah, *Ultrafast spectroscopy of semiconductors and semiconductor nanostructures* (Springer, Berlin, New York, 1999).
8. N.A. Fromer, C. Schüller, D.S. Chemla, T.V. Shahbazyan, I.E. Perakis, K. Maranowski, and A.C. Gossard, Phys. Rev. Lett. **83**, 4646 (1999).
9. N.A. Fromer, C. Schüller, D.S. Chemla, T.V. Shahbazyan, I.E. Perakis, D. Driscoll, and A.C. Gossard, Physica E **12**, 550 (2002).
10. N.A. Fromer, C.W. Lai, D.S. Chemla, I.E. Perakis, D. Driscoll and A.C. Gossard, Phys. Rev. Lett. **89**, 067401 (2002).
11. A.T. Karathanos, I.E. Perakis, N.A. Fromer, and D.S. Chemla, Phys. Rev. B **67**, 035316 (2003).
12. N.A. Fromer, C. Schüller, D.S. Chemla, I.E. Perakis, D. Driscoll, and A.C. Gossard, Phys. Rev. B **66**, 205314 (2002).
13. I.E. Perakis and D.S. Chemla, Phys. Stat. Sol. B **234**, 242 (2002).
14. L.J. Sham, Phys. Rev. **150**, 720 (1966).
15. see, e.g., V.M. Axt and S. Mukamel, Rev. Mod. Phys. **70**, 145 (1998), and references therein.
16. I. Balslev and E. Hanamura, Solid State Commun. **72**, 843 (1989).
17. V.M. Axt and A. Stahl, Z. Phys. B: Condens. Matter **93**, 195 (1994); **93**, 205 (1994); K. Victor, V.M. Axt, and A. Stahl, Phys. Rev. B **51**, 14164 (1995).
18. W. Schäfer, D.S. Kim, J. Shah, T.C. Damen, J.E. Cunningham, K.W. Goosen, L.N. Pfeiffer, and K. Kohler, Phys. Rev. B **53**, 16429 (1996).
19. D.-S. Kim, J. Shah, J.E. Cunningham, T.C. Damen, S. Schmitt-Rink, and W. Schäfer, Phys. Rev. Lett. **68**, 2838 (1992).
20. H. Wang, J. Shah, T.C. Damen, S.W. Pierson, T.L. Reinecke, L.N. Pfeiffer, and K. West, Phys. Rev. B **52**, R17013 (1995).
21. I.E. Perakis and D.S. Chemla, Phys. Rev. Lett. **72**, 3202 (1994); I.E. Perakis, I. Brener, W.H. Knox, and D.S. Chemla, J. Opt. Soc. Am. B **13**, 1313 (1996).
22. I. Brener, W.H. Knox, and W. Schaefer, Phys. Rev. B **51**, 2005 (1995).
23. T.V. Shahbazyan, N. Primozich, I.E. Perakis, and D.S. Chemla, Phys. Rev. Lett. **84**, 2006 (2000); N. Primozich, T.V. Shahbazyan, I.E. Perakis, and D.S. Chemla, Phys. Rev. B **61**, 2041 (2000).
24. I.E. Perakis and T.V. Shahbazyan, Surf. Sci. Reports **40**, 1–74 (2000); Int. J. Mod. Phys. B **13**, 869–893 (1999); I.E. Perakis, Chem. Phys. **210**, 259–277 (1996).
25. A.J. Turberfield, S.R. Haynes, P.A. Wright, R.A. Ford, R.G. Clark, J.F. Ryan, J.J. Harris, and C.T. Foxon, Phys. Rev. Lett. **65**, 637 (1990).
26. B.B. Goldberg, D. Heiman, A. Pinczuk, L. Pfeiffer, and K. West, Phys. Rev. Lett. **65**, 641 (1990).
27. I.V. Kukushkin, R.J. Haug, K. v. Klitzing, and K. Ploog, Phys. Rev. Lett. **72**, 736 (1994); Phys. Rev. Lett. **72**, 3594 (1994).
28. C. Schüller, K.-B. Broocks, Ch. Heyn, and D. Heitmann, Phys. Rev. B **65**, 081301(R) (2002).
29. K.-B. Broocks, P. Schröter, D. Heitmann, Ch. Heyn, C. Schüller, M. Bichler, and W. Wegscheider, Phys. Rev. B **66**, 041309(R) (2002).
30. S.R.E. Yang and L.J. Sham, Phys. Rev. Lett. **58**, 2598 (1987).
31. B.B. Goldberg, D. Heiman, M.J. Graf, D.A. Broido, A. Pinczuk, C.W. Tu, J.H. English, and A.C. Gossard, Phys. Rev. B **38**, 10131 (1988).
32. I.V. Lerner and Yu.E. Lozovik, Zh. Exp. Teor. Fiz. **80**, 1488 (1981) [Sov. Phys. JETP **53**, 763 (1981)].

33. S.M. Girvin, A.H. MacDonald, and P.M. Platzman, Phys. Rev. Lett. **54**, 581 (1985).
34. R. Haussmann, Phys. Rev. B **53**, 7357 (1996).
35. A. Pinczuk, B.S. Dennis, L.N. Pfeiffer, and K. West, Phys. Rev. Lett. **70**, 3983 (1993).
36. M. Potemski, R. Stepniewski, J.C. Maan, G. Martinez, P. Wyder, and B. Etienne, Phys. Rev. Lett. **66**, 2239 (1991); T.A. Vaughan, R.J. Nicholas, C.J.G.M. Langerak, B.N. Murdin, C.R. Pidgeon, N.J. Mason, and P.J. Walker, Phys. Rev. B **53**, 16481 (1996).
37. H. Haug and A.-P. Jauho, *Quantum kinetics in transport and optics in semiconductors*, Springer series in solid state sciences, vol. 123, Springer, Berlin, 1996.
38. S.M. Girvin, A.H. MacDonald, and P.M. Platzman, Phys. Rev. B **33**, 2481 (1986).
39. C. Stafford, S. Schmitt-Rink, and W. Schaefer, Phys. Rev. B **41**, 10000 (1990).
40. P. Kner, S. Bar-Ad, M.V. Marquezini, D.S. Chemla, R. Lövenich, and W. Schäfer, Phys. Rev. B **60**, 4731 (1999); G. Bartels, A. Stahl, V.M. Axt, B. Haase, U. Neukirch, and J. Gutowski, Phys. Rev. Lett. **81**, 5880 (1998); V.M. Axt, S.R. Bolton, U. Neukirch, L.J. Sham, and D.S. Chemla, Phys. Rev. B **63**, 115303 (2001), and references therein.
41. M. Levenson, *Introduction to Nonlinear Laser Spectroscopy* (Academic Press, New York, 1982).
42. N.A. Fromer, P. Kner, D.S. Chemla, R. Lövenich, and W. Schäfer, Phys. Rev. B **62**, 2516 (2000).
43. T.V. Shahbazyan, N. Primozich, and I.E. Perakis, Phys. Rev. B **62**, 15925 (2000).
44. A. Pinczuk, J.P. Valladares, D. Heiman, A.C. Gossard, J.H. English, C.W. Tu, L. Pfeiffer, and K. West, Phys. Rev. Lett. **61**, 2701 (1988).

Ultrafast Formation of Quasiparticles in Semiconductors: How Bare Charges Get Dressed

Markus Betz, Rupert Huber, Florian Tauser, Andreas Brodschelm, Alfred Leitenstorfer, Paul Gartner, Ladislaus Bányai, and Hartmut Haug

Abstract. The femtosecond dressing of bare particles in semiconductors represents a new playground for solid-state physics. Employing ultrafast spectroscopic techniques, the corresponding formation of correlations between charge carriers and collective excitations of the system may be studied in detail. Two important examples are discussed: (i) The transition of bare charge carriers to polarons and (ii) the buildup of screening in a dense electron–hole plasma. Our experimental findings call for a sophisticated quantum kinetic treatment beyond the Kadanoff–Baym ansatz.

1 Introduction

The quasiparticle model is one of the fundamental concepts in many-body physics. It is usually too complicated to regard each isolated ("bare") particle and to take into account its interactions with all the other individual components of the system. More physical insight is gained by introducing new units which are composed of the bare particle plus some average surrounding. Usually, these quasiparticles are assumed to form instantly and to have no internal dynamics. However, this picture turns out to be valid only on timescales which are long compared to the oscillation cycle of the collective mode of the system. Very recently, the quantum dynamical phenomena which occur during the formation of interparticle correlations in systems far from thermal equilibrium have become accessible [1,2].

As a first example, the dynamics of an isolated charge coupled to a highly polar crystal lattice via the Fröhlich interaction is discussed. This scattering mechanism is of central importance for the transport and optical properties of polar semiconductors. Especially, several classes of semiconductors with relatively strong polar optical Fröhlich interaction have obtained special attention as promising candidates for new applications: (i) The III-V nitrides have led to short-wavelength laser diodes [3] and bear large potential for high-speed electronics at elevated power levels and temperatures. (ii) II-VI compounds are currently receiving renewed interest due to their importance for future spintronics [4]. (iii) Organic semiconductors are opening up completely new perspectives such as highly efficient light emitting devices [5]. In this contribution, we demonstrate that the dynamics of non-equilibrium carriers in this so-called regime of intermediate electron-phonon coupling exhibits qualitatively new and important features which are completely unexpected in semiclassical physics and may be understood only on higher levels of quantum kinetic theories [1]. The most striking result is that carriers injected below the threshold for real emission of LO phonons still experience ultrafast relaxation if the Fröhlich coupling is strong enough. For the case of intermediate, the polaron self-energy becomes comparable to the LO phonon energy. As a result, virtual phonons which are generated due to deformation of the crystal lattice during the transition from the bare charge to a polaron lead to significant changes in the carrier distribution functions. Energy conservation holds only within the coupled carrier-phonon system but not for the electronic excitations alone.

The second part of the article deals with carrier–carrier interactions relevant for various microscopic phenomena found in nature. While the interaction of two isolated point charges is described by the bare Coulomb potential, in many-body systems this interaction is modified as a result of the collective response of the screening cloud surrounding each charge carrier. In our experiment, the ultrafast buildup of Coulomb screening and collective behavior in a dense electron–hole plasma photogenerated within 10 fs are demonstrated [2]. In the previous example of polaron formation, the dynamics of the carrier distribution has been monitored which is a result of a coupling process between elementary excitations. Ultrabroadband two-dimensional THz studies add a new dimension: direct access to the details of the interaction process itself. This is achieved by probing the dielectric response of the non-equilibrium system in the frequency regime of its collective mode, in our case a plasma resonance at 15 THz. It turns out that shortly after generation of the plasma, the amplitude and phase distortion of a single-cycle probe transient at 28 THz is almost instantaneous. If the system is probed approximately 100 fs after its excitation, we find a delayed response due to plasma oscillations. These observations are consistent with an ultrafast transition from an undressed state of bare charges far from equilibrium into a correlated many-body ensemble with screening clouds surrounding each carrier.

2 Virtual Carrier–LO Phonon Interaction in the Intermediate Coupling Region: The Quantum Dynamical Formation of Polarons

2.1 Polar Optical Carrier–LO Phonon Interaction

In polar semiconductor materials such as GaAs and CdTe, the polar optical interaction of carriers with LO phonons dominates the energy transfer of hot electrons and holes to the crystal lattice. The coupling strength of this LO phonon scattering for a carrier with effective mass m^* may be characterized by the dimensionless polaron coupling constant α,

$$\alpha = \frac{e^2}{\hbar}\left(\frac{1}{\epsilon_\infty} - \frac{1}{\epsilon_0}\right)\left(\frac{m^*}{2\hbar\omega_{\mathrm{LO}}}\right)^{1/2} , \tag{1}$$

where ϵ_∞ and ϵ_0 denote the high-frequency and static polarizability of the material, respectively. α determines the ratio between the polaron self-energy and the LO phonon energy $\hbar\omega_{\mathrm{LO}}$ ($\hbar\omega_{\mathrm{LO,GaAs}} = 36\,\mathrm{meV}$ and $\hbar\omega_{\mathrm{LO,CdTe}} = 21\,\mathrm{meV}$). Carriers in the Γ-valley of GaAs represent a typical example for weak Fröhlich interaction with $\alpha \ll 1$. In fact, electrons ($m_e^* = 0.067\ m_0$) and heavy holes ($m_{hh}^* \approx 0.4\ m_0$) have polaron coupling constants of $\alpha_e = 0.06$ and $\alpha_{hh} \approx 0.15$, respectively. In contrast, carriers in more polar CdTe fall into the so-called intermediate coupling regime where α is in the order of unity: electrons at the minimum of the conduction band ($m_e^* = 0.09\ m_0$) exhibit a polaron constant of $\alpha_e = 0.33$. Due to their larger effective mass, the heavy holes ($m_{hh}^* \approx 0.8\ m_0$) are even more strongly coupled with $\alpha_{hh} \approx 1.0$.

To provide an intuitive insight into the properties of the polar optical interaction, we start the discussion with a semiclassical description for the weak coupling regime $\alpha \ll 1$. As a consequence, the scattering rate may be calculated analytically for carriers in a parabolic band via Fermi's golden rule. In first order perturbation theory the LO phonon emission rate for an electron with effective mass m^* and kinetic energy E_{kin} is given by

$$\Gamma_{e-\mathrm{LO}} = \sqrt{\frac{m^*}{2E_{\mathrm{kin}}}}\,\frac{e^2\omega_{\mathrm{LO}}}{2\pi\hbar\epsilon_0}\left(\frac{1}{\epsilon_\infty} - \frac{1}{\epsilon_s}\right)\ \operatorname{arcsinh}\left(\sqrt{\frac{E_{\mathrm{kin}}}{\hbar\omega_{\mathrm{LO}}} - 1}\right) . \tag{2}$$

This expression is derived in [6] and is valid for negligible phonon occupation, i.e. for low temperatures $T \ll \hbar\omega_{\mathrm{LO}}/k_B$ and for dispersionless LO phonons. As a result of Eq. (2), all carriers with kinetic energies above $\hbar\omega_{\mathrm{LO}}$ suffer efficient emission of LO phonons. For the material parameters of GaAs and CdTe, the corresponding scattering times for the case of $E_{\mathrm{kin}} \geq 1.5\hbar\omega_{\mathrm{LO}}$ are calculated from Eq. (2) to be 240 fs and 70 fs, respectively, in very good agreement with recent experimental results [7,8]. For charge carriers with kinetic energies below the threshold for LO phonon emission, no direct interaction with LO phonons is expected at least for low temperatures. However, low-energetic charge carriers may interact with lattice vibrations via virtual emission and reabsorption of LO phonons. In the semicassical picture, these processes are accessible in the framework of second-order perturbation theory. As a result, both band gap energy and effective mass of a charge carrier are renormalized due to a deformation of the polar crystal lattice around each slowly moving electron and hole. The renormalized effective polaron mass in the weak-coupling regime is calculated as [9]

$$m_{\mathrm{pol}}^* = m^*\left(1 + \frac{\alpha}{6}\right) \tag{3}$$

and the average number of LO phonons surrounding a charge carrier in the deformed crystal lattice is $\langle N_{\mathrm{LO}}\rangle = \frac{\alpha}{2}$. This screening cloud around each carrier is assumed to have no internal dynamics and to form instantly with the generation of electron–hole pairs.

Comparing the timescales given by Eq. (2) to the LO phonon oscillation periods of $\omega_{\mathrm{LO,GaAs}}^{-1} = 115\,\mathrm{fs}$ and $\omega_{\mathrm{LO,CdTe}}^{-1} = 200\,\mathrm{fs}$, it becomes obvious that the semiclassical picture of instantaneous scattering events is questionable for the interaction dynamics. This is due to the fact that the oscillation period of the emitted quasiparticle may be taken as a measure for the duration of a specific scattering event. As a consequence, wave mechanical features play a dominant role for the interaction. In general, the dynamics of a system is no longer determined by its status at a single point in time but depends on its history since the phase coherence time can no longer be regarded as infinitely short.

Applying femtosecond spectroscopic techniques to the model system of carriers interacting with LO phonons has been proven to be a powerful tool to study these quantum dynamic memory effects in semiconductors. Especially, four-wave-mixing measurements have provided information about the dynamics of interband polarization excited with coherent laser pulses [10–13]. On the other hand, the results of femtosecond transmission experiments [7,14] give direct insight into the temporal evolution of particle distribution functions. Scattering events without energy conservation followed by memory effects restoring the semiclassical limit have been found [7]. Quantum kinetic calculations agree with the experiments [16–19]. However, most of these studies have been restricted to the limit of weak polaron coupling $\alpha \ll 1$, while specific phenomena due to stronger interaction with LO phonons have not been identified.

2.2 Experimental Concept

In the experiments, we create unbound electron–hole pairs with transform limited Gaussian light pulses (durations specified below) of a central photon energy above the band gap energy E_g. The samples are epitaxial layers of high-purity CdTe ($E_g = 1.60\,\mathrm{eV}$) and GaAs ($E_g = 1.52\,\mathrm{eV}$) of a thickness of $d = 370\,\mathrm{nm}$ and $500\,\mathrm{nm}$, respectively. They are anti-reflection coated on both sides, glued to transparent substrates and mounted inside a He cryostat operated at a temperature of $T_L \geq 3\,\mathrm{K}$.

In order to gain insight into the dynamics of the photoexcited carrier distributions, we measure the pump induced transmission changes with a time delayed probe pulse of a duration of 15 fs and a bandwidth of 100 meV. Perfectly synchronized pulses for excitation and probing are provided by a special two-color femtosecond Ti:sapphire laser system [15]. The test pulse is spectrally dispersed with a double monochromator (spectral resolution set to 4 meV) after transmission through the sample. This scheme guarantees uncertainty-limited temporal and spectral resolution for the experimental study of ultrafast carrier dynamics in semiconductors. Moreover, the sensitivity to detect transmission changes $\Delta T/T$ is only limited by the shot noise of the probe photon flux in the order of a few times 10^{-7}.

2.3 Experimental Results and Discussion

Differential transmission spectra (DTS) for various delay times t_D are measured at a photoexcited electron–hole density of $4 \times 10^{14}\,\mathrm{cm}^{-3}$ with cross-linearly polarized pump and probe beams in GaAs (left column of Fig. 1) and CdTe (right column of Fig. 1). The excitation density is kept extremely low to suppress carrier–carrier scattering [7,8]. A lattice temperature of 4.5 K ensures very slow scattering of carriers with acoustic phonons on a time scale of 10 ps. Consequently, the polar-optical interaction with LO phonons is by far the dominant relaxation mechanism in the sub-picosecond regime. The kinetic energies of the excited carriers are determined by the excess energy of the photons above the band gap energy E_g and the ratio of the effective masses in the valence and conduction bands. Exciting GaAs with an 80 fs pulse at 1.65 eV results in a heavy-hole (hh) distribution centered at a kinetic energy of 18 meV which is smaller than $\hbar\omega_{\mathrm{LO}}$, i.e. no scattering with LO phonons is expected. In contrast, the electrons are created with an excess energy of 112 meV and allowed to transfer energy to the crystal lattice via rapid emission of LO phonons [7,14]. At $t_D = 50$ fs in GaAs,

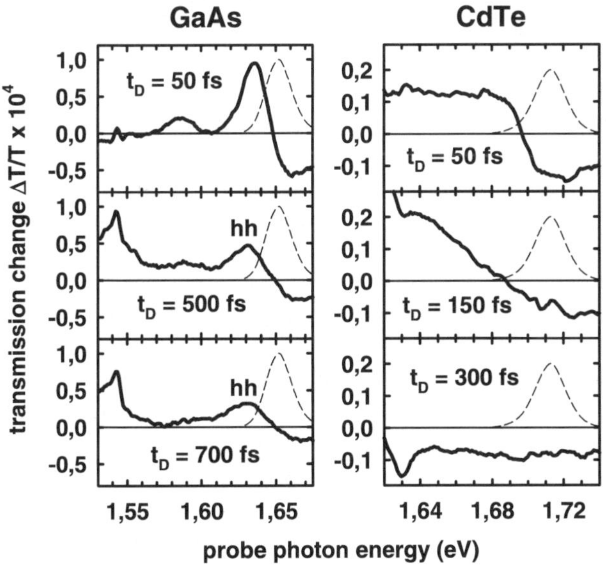

Fig. 1. Spectrally resolved transmission changes $\Delta T/T$ in GaAs (*left*) and CdTe (*right*) for various delay times t_D at a carrier density of $4 \times 10^{14}\,\mathrm{cm}^{-3}$ and $T_L =$ 4.5 K. The excitation spectra are shown as *dashed lines*

a signature of the nonthermal carrier distribution appears near the excitation energy. The spectral hole is slightly redshifted with respect to the excitation spectrum (*dashed*). This phenomenon is related to the excitonic enhancement of the absorption continuum [14]. An additional transmission maximum at a probe photon energy of 1.59 eV is connected to electrons excited from the light hole (lh) band. For a delay of $t_D = 500$ fs, i.e. after twice the electron-LO phonon emission time of approximately 240 fs in GaAs [7,8], most of the electrons have relaxed towards the minimum of the Γ-valley, inducing a transmission increase below a probe photon energy of 1.55 eV. A well resolved bleaching peak due to the generated heavy holes remains at 1.63 eV (indicated by hh in Fig. 1). As late as 700 fs after excitation the increased transmission associated with the hh distribution is still clearly visible in GaAs.

In strong contrast to GaAs, no analogous signature of a hh distribution is found in CdTe (right column of Fig. 1): Excitation with an 80 fs pulse centered at 1.71 eV generates heavy holes with an average kinetic energy of 12 meV. In the semiclassical picture of carrier relaxation the hh distribution should therefore behave similarly as in GaAs. However, at a delay time of $t_D = 150$ fs, approximately twice the electron-LO phonon emission time of 70 fs in CdTe [8], no bleaching peak is observed close to the excitation energy. At 300 fs after excitation, the spectrum high in the absorption continuum is essentially flat. The negative background is related to the renormalization of the fundamental band-gap energy by the pump generated carriers. Apparently, the distribution of heavy holes in CdTe relaxes on a time scale comparable to the electrons even though real emission of LO phonons should be energetically impossible. Nevertheless, the surprisingly fast dynamics can only be related to the increased polaron coupling in CdTe since all other parameters are very similar in GaAs.

In order to analyze our observation of the missing bleaching signature due to heavy holes in CdTe and to illuminate the physical origin of this phenomenon in more detail, DTS are also recorded with co-circularly polarized pump and probe pulses. In this geometry the influence of the lh band on the measured transmission changes is minimized [14,20]. As a result, the experimental findings may be directly compared to simulations based on a two-band model taking into account the hh and conduction bands.

Interband transitions are excited in CdTe with a pump pulse of a duration of 100 fs at 1.72 eV. As a result, we expect an initial electron distribution at 107 meV and a hh distribution around 13 meV. The excitation bandwidth of 16 meV is chosen smaller than $\hbar\omega_{\mathrm{LO}}$ in order to potentially identify LO phonon replicas of the initially prepared electron distribution as verified for the case of weak polaron coupling [14]. Spectrally resolved transmission changes $\Delta T/T$ measured after excitation of 1.3×10^{15} electron–hole pairs per cm^3 are shown in Fig. 2(a).

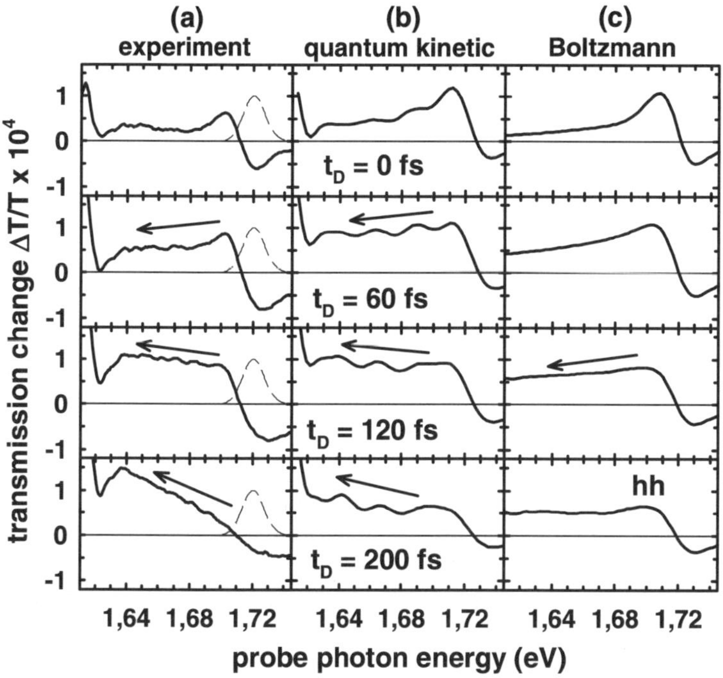

Fig. 2. Differential transmission spectra in CdTe for various delay times t_D at a carrier density of $1.3 \times 10^{15}\,\mathrm{cm}^{-3}$ (**a**) as measured at $T_L = 3\,\mathrm{K}$, (**b**) calculated in a fully two-time dependent quantum kinetic theory and (**c**) obtained via the semiclassical Boltzmann kinetics. The excitation spectrum is shown as a *dashed line* in column (**a**)

2.4 Comparison to Quantum Kinetic Simulations

The theoretical analysis of our experimental data in the intermediate coupling region is difficult since neither the perturbational approaches valid for $\alpha \ll 1$, nor the small polaron model ($\alpha \gg 1$) are expected to produce reliable results. In the equilibrium theory the intermediate-coupling polaron has been treated by unitary transformations and variational calculations [21].

In our approach we consider a quantum kinetic theory based on the Keldysh Green's functions (GF's) with two time arguments. The photoexcited electrons and holes interact with LO phonons. Coulomb collisions between carriers are neglected in this low density regime, but the exciton and excitonic enhancement are included by the Hartree–Fock (HF) approximation. The Dyson equation for the non-equilibrium GF's (see [22] for a detailed presentation) in general matrix notation including band-index, momentum and time indices as well as Keldysh time-contour integrations is

$$G = G_0 + G_0 \Sigma G\,, \tag{4}$$

where G_0 is the free-particle propagator. The self-energy Σ contains instantaneous HF and optical field terms as well as a phonon interaction part:

$$\Sigma_{\mathbf{k}}(t,t') = \imath\hbar \sum_{\mathbf{q}} g_{\mathbf{q}}^2 D_{\mathbf{q}}(t-t') G_{\mathbf{k}-\mathbf{q}}(t,t') \,. \tag{5}$$

The phonons are taken to be in equilibrium and the coupling constant is related to the Fröhlich constant α of electrons or holes via

$$g_{\mathbf{q}}^2 = \alpha \frac{4\pi\hbar \left(\hbar\omega_{\mathrm{LO}}\right)^{3/2}}{\left(2m^*\right)^{1/2} q^2 V} \,. \tag{6}$$

In a previous publication [23] we have developed a straightforward procedure to solve these equations numerically. Our solution for $0.1 < \alpha < 1$ predicted important deviations from the earlier one-time approximations. Vertex corrections are not included, because they are reduced by extra time integrations over short time intervals. For equilibrium this was checked numerically to hold in the relevant time range for intermediate couplings [24]. We take into account time delayed optical test and pump pulses where the coordinate dependence is implemented parametrically through the abstract phases $\phi_P = \mathbf{k}_P\mathbf{x}$ and $\phi_T = \mathbf{k}_T\mathbf{x}$ in order to distinguish the fields (as in [10,17]). The interband polarization is related to the Green's functions by

$$p(t) = \sum_{\mathbf{k}} G^{<}_{\mathbf{k},cv}(t,t) \tag{7}$$

and its part propagating in the test direction is defined by

$$p(t)_T = \frac{1}{2\pi} \int_0^{2\pi} d\phi_T e^{-\imath\phi_T} p(t;\phi_T) \,. \tag{8}$$

The absorption coefficient of the test pulse is

$$\alpha(\omega) = \frac{4\pi\omega_T}{c} \mathrm{Im} \frac{p_T(\omega)}{\mathcal{E}_T(\omega)} \quad \text{with} \quad p_T(\omega) = \int dt e^{\imath\omega t} p_T(t) \,. \tag{9}$$

The electron and hole populations may also be expressed by the Green's functions

$$f^e_{\mathbf{k}}(t) = -\imath\hbar G^{<}_{\mathbf{k},cc}(t,t) \quad \text{and} \quad f^h_{\mathbf{k}}(t) = 1 + \imath\hbar G^{<}_{\mathbf{k},vv}(t,t) \,. \tag{10}$$

Numerical calculations are performed for the parameters of CdTe and GaAs at $T_L = 3\,\mathrm{K}$ considering electrons and heavy holes. The bare band gap used in the theory was deduced from the experimental gap energy taking into account the calculated equilibrium polaron shift. This correction is especially important in the case of more polar CdTe.

As it may be seen from Fig. 2(a) and 2(b) the most important common feature of the experimental and the quantum kinetic results is a reversal of the slope of the DTS between $t_D = 60\,\mathrm{fs}$ and 200 fs, as indicated by the arrows. This is a new feature absent for low LO phonon coupling as in GaAs, which is dominated by the phonon replicas in the electron distribution and a stationary bleaching peak due to the hole population [14,17].

To demonstrate the quantum kinetic origin of our findings in CdTe, a comparison is made with a simplified, semiclassical theory based on the semiconductor Bloch equation (with HF effects). Conventional Boltzmann collision terms are assumed for LO phonon assisted transitions of the populations, according to Fermi's golden Rule. For the polarization a phenomenological dephasing time of $T_2 = 120\,\mathrm{fs}$ is adopted. Due to the strict energy conservation the heavy holes cannot relax. The results are shown in Fig. 2(c): In contrast to the experiment and the quantum kinetic simulation [Fig. 2(a) and (b)], the semiclassically calculated DTS spectra never reverse the slope and a transmission maximum induced by the heavy holes [see Fig. 2(c)] remains prominent at $t_D = 200\,\mathrm{fs}$.

The hh energy distributions computed with the two-time quantum kinetics (thick lines) and the semiclassical Boltzmann kinetics (thin lines) are depicted in Fig. 3 for GaAs (left column) and CdTe (right column). In GaAs, both models result in practically identical hh populations for all delay times. A strongly peaked distribution is conserved on a sub-picosecond time scale. In contrast, the populations obtained with the Dyson equation for CdTe show a significant relaxation of the holes, resulting in a broad background already after a time delay t_D as short as 60 fs. Even at a delay time of $t_D = 200\,\mathrm{fs}$

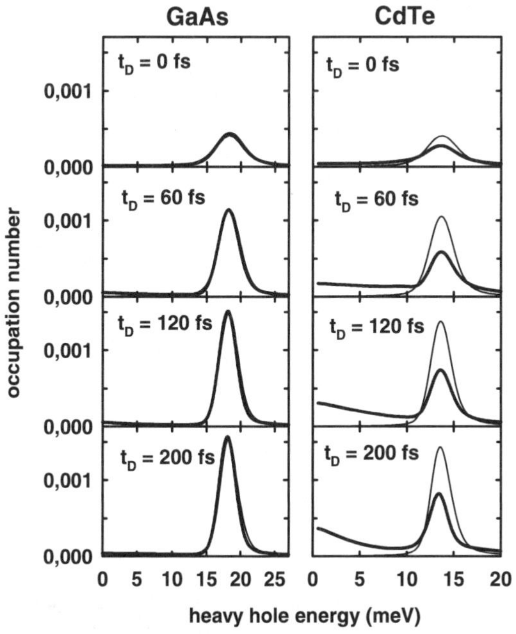

Fig. 3. Heavy-hole energy distributions in GaAs (*left*) and CdTe (*right*), as calculated in the two-time quantum kinetic simulation (*thick lines*) and with the semiconductor Bloch equations including Boltzmann scattering terms (*thin lines*) for various delay times t_D

the background component in CdTe experiences further relaxation indicating that the phenomenon is not linked exclusively to the energy uncertainty during the ultrafast carrier generation process. These effects are purely quantum kinetic in nature: If the sub-threshold hole dynamics in CdTe is simulated semiclassically (thin lines in the right column of Fig. 3), the distribution functions undergo no relaxation.

In order to check that the quantum kinetic features in the DTS originate alone from the modification of the hole distribution, we introduced the pump generated populations obtained with the Dyson equation directly into the simplified Bloch equation for the polarization. As in the quantum kinetic treatment, the DTS reverses its slope at the same delay time as in the experiment (not shown).

To understand the unexpected dynamics of the heavy holes, one has to take into account the fact that within the two-time quantum kinetics with a stronger coupling constant the energy of the free particles is no longer conserved. The interaction energy plays an important role and allows transitions that are forbidden in the Boltzmann picture. This non-conservation is related to the finite lifetime of the states and increases with the strength of the interaction. The unusual hole distribution may also be interpreted as the buildup of a polaronic state that is characterized by virtual emission and reabsorption of LO phonons. The typical timescale of this polaron formation is given by the oscillation period of the collective mode, i.e. $\omega_{\mathrm{LO}}^{-1} = 200\,\mathrm{fs}$ in CdTe.

In the measured DTS [Fig. 2(a)] a sharp maximum at 1.609 eV (one LO energy above a strong exciton bleaching at 1.588 eV) and a transmission minimum at 1.615 eV are found. These features result from a combination of excitonic nonlinearities and polaron effects. Their position and shape is reproduced by the quantum kinetics [Fig. 2(b)]. In contrast, only a shift of the band edge and the exciton (not shown) appears in the DTS calculated via the Boltzmann equation while the structure in the absorption continuum is absent [Fig. 2(c)].

At intermediate electron-phonon coupling, the phonon replicas of the electron distribution in the DTS are strongly smeared out [Fig. 2(a)] in contrast to previous experimental results for the case of weakly polar GaAs [14]. This effect is well understood as a result of time-energy uncertainty in the generation process [25] and is reproduced also in the framework of the simplified Boltzmann kinetics [Fig. 2(c)]. In the full quantum kinetic calculations [Fig. 2(b)] one still sees a trace of the phonon satellite structure which is not resolved in the experiment. Possible explanations for this slight discrepancy are spurious contributions due to the light holes as well as the anisotropy of the hh band which are not included in the simulations.

2.5 Conclusion

In conclusion, we have demonstrated an ultrafast dynamics of low-energy heavy holes interacting with unoccupied polar-optical modes in CdTe. This unexpected dynamics may be interpreted as the femtosecond buildup of a polaronic state from a photogenerated bare charge. A theoretical description of the sub-threshold scattering calls for a sophisticated quantum kinetic treatment beyond the Kadanoff–Baym ansatz. This phenomenon represents a typical many-body effect: The free-particle energy ceases to be a constant of motion in systems where the coupling between electronic and lattice degrees of freedom can no longer be regarded as a weak perturbation. In general, our findings are relevant for a detailed understanding of nonequilibrium carrier dynamics in all materials with a somewhat stronger polaron coupling.

3 Femtosecond Buildup of Coulomb Screening in GaAs Probed via Ultrabroadband THz Spectroscopy

3.1 Introduction

The earliest stage in the dynamics of interacting many-body excitations in solids is the buildup of inter-particle correlations. A photogenerated electron–hole plasma in a semiconductor constitutes an ideal playground to study such ultrafast transient phenomena: In a plasma each charge carrier attracts a polarization cloud of opposite sign to form a dressed quasiparticle. This screening cloud modifies the interaction potential of the charges with respect to the bare Coulomb potential V_q:

$$V_q = \frac{4\pi e^2}{q^2}\,, \tag{11}$$

q denotes the momentum exchange in the interaction process in reciprocal space, e is the elementary charge. The effective interaction in the carrier plasma may be described by the potential $W_q(\omega, t_D)$ which is defined as

$$W_q(\omega, t_D) = \frac{V_q}{\epsilon(\omega, t_D)}\,. \tag{12}$$

The longitudinal dielectric function $\epsilon(\omega, t_D)$ renormalizes the bare Coulomb potential V_q (Eq. (11)). While V_q obviously mediates an instantaneous interaction with a white frequency spectrum, $W_q(\omega, t_D)$ may exhibit a pronounced dependence on the frequency ω corresponding to the energy exchanged in a collision. This retardation in the many-body system is caused by collective effects such as plasma oscillations at a frequency ω_{pl}. In the nonequilibrium stage immediately after femtosecond photogeneration of a plasma, $W_q(\omega, t_D)$ might also depend on the time delay t_D with respect to the event of the photoinjection. The buildup of screening and collective phenomena in an electron–hole plasma has been studied intensely by quantum kinetic theories [26–30]. Especially, a sophisticated quantum kinetic theory for the dynamics of $W_q(\omega, t_D)$ has been developed [28,30]. Experiments investigating the regime of Coulomb quantum kinetics [31–34] have been sensitive to the dynamics of interband polarizations and particle distributions. However, they have not provided access to the interaction potential itself. In contrast, we present a direct observation of the buildup of screening and the formation of dressed quasiparticles in a dense electron–hole plasma photogenerated in GaAs within 10 fs. Especially, the low-energy excitations of the plasma and the resulting resonances in $\epsilon(\omega, t_D)$ are mapped out with femtosecond temporal resolution.

3.2 Experimental Setup

The experimental setup is shown in Fig. 4. We start with 10 fs laser pulses (center wavelength: 780 nm) at a repetition rate of 64 MHz and an average output power of 1 W from a mode-locked Ti:sapphire laser system. The near infrared light is split up into three beams. The major part of the laser intensity is directly focussed onto a GaAs sample. The specimen consists of a 200-nm-thin epitaxial layer of high-purity GaAs attached to a transparent diamond substrate by van-der-Waals forces. Via resonant interband absorption of a laser pulse, free electron–hole pairs are created in GaAs. The photoinduced particle pair density is $N = 2 \times 10^{18}\,\mathrm{cm}^{-3}$, resulting in a plasma frequency of $\omega_{\mathrm{pl}} \approx 15\,\mathrm{THz}$.

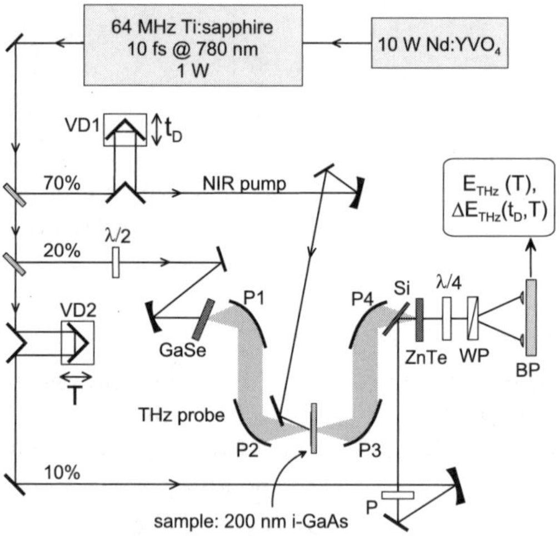

Fig. 4. Experimental setup for ultrafast near infrared pump – THz probe measurements: VD1 and VD2, variable delay lines; $\lambda/2$, half-wave plate; GaSe, nonlinear optical crystal (thickness: 30 µm); P1, P2, P3, P4, parabolic gold mirrors; P, polarizer; Si, high-resistivity silicon window; ZnTe, electro-optic crystal (ZnTe, ⟨110⟩-oriented, thickness: 10 µm); $\lambda/4$, quarter-wave plate; WP, Wollaston prism; BP, balanced pair of photodiodes

At a delay time t_D after photoexcitation, the mid infrared polarizability of this nonequilibrium system is tested in the long-wavelength limit, i.e. $q \approx 0$, by a single-cycle electric field transient of a duration of 27 fs (FWHM of the intensity, Fig. 5(a)) and a center frequency of 28 THz (Fig. 5(b)). Phase matched optical rectification of another portion of the 10 fs laser pulse in a GaSe nonlinear crystal as thin as 30 μm [35] is exploited to generate this THz probe. Any resonance in the system will lead to a distortion of the wave form and a retarded tail to the probe transient transmitted through the sample. The changes ΔE_{THz} induced in the probe electric field E_{THz} are directly measured in the time domain via ultrabroadband electro-optic sampling [36,37]: The transmitted THz transient is focussed into a $\langle 110 \rangle$ oriented ZnTe crystal of a thickness of 10 μm. The electro-optic effect results in a birefringence of ZnTe which is proportional to the THz electric field amplitude present in the crystal. A time delayed third part of the 10 fs laser pulse reads out the birefringence thereby sampling the THz wave form as a function of a second delay time T. Interestingly, our setup allows to detect birefringence induced modifications of the photon flux as low as 5×10^{-9} limited by the shot noise of the laser light used to sample the THz wave form.

Figure 5(a) depicts the electric field amplitude E_{THz} of the THz probe transmitted through the unexcited GaAs sample versus delay time T. The Fourier transform of this THz transient is shown in Fig. 5(b) and (c). The field amplitude (Fig. 5(b)) exhibits frequency components between 1 THz and 100 THz, covering the entire mid and far infrared wavelength region from $\lambda =$ 300 μm to 3 μm. The phase spectrum in Fig. 5(c) is flat between 10 THz and

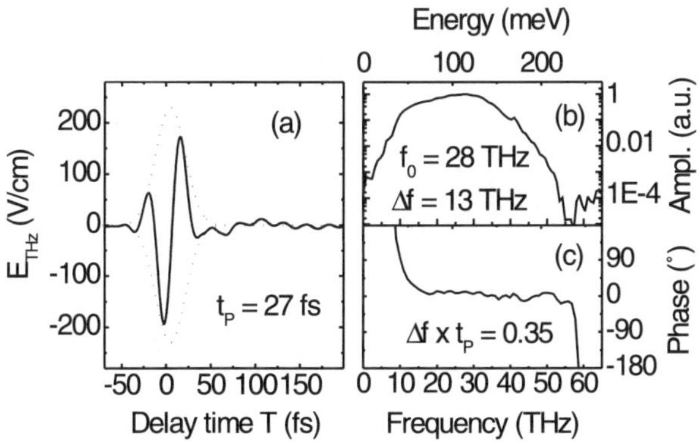

Fig. 5. (**a**) Electric field amplitude E_{THz} of the single-cycle probe versus time T. The pulse duration is $t_p = 27$ fs (FWHM of the intensity envelope). (**b**) Amplitude and (**c**) phase spectra of the single-cycle probe transient versus frequency and photon energy. $f_{\max}$ denotes the center frequency. Δf represents the FWHM of the spectral intensity

55 THz giving rise to a time-bandwidth product as small as $t_p \times \Delta f = 0.35$. In this wavelength regime, our single-cycle pulses represent the ultimate probe with a temporal resolution limited only by the uncertainty principle.

3.3 Experimental Results

In a two-dimensional spectroscopy, we measure the THz electric field change ΔE_{THz} due to carrier injection with the 10 fs pump as a function of pump-probe delay t_D and THz sampling delay T [2]. The electric waveform E_{THz} of the test transient is shown again in Fig. 6(a). The THz electric field change ΔE_{THz} due to carrier excitation with the 10 fs pump is displayed in a grayscale map versus pump-probe delay t_D (vertical) and THz sampling delay T (horizontal) in Fig. 6(b). The diagonal dotted line denotes the position of the maximum of the 10 fs pump pulse. For pump-probe delays between

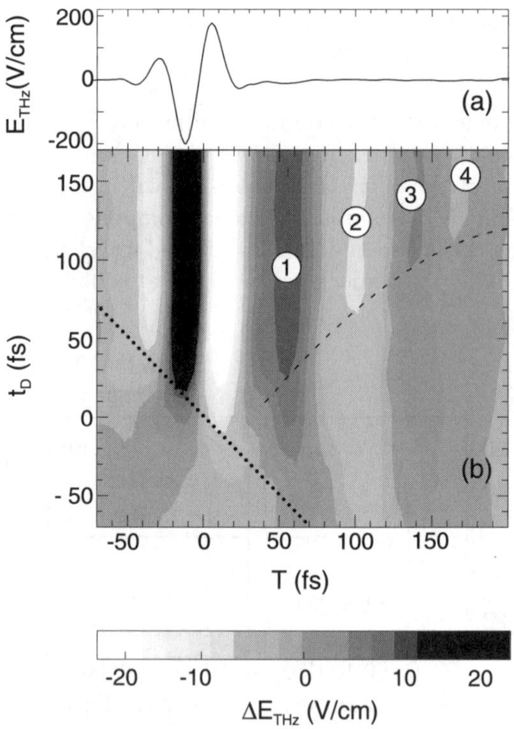

Fig. 6. (**a**) Electric field amplitude E_{THz} of the single-cycle probe versus time T. (**b**) The polarization response of the plasma depends on two delay times, t_D and T. The electric field change ΔE_{THz} induced by 10 fs photoexcitation of $2 \times 10^{18}\,\mathrm{cm}^{-3}$ electron–hole pairs in GaAs is shown as a grey scale map versus t_D and T. The *dotted diagonal line* denotes the position of the 800 nm pump pulse and the *dashed curve* serves to emphasize the buildup of the retarded plasmon response

$t_D = -20$ fs and $+20$ fs, the excitation pulse overlaps with the electric field of the THz and a negative (white) and positive (black) half cycle in ΔE_{THz} appear along the excitation diagonal. These striking features in the vertical region around $T = 0$ fs correspond to an instantaneous perturbation of the single-cycle probe which appears as soon as the plasma is injected. A retarded oscillatory response of the plasma starts to develop at $t_D = 40$ fs, visualized by the dashed black line in Fig. 6(b) to guide the eye: An additional half cycle in ΔE_{THz} builds up, represented by a vertical dark gray column labeled **1** at $T = 60$ fs. In this region no probe field is present without excitation (see Fig. 6(a)). For even longer pump-probe delays $t_D \geq 70$ fs, a second half wave (label **2**) appears in the retarded response at $T = 100$ fs. Another maximum (**3**) shows up at $T = 140$ fs for $t_D \geq 100$ fs and finally a last minimum (**4**) appears at $T = 170$ fs for $t_D \geq 120$ fs. For pump-probe delays beyond $t_D = 150$ fs the electric field change ΔE_{THz} versus T becomes stationary on a picosecond time scale given by the carrier recombination time in the sample.

In order to obtain a more quantitative description of the experimental data in Fig. 2 and to extract $\epsilon_{q=0}(\omega, t_D)$, we transfer our results into the frequency domain [27,28,30]. Since our detection scheme is sensitive to both amplitude and phase of the transmitted probe field, we are able to access the full complex dielectric function in the long-wavelength limit, i.e. $\epsilon_{q=0}(\omega, t_D)$, via the following procedure similar to the one described in [38]: Both sets of real time data $E_{\text{THz}}(T)$ and $\Delta E_{\text{THz}}(t_D, T)$ are Fourier transformed along the electro-optic sampling axis T. In addition, only precisely known information enters the calculation: The layer structure of the sample is accounted for by a transfer matrix formalism which includes reflexion and absorption losses as well as phase shifts upon propagation through the interfaces and the layer material, respectively [39]. The dielectric function of intrinsic GaAs without optical excitation is parameterized by a dielectric oscillator model where $\epsilon_{q=0}(\omega, t_D \ll 0)$ is represented by the following equation, with ω_{pl} set to zero:

$$\epsilon(\omega) = \epsilon_\infty \times \left(1 + \frac{\omega_{\text{LO}}^2 - \omega_{\text{TO}}^2}{\omega_{\text{TO}}^2 - \omega^2 - i\gamma\omega} - \frac{\omega_{\text{pl}}^2}{\omega^2 + i\omega/\tau}\right), \tag{13}$$

ω_{LO} and ω_{TO} are the longitudinal and transverse optical phonon frequencies in the center of the Brillouin zone ($\omega_{\text{LO}}/2\pi = 8.8$ THz and $\omega_{\text{TO}}/2\pi = 8.1$ THz in GaAs). The lattice damping is described by $\gamma = 0.2\,\text{ps}^{-1}$ and the nonresonant background polarizability of the bound electrons is accounted for by $\epsilon_\infty = 11.0$.

Figure 7(a) and (b) shows the imaginary and real parts of the inverse dielectric function versus frequency (or equivalently, versus energy exchanged in a long-range Coulomb collision) for various pump-probe delay times t_D. The negative imaginary part of $1/\epsilon_{q=0}$ (Fig. 7(a)) reflects the buildup of dissipation in the plasma: 25 fs after carrier generation a wide range of energies may be exchanged between the particles, indicating an uncorrelated

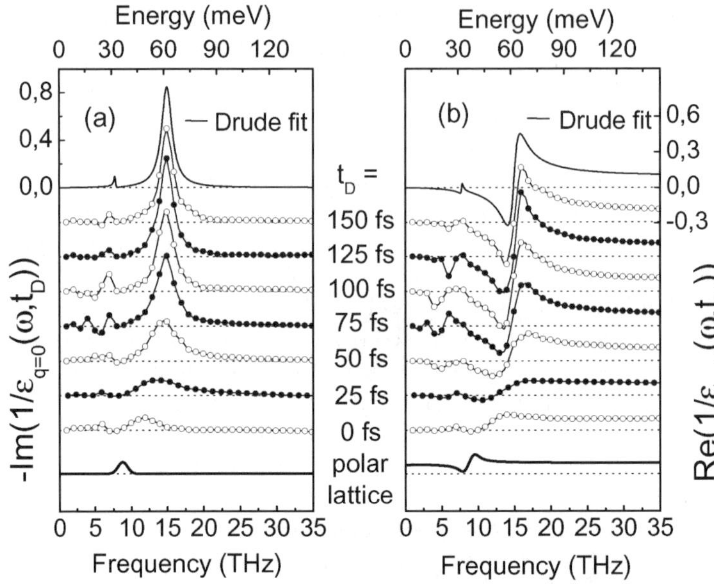

Fig. 7. (**a**) Imaginary and (**b**) real part of the long-wavelength limit of the inverse dielectric function of GaAs versus frequency for different pump-probe delays t_D. The *lower solid lines* result from the dielectric oscillator model of Eq. (3) for unexcited GaAs. A Drude-fit of the data at $t_D = 150\,\mathrm{fs}$ is represented by the *upper solid line*

state in an extreme nonequilibrium situation. Within 100 fs a sharp maximum at the plasma frequency of $\omega_{\mathrm{pl}} = 14.4\,\mathrm{THz}$ appears due to the transition to collective plasmon scattering. In the unexcited sample (*lower solid line* in Fig. 7(a)) $-\mathrm{Im}(1/\epsilon_{q=0})$ vanishes exactly, except for a small peak at $\omega_{\mathrm{pl}}/2\pi = 8.8\,\mathrm{THz}$ which is due to energy exchange with the crystal lattice via polar-optical scattering with LO phonons as discussed in Sect. 2.1. The real part of $1/\epsilon_{q=0}$ (Fig. 7(b)) describes the buildup of Coulomb screening. The quantity $\mathrm{Re}(1/\epsilon_{q=0})$ renormalizes the effective charge that an electron feels in the interaction process with a quasiparticle in the plasma exchanging an energy (see Eqs. (11) and (12)). For late pump-probe delays of $t_D \geq 150\,\mathrm{fs}$, a resonant dispersive feature of over- and antiscreening [2] is found around the plasma frequency. This phenomenon is consistent with a fully developed dressed interaction. Interestingly, the resonance extrema do not emerge instantaneously with carrier generation: At $t_D = 25\,\mathrm{fs}$, the spectrum is completely flat above $\hbar\omega_{\mathrm{pl}}$, indicating bare Coulomb collisions. However, $\mathrm{Re}(1/\epsilon_{q=0})$ drops close to zero in the low-energy region with a step at $\hbar\omega_{\mathrm{pl}}$. This behavior means a very large polarizability at low frequencies, consistent with an onset of the free-carrier conductivity immediately after carrier generation at $t_D = 0$. In contrast, $\mathrm{Re}(1/\epsilon_{q=0})$ is always finite for ground-state GaAs (*lower solid line* in Fig. 7(b)), in agreement with the insulating properties of the unexcited semiconductor.

The classical Drude formula of Eq. (13) describes the long-wavelength and high-frequency limit of the dielectric function of an electron gas assuming an exponential plasmon damping with a time constant τ. We have performed least-square fits to our data allowing ω_{pl} and τ in the Drude response as free parameters while we kept the lattice part fixed. Good agreement is found for late delay times. An example is given by the uppermost curves in Fig. 7(a) and (b) with $\omega_{\mathrm{pl}} = 14.4\,\mathrm{THz}$ and $\tau = 85\,\mathrm{fs}$ for $t_D = 150\,\mathrm{fs}$. Interestingly, the dielectric functions measured in the non-Markovian quantum regime at early times t_D show clear deviations from a Drude shape. We want to emphasize that our experimental findings strongly support quantum kinetic theories for the nonequilibrium dynamics of the Coulomb interaction [28,30]. In fact, these models do predict a strongly broadened plasmon pole and a delayed buildup of Coulomb screening at early times after a strong perturbation. The typical time scale for these phenomena is of the order of the duration of a plasma oscillation period, i.e. $2\pi/\omega_{\mathrm{pl}} = 70\,\mathrm{fs}$ in our case.

3.4 Conclusion

In conclusion, we have reported the direct observation of the ultrafast buildup of Coulomb correlation effects and the formation of dressed quasiparticles in a many-body system. We demonstrate that plasmon scattering and Coulomb screening are not present instantaneously after 10 fs photogeneration of a dense electron–hole plasma in GaAs. However, they evolve on a time scale which is linked to the inverse plasma frequency. Conductivity of the plasma is shown to set in immediately with carrier generation. These findings are in agreement with recent simulations in the framework of Coulomb quantum kinetics. The results are obtained utilizing a novel technique based on ultrabroadband THz spectroscopy that allows us to resolve the polarization response of the system with sub-cycle resolution of the electric field amplitude and phase.

We want to point out that the present scheme may be employed to study the dynamics of various elementary excitations in the mid to far infrared spectral region with a time resolution close to the ultimate limit. As an example, new perspectives arise for investigations in systems such as magnons in high-T_c superconductors, lattice dynamics in organic semiconductors and vibrational dynamics in large molecules and biological complexes.

Acknowledgements

We gratefully acknowledge stimulating discussions and continuous support by A. Laubereau, W. Kaiser and L.V. Keldysh. The high-quality GaAs samples have been grown by G. Böhm, M. Bichler and G. Abstreiter. We thank

K. Ortner and C.R. Becker for providing the high-quality CdTe specimen. The work has been supported by the DFG within the project "Quantum Coherence in Semiconductors".

References

1. M. Betz, G. Göger, A. Laubereau, P. Gartner, L. Bányai, H. Haug, K. Ortner, C.R. Becker, and A. Leitenstorfer, Phys. Rev. Lett. **86**, 4684 (2001)
2. R. Huber, F. Tauser, A. Brodschelm, M. Bichler, G. Abstreiter, and A. Leitenstorfer, Nature **416**, 286 (2001)
3. S. Nakamura, G. Fasol, *The Blue Laser Diode* (Springer, Berlin, 1997)
4. R. Fiederling, M. Keim, G. Reuscher, W. Ossau, G. Schmidt, A. Waag, and L.W. Molenkamp, Nature **402**, 787 (1999)
5. M.A. Baldo, M.E. Thompson, and S.R. Forrest, Nature **403**, 750 (2000)
6. B.K. Ridley: *Quantum Processes in Semiconductors*, Clarendon Press, Oxford (1993)
7. A. Leitenstorfer, C. Fürst, A. Laubereau, W. Kaiser, G. Tränkle, and G. Weimann, Phys. Rev. Lett. **76**, 1545 (1996)
8. M. Betz, G. Göger, A. Leitenstorfer, K. Ortner, C.R. Becker, G. Böhm, and A. Laubereau, Phys. Rev. B **60**, R11265 (1999)
9. J. Callaway, *Quantum Theory of the Solid State*, Academic Press (London), 2nd ed. (1991)
10. L. Bányai, D.B. Tran Thoai, E. Reitsamer, H. Haug, D. Steinbach, M.U. Wehner, M. Wegener, T. Marschner, and W. Stolz, Phys. Rev. Lett. **75**, 2188 (1995)
11. M.U. Wehner, M.H. Ulm, D.S. Chemla, and M. Wegener, Phys. Rev. Lett. **80**, 1992 (1998)
12. D. Steinbach, G. Kocherscheidt, M.U. Wehner, H. Kalt, M. Wegener, K. Ohkawa, D. Hommel, and V.M. Axt, Phys. Rev. B **60**, 12079 (1999)
13. U. Woggon, F. Gindele, W. Langbein, and J.M. Hvam, Phys. Rev. B **61**, 1935 (2000)
14. C. Fürst, A. Leitenstorfer, A. Laubereau, and R. Zimmermann, Phys. Rev. Lett. **78**, 3733 (1997)
15. C. Fürst, A. Leitenstorfer, and A. Laubereau, IEEE J. of Selected Topics in Quantum Electronics **2**, 473 (1996)
16. R. Zimmermann, J. Wauer, A. Leitenstorfer, and C. Fürst, J. Lumin. **76 & 77**, 34 (1998)
17. A. Schmenkel, L. Bányai, and H. Haug, J. Lumin. **76 & 77**, 134 (1998)
18. K. Hannewald, S. Glutsch, and F. Bechstedt, Phys. Rev. B **61**, 10792 (2000)
19. H. Castella and J.W. Wilkins, Phys. Rev. B **61**, 15827 (2000)
20. D.N. Mirlin, in F. Meier, and B.P. Zakharchenya (eds.): *Optical Orientation*, Chapter 4, North Holland, Amsterdam (1984)
21. T.D. Lee, F. Low and D. Pines, Phys. Rev. **90**, 297 (1953)
22. H. Haug, A.-P. Jauho, *Quantum Kinetics in Transport and Optics of Semiconductors*, Springer Series in Solid-State Sciences Vol. 123 (Springer, Berlin, 1996)
23. P. Gartner, L. Bányai, and H. Haug, Phys. Rev. B **60**, 14234 (1999)
24. P. Gartner, L. Bányai and H. Haug, Phys. Rev. B **66**, 07505 (2002)

25. A. Leitenstorfer, A. Lohner, T. Elsässer, S. Haas, F. Rossi, T. Kuhn, W. Klein, G. Böhm, G. Tränkle, and G. Weimann, Phys. Rev. Lett. **73**, 1687 (1994)
26. M. Hartmann, H. Stolz and R. Zimmermann, phys. stat. sol. (b) **159**, 35 (1990)
27. K. El Sayed, S. Schuster, H. Haug, F. Herzel, and K. Henneberger, Phys. Rev. B **49**, 7337 (1994)
28. L. Bányai, Q.T. Vu, B. Mieck, and H. Haug, Phys. Rev. Lett. **81**, 882 (1998)
29. N.-H. Kwong and M. Bonitz, Phys. Rev. Lett. **84**, 1768 (2000)
30. Q.T. Vu and H. Haug, Phys. Rev. B **62**, 7179 (2000)
31. F.X. Camescasse, A. Alexandrou, D. Hulin, L. Banyai, D.B. Tran Thoai, and H. Haug, Phys. Rev. Lett. **77**, 5429 (1996)
32. W.A. Hügel, M.F. Heinrich, M. Wegener, Q.T. Vu, L. Banyai, and H. Haug, Phys. Rev. Lett. **83**, 3313 (1999)
33. M. Bonitz, J.F. Lampin, F.X. Camescasse, A. Alexandrou, Phys. Rev. B 62, 15724 (2000)
34. Q.T. Vu, H. Haug, W.A. Hügel, S. Chatterjee, and M. Wegener, Phys. Rev. Lett. **85**, 3508 (2000)
35. R. Huber, A. Brodschelm, F. Tauser, and A. Leitenstorfer, Appl. Phys. Lett. **76**, 3191 (2000)
36. Q. Wu and X.-C. Zhang, Appl. Phys. Lett. **71**, 1285 (1997)
37. A. Leitenstorfer, S. Hunsche, J. Shah, M.C. Nuss, and W.H. Knox, Appl. Phys. Lett. **74**, 1516 (1999)
38. M. Schall and P.U. Jepsen, Opt. Lett. **25**, 13 (2000)
39. M. Born and E. Wolf, Principles of Optics, 7th ed., Cambridge University Press, Cambridge (1999)

Self-consistent Projection Operator Theory of Intersubband Absorbance in Semiconductor Quantum Wells

Inès Waldmüller, Jens Förstner and Andreas Knorr

1 Introduction

Due to their many-particle character and their application in quantum cascade lasers, optical intersubband excitations in semiconductor quantum wells have become the focus of many recent publications [1,2]. In samples of high quality, intrinsic processes like electron–electron and electron-phonon many particle correlations determine the basic optical and transport properties such as lineshape and ultrafast dynamics. At the same time, intersubband excitations allow the direct investigation of dynamical properties of an important model system of many particle physics – the two-dimensional electron gas. We here present a microscopic theory for the intersubband dynamics and absorption.

The calculation of absorption spectra of MQW systems is in principle composed of two parts: the determination of the polarization in a single quantum well within a density matrix approach as the source of electromagnetic radiation (Fig. 1a) and the calculation of the generated fields in the geometry of interest (Fig. 1b) within a Green's function approach [3,4]. We will here focus on the so-called single-pass geometry (cf. Fig. 1b, [5]).

2 Optical Polarization and many-particle effects

In order to determine the generated field in a multi-quantum well (MQW) system (Fig. 1: 1b), we first calculate its source, i.e. the two-dimensional optical polarization $\mathbf{P}$ (dipole density) due to intersubband transitions in a single quantum well:

$$\mathbf{P}(z,t) = \frac{1}{A} \sum_{ab} \{ \sigma_{ab}(t)\, \mathbf{d}_{ba}(z) + c.c. \} \,. \tag{1}$$

$\sigma_{ab} = \langle a_a^\dagger\, a_b \rangle$ denotes the density matrix elements (intersubband coherence), being the expectation value for the simultaneous generation ($a_a^\dagger = a^{\dagger a}_{\mathbf{k_a}} = a^{\dagger a}_{\mathbf{k_a},s_a}$) of an electron with charge $(-e)$ in subband a with wavenumber k_a and spin s_a and destruction ($a_b = a^b_{\mathbf{k_b}} = a^b_{\mathbf{k_b},s_b}$) of an electron in subband b with wavenumber k_b and spin s_b (cf. Fig. 1: 1a), [6], $\mathbf{d}_{ba}(z) = -e\zeta_b(z)\mathbf{z}\zeta_a(z)$

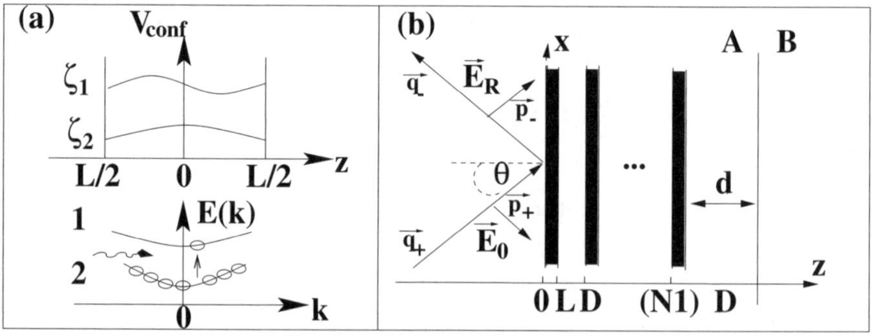

Fig. 1. Theoretical model system: (**a**) source calculation: single quantum well with 2-band in-plane dispersion $E(k)$ (confined in growth direction z), (**b**) field calculation: N quantum wells (well width L) in single-pass geometry. Here $\mathbf{E}_0 = E_0 \mathbf{p}_+ e^{i\mathbf{q}_+ \mathbf{r}}$ ($\mathbf{E}_R = E_R \mathbf{p}_- e^{i\mathbf{q}_- \mathbf{r}}$) denotes the incident (reflected) p-polarized field. The angle of incidence is chosen to obtain first a large density oscillation strength and second total reflection at the medium A/medium B interface to enhance the absorbance (standing-wave effect)

and ζ_i the quantum well wavefunctions of the conduction subbands in effective mass approximation[1]. Thus, in order to determine the polarization we start with calculating the intersubband coherence by using the density operator σ of the system:

$$\langle a_a^\dagger \, a_b \rangle = \mathrm{tr}(a_a^\dagger \, a_b \sigma) \,. \tag{2}$$

2.1 The Density Operator

In a many-particle system, only the expectation values of a certain set of observables $\{O\}$, the so called observation level, are of interest. To determine the polarization, we choose $O = a_a^\dagger a_b$. The observation level $\{O\}$, is typically not a complete system in the space of observables. Therefore, the knowledge of the true density operator $\sigma(t)$ of the system contains more information than necessary and it is possible to decompose the density operator with respect to the observables of interest into a relevant part, $\sigma_{\mathrm{rel}}(t)$, and an irrelevant part, $\sigma_{\mathrm{irr}}(t)$, with the properties[2]:

$$\sigma(t) = \sigma_{\mathrm{rel}}(t) + \sigma_{\mathrm{irr}}(t) \,, \quad \mathrm{tr}(O\sigma(t)) = \mathrm{tr}(O\sigma_{\mathrm{rel}}(t)) \,, \quad \mathrm{tr}(O\sigma_{\mathrm{irr}}(t)) = 0 \,. \tag{3}$$

When t_0 is the switch-on time of the interaction, we assume in the following a Hartree–Fock ground-state, i.e. $\sigma(t_0) = \sigma_{\mathrm{rel}}(t_0)$. Thus, the initial states

[1] For a single symmetric quantum well with infinitely high barriers and well width L the wavefunctions are given by $\zeta_1 = \sqrt{2/L}\sin((2\pi z)/L), \zeta_2 = \sqrt{2/L}\cos((\pi z)/L)$, cf. Fig. 1.

[2] A detailed introduction into the theory of the relevant density operator can be found in [7,8].

under consideration, $\sigma(t_0)$ shall have no irrelevant part. The relevant part, $\sigma_{\mathrm{rel}}(t)$, is a mapping of the density operator $\sigma(t)$, i.e. $\sigma(t)$ and $\sigma_{\mathrm{rel}}(t)$ are related by a functional f:

$$\sigma_{\mathrm{rel}}(t) = f[\sigma(t)] \,. \tag{4}$$

For a differential variation of $\sigma(t)$, the variation of $f[\sigma(t)]$ defines a superoperator $\mathcal{P}[\sigma(t)]$ in Liouville space[3]

$$\mathrm{d}\, f[\sigma(t)] = \mathcal{P}[\sigma(t)]\, \mathrm{d}\sigma(t) = \mathcal{P}[\sigma_{\mathrm{rel}}(t)]\, \mathrm{d}\sigma(t) \,. \tag{5}$$

To derive an equation of motion for $\sigma_{\mathrm{rel}}(t)$ one starts with the von Neumann-equation (Liouville equation) for the full density operator:

$$\frac{\mathrm{d}}{\mathrm{d}t}\sigma(t) = -\mathrm{i}\mathcal{L}\,\sigma(t) \,, \tag{6}$$

where $\mathcal{L}$ is the so-called Liouville operator, mediating the mapping of an element G of the Liouville space via the commutation $(1/\hbar)[H(t), G]$. Here $H(t)$ is assumed to be a many particle Hamiltonian including time-dependent external fields. Using (5) and (6) the time derivative of $\sigma_{\mathrm{rel}}(t)$ reads

$$\frac{\mathrm{d}}{\mathrm{d}t}\sigma_{\mathrm{rel}}(t) = \mathcal{P}[\sigma_{\mathrm{rel}}(t)]\,\frac{\mathrm{d}}{\mathrm{d}t}\sigma(t) = -\mathrm{i}\,\mathcal{P}[\sigma_{\mathrm{rel}}(t)]\,\mathcal{L}\,\sigma(t) \,. \tag{7}$$

In order to get a closed equation of motion for $\sigma_{\mathrm{rel}}(t)$ the full density operator σ in (7) is now related to σ_{rel}. For this purpose, the superoperator $\mathcal{Q}[\sigma_{\mathrm{rel}}(t)]$ is introduced by using the definition:

$$\mathcal{Q}[\sigma_{\mathrm{rel}}(t)] + \mathcal{P}[\sigma_{\mathrm{rel}}(t)] = 1 \,. \tag{8}$$

Using (8), the Liouville operator $\mathcal{L}$ is decomposed into a relevant and a irrelevant part, $\mathcal{L} = \mathcal{L}_{\mathcal{P}} + \mathcal{L}_{\mathcal{Q}}$:

$$\mathcal{L}_{\mathrm{rel}}(t) = \mathcal{L}_{\mathcal{P}}(t) = \mathcal{P}[\sigma_{\mathrm{rel}}(t)]\,\mathcal{L}, \quad \mathcal{L}_{\mathrm{irr}}(t) = \mathcal{L}_{\mathcal{Q}}(t) = \mathcal{Q}[\sigma_{\mathrm{rel}}(t)]\,\mathcal{L} \,. \tag{9}$$

Applying (7) and the operator identity

$$\mathcal{U}(t,t_0) = \mathcal{U}_{\mathcal{Q}}(t,t_0) - \mathrm{i}\int_{t_0}^{t} \mathrm{d}t'\,\mathcal{U}_{\mathcal{Q}}(t,t')\,\mathcal{L}_{\mathcal{P}}(t')\,\mathcal{U}(t',t_0) \,, \tag{10}$$

where $\mathcal{U}$ and $\mathcal{U}_{\mathcal{Q}} = \mathcal{Q}\mathcal{U}$ denote the time evolution operator for $\mathcal{L}$ and $\mathcal{L}_{\mathcal{Q}}$ [7], the density operator can be written as:

$$\begin{aligned} \sigma(t) &= \mathcal{U}(t,t_0)\,\sigma(t_0) \\ &= \mathcal{U}_{\mathcal{Q}}(t,t_0)\,\sigma(t_0) + \int_{t_0}^{t} \mathrm{d}t'\,\mathcal{U}_{\mathcal{Q}}(t,t')\,\frac{\mathrm{d}}{\mathrm{d}t'}\sigma_{\mathrm{rel}}(t') \,. \end{aligned} \tag{11}$$

3 Here we used $\mathcal{P}^2 = \mathcal{P}$.

A partial integration[4] leads to the desired decomposition of the full density operator into a relevant and an irrelevant part:

$$\begin{aligned}\sigma(t) &= \sigma_{\mathrm{rel}}(t) - \mathrm{i}\int_{t_0}^{t} \mathrm{d}t'\, \mathcal{U}_{\mathcal{Q}}(t,t')\mathcal{Q}[\sigma_{\mathrm{rel}}(t')]\,\mathcal{L}\,\sigma_{\mathrm{rel}}(t') \\ &= \sigma_{\mathrm{rel}}(t) + \sigma_{\mathrm{irr}}(t)\,. \end{aligned} \tag{12}$$

Inserting (12) into (7) one finally obtains a closed equation of motion for σ_{rel}:

$$\begin{aligned}\frac{\mathrm{d}}{\mathrm{d}t}\sigma_{\mathrm{rel}}(t) &= -\mathrm{i}\mathcal{L}_{\mathcal{P}}(t)\sigma(t) = -\mathrm{i}\mathcal{L}_{\mathcal{P}}(t)(\sigma_{\mathrm{rel}}(t) + \sigma_{\mathrm{irr}}(t)) \\ &= -\mathrm{i}\mathcal{L}_{\mathcal{P}}(t)\sigma_{\mathrm{rel}}(t) - \int_{t_0}^{t} \mathrm{d}t'\, \mathcal{L}_{\mathcal{P}}(t)\,\mathcal{U}_{\mathcal{Q}}(t,t')\,\mathcal{L}_{\mathcal{Q}}(t')\sigma_{\mathrm{rel}}(t')\,. \end{aligned} \tag{13}$$

The expectation values of the observables of interest $O_n \in \{O\}$, e.g. σ_{ab} for the dipole density of the intersubband transition, can be calculated by using (3):

$$\begin{aligned}\frac{\mathrm{d}}{\mathrm{d}t}\langle O_n\rangle &= \frac{\mathrm{d}}{\mathrm{d}t}\mathrm{tr}(O_n\,\sigma_{\mathrm{rel}}(t)) \\ &= -\mathrm{i}\,\mathrm{tr}(O_n\,\mathcal{L}_{\mathcal{P}}(t)\sigma_{\mathrm{rel}}(t)) \\ &\quad - \int_{t_0}^{t} \mathrm{d}t'\,\mathrm{tr}(O_n\,\mathcal{L}_{\mathcal{P}}(t)\,\mathcal{U}_{\mathcal{Q}}(t,t')\,\mathcal{L}_{\mathcal{Q}}(t')\sigma_{\mathrm{rel}}(t'))\,. \end{aligned} \tag{14}$$

Equation (14) shows, that even though $\langle O_n\rangle = \mathrm{tr}(O_n\sigma_{\mathrm{rel}})$ depends only on the relevant part of the density operator, the equation of motion for $\langle O_n\rangle$ depends on both the relevant and the irrelevant part of the density matrix.

The Relevant Part of the Density Operator. In order to use (14), the relevant density matrix operator has to be determined. As known from non-equilibrium statistical physics, the generalized canonical density operator

$$\sigma_{\mathrm{can}}(t) = \frac{\mathrm{e}^{-\sum_n \lambda_n(t)O_n}}{\mathrm{tr}(\mathrm{e}^{-\sum_n \lambda_n(t)O_n})} \tag{15}$$

is the density operator which has the maximum uncertainty measure within a fixed set of observables. As can be found in textbooks [7], the mapping $\sigma(t)$ onto $\sigma_{\mathrm{can}}(t)$ has the required properties for the relevant density matrix

[4] Here the following two properties of the time evolution operator have been used:

$$\frac{\partial}{\partial t'}\mathcal{U}_{\mathcal{Q}}(t,t') = \mathrm{i}\,\mathcal{U}_{\mathcal{Q}}(t,t')\,\mathcal{Q}[\sigma_{\mathrm{rel}}(t')]\,\mathcal{L}\,, \quad \mathcal{U}_{\mathcal{Q}}(t,t) = 1$$

operator (3). Thus, in the following we choose[5] $\sigma_{\rm rel} = \sigma_{\rm can}$. Note that the expression for the canonical density operator does not imply a restriction to equilibrium processes, the full time dependence is included in the Lagrange parameters λ_n.

The Irrelevant Part of the Density Operator. Starting from (12) we can rewrite the irrelevant density operator in a time-convolutionless expression[6] [9]:

$$\begin{aligned}\sigma_{\rm irr}(t) &= -{\rm i}\int_{-\infty}^{t} {\rm d}t'\, \mathcal{T}_{+}{\rm e}^{-{\rm i}\int_{t'}^{t}{\rm d}t''\mathcal{L}_{\mathcal{Q}}(t'')}\mathcal{L}_{\mathcal{Q}}(t')\,\mathcal{T}_{-}{\rm e}^{{\rm i}\int_{t'}^{t}{\rm d}t''\mathcal{L}_{\mathcal{P}}(t'')}(\sigma_{\rm rel}(t)+\sigma_{\rm irr}(t))\\ &= (1-\Xi(t))^{-1}\Xi(t)\sigma_{\rm rel}(t)\,, \qquad (16)\\ \Xi(t) &= -{\rm i}\int_{-\infty}^{t} {\rm d}t'\mathcal{T}_{+}{\rm e}^{-{\rm i}\int_{t'}^{t}{\rm d}t''\mathcal{L}_{\mathcal{Q}}(t'')}\mathcal{L}_{\mathcal{Q}}(t')\,\mathcal{T}_{-}{\rm e}^{{\rm i}\int_{t'}^{t}{\rm d}t''\mathcal{L}_{\mathcal{P}}(t'')}\,. \qquad (17)\end{aligned}$$

Often, this formulation is a good starting point for approximation schemes to many particle correlations.

2.2 Equations of Motion for Quantum Well Excitations

We now derive the equation of motion for the intersubband coherence by applying (14):

$$\frac{\rm d}{{\rm d}t}\sigma_k^{12} = -{\rm i}\,{\rm tr}(a_{1k}^{\dagger}a_{2k}\,\mathcal{L}_{\mathcal{P}}(t)\sigma_{\rm rel}(t)) - {\rm i}\,{\rm tr}(a_{1k}^{\dagger}a_{2k}\,\mathcal{L}_{\mathcal{P}}(t)\sigma_{\rm irr}(t))\,. \qquad (18)$$

This equation describes the optical response of intersubband transitions and can be decomposed into a mean-field part [10] dependent on $\sigma_{\rm rel}$ and correlation contributions (dephasing of the macroscopic polarization, [11]) dependent on $\sigma_{\rm irr}$.

Hamiltonian. To evaluate (18), the Hamiltonian of the intersubband quantum well system is necessary. In our model, the total Hamiltonian in second quantization is given by $H = H_0 + H_{\rm cf} + H_{\rm cc} + H_{\rm cp}$, where the Hamiltonian

[5] It can be shown that $\mathcal{P}$ is the Kawasaki–Gunton operator [7]:

$$\mathcal{P}[\sigma_{\rm rel}(t)]Y = \left(\sigma_{\rm rel}(t) - \sum_{\nu=1}^{n}\frac{\partial\sigma_{\rm rel}(t)}{\partial\langle O_\nu\rangle(t)}\langle O_\nu\rangle(t)\right){\rm tr}(Y) + \sum_{\nu=1}^{n}\frac{\partial\sigma_{\rm rel}(t)}{\partial\langle O_\nu\rangle(t)}{\rm tr}(O_\nu Y)$$

[6] On the basis of $\frac{\rm d}{{\rm d}t}\mathcal{U}_{\mathcal{Q}}(t,t') = -{\rm i}\mathcal{L}_{\mathcal{Q}}(t)\mathcal{U}_{\mathcal{Q}}(t,t')$, $\frac{\rm d}{{\rm d}t}\mathcal{U}(t,t') = -{\rm i}\mathcal{L}(t)\mathcal{U}(t,t')$ and (6), (7) we derive ($\mathcal{T}_{\pm}$ denotes the time/anti-time ordering operators)

$$\begin{aligned}\mathcal{U}(t,t') &= \mathcal{T}_{+}{\rm e}^{-{\rm i}\int_{t'}^{t}{\rm d}t''\mathcal{L}(t'')} \rightarrow \mathcal{U}_{\mathcal{Q}}(t,t') = \mathcal{T}_{+}{\rm e}^{-{\rm i}\int_{t'}^{t}{\rm d}t''\mathcal{L}_{\mathcal{Q}}(t'')}\,,\\ \sigma(t') &= \mathcal{T}_{-}{\rm e}^{{\rm i}\int_{t'}^{t}{\rm d}t''\mathcal{L}(t'')}\sigma(t) \rightarrow \sigma_{\rm rel}(t') = \mathcal{T}_{-}{\rm e}^{{\rm i}\int_{t'}^{t}{\rm d}t''\mathcal{L}_{\mathcal{P}}(t'')}\sigma(t)\end{aligned}$$

of the noninteracting Bloch electrons and phonons is represented by H_0, the field-carrier interaction by H_{cf}, the Coulomb interaction of the electronic system by $H_{\rm cc}$ and the electron-phonon interaction by $H_{\rm cp}$ [12][7]:

$$H_0 = \sum_a \epsilon_a\, a_a^\dagger\, a_a + \sum_{\mathbf{q}} \hbar\omega_{\rm LO} b_{\mathbf{q}}^\dagger b_{\mathbf{q}} \tag{19}$$

$$H_{\rm cf} = \sum_{a,b} \int {\rm d}z\, \mathbf{d}_{ab} \cdot \mathbf{E}(z,t)\, a_a^\dagger\, a_b \tag{20}$$

$$H_{\rm cc} = \frac{1}{2} \sum_{abcd} V_{abcd}\, a_a^\dagger a_b^\dagger a_d a_c \tag{21}$$

$$H_{\rm cp} = \sum_{a,b,\mathbf{k}} \sum_{\mathbf{q}} \left[g_{\mathbf{q}}^{ab} a_{\mathbf{k}}^{\dagger a} b_{\mathbf{q}} a_{\mathbf{k}-\mathbf{q}_{||}}^{b} + g_{\mathbf{q}}^{*ab} a_{\mathbf{k}-\mathbf{q}_{||}}^{\dagger b} b_{\mathbf{q}}^\dagger a_{\mathbf{k}}^{b} \right] \tag{22}$$

Here, LO-phonons as the dominant part of the electron-phonon interaction have been taken into account.

2.3 Mean-field Contributions

Using footnote (5) we find[8] the identity ${\rm tr}(a_1^\dagger a_2 \mathcal{L}_\mathcal{P} \sigma_{\rm rel}) = {\rm tr}(a_1^\dagger a_2 \mathcal{L} \sigma_{\rm rel})$ and can calculate the mean-field contributions:

$$\begin{aligned} \frac{\rm d}{{\rm d}t} \sigma_k^{12}|_{\rm MF} &= -{\rm i}\, {\rm tr}(a_k^{\dagger 1} a_k^2\, \mathcal{L}_\mathcal{P}(t) \sigma_{\rm rel}(t)) = {\rm i}\, {\rm tr}(\mathcal{L}(t) a_k^{\dagger 1} a_k^2 \sigma_{\rm rel}(t)) \\ &= \frac{\rm i}{\hbar}\, {\rm tr}([H, a_k^{\dagger 1} a_k^2] \sigma_{\rm rel}(t))\,. \end{aligned} \tag{23}$$

With (19) and the commutation relations for fermions this yields:

$$\frac{\rm d}{{\rm d}t} \sigma_k^{12}|_{\rm MF}^{0} = {\rm i}\,(\omega_k^1 - \omega_k^2)\, \sigma_k^{12} \tag{24}$$

$$\frac{\rm d}{{\rm d}t} \sigma_k^{12}|_{\rm MF}^{\rm cf} = \frac{\rm i}{\hbar} \int {\rm d}z\, \mathbf{E}(z,t) \cdot \mathbf{d}_{21}\, (\sigma_k^{11} - \sigma_k^{22}) \tag{25}$$

---

[7] $\epsilon_a = \hbar\omega_a$ is the energy of an electron in subband a with wave vector $\mathbf{k_a}$, $\hbar\omega_{\rm LO}$ the LO-phonon energy. The Coulomb (V_{abcd}) and the Fröhlich ($g_{\mathbf{q}}^{ab}$) coupling matrix elements are given by:

$$\begin{aligned} V_{abcd} &= \frac{e^2}{2\varepsilon_0 A} V_{|\mathbf{k}_a - \mathbf{k}_c|}^{abcd} \delta_{k_a + k_b, k_c + k_d} \delta_{s_a, s_c} \delta_{s_b, s_d},\ g_{\mathbf{q}}^{ab} \\ &= -{\rm i} \sqrt{\frac{e^2 \hbar\omega_{\rm LO}}{2\varepsilon_0 V} \left(\frac{1}{\epsilon_\infty} - \frac{1}{\epsilon_s} \right)} \frac{1}{q} \mathcal{F}_{\mathbf{q}_\perp}^{ab}, \\ V_{|\mathbf{k}|}^{abcd} &= \int {\rm d}z \zeta_a^*(z) \zeta_c(z) \int {\rm d}z' \zeta_b^*(z') \zeta_d(z') \frac{{\rm e}^{-|\mathbf{k}||z-z'|}}{|\mathbf{k}|},\ \mathcal{F}_{\mathbf{q}_\perp}^{ab} \\ &= \int {\rm d}z \zeta_a^*(z) \zeta_b(z) {\rm e}^{{\rm i}q_\perp z}\,. \end{aligned}$$

Here, $V = A\ L$ denotes the normalization volume and $\varepsilon_s/\varepsilon_\infty$ the static/optical dielectric constant.

[8] ${\rm tr}(O_\mu \mathcal{P} \mathcal{L} \sigma_{\rm rel}) = \{\langle O_\mu \rangle(t) - \sum_{\nu=1}^n \delta_{\mu\nu} \langle O_\nu \rangle(t)\} {\rm tr}(\mathcal{L}\sigma_{\rm rel}) + \sum_{\nu=1}^n \delta_{\mu\nu} {\rm tr}(O_\nu \mathcal{L} \sigma_{\rm rel})$

$$\frac{\mathrm{d}}{\mathrm{d}t}\sigma_k^{12}|_{\mathrm{MF}}^{\mathrm{cc}} = \frac{\mathrm{i}}{\hbar}\sum_{a,b,c}(V_{ab1c}\,\mathrm{tr}(a_a^\dagger a_b^\dagger a_c a_2 \sigma_{\mathrm{rel}}) - V_{2abc}\,\mathrm{tr}(a_1^\dagger a_a^\dagger a_c a_b \sigma_{\mathrm{rel}})) \quad (26)$$

$$\frac{\mathrm{d}}{\mathrm{d}t}\sigma_k^{12}|_{\mathrm{MF}}^{\mathrm{cp}} = \frac{\mathrm{i}}{\hbar}\sum_{a,\mathbf{q}}[g_{\mathbf{q}}^{a1}\,\mathrm{tr}(a_{\mathbf{k}+\mathbf{q}_{||}}^{\dagger a} b_{\mathbf{q}} a_{\mathbf{k}}^2 \sigma_{\mathrm{rel}}) + g_{\mathbf{q}}^{*1a}\,\mathrm{tr}(a_{\mathbf{k}-\mathbf{q}_{||}}^{\dagger a} b_{\mathbf{q}}^\dagger a_{\mathbf{k}}^2 \sigma_{\mathrm{rel}})$$
$$- g_{\mathbf{q}}^{2a}\,\mathrm{tr}(a_{\mathbf{k}}^{\dagger 1} b_{\mathbf{q}} a_{\mathbf{k}-\mathbf{q}_{||}}^{a} \sigma_{\mathrm{rel}}) - g_{\mathbf{q}}^{*a2}\,\mathrm{tr}(a_{\mathbf{k}}^{\dagger 1} b_{\mathbf{q}}^\dagger a_{\mathbf{k}+\mathbf{q}_{||}}^{a} \sigma_{\mathrm{rel}})]\,. \quad (27)$$

Mean-field Factorization. Regarding (26) and (27) it becomes obvious, that the system of equations does not close and thus we have to calculate 3-point and 4-point expectation values in order to obtain $\frac{\mathrm{d}}{\mathrm{d}t}\sigma_k^{21}|_{\mathrm{MF}}^{\mathrm{cc/cp}}$. We assume that the phonons can be treated as a bath for the dynamical electronic system and choose

$$\sigma_{\mathrm{can}} = \frac{\mathrm{e}^{-\sum_{i,j}\lambda_{ij}(t)a_i^\dagger a_j - \hbar k_B T\sum_{\mathbf{q}}\omega_{\mathbf{q}} b_{\mathbf{q}}^\dagger b_{\mathbf{q}}}}{\mathrm{tr}(\mathrm{e}^{-\sum_{i,j}\lambda_{ij}(t)a_i^\dagger a_j - \hbar k_B T\sum_{\mathbf{q}}\omega_{\mathbf{q}} b_{\mathbf{q}}^\dagger b_{\mathbf{q}}})}\,. \quad (28)$$

In order to facilitate calculations, we first diagonalize the electronic part of the relevant density operator. In this representation[9], we can calculate $\sigma_{\mathrm{rel}}^{\mathrm{el}}$ to be

$$\sigma_{\mathrm{rel}}^{el} = \frac{\mathrm{e}^{-\sum_i \lambda_i d_i^\dagger d_i}}{\mathrm{tr}(\mathrm{e}^{-\sum_i \lambda_i d_i^\dagger d_i})} = \frac{\prod_i \mathrm{e}^{-\lambda_i n_i}}{\sum_{\{n_\alpha\}}\langle\{n_\alpha\}|\prod_i \mathrm{e}^{-\lambda_i n_i}|\{n_\alpha\}\rangle}\,. \quad (31)$$

Here $|\{n_\alpha\}\rangle = |n_1,\ldots,n_\alpha,\ldots\rangle$ with $n_\alpha \in \{0,1\}$, since we are considering fermions. Taking the relevant set of observables to be $a_i^\dagger a_j$ we find[10]

$$\mathrm{tr}(a_1^\dagger a_2^\dagger a_3 a_4 \sigma_{\mathrm{rel}}) = -\sum_{\{n_\alpha\}}\sum_{a,b} U_{a1}U_{b2}U_{a3}^* U_{b4}^* \langle\{n_\alpha\}|n_a\ n_b \prod_i \frac{\mathrm{e}^{-\lambda_i n_i}}{1+\mathrm{e}^{-\lambda_i}}|\{n_\alpha\}\rangle$$
$$+\sum_{\{n_\alpha\}}\sum_{a,b} U_{a1}U_{b2}U_{b3}^* U_{a4}^* \langle\{n_\alpha\}|n_a\ n_b \prod_i \frac{\mathrm{e}^{-\lambda_i n_i}}{1+\mathrm{e}^{-\lambda_i}}|\{n_\alpha\}\rangle$$
$$= \sum_a \frac{U_{a1}U_{a4}^*}{1+\mathrm{e}^{\lambda_a}}\sum_b \frac{U_{b2}U_{b3}^*}{1+\mathrm{e}^{\lambda_b}} - \sum_a \frac{U_{a1}U_{a3}^*}{1+\mathrm{e}^{\lambda_a}}\sum_b \frac{U_{b2}U_{b4}^*}{1+\mathrm{e}^{\lambda_b}}\,. \quad (32)$$

---

[9] We can rewrite the elements of the hermitian matrix λ with respect to a complete set of orthonormal functions, $\{\phi_n\}$ (with $U_{ik} = \langle\lambda_i|\phi_k\rangle$, $\langle\lambda_i|\lambda|\lambda_j\rangle = \lambda_j\delta_{i,j}$, $n_i = d_i^\dagger d_i$),

$$\lambda_{nm} = \langle\phi_n|\lambda|\phi_m\rangle = \sum_{i,j}\langle\phi_n|\lambda_i\rangle\langle\lambda_i|\lambda|\lambda_j\rangle\langle\lambda_j|\phi_m\rangle = \sum_i U_{in}^*\lambda_i U_{im}\ , \quad (29)$$

thus we introduce new operators $d_i, d_i^\dagger$:

$$d_i = \sum_j U_{ij}a_j\,,\ \left(a_i = \sum_j U_{ji}^* d_j\right)\,,\ d_i^\dagger = \sum_j U_{ij}^* a_j^\dagger\,,\ \left(a_i^\dagger = \sum_j U_{ji} d_j^\dagger\right)\,. \quad (30)$$

[10] Note that as we are dealing with fermions we have to consider that the states are antisymmetric. Thus, for simplicity we use the commutation relations in order to be able to switch to number operators n_i.

If we calculate $\langle a_1^\dagger a_2 \rangle = \mathrm{tr}(a_1^\dagger a_2 \sigma_{\mathrm{rel}})$ the same way, we find $\langle a_1^\dagger a_2 \rangle = \sum_a \frac{U_{a1}U_{a2}^*}{1+\mathrm{e}^{\lambda_a}}$ and therewith obtain the so-called mean-field factorization for four-particle correlations:

$$\mathrm{tr}(a_1^\dagger a_2^\dagger a_3 a_4 \sigma_{\mathrm{rel}}) = \langle a_1^\dagger a_4 \rangle \langle a_2^\dagger a_3 \rangle - \langle a_1^\dagger a_3 \rangle \langle a_2^\dagger a_4 \rangle \,. \tag{33}$$

The expression *mean-field factorization* for (33) results from the fact, that the decomposition of the four-particle correlations in (26) leads to the same result as approximating H_{cc} by the one-particle Hamiltonian $H_{\mathrm{MF}}^{\mathrm{cc}}$:

$$H_{\mathrm{MF}}^{\mathrm{cc}} = \sum_{ab} \left(\sum_{cd} [V_{acbd} - V_{cabd}] \langle a_c^\dagger a_d \rangle \right) a_a^\dagger \; a_b \,. \tag{34}$$

Applying (33) and (26) or instead (34) yields in rotating wave approximation:

$$\begin{aligned} \frac{\mathrm{d}}{\mathrm{d}t}\sigma_k^{12}|_{\mathrm{MF}}^{\mathrm{cc}} = \frac{\mathrm{i}}{\hbar} \Bigg\{ & \sum_{q\neq 0} [(V_q^{2112} - V_q^{1111})\sigma_{k-q}^{11}\sigma_k^{12} - (V_q^{2112} - V_q^{2222})\sigma_{k-q}^{22}\sigma_k^{12}] \\ & + \sum_{q \neq 0} V_q^{1212}(\sigma_k^{11} - \sigma_k^{22})\sigma_{k-q}^{12} - 2V_0^{2112}(\sigma_k^{11} - \sigma_k^{22}) \sum_q \sigma_q^{12} \Bigg\} \,. \end{aligned} \tag{35}$$

Using σ_{can} from (28) we obtain

$$\begin{aligned} \mathrm{tr}(a_{\mathbf{k}+\mathbf{q}_{||}}^{\dagger a} b_{\mathbf{q}} a_{\mathbf{k}}^{b} \sigma_{\mathrm{rel}}) &= \sum_{n_\alpha, \tilde{n}_{\mathbf{q}}} \langle \{\tilde{n}_{\mathbf{q}}\} | \langle \{n_\alpha\} | \sum_{i,j} U_{ia} U_{jb}^* d_{\mathbf{k}+\mathbf{q}_{||}}^{\dagger i} b_{\mathbf{q}} d_{\mathbf{k}}^{j} \sigma_{\mathrm{can}} | \{n_\alpha\} \rangle | \{\tilde{n}_{\mathbf{q}}\} \rangle \\ &= 0 \,, \end{aligned} \tag{36}$$

and thus the electron-phonon mean-field contribution for the here considered phonon bath vanishes:

$$\frac{\mathrm{d}}{\mathrm{d}t}\sigma_k^{12}|_{\mathrm{MF,cp}} = 0 \,. \tag{37}$$

This is consistent with the assumption that treating the phonons as a bath does not allow coherent phonon fields. To obtain an interpretation of the mean-field effects, we present spectra (Fig. 2) showing the susceptibility [Im($\chi_k(\omega)$), cf. (55)] of an InAs/AlSb single quantum well with respect to different mean-field contributions[11]. As can be seen in Fig. 2a, the exchange energy, the first term in (35), renormalizes the transition energy (blue shift) which is decreasing with increasing well width. The excitonic contribution, the second term in (35), causes a renormalization of the electric-dipole interaction energy (hydrogen-like resonance line). In principle, the electrons couple via Coulomb interaction to the hole, which is left in the lower subband, when

[11] Note, that we here approximate the microscopic dephasing due to electron–electron and electron-phonon interaction with a constant phenomenological dephasing. These spectra were calculated by solving the equation of motion (18) in time domain and then performing a time integration (e.g. [13]).

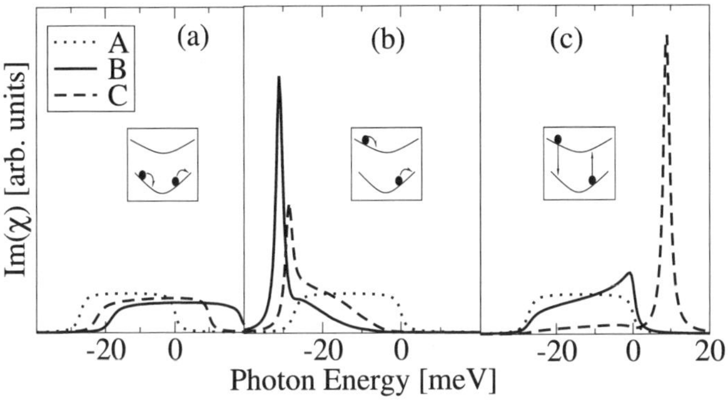

Fig. 2. Susceptibility, $\mathrm{Im}(\sum_k \chi_k(\omega))$, cf. (55), of an InAs/AlSb single quantum well [temperature: 15 Kelvin, electron density: $1 \times 10^{12}\,\mathrm{cm}^{-2}$, well width $L = 5\,\mathrm{nm}$ (B), 10 nm (C)] with regard to different mean-field contributions in comparison to non-interacting single particle excitation (A): (**a**) exchange shift; (**b**) excitonic contribution; (**c**) depolarization; In the *inset* we show the dominant Coulomb processes between the electrons. Material parameters are given in [17]

the electrons are lifted by the optical field to the higher subband [10], similar to the electron-hole coupling in interband transitions [13]. The last term, the depolarization, is caused by the macroscopic distribution of electrons and is proportional to the density and well width. Note that the depolarization term is a Hartree-contribution and equivalent to the classical longitudinal field $(\mathbf{M}_{ab} \cdot \mathbf{E}_L)$.

2.4 Correlation Contributions

Next, we want to determine the correlation contributions which describe the microscopic dephasing, i.e. the part of the expectation value dependent on σ_{irr}. Using footnote (5) in a way similar to footnote (8) we find

$$\frac{\mathrm{d}}{\mathrm{d}t}\sigma_k^{12}|_{\mathrm{Corr}} = -\mathrm{i}\,\mathrm{tr}(a_k^{\dagger 1} a_k^2 \mathcal{L}_{\mathcal{P}}(t)\sigma_{\mathrm{irr}}) = \mathrm{i}\,\mathrm{tr}(\mathcal{L}(t) a_k^{\dagger 1} a_k^2 \sigma_{\mathrm{irr}})\,. \tag{38}$$

Here we can restrict the Liouville operator $\mathcal{L}$ to the many particle (MP) Liouvillian, since only $\mathcal{L}_{\mathrm{MP}}$ yields non-vanishing contributions to the expectation value with the irrelevant part, cf. (3). Assuming that the relevant and irrelevant Liouvillians depend only weakly on time (Markovian approximation), we approximate the integral kernels in (16) by:

$$\mathcal{T}_{+}\mathrm{e}^{-\mathrm{i}\int_{t'}^{t}\mathrm{d}t''\,\mathcal{L}_{\mathcal{Q}}(t'')} \approx \mathrm{e}^{-\mathrm{i}(t-t')\mathcal{L}_{\mathcal{Q}}}, \mathcal{T}_{-}\mathrm{e}^{\mathrm{i}\int_{t'}^{t}\mathrm{d}t''\,\mathcal{L}_{\mathcal{P}}(t'')} \approx \mathrm{e}^{\mathrm{i}(t-t')\mathcal{L}_{\mathcal{P}}} \tag{39}$$

and rewrite (17) by changing the integration variable:

$$\Xi = -\mathrm{i}\int_0^\infty \mathrm{d}s\,\mathrm{e}^{-\mathrm{i}\mathcal{L}_{\mathcal{Q}}s}\,\mathcal{L}_{\mathcal{Q}}\,\mathrm{e}^{\mathrm{i}\mathcal{L}_{\mathcal{P}}s}\,. \tag{40}$$

Since (40) contains an exponential of $\mathcal{L}_Q$, we approximate $\mathcal{L}_Q$ by the free-particle Liouvillian $\mathcal{L}_0$ and a correction $\mathcal{L}_C$ ($\mathcal{L}_Q = \mathcal{L}_0 + \mathcal{L}_{\mathrm{MF}} + \mathcal{L}_C \approx \mathcal{L}_0 + \mathcal{L}_C$) and derive a perturbation series of the term $\mathrm{e}^{-\mathrm{i}\mathcal{L}_Q s}$ with respect to $\mathcal{L}_0$. We rewrite (40)[12] by introducing the operator $\mathcal{X}$:

$$\Xi = -\mathrm{i}\int_0^\infty \mathrm{d}s \mathcal{X}(s)\mathcal{L}_Q \mathrm{e}^{\mathrm{i}\mathcal{L}_0 s}\,, \tag{41}$$

$$\mathcal{X}(s) = \left(1 + \int^s \mathrm{d}s' \mathrm{e}^{-\mathrm{i}(\mathcal{L}_0 + \mathcal{Q}\mathcal{L}_C)s'}[-\mathrm{i}\mathcal{Q}\mathcal{L}_C]\mathrm{e}^{\mathrm{i}\mathcal{L}_0 s'}\right)\mathrm{e}^{-\mathrm{i}\mathcal{L}_0 s}\,, \tag{42}$$

$$\frac{\mathrm{d}}{\mathrm{d}s}(\mathcal{X}(s)) = \mathcal{X}(s)[-\mathrm{i}\mathcal{Q}\mathcal{L}_C] + \mathcal{X}(s)[-\mathrm{i}\mathcal{L}_0] \tag{43}$$

and use the Laplace-transformation $\tilde{\mathcal{X}}(z) = L(\mathcal{X}(s))$ for the operator $\mathcal{X}(s)$ to derive the homogeneous equation

$$\tilde{\mathcal{X}}(z)(z + \mathrm{i}\mathcal{L}_0) = \tilde{\mathcal{X}}(z)(-\mathrm{i}\mathcal{Q}\mathcal{L}_C)\,, \tag{44}$$

which can be solved by iteration (with $\mathcal{X}_0 = 1$, $\tilde{\mathcal{X}}_0 = (z + \mathrm{i}\mathcal{L}_0)^{-1}$ as initial values):

$$\tilde{\mathcal{X}}(z) = (1 - \mathcal{U}_0(z)[-\mathrm{i}\mathcal{Q}\mathcal{L}_C])^{-1}\mathcal{U}_0(z), \quad \mathcal{U}_0(z) = (z + \mathrm{i}\mathcal{L}_0)^{-1}\,. \tag{45}$$

Therewith we rewrite Ξ:

$$\Xi = -\mathrm{i}\lim_{z\to 0}\int_0^\infty \mathrm{d}s\, \mathrm{e}^{-sz}\mathcal{X}(s)\mathrm{e}^{-\mathrm{i}\mathcal{L}_0 s}\mathcal{L}_Q \mathrm{e}^{\mathrm{i}\mathcal{L}_0 s}\,, \tag{46}$$

$$= -\mathrm{i}\lim_{z\to 0}(1 - \mathcal{U}_1(z)[-\mathrm{i}\mathcal{Q}\mathcal{L}_C])^{-1}\mathcal{U}_1(z)\mathcal{L}_Q(2)\,, \tag{47}$$

$$\mathcal{U}_1(z) = (z + \mathrm{i}\mathcal{L}_0(3) - \mathrm{i}\mathcal{L}_0(1))^{-1}\,. \tag{48}$$

Here the arguments in the operators denote the order of application to the right. Collecting all results, the electron–electron correlation part of the equations of motion can be cast into:

$$\frac{\mathrm{d}}{\mathrm{d}t}\sigma_k^{12}|_{\mathrm{Corr,cc}} = -\mathrm{tr}(\epsilon^{-1}\mathcal{L}_{\mathrm{MP}}a_k^{\dagger 1}a_k^2\epsilon^{-1}\mathcal{Q}\mathcal{L}_{\mathrm{MP}}(2)\zeta(\mathcal{L}_0(1) - \mathcal{L}_0(3))\sigma_{\mathrm{rel}})\,, \tag{49}$$

$$= -\mathrm{tr}\left(\epsilon^{-1}\sum_{abc}(V_{ab1c}a_a^\dagger\, a_b^\dagger a_c a_2 - V_{2abc}a_1^\dagger a_a^\dagger\, a_c a_b)\right.$$
$$\left.\epsilon^{-1}\mathcal{Q}\mathcal{L}_{\mathrm{MP}}(2)\zeta(\mathcal{L}_0(3) - \mathcal{L}_0(1))\sigma_{\mathrm{rel}}\right)\,, \tag{50}$$

---

12 Here we use a disentangling formula for superoperators [7] to decompose the exponential $\mathrm{e}^{-\mathrm{i}\mathcal{L}_Q s}$:

$$\mathrm{e}^{\mathrm{i}(A+B)t} = \mathrm{e}^{\mathrm{i}At} + \int^t \mathrm{d}t'\mathrm{e}^{\mathrm{i}(A+B)t'}\mathrm{i}B\mathrm{e}^{\mathrm{i}A(t-t')}$$

$$\to \mathrm{e}^{-\mathrm{i}\mathcal{Q}(\mathcal{L}_0+\mathcal{L}_C)s} \approx \mathrm{e}^{-\mathrm{i}(\mathcal{L}_0+\mathcal{Q}\mathcal{L}_C)s} = \left(1 + \int^s \mathrm{d}s'\mathrm{e}^{-\mathrm{i}(\mathcal{L}_0+\mathcal{Q}\mathcal{L}_C)s'}[-\mathrm{i}\mathcal{Q}\mathcal{L}_C]\mathrm{e}^{\mathrm{i}\mathcal{L}_0 s'}\right)\mathrm{e}^{-\mathrm{i}\mathcal{L}_0 s}\,.$$

where we have introduced Heitler's zeta-function ζ and the superoperator

$$\epsilon = 1 + \mathcal{L}_{\mathcal{Q}}(\mathcal{L}_0(3) - \mathcal{L}_0(1))^{-1} \,. \tag{51}$$

Equation (49) shows that the cc-correlations have to be calculated with many particle Liouvillianes screened by the operator ϵ, which in its simplest approximation contributes via the Lindhard screening of the Coulomb interaction matrix elements V_{abcd}. To show this, we have to calculate the action of ϵ on the 4-particle operator products, cf. (50). In order to apply ϵ as a whole to the 4-particle functions, we restrict to its eigenfunctions by choosing the index combinations which conserve the 4-particle function and find:

$$\begin{aligned}\epsilon^{-1}\sum_{abc} V_{ab1c}a_a^\dagger\, a_b^\dagger a_c a_2 &= \langle\epsilon\rangle^{-1}\sum_{abc} V_{ab1c}a_a^\dagger\, a_b^\dagger a_c a_2 \\ &= \left(1 - V_q \sum_{a,\mathbf{k_a}} \frac{\sigma^{aa}_{\mathbf{k}_a} - \sigma^{aa}_{\mathbf{k}_a-\mathbf{q}}}{\varepsilon^{a}_{\mathbf{k}_a} - \varepsilon^{a}_{\mathbf{k}_a-\mathbf{q}}}\right)\sum_{abc} V_{ab1c}a_a^\dagger\, a_b^\dagger a_c a_2 \end{aligned} \tag{52}$$

Then, by applying the second screening operator ϵ and the remaining superoperators ($\mathcal{Q}\mathcal{L}_{\mathrm{MP}} \approx \mathcal{L}_{\mathrm{MP}} - \mathcal{L}_0$) and using a similar treatment for the correlation contribution of the electron-phonon interaction, we finally obtain a nonlinear quantum master equation:

$$\frac{\mathrm{d}}{\mathrm{d}t}\sigma_k^{12}\Big|^{\mathrm{cc,cp}}_{\mathrm{Corr}} = -\{\, \Sigma^{\mathrm{cc}}_{\mathrm{d}}(k) + \Sigma^{\mathrm{cp}}_{\mathrm{d}}(k)\}\,\sigma^{12}_{k} + \sum_{k'}\{\Sigma^{\mathrm{cc}}_{\mathrm{nd}}(k,k') + \Sigma^{\mathrm{cp}}_{\mathrm{nd}}(k,k')\}\,\sigma^{12}_{k'}\,, \tag{53}$$

which yields the dephasing dynamics[13]. Thus, by solving (18) we can now calculate the imaginary part of the susceptibility (often taken as a measure for the absorption) without any phenomenological damping, cf. (1) and (2).

A detailed analysis shows that the non-diagonal dephasing contributions compensate the influence of the diagonal part by one order of magnitude, i.e. even stronger than for the interband case [13]. Considering the special case of an ideal-2d quantum well and neglecting nonparabolicity effects (i.e. equal

[13] The diagonal and non-diagonal correlation contributions read:

$$\begin{aligned}\Sigma^{\mathrm{cc}}_{d} = \frac{\pi}{\hbar}\sum_{q,k'}^{\substack{a\neq b\\ a,b=1,2}} \{&\delta^{abab}_{k,k',k'-q,k+q} W^{baab}_{q} W^{baab,abab}_{q,k'-k-q}\Gamma^{b,b,a}_{k',k+q,k'-q} \\ +\, &\delta^{abba}_{k,k',k'-q,k+q} W^{abab}_{q} W^{abab,baab}_{q,k'-k-q}\Gamma^{b,a,b}_{k',k+q,k'-q} \\ +\, &\delta^{aabb}_{k,k',k'-q,k+q} W^{aabb}_{q} W^{aabb,aabb}_{q,k'-k-q}\Gamma^{a,b,b}_{k',k+q,k'-q} \\ +\, &\delta^{bbbb}_{k,k',k'-q,k+q} W^{bbbb}_{q} W^{bbbb,bbbb}_{q,k'-k-q}\Gamma^{b,b,b}_{k',k+q,k'-q}\}\,,\end{aligned}$$

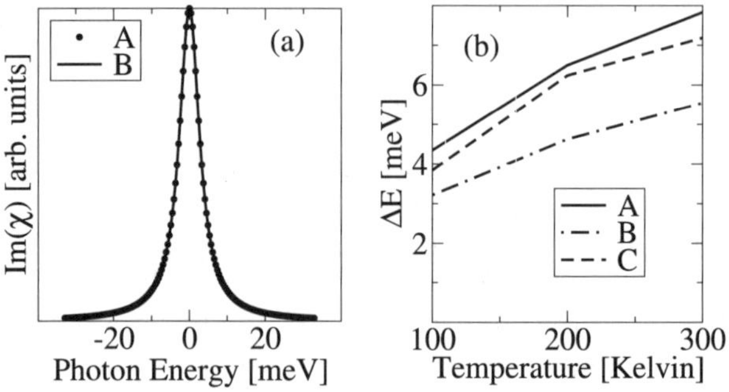

Fig. 3. (**a**) Susceptibility, $\mathrm{Im}(\sum_k \chi_k(\omega))$, cf. (55), of an ideal-2d artificial quantum well *neglecting parabolicity effects*(i.e. equal effective masses in both subbands): (I) spectrum of noninteracting single-particle excitations, (II) inclusion of MF and cc-/cp contributions, (**b**) Linewidth of a GaAs/AlGaAs single quantum well with regard to different *non-additive* material dephasing contributions: (I) cc-correlations (II) cp -correlations (III) cc- and cp-correlations, material parameters given in [17]

effective in-plane masses for both subbands), all mean-field (35) and correlation contributions (53) cancel and the spectrum is that of noninteracting

---

$$
\begin{aligned}
\Sigma^{\mathrm{cc}}_{\mathrm{nd}} = \frac{\pi}{\hbar} \sum_{q}^{\substack{a\neq b\\ a,b=1,2}} \{ & \delta^{abab}_{k,k'+q,k',k+q} W^{bbbb}_q \, W^{baab,abab}_{q,k'-k} \Gamma^{b,b,a}_{k+q,k'+q,k} \\
& + \delta^{bbbb}_{k,k'+q,k',k+q} W^{baab}_q W^{bbbb,bbbb}_{q,k'-k} \Gamma^{b,b,b}_{k+q,k'+q,k} \\
& + \delta^{abba}_{k,q,q-k'+k,k'} W^{bbbb}_{k'-k} W^{abab,baab}_{k'-k,q-k'} \Gamma^{b,a,b}_{q-k'+k,k,q} \\
& + \delta^{bbbb}_{k,q,q-k'+k,k'} W^{abab}_{k'-k} \, W^{bbbb,bbbb}_{k'-k,q-k'} \Gamma^{b,b,b}_{q-k'+k,k,q} \\
& - \delta^{baba}_{k,k',k'-q,k+q} W^{abab}_q W^{baab,abab}_{q,k'-k-q} \Gamma^{b,b,a}_{k,k'-q,k+q} \\
& - \delta^{baab}_{k,k',k'-q,k+q} W^{baab}_q W^{abab,baab}_{q,k'-k-q} \Gamma^{b,a,b}_{k,k'-q,k+q} \} ,
\end{aligned}
$$

$$
\Sigma^{\mathrm{cp}}_{\mathrm{d}} = \frac{\pi}{\hbar} \sum_{k',\pm}^{\substack{a\neq b\\ a,b=1,2}} \{ \delta_{bb} |g^{bb}_{k+k'}|^2 + \delta_{ab} \, |g^{ab}_{k+k'}|^2 \} \Gamma^{+,\mp,b}_{k',k+k'} ,
$$

$$
\Sigma^{\mathrm{cp}}_{\mathrm{nd}} = \frac{\pi}{\hbar} \sum_{\pm}^{\substack{a\neq b\\ a,b=1,2}} \delta_{aa} \, g^{*aa}_{k+k'} g^{bb}_{k+k'} \Gamma^{+,\pm,a}_{k,k+k'}
$$

Here $\Gamma^{a,b,c}_{k_1,k_2,k_3} = (1-\sigma^a_{k_1})\, \sigma^b_{k_2} \sigma^c_{k_3} + \sigma^a_{k_1} \, (1-\sigma^b_{k_2})\, (1-\sigma^c_{k_3})$, $\Gamma^{+,\mp,b}_{k,q} = [n_q + \frac{1}{2} \mp \frac{1}{2}]$ $(1 - \sigma^{bb}_k) + [n_q + \frac{1}{2} \pm \frac{1}{2}] \, \sigma^{bb}_k$ denote the in- and outscattering rates, $W^{abcd}_k = V^{abcd}_k / \langle \epsilon \rangle$ and $W^{abcd,ABCD}_{q,k'-k-q} = 2W^{abcd}_q - W^{ABCD}_{k'-k-q}$, g^{ab}_q the Coulomb (Lindhard screening)/Fröhlich matrix elements and $\delta^{abcd}_{k_1,k_2,k_3,k_4} = \delta(\epsilon^a_{k_1} + \epsilon^b_{k_2} - \epsilon^c_{k_3} - \epsilon^d_{k_4})$, $\delta_{ab} = \delta(-\epsilon^a_k + \epsilon^b_{k'} \mp \hbar\omega_{\mathrm{LO}})$ conservation of energy between the first and final states.

single-particle excitations (free carrier) with a constant dephasing (Fig. 3a, Kohns Theorem). Thus, nonparabolicity [14] and non-diagonal correlation contributions have to be considered properly, i.e. considering the different electron–electron and electron-phonon correlations separately can lead to wrong results, since due to non-diagonal correlation contributions, the dephasing contributions are non-additive (Fig. 3b).

3 Optical Fields and Cooperative Phenomena

Based on the density matrix approach we have so far determined the source of electromagnetic radiation due to intersubband transitions. Thus, now the electromagnetic fields generated by the polarization in the m-th quantum well[14]

$$\mathbf{P}_{\omega}^{(m)}(z) = \int_{(m)} \mathrm{d}z' \; \overset{\leftrightarrow}{\chi}{}_{\omega}^{(m)}(z,z')\, \mathbf{E}_{\omega}^{(m)}(z') \tag{54}$$

has to be determined as a function of the self-consistent field. Here, we have introduced the nonlocal susceptibility tensor[15].

$$\begin{aligned}
\overset{\leftrightarrow}{\chi}{}_{\omega}^{(m)}(z,z') &= \frac{1}{A\hbar} \sum_{\substack{n_1,n_2 \\ s_1,s_2 \\ k}} \chi_k(\omega)\, \mathbf{d}_{ab}^{(m)}(z)\, \mathbf{d}_{ab}^{(m)}(z')\,, \\
\chi_k(\omega) &= \frac{\sigma_k^{ab}(\omega)}{\frac{1}{\hbar}\int_{QW} \mathrm{d}z'\, \mathbf{E}_{\omega}(z') \cdot \mathbf{d}_{ab}(z')} = \frac{\tilde{\sigma}_k^{ab}(\omega)}{\frac{1}{\hbar}\mathbf{E}_{\omega}(z_{QW}) \cdot \int_{QW} \mathrm{d}z' \mathbf{d}_{ab}(z')}\,.
\end{aligned} \tag{55}$$

3.1 Calculation of the Generated Field

In order to calculate the fields generated by a multiple quantum well system, cf. Fig. 1, first the solution of the wave equation is derived[16]. Assuming translational invariance parallel to the wells and considering stationary fields

---

[14] The "(m)" under the integral denotes that the integration runs over the m-th quantum well, i.e. $\int_{(m-1)D}^{(m-1)D+L}$. Note that here the m-th well is going from $(m-1)D$ to $(m-1)D+L$ in contrast to the single-quantum-well-model, Fig. 1, and that this has to be considered when using the wavefunctions and dipole moment defined in Sect.(2).

[15] Note, that $\chi_k(\omega)$ is independent of the electromagnetic field, since $\sigma_k^{ab}(\omega)$ is determined by an equation of motion linear in $\frac{1}{\hbar}\int_{QW} \mathrm{d}z'\, \mathbf{E}_{\omega}(z') \cdot \mathbf{d}_{ab}(z')$. Here $\tilde{\sigma}_k^{ab}(\omega)$ denotes that the reduced density matrix is calculated numerically by approximating the electric field in the equation of motion with $\mathbf{E}(z) \approx \mathbf{E}(z_{QW})$ (z_{QW} is the center of the quantum well).

[16] We will first neglect the medium1/medium2 interface.

$\mathbf{F}(\mathbf{r},t)$ monochromatic with a frequency ω, it is convenient to make the following ansatz:

$$\mathbf{F}(\mathbf{r},t) = \mathbf{F}(\mathbf{r})\,\mathrm{e}^{-\mathrm{i}\,\omega\,t} = \mathbf{F}_\omega(z)\,\mathrm{e}^{\mathrm{i}(q_{||}\,x-\omega\,t)}\,. \tag{56}$$

For an uncharged, unmagnetic material ($\rho = 0, \mu_r = 1, \mathbf{M} = 0, \mathbf{j} = 0$) Maxwell's Equations [15] and the material equation $\mathbf{D} = \epsilon_0\,\epsilon_b\,\mathbf{E} + \mathbf{P}$ yield the wave equation for $\mathbf{E}$

$$\overset{\leftrightarrow}{\mathcal{L}}\cdot\mathbf{E}(\mathbf{r}) = -\omega^2\,\mu_0\mathbf{P}(\mathbf{r})\,. \tag{57}$$

$\mathbf{P}$ denotes the resonant intersubband part of the polarization. $\overset{\leftrightarrow}{\mathcal{L}}$ in dyadic form ($\overset{\leftrightarrow}{\mathcal{U}}$ is the unit tensor) is given by

$$\overset{\leftrightarrow}{\mathcal{L}} = \left(\frac{\omega^2\epsilon_b}{c^2} - q_{||}^2 + \frac{\partial^2}{\partial z^2}\right)\overset{\leftrightarrow}{\mathcal{U}} - \left(\mathrm{i}\,\mathbf{q}_{||} + \hat{\mathbf{z}}\,\frac{\partial}{\partial z}\right)\left(\mathrm{i}\,\mathbf{q}_{||} + \hat{\mathbf{z}}\,\frac{\partial}{\partial z}\right)\,. \tag{58}$$

In order to solve (57) and therewith calculate the fields generated by the polarization $\mathbf{P}$, first the homogeneous wave equation is solved.

Homogeneous Equation. We start by subdividing the fields in left- and rightward propagating waves by making the Fourier ansatz:

$$\mathbf{E}_+(\mathbf{r}) = \mathbf{E}_+\mathrm{e}^{\mathrm{i}\,\mathbf{q}_+\cdot\mathbf{r}}\,,\quad \mathbf{E}_-(\mathbf{r}) = \mathbf{E}_-\mathrm{e}^{\mathrm{i}\,\mathbf{q}_-\cdot\mathbf{r}} \tag{59}$$

where $\mathbf{q}_\pm$ (without loss of generality: $\mathbf{q}_\pm = (q_{||}, 0, \pm q_\perp)$) lies in the plane of incidence spanned by $\hat{\mathbf{q}}_{||}$ and $\hat{\mathbf{z}}$. In the geometry considered here (Fig. 1), we find $\hat{\mathbf{q}}_{||} = \hat{\mathbf{x}}$. Next, we introduce a unit vector perpendicular to the plane of incidence $\hat{\mathbf{s}} = \hat{\mathbf{q}}_{||} \times \hat{\mathbf{z}} = \hat{\mathbf{x}} \times \hat{\mathbf{z}} = -\hat{\mathbf{y}}$. Considering that the solution of the homogeneous wave equation must be transverse ($\nabla\cdot\mathbf{E} = 0 \rightarrow \mathbf{q} \perp \mathbf{E}$), it is convenient to introduce two vectors perpendicular to $\mathbf{q}_\pm$ that span the possible $\mathbf{E}$ [3]. One of these vectors can obviously be taken as the unit vector $\hat{\mathbf{s}}$, the other can be taken as (cf. Fig. 1b)

$$\hat{\mathbf{p}}_\pm = q^{-1}\,(q_{||}\,\hat{\mathbf{z}} \mp q_\perp\,\hat{\mathbf{q}}_{||}) = q^{-1}\,(q_{||}\,\hat{\mathbf{z}} \mp q_\perp\,\hat{\mathbf{x}})\,. \tag{60}$$

Returning to (59) and having in mind that the wave must be transverse, it is clear that choosing $\mathbf{q}_\pm$ as the wave vector, $\mathbf{E}$ can have only $\hat{\mathbf{s}}$ and $\hat{\mathbf{p}}_\pm$ components:

$$\mathbf{E}_+(\mathbf{r}) = (E_{s+}\,\hat{\mathbf{s}} + E_{p+}\,\hat{\mathbf{p}}_+)\,\mathrm{e}^{\mathrm{i}\,\mathbf{q}_+\cdot\mathbf{r}}\,,\mathbf{E}_-(\mathbf{r}) = (E_{s-}\,\hat{\mathbf{s}} + E_{p-}\,\hat{\mathbf{p}}_-)\,\mathrm{e}^{\mathrm{i}\,\mathbf{q}_-\cdot\mathbf{r}}\,. \tag{61}$$

Once $\mathbf{E}(\mathbf{r})$ is specified, $\mathbf{B}(\mathbf{r})$ follows from $\nabla\times\mathbf{E} = \mathrm{i}\omega\mathbf{B}$ and one obtains

$$\mathbf{B}_+(\mathbf{r}) = \frac{\sqrt{\epsilon_b}}{c}\,(E_{p+}\,\hat{\mathbf{s}} - E_{s+}\,\hat{\mathbf{p}}_+)\,\mathrm{e}^{\mathrm{i}\,\mathbf{q}_+\cdot\mathbf{r}}\,,\mathbf{B}_-(\mathbf{r}) = \frac{\sqrt{\epsilon_b}}{c}\,(E_{p-}\,\hat{\mathbf{s}} - E_{s-}\,\hat{\mathbf{p}}_-)\,\mathrm{e}^{\mathrm{i}\,\mathbf{q}_-\cdot\mathbf{r}}\,. \tag{62}$$

Inhomogeneous Equation. Knowing the solutions of the homogeneous equation, we now want to solve (57). Here the source is, within the plane-wave approximation, of the form [cf. (1)]

$$\mathbf{P}(\mathbf{r}) = \mathbf{P}(z)\,\mathrm{e}^{\mathrm{i}\,\mathbf{q}_{||}\cdot\mathbf{r}} = \sum_{m=1}^{N} \int_{(m)} \mathrm{d}z'\,\mathbf{P}^{(m)}\,\delta(z-z')\,\mathrm{e}^{\mathrm{i}\,\mathbf{q}_{||}\cdot\mathbf{r}}\,. \tag{63}$$

Since Maxwell's equations are linear in $\mathbf{P}(\mathbf{r})$, it is possible to decompose the electromagnetic field in the n-th quantum well as follows:

$$\mathbf{E}^{(n)}(\mathbf{r}) = \sum_{m=1}^{N} \int_{(m)} \mathrm{d}z'\,\tilde{\mathbf{E}}^{(m)}(\mathbf{r})\,, \tag{64}$$

where $\tilde{\mathbf{E}}^{(m)}$ denotes the field in the m-th quantum well, generated by the source $\tilde{\mathbf{P}}^{(m)}(\mathbf{r}) = \mathbf{P}^{(m)}\,\delta(z-z')\,\mathrm{e}^{\mathrm{i}\,\mathbf{q}_{||}\cdot\mathbf{r}}$. Knowing that for $z' \neq z$ the solutions of homogenous equation satisfy (57), leads to [17]:

$$\tilde{\mathbf{E}}(\mathbf{r}) = \tilde{\mathbf{E}}_{+}(\mathbf{r})\mathrm{e}^{-\mathrm{i}\,q_\perp\,z'}\Theta(z-z') + \tilde{\mathbf{E}}_{-}(\mathbf{r})\mathrm{e}^{\mathrm{i}\,q_\perp\,z'}\Theta(z'-z) + \tilde{\mathcal{E}}\delta(z-z')\mathrm{e}^{\mathrm{i}\,q_{||}\,x} \tag{65}$$

with $\tilde{\mathcal{E}} = \tilde{\mathcal{E}}_s\,\hat{\mathbf{s}} + \tilde{\mathcal{E}}_{q_{||}}\,\hat{\mathbf{q}}_{||} + \tilde{\mathcal{E}}_z\,\hat{\mathbf{z}} = \tilde{\mathcal{E}}_y\,\hat{\mathbf{y}} + \tilde{\mathcal{E}}_x\,\hat{\mathbf{x}} + \tilde{\mathcal{E}}_z\,\hat{\mathbf{z}}$. To ensure physically reasonable behavior for $z \to \infty$ and $z \to -\infty$, we choose here a transmitted wave for $z > z'$ and a reflected wave for $z < z'$. The factors $\mathrm{e}^{-\mathrm{i}\,q_\perp\,z'}$, $\mathrm{e}^{\mathrm{i}\,q_\perp\,z'}$ are included for convenience [3]. Inserting the ansatz (65) into the wave equation (57), the unknown coefficients $\tilde{E}_i$, $\tilde{\mathcal{E}}_i$ can be determined[18]

$$\begin{aligned}
& \overset{\leftrightarrow}{\mathcal{L}}\cdot\tilde{\mathbf{E}}(\mathbf{r}) &&= -\omega^2\,\mu_0\,\tilde{\mathbf{P}}(\mathbf{r}) \\
\Rightarrow I. \quad & \left(\left(q^2 + \frac{\partial^2}{\partial z^2}\right)\tilde{E}_x - \mathrm{i}\,q_{||}\,\frac{\partial}{\partial z}\tilde{E}_z\right) &&= -\omega^2\,\mu_0\,P_x\,\delta(z-z') \\
\Rightarrow II. \quad & \left(\left(q_\perp^2 + \frac{\partial^2}{\partial z^2}\right)\tilde{E}_y\right) &&= -\omega^2\,\mu_0\,P_y\,\delta(z-z') \\
\Rightarrow III. \quad & \left(q_\perp^2\,\tilde{E}_z - \mathrm{i}\,q_{||}\,\frac{\partial}{\partial z}\tilde{E}_x\right) &&= -\omega^2\,\mu_0\,P_z\,\delta(z-z')
\end{aligned}$$

[17] for brevity we will suppress the index $^{(m)}$ indicating the m-th quantum well during the following calculation

[18] Here we used

$$\begin{aligned}
\frac{\partial^2}{\partial z^2}\mathrm{e}^{\mathrm{i}\,q_\perp\,(z-z')}\,\Theta(z-z') &= -q_\perp^2\Theta(z-z')\mathrm{e}^{\mathrm{i}\,q_\perp\,(z-z')} + \mathrm{i}\,q_\perp\delta(z-z') + \delta'(z-z')\,, \\
\frac{\partial^2}{\partial z^2}\mathrm{e}^{-\mathrm{i}\,q_\perp\,(z-z')}\,\Theta(z'-z) &= -q_\perp^2\Theta(z'-z)\mathrm{e}^{-\mathrm{i}\,q_\perp\,(z-z')} + \mathrm{i}\,q_\perp\delta(z-z') - \delta'(z-z')
\end{aligned}$$

with $\frac{\mathrm{d}}{\mathrm{d}x}\Theta(\pm x) = \pm\delta(x)$ and $\delta(x-x')\,f(x) = \delta(x-x')\,f(x')$.

$$
\begin{aligned}
\Leftrightarrow I. \quad & \left(-\mathrm{i}q\,(\tilde{E}_{p+} - \tilde{E}_{p-}) + \tilde{\mathcal{E}}_x\, q^2 + \omega^2\,\mu_0\, P_x\right)\ \delta(z-z') \\
+ \quad & \left(-\tfrac{q_\perp}{q}\,(\tilde{E}_{p+} + \tilde{E}_{p-}) - \tilde{\mathcal{E}}_z\,\mathrm{i}\,q_{||}\right)\ \delta'(z-z') \\
+ \quad & \tilde{\mathcal{E}}_x\ \delta''(z-z') = 0 \\
\Leftrightarrow II. \quad & \left(-\mathrm{i}q_\perp\,(\tilde{E}_{s+} + \tilde{E}_{s-}) + \tilde{\mathcal{E}}_y\, q_\perp^2 + \omega^2\,\mu_0\, P_y\right)\ \delta(z-z') \\
- \quad & (\tilde{E}_{s+} - \tilde{E}_{s-})\ \delta'(z-z') \\
+ \quad & \tilde{\mathcal{E}}_y\ \delta''(z-z') = 0 \\
\Leftrightarrow III. \quad & \left(\mathrm{i}\tfrac{q_{||}\,q_\perp}{q}\,(\tilde{E}_{p+} + \tilde{E}_{p-}) + \tilde{\mathcal{E}}_z\, q_\perp^2 + \omega^2\,\mu_0\, P_z\right)\ \delta(z-z') \\
- \quad & \mathrm{i}\,q_{||}\,\tilde{E}_x\ \delta'(z-z') = 0\,. \qquad (66)
\end{aligned}
$$

Taking into account that the different orders of singularities must vanish separately yields finally

$$
\tilde{\mathcal{E}}_{x,y} = 0,\ \tilde{\mathcal{E}}_z = -\frac{c^2\,\mu_0}{\epsilon_b}\,P_z,\ \tilde{E}_{s\pm} = -\frac{\omega^2\,\mu_0}{2\,\mathrm{i}\,q_\perp}\,\hat{\mathbf{s}}\cdot\tilde{\mathbf{P}},\ \tilde{E}_{p\pm} = -\frac{\omega^2\,\mu_0}{2\,\mathrm{i}\,q_\perp}\,\hat{\mathbf{p}}_\pm\cdot\tilde{\mathbf{P}}\,, \tag{67}
$$

and thus the electric field in the n-th quantum well is given by

$$
\begin{aligned}
\mathbf{E}^{(n)}(\mathbf{r}) &= \sum_{m=1}^{N}\int_{(m)} \mathrm{d}z'\,\tilde{\mathbf{E}}^{(m)}(z')\,\mathrm{e}^{\mathrm{i}\,q_{||}\,x} \\
&= -\mu_0\,\omega^2 \sum_{m=1}^{N}\int_{(m)} \mathrm{d}z'\ \overset{\leftrightarrow}{\mathcal{G}}(z,z')\,\tilde{\mathbf{P}}^{(m)}(z')\,\mathrm{e}^{\mathrm{i}\,q_{||}\,x}\,.
\end{aligned} \tag{68}
$$

Here $\overset{\leftrightarrow}{\mathcal{G}}(z,z')$ is the retarded Green's function tensor in dyadic form

$$
\overset{\leftrightarrow}{\mathcal{G}}(z,z') = \frac{\mathrm{e}^{\mathrm{i}\,q_\perp\,|z-z'|}}{2\,\mathrm{i}\,q_\perp}[\hat{\mathbf{s}}\hat{\mathbf{s}} + \Theta(z-z')\hat{\mathbf{p}}_+\hat{\mathbf{p}}_+ + \Theta(z'-z)\hat{\mathbf{p}}_-\hat{\mathbf{p}}_-] + \frac{c^2}{\omega^2\,\epsilon_b}\delta(z-z')\hat{\mathbf{z}}\hat{\mathbf{z}}. \tag{69}
$$

Generated Field in the Presence of Interfaces. Next, the Green's function approach shall be generalized to a single-pass geometry, cf. Fig. 1. With respect to this experimental setup, the special case of total reflection at a medium 1/medium 2 interface will be considered in the following (see e.g. [16]). Thus, in order to calculate the absorbance, the field in medium 2 can be neglected and only the electric field inside the n-th quantum well has to be calculated.

Therefore we neglect, as a a starting point, any sources and assume the electric fields to be of the form

$$
\mathbf{E_1}(\mathbf{r}) = (E_{s_1+}\,\hat{\mathbf{s}} + E_{p_1+}\,\hat{\mathbf{p}}_{1+})\,\mathrm{e}^{\mathrm{i}\,\mathbf{q}_{1+}\cdot\mathbf{r}} + (E_{s_1-}\,\hat{\mathbf{s}} + E_{p_1-}\,\hat{\mathbf{p}}_{1-})\,\mathrm{e}^{\mathrm{i}\,\mathbf{q}_{1-}\cdot\mathbf{r}}\,, \tag{70}
$$

$$
\mathbf{E_2}(\mathbf{r}) = (E_{s_2+}\,\hat{\mathbf{s}} + E_{p_2+}\,\hat{\mathbf{p}}_{2+})\,\mathrm{e}^{\mathrm{i}\,\mathbf{q}_{2+}\cdot\mathbf{r}} + (E_{s_2-}\,\hat{\mathbf{s}} + E_{p_2-}\,\hat{\mathbf{p}}_{2-})\,\mathrm{e}^{\mathrm{i}\,\mathbf{q}_{2-}\cdot\mathbf{r}}\,, \tag{71}
$$

where "1" denotes medium 1 and "2" medium 2, respectively. In that case, simply the effect of the medium 1/medium 2 interface on the $\hat{\mathbf{s}}$- and $\hat{\mathbf{p}}$-polarized waves has to be calculated. Considering a wave incident upon the medium 1/medium 2 interface from medium 1 the boundary condition at $z = \infty$ demands that $E_{s_2-} = E_{p_2-} = 0$. Furthermore applying the Maxwell boundary conditions yields $E_{s_1-} = r^s_{12} E_{s_1+}$ and $E_{p_1-} = r^p_{12} E_{p_1+}$ with $r^{s/p}_{12}$ being the Fresnel coefficients for $\hat{\mathbf{s}}/\hat{\mathbf{p}}$-polarized waves[19]. With this background, the Green's function approach can be easily generalized to a geometry with interfaces. Without an interface located at $z = (N-1)D + L + d$, each sheet of polarization at a given z would produce – besides the local contribution – a leftward and a rightward wave, cf. (68) and (69). Since in the geometry considered here, the interface is located on the right side of the MQW system (Fig. 1), the interface does not affect the leftward wave. Thus in medium 1 we find that at a given z there is (I.) the local contribution, (II.) rightward-wave contributions from all planes $z' < z$, (III.) leftward-wave contributions from the reflection of all the rightward waves at the interface and (IV.) directly generated leftward waves from the planes $z' > z$. Therefore, the electric field in the n-th quantum well in medium 1 is given by

$$\begin{aligned}\mathbf{E_1}^{(n)}(\mathbf{r}) &= (E_{s_1+}\,\hat{\mathbf{s}} + E_{p_1+}\,\hat{\mathbf{p}}_{1+})\,\mathrm{e}^{\mathrm{i}\,\mathbf{q}_{1+}\cdot\mathbf{r}} + (E_{s_1-}\,\hat{\mathbf{s}} + E_{p_1-}\,\hat{\mathbf{p}}_{1-})\,\mathrm{e}^{\mathrm{i}\,\mathbf{q}_{1-}\cdot\mathbf{r}} \\ &\quad -\mu_0\,\omega^2 \sum_{m=1}^{N} \int_{(m)} \mathrm{d}z'\, \frac{c^2}{\omega^2\,\epsilon_b}\delta(z-z')\,\hat{\mathbf{z}}\hat{\mathbf{z}}\,\tilde{\mathbf{P}}^{(m)}(z')\,\mathrm{e}^{\mathrm{i}\,q_{\|}\,x} \\ &= -\mu_0\,\omega^2 \sum_{m=1}^{N} \int_{(m)} \mathrm{d}z'\,[\overleftrightarrow{\mathcal{G}}(z,z') + \overleftrightarrow{\mathcal{G}}_1(z,z')]\,\tilde{\mathbf{P}}^{(m)}(z')\,\mathrm{e}^{\mathrm{i}\,q_{\|}\,x} \end{aligned} \tag{72}$$

with

$$\overleftrightarrow{\mathcal{G}}_1(z,z') = \frac{\mathrm{e}^{-\mathrm{i}\,q_\perp\,(z+z')}}{2\,\mathrm{i}\,q_\perp}[r^s_{12}\hat{\mathbf{s}}\hat{\mathbf{s}} + r^p_{12}\hat{\mathbf{p}}_{1+}\hat{\mathbf{p}}_{1-}] \tag{73}$$

and $\overleftrightarrow{\mathcal{G}}(z,z')$ being the the retarded Green's function tensor in dyadic form for a geometry without interface, cf. (69).

3.2 Absorbance

We will now determine the absorbance in the geometry considered. Therefore we consider a monochromatic p-polarized plane wave $\mathbf{E}_0$ incident at

[19] remember that the interface is located at $z = (N-1)D + L + d$

$$r^s_{12} = \frac{q_{1\perp} - q_{2\perp}}{q_{1\perp} + q_{2\perp}}\,\mathrm{e}^{2\,\mathrm{i}\,q_\perp((N-1)D+L+d)}, \quad r^p_{12} = \frac{q_{1\perp}\,\epsilon_{2b} - q_{2\perp}\,\epsilon_{1b}}{q_{1\perp}\,\epsilon_{2b} + q_{2\perp}\,\epsilon_{1b}}\,\mathrm{e}^{2\,\mathrm{i}\,q_\perp((N-1)D+L+d)}$$

$$q_{1\perp} = \sqrt{\frac{\omega^2\,\epsilon_{1b}}{c^2} - q^2_{1\|}}\,, \quad q_{2\perp} = \sqrt{\frac{\omega^2\,\epsilon_{2b}}{c^2} - q^2_{1\|}}\,.$$

an (internal) angle θ on a multiple-quantum well structure consisting of N quantum wells embedded in an infinite medium with the dielectric constant $\epsilon_b(\omega)$, Fig. 1. The barrier layer in the MQW structure is assumed to be so thick that the quantum wells are electronically uncoupled, i.e. the overlap of the wave functions of the electrons belonging to different quantum wells can be neglected. Consequently, the electronic properties (wave functions and eigenenergies) of the MQW structure are determined essentially from a single quantum well and the field inside the n-th quantum well is given by

$$\begin{aligned}\mathbf{E}^{(n)}(z) &= E_0\,\hat{\mathbf{p}}_+\,\mathrm{e}^{\mathrm{i}\,q_\perp\,z} + r_{12}^p\,E_0\,\hat{\mathbf{p}}_-\,\mathrm{e}^{-\mathrm{i}\,q_\perp\,z}\\ &\quad -\mu_0\,\omega^2\sum_{m=1}^{N}\int_{(m)}\mathrm{d}z'\,[\overset{\leftrightarrow}{\mathcal{G}}(z,z')+\overset{\leftrightarrow}{\mathcal{G}}_1(z,z')]\,\tilde{\mathbf{P}}^{(m)}(z')\\ &= E_0\,\hat{\mathbf{p}}_+\,\mathrm{e}^{\mathrm{i}\,q_\perp\,z} + r_{12}^p\,E_0\,\hat{\mathbf{p}}_-\,\mathrm{e}^{-\mathrm{i}\,q_\perp\,z}\\ &\quad -\mu_0\,\omega^2\,\tilde{N}\sum_{m=1}^{N}\int\int_{(m)}\mathrm{d}z'\,\mathrm{d}z''\,[\overset{\leftrightarrow}{\mathcal{G}}(z,z')+\overset{\leftrightarrow}{\mathcal{G}}_1(z,z')]\\ &\quad \mathbf{d}_{12}^{(m)}(z')\,(\mathbf{E}^{(m)}(z'')\cdot\mathbf{d}_{ab}^{(m)}(z''))\,,\end{aligned} \tag{74}$$

where for brevity we have rewritten the nonlocal susceptibility tensor by introducing the quantity $\tilde{N}$:

$$\overset{\leftrightarrow}{\chi}{}_\omega^{(m)}(z,z') = \frac{1}{A\hbar}\sum_{\substack{n_1,n_2\\ s_1,s_2\\ k}}\chi_k(\omega)\,\mathbf{d}_{ab}^{(m)}(z)\,\mathbf{d}_{ab}^{(m)}(z') = \tilde{N}\,\mathbf{d}_{ab}^{(m)}(z)\,\mathbf{d}_{ab}^{(m)}(z')\,. \tag{75}$$

Inserting $\overset{\leftrightarrow}{\mathcal{G}}(z,z')$ and $\overset{\leftrightarrow}{\mathcal{G}}_1(z,z')$, cf. (69) and (73), yields

$$\begin{aligned}E_x^{(n)}(z) &= -E_0\frac{q_\perp}{q}(\mathrm{e}^{\mathrm{i}\,q_\perp\,z} - r_{12}^p\,\mathrm{e}^{-\mathrm{i}\,q_\perp\,z}) - \mu_0\,\omega^2\,\tilde{N}\,\frac{e^2\,q_\perp\,q_{||}}{2\,\mathrm{i}\,q_\perp\,q^2}\sum_{m=1}^{N}\Gamma_z^{(m)}\\ &\quad \int_{(m)}\mathrm{d}z'\,d_{12}^{(m)}(z')\left[\mathrm{e}^{\mathrm{i}\,q_\perp|z-z'|}(\Theta(z'-z)-\Theta(z-z'))\right.\\ &\quad \left.+r_{12}^p\,\mathrm{e}^{-\mathrm{i}\,q_\perp(z+z')}\right]\,,\end{aligned} \tag{76}$$

$$\begin{aligned}E_z^{(n)}(z) &= E_0\frac{q_{||}}{q}(\mathrm{e}^{\mathrm{i}\,q_\perp\,z} + r_{12}^p\,\mathrm{e}^{-\mathrm{i}\,q_\perp\,z}) - \mu_0\,\omega^2\,\tilde{N}\,e^2\sum_{m=1}^{N}\Gamma_z^{(m)}\\ &\quad \int_{(m)}\mathrm{d}z'\,d_{12}^{(m)}(z')\left[\frac{q_{||}^2}{2\,\mathrm{i}\,q_\perp\,q^2}\mathrm{e}^{\mathrm{i}\,q_\perp|z-z'|}(\Theta(z'-z)+\Theta(z-z'))\right.\\ &\quad \left.+\frac{c^2}{\omega^2\,\epsilon_b}\delta(z-z') + \frac{q_{||}^2}{2\,\mathrm{i}\,q_\perp\,q^2}\,r_{12}^p\,\mathrm{e}^{-\mathrm{i}\,q_\perp(z+z')}\right]\,.\end{aligned} \tag{77}$$

Here we have applied

$$\Gamma_z^{(m)} = \int_{(m)} dz''\, d_{12}^{(m)}(z'')\, E_z^{(m)}(z'') \,. \tag{78}$$

In order to determine the unknown quantity $\Gamma_z^{(m)}$, (77) is multiplied with the dipole moment $d_{12}^{(n)}(z)$ and the resulting equation is integrated over z across the n-th quantum well [4] leading to

$$\begin{aligned}\Gamma_z^{(n)} &= E_0 \frac{q_{||}}{q} \int_{(n)} dz\, (e^{i\, q_\perp\, z} + r_{12}^p e^{-i\, q_\perp\, z})\, d_{12}^{(n)}(z) \\ &\quad -\mu_0\, \omega^2\, \tilde{N}\, e^2 \sum_{m=1}^{N} \Gamma_z^{(m)} \int_{(n)} dz\, d_{12}^{(n)}(z) \Bigg(\int_{(m)} dz'\, d_{12}^{(m)}(z') \frac{q_{||}^2}{2\, i\, q_\perp\, q^2} \\ &\quad \Bigg[e^{i\, q_\perp |z-z'|} (\Theta(z'-z) + \Theta(z-z')) + \frac{c^2}{\omega^2\, \epsilon_b} \delta(z-z') \\ &\quad + r_{12}^p\, e^{-i\, q_\perp (z+z')} \Bigg] \Bigg) \,. \end{aligned} \tag{79}$$

By letting the observation point z be located in space we finally determine the reflected field

$$\begin{aligned}\mathbf{E}^R(z) &= -E_0\, r_{12}^p\, e^{-i\, q_\perp\, z} \begin{pmatrix} \cos(\theta) \\ 0 \\ \sin(\theta) \end{pmatrix} \\ &\quad -\mu_0\, \omega^2\, \tilde{N}\, \frac{e^2\, q_{||}}{2\, i\, q^2} e^{-i\, q_\perp\, z} \sum_{m=1}^{N} \Gamma_z^{(m)} K_R^{(m)} \begin{pmatrix} 1 \\ 0 \\ \tan(\theta) \end{pmatrix} , \\ K_R^{(m)} &= \int_{(m)} dz'\, d_{12}^{(m)}(z')\, (e^{i\, q_\perp\, z'} + r_{12}^p\, e^{-i\, q_\perp\, z'}) \end{aligned} \tag{80}$$

and therewith the absorbance given by:

$$A = 1 - R - T = 1 - |\mathbf{E}^R/\mathbf{E}_0|^2 - |\mathbf{E}^T/\mathbf{E}_0|^2 \,. \tag{81}$$

Thus, we are now able to calculate the absorbance including radiative coupling and many particle effects[20]. In Fig. 4 we present the dependence of the absorbance of a GaAs/AlGaAs multiple-quantum well (parameters in [17]) system in the single-pass geometry on the number of wells, N, and the distance d between wells and interface (cf. Fig. 1b). The inset shows the sum over

[20] Note that in the field calculation we did not distinguish between longitudinal and transversal field components, in contrast to the source calculation. Thus, in order to calculate the absorbance the depolarization term (last term in (35), identical with the longitudinal field) in the source calculation has to be neglected to avoid double counting.

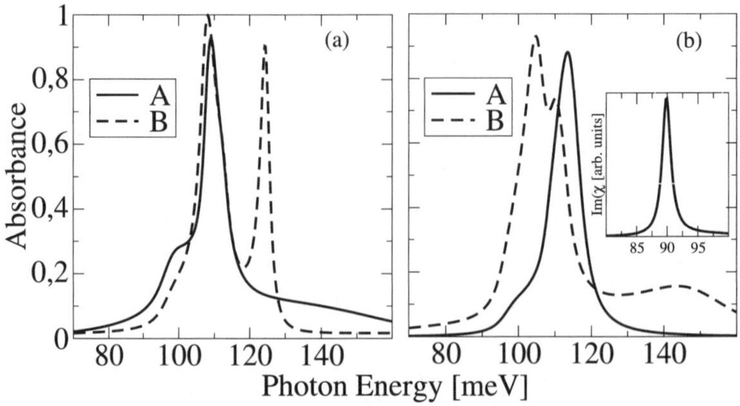

Fig. 4. Absorbance of a GaAs/AlGaAs MQW system (15 K, density: $2 \times 10^{12}\,\mathrm{cm}^{-2}$, $L = 10\,\mathrm{nm}$, $\theta = 63°$) with regard to influence of radiative broadening effects: (**a**) different distance d, i.e. different intensities at the center of the MQW system (51 wells) I: (A) $I \approx I_{\max}$, (B) $I \approx 0.5\, I_{\max}$; (**b**) different number N of quantum wells (d is chosen to obtain $I = I_{\max}$): (A) 5 wells, (B) 80 wells; *inset*: susceptibility, $\mathrm{Im}(\sum_k \chi_k(\omega))$, cf. (55), of a single quantum well, material parameters given in [17]

the imaginary part of the susceptibility, $\mathrm{Im}(\sum_k \chi_k(\omega))$, cf. (55), of a single quantum well, which enters (54) as source. The absorbance shows cooperative phenomena dependent on the geometric parameters. By varying the distance d, the center of the MQW system can be shifted to different intensities of the standing wave in the single-pass geometry, $I = |E(z)|^2$. Whereas for a small number of wells, increasing the intensity leads only to higher absorbance, for a higher number of wells, the lineshape can change significantly. If the center of the wells is positioned at half of the intensity (cf. Fig. 4a B), the coupling of the wells even leads to two distinguished peaks due to very different intensities at the outer wells of the sample. Increasing the number of wells, Fig. 4b, the spectrum becomes more asymmetric and the peak position is shifted to lower energies. The appearance of the second peak is due to increased broadening and distortion effects caused by stronger electromagnetic coupling [4].

4 Conclusion

We have presented a self-consistent theory for the interplay of basic many-particle interactions of intersubband transitions in semiconductor quantum wells including microscopic dephasing dynamics via electron–electron and electron-phonon interaction and radiative coupling effects.

Acknowledgements

We acknowledge stimulating discussions with Michael Woerner and S.C. Lee. Financial support was given by the "Berliner Programm zur Förderung von Frauen in Forschung und Lehre" and the Berliner DFG-Forschergruppe 394.

References

1. M. Helm: The basic physics of intersubband transitions: In *Semiconductor and Semimetals*, volume 62, chapter 1, pages 1–99. Academic Press (2000)
2. T. Elsaesser and M. Woerner: Physics Reports, **321**, 253 (1999)
3. J.E. Sipe: J. Opt. Soc. Am. B, **4**, 481 (1987)
4. A. Liu: Phys. Rev. B, **50**, 8569 (1994)
5. R.A. Kaindl et al.: Phys. Rev. B, **63**, 161308 (2001)
6. H. Haug and S.W. Koch: *Quantum Theory of the Optical and Electronic Properties of Semiconductors*, World Scientific Publishing Co. Pte. Ltd., Singapore, 3rd edition (1994)
7. E. Fick and G. Sauermann: *The Quantum Statistics of Dynamical Processes*, Springer Verlag, Berlin (1990)
8. H.-P. Breuer and F. Petruccione: *The Theory of Open Quantum Systems*, Oxford, New York (2002)
9. D. Ahn: Phys. Rev. B, **50**, 8310 (1994)
10. D. Nikonov et al.: Phys. Rev. Lett., **79**, 4633 (1997)
11. I. Waldmüller, J. Förstner and A. Knorr: Theory of ultrafast dynamics and lineshape of semiconductor quantum well intersubband emitters: In *OSA Trends in Optics and Photonics (TOPS) Vol. 79*, pages 301–303, Washington DC, 2002. OSA Technical Digest
12. T. Kuhn: Density matrix theory of coherent ultrafast dynamics: In *Theory of Transport Properties of Semiconductor Nanostructures*, volume 4 of *Electronic Materials*, chapter 6, pages 173–214. Chapman & Hall (1998)
13. W.W. Chow and S.W. Koch: *Semiconductor–Laser Fundamentals – Physics of the Gain Material*, Springer Verlag, Berlin (1999)
14. U. Ekenberg: Phys. Rev. B, **40**, 7714 (1989)
15. J.D. Jackson: *Classical Electrodynamics*, John Wiley & Sons, Inc., New York (1975)
16. M. Born and E. Wolf: *Principles of Optics*, Pergamon Press, Oxford (1959)
17. Parameter used in calculations: Fig. 2: InAs/AlSb, infinite-well calculation, $n = 1 \times 10^{12}\,\mathrm{cm}^{-2}$, $T = 15\,\mathrm{K}$, $L = 5\,\mathrm{nm}$, $30\,\mathrm{nm}$, $m_1 = 0.039\,m_0$, $m_2 = 0.027\,m_0$ [10], $\tau = 0.2\,\mathrm{ps}$; Fig. 3: (a) GaAs/AlGaAs, infinite-well calculation, $n = 5 \times 10^{10}\,\mathrm{cm}^{-2}$, $T = 15\,\mathrm{K}$, $L = 0$, $m_1 = m_2 = 0.0665\,m_0$, $\tau = 0.2\,\mathrm{ps}$; (b) GaAs/$Al_{0.35}$GaAs, finite-well calculation, $n = 6 \times 10^{11}\,\mathrm{cm}^{-2}$, $L = 9\,\mathrm{nm}$, $m_1 = 0.092\,m_0$, $m_2 = 0.072\,m_0$ calculated with [14], Fig. 4: GaAs/$Al_{0.35}$GaAs, finite-well calculation, $n = 6 \times 10^{11}\,\mathrm{cm}^{-2}$, $L = 10\,\mathrm{nm}$, $T = 15\,\mathrm{K}$, $m_1 = 0.088 m_0$, $m_2 = 0.071\,m_0$ calculated with [14], (a): (A) $d = 1.26\,\mu\mathrm{m}$, (B) $d = 2.4\,\mu\mathrm{m}$, $N = 51$, (b): (A) $N = 5$, $d = 1.86\,\mu\mathrm{m}$, (B) $N = 80$, $d = 0.8\,\mu\mathrm{m}$

Disorder Induced e–h Correlation in Photoexcited Transients in Semiconductors

Anděla Kalvová and Bedřich Velický

Abstract. In mixed semiconductor crystals, the alloy disorder in the valence and conduction bands is statistically correlated. This leads to kinematical correlations in the motion of electrons and holes analogous to exciton effects. This mechanism long known to affect the linear optical response is presently shown to act also in the non-linear regime induced by strong short light pulses. A direct numerical solution of the Kadanoff–Baym equations for non-equilibrium Green functions (employing a self-consistent and conserving single-site approximation) is used to demonstrate the onset and a ripe stage of correlations between the electron and hole photoexcited populations, and the influence of the transient light induced band hybridization.

1 Introduction

Short time optical transients of electrons in semiconductors as induced by sub-picosecond light pulses are one of the dynamically developing areas of research, because of their importance for semiconductor physics, but also as a prototype case of fully quantum transient phenomena far from equilibrium [1,2]. Improved experimental and theoretical techniques provide an access to rather subtle phenomena, including the formation of correlations in many-electron systems. A sample recent work concerning transient exciton phenomena, that is the time evolution of the electron–hole correlations, is presented in [3].

Our previous research in this area was oriented on the sub-problem of optical transients in disordered semiconductors, in which, in particular at the early stages of the transient process, scattering on the static random potential in the disordered sample may be dominant. This leads to important modifications of the physical picture and formal tools for its description as compared with the crystal case. The necessary formal tools and numerical methods based on the use of non-equilibrium Green's functions are described and illustrated on a few case studies in [4]. The specific systems we have in mind are random III-V semiconductor alloys, materials of special importance in optoelectronics [5]; the actual models used are simplified, however, in that it is assumed that the valence and conduction wave functions are not mutually mixed by disorder [6]. The alloy case is better susceptible to a theoretical treatment than the other important class of disordered materials, the amorphous semiconductors, because of the availability of effective approximations of the CPA family, as discussed in the references given above.

The specific problem addressed in this lecture is the effective electron–hole "interaction" representing the spatial correlation in the propagation of an e–h pair moving in the same random environment. This phenomenon is known from linear optics, and it has been found as crucial for understanding the absorption edges of amorphous semiconductors, in particular their exponential ("Urbach") tails. There, it combines with the excitonic interactions proper, and a quasiclassical theory of the doubly correlated e–h motion was worked out decades ago. Summaries can be found in two chapters, 2 and 8, of [7]. In Chap. 8 of this book, a comparison is made also with the case of short range disorder of the alloy type, which is similar on the one hand, but quite different on the other one, because of its fully quantum nature. In the linear theory, the e–h correlation was expressed in terms of the so-called optical vertex defined in terms of the configuration average of two propagators. Schematically, $\langle G^{(e)} \otimes G^{(h)} \rangle = \langle G^{(e)} \rangle \otimes \langle G^{(h)} \rangle + \langle G^{(e)} \rangle \times \text{vertex} \times \langle G^{(h)} \rangle$. A number of papers devoted to an actual construction of this vertex within the CPA formalism could be given, but we single out just one paper [8], to which it will be frequently referred below.

These authors transfer the notion of a correlated and anticorrelated disorder, introduced originally for the quasiclassical disorder in amorphous semiconductors, to the case of a disorder on the atomic scale and obtain the corresponding optical response. The atomic disorder in two bands, conduction and valence, is said to be correlated/anticorrelated, if the underlying quasi-atomic levels fluctuate in parallel, or against each other. Each case leads to a different type of the e–h correlation.

We generalize all these notions to the transient non-linear behavior. The effect of the kinematical e–h correlation is predicted to be perhaps even more pronounced than in the case of linear response. At the same time, it is found that this correlation develops gradually. Thus, we find that it has the nature of a final state interaction, just like the genuine excitons. Changing the pulse strength (measured by a compound characteristic, the Rabi phase), we obtain a marked variation of the degree of correlation in the e–h transient. Altogether, it is found that the kinematic e–h correlation is a basic phenomenon in the optical transients in semiconductor alloys.

Section 2 serves to give a qualitative picture of the optical transitions in disordered systems. In Sect. 3, we introduce our alloy model and contrast the case of the linear response and of the non-linear transients employing the simple density matrix language. Next Sect. 4 gives a brief overview of the nonequilibrium Green's function (NGF) technique used, and the final Sect. 5 shows the computed response of our model to two pulses, one comparatively weak, the other moderately strong, both lasting about 100 fs.

2 Qualitative Picture of Disorder Effects in the Optical Response of Semiconductors

The format of lectures at this meeting allows to present the background motivation of the reported research in more detail and more deeply than is usually possible, and we take the liberty to place our problem into the context of several broader areas: linear optical properties of crystalline semiconductors, non-linear response to short light pulses, and disordered systems. Linear optics of semiconductors is a classical field, and it is reviewed in advanced text-books [9]. A point of departure for us will be the simplest case of the interband transitions across the gap between just two ideal bands $\epsilon_c(\mathbf{k})$, $\epsilon_v(\mathbf{k})$. The absorptive part of complex permittivity is in this case given by a Golden Rule expression as

$$\varepsilon_2(\omega) = 4\pi(N\Upsilon_0\epsilon_0)^{-1} \sum_{\mathbf{k}\in\mathrm{BZ}} |ex_{vc}|^2 \delta(\hbar\omega - [\epsilon_c(\mathbf{k}) - \epsilon_v(\mathbf{k})])\,. \tag{1}$$

Here, N is the number of primitive cells, Υ_0 the cell volume, the dipole matrix element ex_{vc} should in general depend on $\mathbf{k}$. For simplicity, we assume that it is constant throughout the paper.

If, first, we want to generalize this expression to incorporate the effect of phonon emission and absorption, it is necessary to consider the evolution of the photoexcited e–h pair as a motion of two particles which interact with the phonons independently. This interaction leads to relaxing the vertical selection rule $\mathbf{k}_c = \mathbf{k}_v = \mathbf{k}$ (indirect transitions) and to a broadening in the region of direct transitions. Both these effects amount to the transformation of the bare particles to the respective polarons and can be taken into account by introducing the polaron spectral functions $A_b(\mathbf{k},\omega)$, $b = c, v$. With them, $\varepsilon_2(\omega)$ becomes a convolution,

$$\varepsilon_2(\omega) = 4\pi(N\Upsilon_0\epsilon_0)^{-1} \int \mathrm{d}\eta \sum_{\mathbf{k}\in\mathrm{BZ}} |ex_{vc}|^2 A_v(\mathbf{k},\eta) A_c(\mathbf{k},\eta+\hbar\omega)\,. \tag{2}$$

If, now, the electron-phonon interaction is turned off, the spectral functions have the limit $A_b(\mathbf{k},\omega) \to \delta(\hbar\omega - \epsilon_b(\mathbf{k}))$ and (2) reduces to the ideal crystal case (1).

Excitonic interactions represent the other limiting type of modification of the interband formula (1). Again, a bare e–h pair is photogenerated, but now the mutual interaction between both particles, as they fly off, is important, and the final state of the process is modified; this is the so-called final state interaction. The correlated e–h motion leads to a change in the absorption strength, while the energy and total momentum of the final scattering states does not change. As shown first in [10], the simple interband formula (1) is modified twofold: there may exist bound excitonic states, and the continuous part of the ε_2 spectrum differs from (1) by the appearance of the so-called

Elliott factor $(1 + \Gamma(\mathbf{k}))$:

$$\varepsilon_2(\omega) = 4\pi(N\Upsilon_0\epsilon_0)^{-1}\Big\{[\text{BOUND STATES}] + \sum_{\mathbf{k}\in\mathrm{BZ}} |ex_{vc}|^2(1+\Gamma(\mathbf{k}))\,\delta(\hbar\omega - [\epsilon_c(\mathbf{k}) - \epsilon_v(\mathbf{k})])\Big\}\,. \qquad (3)$$

The Elliott factor is seen to modify, appropriately enough, the effective transition matrix element, while the energies of the two-particle excitations remain without change.

We will not continue to review this, leaving thus aside the general case of combined exciton-phonon processes, interband phonon induced correlations, etc.

Our main goal is to characterize the effects of disorder on the electrons in semiconductors, and to draw an analogy with the genuine phonon and exciton effects. Disorder affects the electrons through many channels, but we have in mind the modification of the one-electron orbitals caused by a random potential, in which the electrons move. We will consider a moderately pronounced disorder, for which it will be productive to think about its effect in a perturbative manner, with a reference crystal in mind. This approach makes sense for both major classes of disordered semiconductors, amorphous and glassy semiconductors on the one hand, mixed crystals, i.e., random alloys, in which crystal nodes are occupied by different atoms in an irregular way, on the other hand. Presently, we are interested in the latter, alloy, case. We have to mention the amorphous semiconductors, however, for several reasons. Firstly, the exciton-like disorder-induced kinematical e–h correlations in the linear optical response of electrons in amorphous semiconductors were, in their time, a seminal problem for the whole disorder area. These phenomena were then studied in detail, in connection with the problem of the optical absorption edge in disordered amorphous semiconductors. We refer the reader to two chapters, 2 and 8, of the book [7]. The disorder in question was a consequence of smooth random fields, due to either space charges, or internal strains, both typical for amorphous semiconductors. This case can be treated using the effective mass picture combined with a quasiclassical approximation. This is easy to visualize, and we will use it for illustrative purposes. We must stress that near the absorption edge the kinematic effect combines with a true screened Coulomb interaction between both particles, and this makes the problem substantially more difficult. At higher excitation energies, however, the dynamic effect becomes negligible and the kinematic coupling prevails. This is especially true for a sufficiently strong disorder, and we will refer to this remark to justify that we work later in a purely one-electron approximation.

Let us gradually introduce disorder into the reference crystal. The crystal symmetry is lowered, so that none of the quantum numbers b, $\mathbf{k}$ is preserved. In particular, the electrons do not have a sharp k-vector, and their eigenfunctions are inhomogeneously distributed in space. In general, also the bands are intermixed. For simplicity, we neglect this latter blurring of the band index b. This defines the so-called model of independent bands. Most of the existing work on linear optics of disordered semiconductors uses this model expressly or tacitly. A general expression for ε_2 then reads

$$\varepsilon_2(\omega) = 4\pi(N\Upsilon_0\epsilon_0)^{-1}\sum_i\sum_f |\langle vi|e\hat{x}_{vc}|cf\rangle|^2\delta(\hbar\omega - [\epsilon_{cf} - \epsilon_{vi}])\,. \quad (4)$$

The random eigenfunctions have the form

$$\begin{aligned} |vi\rangle &= \sum_{\mathbf{k}\in\mathrm{BZ}} a_{vi}|v\mathbf{k}\rangle \\ |cf\rangle &= \sum_{\mathbf{k}\in\mathrm{BZ}} a_{cf}|v\mathbf{k}\rangle \end{aligned} \quad (5)$$

by assumption, while the transition operators look like

$$\begin{aligned} e\hat{x}_{vc} &= ex_{vc}\sum_{\mathbf{k}\in\mathrm{BZ}}|v\mathbf{k}\rangle\langle c\mathbf{k}| \\ &\equiv ex_{vc}\Pi_{vc}\,. \end{aligned} \quad (6)$$

The eigenfunction expansions may be wild, in particular for energies at the band edge, where the Anderson localization may be expected. Deeper in the band, however, we assume that a signature of the parent Bloch function will be apparent. We use this remark to illustrate schematically the origins of the analogy of the polaron effect. Physically speaking, the random potential may be considered as a distortion of an original periodic potential due to immobile, "frozen phonons". Assuming that the valence band states are practically undistorted and the disorder only smeared the conduction states, we may rewrite (4) with the use of (5) and (6) in the form

$$\varepsilon_2(\omega) = 4\pi(N\Upsilon_0\epsilon_0)^{-1}|ex_{vc}|^2\int \mathrm{d}\eta\sum_f\sum_{\mathbf{k}\in\mathrm{BZ}}\delta(\eta - \varepsilon_v(\mathbf{k}))\delta(\hbar\omega + \eta - \epsilon_{cf})|a_{cf}(\mathbf{k})|^2\,. \quad (7)$$

Figure 1a suggests an interpretation of the formula. In the crystal, along the depicted line in the k-space, only one sharply defined wave vector at the crossing of ϵ_c and of $\epsilon_v + \hbar\omega$ led to an interband transition. In the disordered case, this sharp selection rule is relaxed. In principle, for a selected $\hbar\omega$, there always exists a transition between any of the cf states and some $v\mathbf{k}^\star$ state whose energy is $\epsilon_v(\mathbf{k}^\star) = \hbar\omega - \epsilon_{cf}$. The transition strength is weighted by $|a_{cf}(\mathbf{k}^\star)|^2$. This is a clear picture of "non-direct", or non-vertical transitions. For a systematic theory, it is preferable to make another rearrangement,

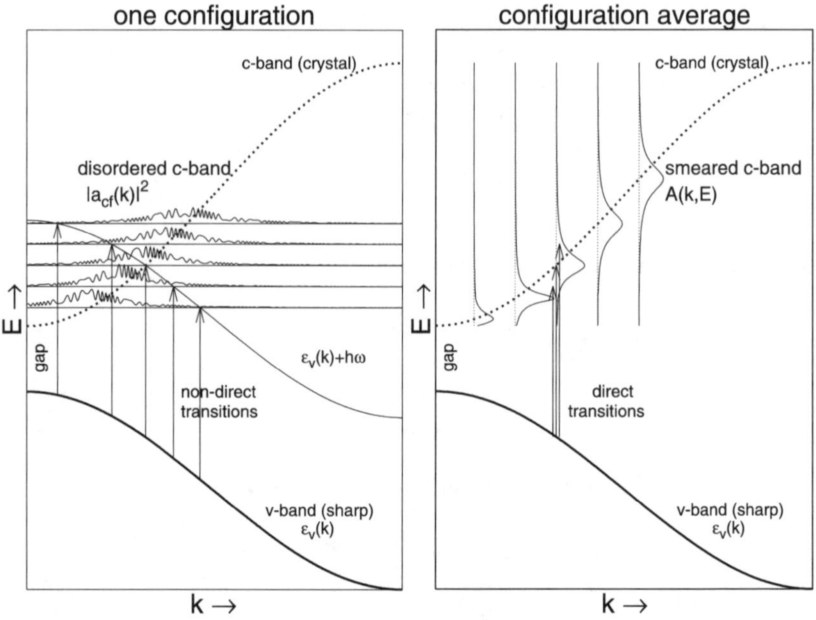

Fig. 1. (**a**) *Left panel*: **k** is not conserved in non-direct transitions between sharp energy levels for one alloy configuration (**b**) *Right panel*: **k** conservation is restored for direct transitions to a broadened ("complex") quasiparticle band after configuration average

summing first over the final states. This internal sum is

$$\begin{aligned}&\sum_f \delta(\hbar\omega + \eta - \epsilon_{cf})|a_{cf}(\mathbf{k})|^2 \\ &= \langle c\mathbf{k}| \left\{ \sum_f |cf\rangle \delta(\hbar\omega + \eta - \epsilon_{cf}) \langle c\mathbf{k}| \right\} |c\mathbf{k}\rangle \\ &\equiv A_c(\mathbf{k})\,. \end{aligned} \tag{8}$$

By these rearrangements, we introduce the spectral density function $A_c(\mathbf{k})$ for the conduction band. This is a random quantity, of course, but, unlike the individual orbitals and energies, it is self-averaging. This term means that for very large samples of the random medium, the value of the spectral density will have a vanishingly small statistical dispersion, so that it will be practically independent of the particular choice of the configuration, that is a realization of the randomly generated sample. In that case, these individual values can be replaced by their statistical, or, as the term goes, configuration average. The dielectric function then assumes a form fully resembling the polaron case (2):

$$\varepsilon_2(\omega) = 4\pi(N\Upsilon_0\epsilon_0)^{-1}\int \mathrm{d}\eta \sum_{\mathbf{k}\in\mathrm{BZ}} |ex_{vc}|^2 A_v(\mathbf{k},\eta)A_c(\mathbf{k},\eta+\hbar\omega)\,. \tag{9}$$

Here, of course, the valence band spectral function is sharp: $A_v(\mathbf{k},\eta) = \delta(\eta - \epsilon_v(\mathbf{k}))$. The physical picture obtained is sketched in Fig. 1b. The configuration average implicit in the procedure restores the translational invariance of the sample. The $\mathbf{k}$-vector becomes sharp again, and the interband transitions can be visualized as vertical. The transition energies become blurred, however, so as to capture the disorder effect.

Finally, we come to the basic situation studied in this paper. Disorder causes a random potential acting in each of the bands separately and differently in general. The assumed dynamical decoupling of both bands does not preclude a statistical coupling; the two random fields are a result of the same random environment in the sample, so that they should show some degree of statistical correlation. While this feature does not reflect itself in statical properties of the electrons, it plays an important role for the interband transitions, where the statistically correlated initial and final states are linked through the transition matrix element $\langle ci|\Pi_{vc}|vf\rangle$. Without entering into details, we quote [6,8,11,12] the general configuration averaged form of ε_2:

$$\begin{aligned}\varepsilon_2(\omega) &= 4\pi(N\Upsilon_0\epsilon_0)^{-1}\\ &\times \int \mathrm{d}\eta \sum_{\mathbf{k}} |ex_{vc}|^2 A_v(\mathbf{k},\eta)A_c(\mathbf{k},\eta+\hbar\omega)(1+\Gamma(\eta,\eta+\hbar\omega,\mathbf{k}))\,.\end{aligned} \tag{10}$$

As compared to (9), a new $1+\Gamma$ factor appears. For an uncorrelated disorder in the two bands, $\Gamma \to 0$, and only the two convoluted spectral functions remain. Thus, the polaron-like disorder broadening is always present. The additional correlation related factor is, in the Green's function language, a vertex part. Very clearly, it stems from the statistical correlations in the transition matrix element, and it also modifies the transition strength multiplicatively. A comparison with (3) shows that it plays exactly the role of the Elliott factor for excitons, only now it depends, in addition to the $\mathbf{k}$ dependence, also on the energies of both one-particle excitations.

The analogy with the Elliott factor is not only formal. It may be said that the statistical correlation of the potentials leads to a kinematic correlation between the electron and the hole whose appearance resembles an effective interaction. Depending on the type of correlation, this interaction may be attractive or repulsive, as sketched in Fig. 2 for the case of smooth random fields. Two situations are sketched. In the first case, the potentials fluctuate in parallel (as corresponds to a fluctuating electric field). These are the so-called correlated potentials [8]. In the other case, the potentials fluctuate against each other (…elastic strains), the anti-correlated potentials. For energies moderately remote from the band edges, the wave function can be found

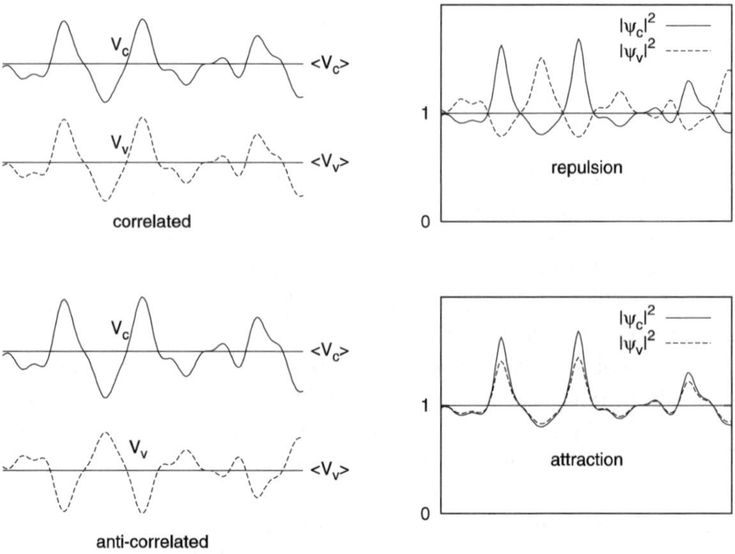

Fig. 2. Smooth random potentials acting in the conduction and the valence bands. *Upper row*: the two potentials positively correlated (*left panel*). A negative correlation of the e and h WKB wave functions results (*right panel*). *Lower row*: for negatively correlated potentials, the wave functions are correlated positively

using the WKB approximation easily, $\psi(x) \sim \sqrt{k(x)}\exp(\pm\mathrm{i}\int \mathrm{d}x' k(x'))$ with $k(x) = \hbar^{-1}\sqrt{2m_b(E - V_b(x))}$. We only have to recall that the valence band effective mass is negative. This is just like to say that for the hole its mass and energy are positive, while the potential changes sign. Then, it is easily found that, in the correlated case, the two particles prefer to "meet" at the common minima of their potentials. This means an effective attraction. For the anticorrelated potential, the behavior is the opposite, and an effective repulsion results. By analogy with the exciton case, we may expect that this will lead to an enhancement of the optical absorption in the correlated case, to a reduction in the other one.

It is interesting to compare this simple picture with the calculations of Abe and Toyozawa [8] in more detail. The situation they consider is far from being quasiclassical, but quite close to our model alloy, as the randomness in their two-band model is represented by a Gaussian distribution of the "atomic" levels in the Wannier representation. The random levels may be positively correlated, uncorrelated, or negatively correlated, as described by a parameter γ, which assumes a corresponding value 1, 0, and -1. The authors demonstrate, using the CPA, that the sharp crystal absorption edge is para-exponentially smeared and they find that this so-called Urbach tail of the absorption edge is enhanced/reduced for an anticorrelated/correlated potential. This appears like the exact opposite of the above qualitative con-

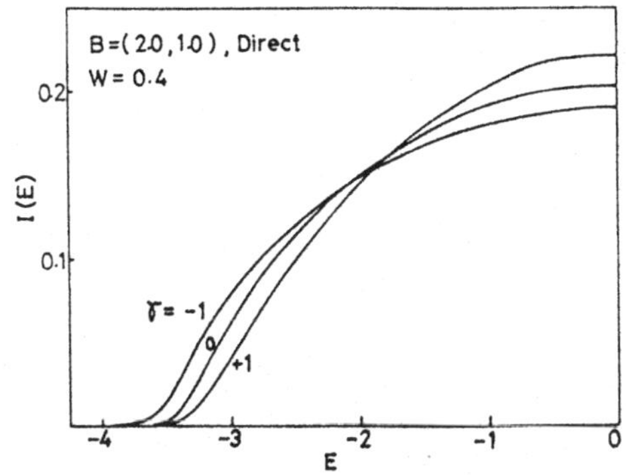

Fig. 3. Reproduction of Fig. 6 of the paper [8] with kind permission of the Publication Committee of the Phys. Soc. Jap. $I(E)$ is the absorption profile for photon energies around the direct optical gap of a reference crystal. B's and W are quantitative parameters of the model. Important are the labels $\gamma = 0, +1, -1$ which specify the uncorrelated, correlated, and anticorrelated disorder in both bands. The crystal absorption edge is at $E = 0$. The part of the spectrum relevant to the present discussion is not the tail region (negative E), but the main absorption band, $E \geq 0$, as discussed in the main text

clusion. We reproduce their Fig. 6 to clarify this seeming contradiction. In this picture, see Fig. 3, the optical absorption $I(E)$ is shown as a function of energy E measured from the crystal energy gap. The tails then appear for negative E, and they behave precisely as the authors describe. The ordering of the absorption curves becomes reverted, however, already around $E = 0$. There, it agrees with the prediction resulting from Fig. 2, which also holds only for energies somewhat above the gap. Thus, the quasiclassical reasoning has some bearing also on the fully quantum case.

3 Model Alloy and its Linear and Non-linear Response

In this section, we formalize the qualitative model outlined above and define the physical questions we will address in a straightforward manner employing the language of the one-electron density matrix.

3.1 Model Semiconductor Alloy

The model used is schematic, but possessing most of the salient features of the realistic systems we mimic here, namely the mixed semiconductor crystals thoroughly characterized in [5]. In these systems, one of the sublattices is randomly occupied by atoms which are isoelectronic and chemically similar. In the basic approximation, such alloying can best be described in an LCAO language by random shifts of the corresponding quasiatomic levels. Our model is even cruder, and its justification is outlined in [4].

Following [4], we consider a completely random binary alloy $A_{c^A}B_{c^B}$ with $c^A + c^B = 1$. The semiconductor has a two-band electron structure with the gap between two isotropic parabolic band edges at the center of the Brillouin zone (*"standard band structure"*). All many-body interactions are ignored. The disorder potential acts in each band separately, it is not mixing states of both bands (*"independent bands"*). The effect of the optical pulse is restricted to a non-random interband dipole coupling treated in the rotating wave approximation (RWA). The initial condition is a fully occupied valence band and a completely empty conduction band before the arrival of light.

The notion of independent bands is formalized by introducing the band projectors which are non-random and diagonal in both the Bloch and the Wannier (site) basis, labeled by $\mathbf{k}$ and i, respectively:

$$P_b = \sum_{\mathsf{BZ}} |b\mathbf{k}\rangle\langle b\mathbf{k}| = \sum_{\text{lattice}} |bi\rangle\langle bi| \ , \quad b = c, v. \tag{11}$$

The full one-electron Hamiltonian for one configuration of our alloy has the following structure:

$$\mathcal{H} = W_v + W_c + \mathcal{V}_c + \mathcal{V}_v + U(t) \equiv \mathcal{H}_c + \mathcal{H}_v + U(t) \equiv \mathcal{H}_{\text{DARK}} + U(t). \tag{12}$$

The non-random configuration independent or averaged quantities are denoted by italic, while the configuration dependent ones by calligraphic characters. The dark Hamiltonian $\mathcal{H}_{\text{DARK}}$ is band diagonal, as both the Bloch Hamiltonians W_b and the random potentials $\mathcal{V}_b$ are assumed to obey the rule $\mathcal{O}_b = P_b \mathcal{O}_b P_b$ with $b = c,\ v$. The interaction with the pulse is taken as purely interband, that is off-diagonal; with the linear polarization in the x direction, the basic frequency Ω, and in the Rotating Wave Approximation, it reads

$$\begin{aligned} U(t) &= U_{cv}(t) + U_{vc}(t) \\ &= -Q \cdot \Phi(t) \left\{ \sum_k |c\mathbf{k}\rangle \mathrm{e}^{-i\Omega t} \langle v\mathbf{k}| + \sum_k |v\mathbf{k}\rangle \mathrm{e}^{i\Omega t} \langle c\mathbf{k}| \right\} \\ &\equiv -Q \cdot \Phi(t) \{ \Pi_{cv} \mathrm{e}^{-i\Omega t} + \Pi_{vc} \mathrm{e}^{+i\Omega t} \} , \end{aligned} \tag{13}$$

where $Q = ex_{cv}\mathrm{E}_{\mathrm{M}}$ is the strength of the pulse, ex_{cv} a real constant electric dipole matrix element and E_{M} the peak electric field of the pulse; $\mathcal{E}_x(t) = \mathrm{E}_{\mathrm{M}}\Phi(t) \geq 0$ is the pulse envelope. The $\Pi_{bb'}$ operators were already defined in (6). The RWA is not fully quantitative for short pulses, but it is still well justified in our model, where the transition energies across the gap are about 1.5 eV corresponding to periods about 4 fs, while the pulse duration is of the order of 100 fs.

3.2 Transient Optical Response

Our model treats electrons as independent particles. In such case, the natural quantity describing time evolution of the photoexcited transient is the one-electron density matrix ϱ of a single random alloy configuration. It evolves according to a Liouville equation

$$\begin{aligned} \mathrm{i}\hbar\dot{\varrho} &= [\mathcal{H}(t), \varrho(t)] \\ &= [\mathcal{H}_{\mathrm{DARK}}, \varrho(t)] + [U(t), \varrho(t)] \,. \end{aligned} \tag{14}$$

The *initial condition* is $\varrho(t_0) = P_v$ for t_0 in a distant past; this initial condition is non-random. A formal solution of (14) is expressed in terms of the usual evolution operator $\mathcal{S}(t, t')$ for the full Hamiltonian (12):

$$\begin{aligned} \varrho &= \mathcal{S}(t, t_0)\varrho(t_0)\mathcal{S}(t_0, t) \\ &= \mathcal{S}(t, t_0)P_v\mathcal{S}(t_0, t) \,. \end{aligned} \tag{15}$$

All effect of the external optical field is contained in the evolution operator in a closed non-linear fashion.

In general, the macroscopic optical response is given by the electric polarization $\boldsymbol{P}$ induced by the electric field $\boldsymbol{E}$ of incident light. In our isotropic case, we need

$$P_x(t) = (N\Upsilon_0\epsilon_0)^{-1}\mathrm{Tr}\langle e\hat{x} \cdot \varrho(t)\rangle \,, \tag{16}$$

with $\hat{x} = x_{cv}\{\Pi_{cv} + \Pi_{vc}\}$. Macroscopic observables should not depend on a particular configuration; hence they should be given by a configuration average. The symbol $\langle\cdots\rangle$ indicates such averaging over all random configurations. N is the number of primitive cells, Υ_0 the cell volume. The electric field enters through the time dependence of ϱ. Clearly, for $\varrho(t_0) = P_v$ in the distant past, $P_x(t_0) = 0$. All polarization is an induced transient.

Linear response. Some understanding of (16) can be gained in the linear case. Then, $\boldsymbol{P}$ and $\boldsymbol{E}$ are linked by the dielectric function ε defined by a symbolic equation $\boldsymbol{P} = \epsilon_0(\varepsilon - 1) \cdot \boldsymbol{E}$. An explicit expression is obtained by expanding the field-dependent evolution operator $\mathcal{S}(t, t')$ in terms of the dark evolution operator $\mathcal{S}_{\mathrm{D}}(t, t')$ and of the external field to the lowest order:

$$\begin{aligned} P_x(t) = 2\,\mathrm{Re}\int_{t_0}^{t} \mathrm{d}\bar{t}\,(ex_{cv})^2(N\Upsilon_0)^{-1} \times \\ \mathrm{Tr}\langle\mathcal{S}_{\mathrm{D};vv}(\bar{t} - t)\Pi_{vc}\mathcal{S}_{\mathrm{D};cc}(t - \bar{t})\Pi_{cv}\rangle \underbrace{\mathrm{e}^{-\mathrm{i}\Omega\bar{t}}\,\mathcal{E}_x(\bar{t})}_{E_x(\bar{t})} \,. \end{aligned} \tag{17}$$

As indicated, the dark evolution operator only depends on the time difference $t - t'$, so the integral is, in fact, a convolution, and it neatly defines a linear response function. Its Fourier transform is the complex polarizability, and using the quoted macroscopic relations, the absorptive part of the complex permittivity for our model can be rewritten [6] in the form

$$\varepsilon_2(\omega) = 4\pi(N\Upsilon_0\epsilon_0)^{-1}\int \mathrm{d}\eta \mathrm{Tr}\langle\delta(\eta-\mathcal{H}_v)\Pi_{vc}\delta(\eta+\hbar\omega-\mathcal{H}_c)\Pi_{cv}\rangle\,. \quad (18)$$

It is easy to see that this is a configuration average of (4) written in a compact operator form. The last two equations (17), (18) are equivalent and lead to the following interpretation of the vertex corrections, that is the effective Elliott factors in (10). Before averaging, the electron and the hole move independently in their respective bands. Configuration average of the product of two δ-functions would not factorize in general, only if the randomness in both bands were uncorrelated. In a chemical environment common for both particles, a disorder induced optical vertex Γ_{vc} modifies the product of the averaged δ-functions $A_b(E) = \langle\delta(E-\mathcal{H}_b)\rangle$. Quantities $A_b(E)$ have the meaning of spectral densities in the individual bands. They describe the band broadening and shift, in other words the polaron-like effects of the disorder. In the time variable representation, the quantities to be factorized are the band diagonal components of the dark propagator. Again, correction factors will compensate for the simple factorization.

Generalization to the non-linear response. In the non-linear case, the induced polarization $\boldsymbol{P}$ cannot be expressed in terms of the electric field $\boldsymbol{E}$ of incident light and of a response function; still, it is not difficult to obtain

$$P_x(t) = 2\,\mathrm{Re}\int_{t_0}^{t} \mathrm{d}\bar{t}\,(ex_{cv})^2(N\Upsilon_0)^{-1}$$
$$\times \mathrm{Tr}\langle\mathcal{S}_{vv}(\bar{t},t)\Pi_{vc}\mathcal{S}_{cc}(t,\bar{t})\Pi_{cv} + \mathcal{S}_{vc}(\bar{t},t)\Pi_{cv}\mathcal{S}_{vc}(t,\bar{t})\Pi_{cv}\rangle \mathrm{e}^{-i\Omega\bar{t}}\,\mathcal{E}_x(\bar{t})\,. \quad (19)$$

Equations (19) and (17) are quite similar. In both equations, the external field amplitude is singled out in the integral. The other factor is thus of zeroth order, that is field independent, in the linear response limit. The full expression (19) then reduces exactly to (17). There are also qualitative differences. The first term in (19) represents an e–h correlation function, similarly to (17). The band diagonal blocks $\mathcal{S}_{vv}$, $\mathcal{S}_{cc}$ are dressed by the light, however. This can be interpreted simply as a non-linear effect. At the same time, it may be said that each of the particles is virtually excited into the other band and their identity is thus spurious. Even more striking is the appearance of the off-diagonal blocks $\mathcal{S}_{vc}$ in the second term in (17). Here, the band mixing is real, unlike the virtual mixing in the diagonal part. The phenomenon involved is the light induced hybridization of both bands and cannot be interpreted as a propagation of a well defined e–h pair at all. The excitations in question are reminiscent of the Bogolyubov–Valatin anomalous propagators.

While these results are physically revealing, they offer no easy way for an actual evaluation, as it is typically the case with a density matrix formulation. It is then preferable to turn to the NGF formulation.

4 Use of the Nonequilibrium Green's Functions

This section is going to be rather terse for two reasons. We use here the NGF technique in the form developed and explained at length in [4]. It is thus sufficient to recapitulate here several essential points. Besides, at this meeting, at the center of attention is physics rather than techniques. We work in the standard LW matrix form of NGF [13]. For example:

$$\mathbf{G} = \begin{Vmatrix} G^R & G^< \\ 0 & G^A \end{Vmatrix}; \quad \mathbf{\Sigma} = \begin{Vmatrix} \Sigma^R & \Sigma^< \\ 0 & \Sigma^A \end{Vmatrix}; \quad \boldsymbol{\mathcal{H}} = \begin{Vmatrix} \mathcal{H} & 0 \\ 0 & \mathcal{H} \end{Vmatrix}. \tag{20}$$

Among the reasons for using this variant is that, for the elastic alloy scattering, the equations for propagators decouple from the equations for the particle correlation function, a substantial advantage.

Virtual crystal and dark GF. The Hamiltonians are split into their averaged (mean field, virtual crystal) and fluctuating parts:

$$\left.\begin{aligned} \boldsymbol{\mathcal{H}} &= \langle \boldsymbol{\mathcal{H}} \rangle + \boldsymbol{\mathcal{D}} \\ \boldsymbol{\mathcal{H}}_{\text{DARK}} &= \langle \boldsymbol{\mathcal{H}}_{\text{DARK}} \rangle + \boldsymbol{\mathcal{D}} \end{aligned}\right\} \boldsymbol{\mathcal{D}} = \boldsymbol{\mathcal{D}}_v + \boldsymbol{\mathcal{D}}_c \equiv (\boldsymbol{\mathcal{V}}_v - \langle \boldsymbol{\mathcal{V}}_v \rangle) + (\boldsymbol{\mathcal{V}}_c - \langle \boldsymbol{\mathcal{V}}_c \rangle). \tag{21}$$

We might introduce the NGF $\boldsymbol{\mathcal{G}}$ for each random configuration. For example, $\mathcal{G}^R = (\mathrm{i}\hbar)^{-1}\mathcal{S}(t,t')\vartheta(t-t')$, etc. We will only need the averaged Green's functions, however, like $\mathbf{G} = \langle \boldsymbol{\mathcal{G}} \rangle$. The self-energies are then defined with respect to the virtual crystal Green's functions. The latter GF are eliminated, however, as the Dyson equation for the full $\mathbf{G}$ is rewritten in terms of the dark Green's function $\mathbf{G}_{\text{DARK}}$:

$$\begin{aligned} \mathbf{G}_{\text{DARK}}^{-1} &= i\hbar\partial_t \mathbf{1} - \langle \boldsymbol{\mathcal{H}}_{\text{DARK}} \rangle - \mathbf{\Sigma}_{\text{DARK}} \\ \mathbf{G}^{-1} &= i\hbar\partial_t \mathbf{1} - \langle \boldsymbol{\mathcal{H}}_{\text{DARK}} \rangle - \mathbf{U} - \mathbf{\Sigma} \\ &= \mathbf{G}_{\text{DARK}}^{-1} - \mathbf{U} - \underbrace{(\mathbf{\Sigma} - \mathbf{\Sigma}_{\text{DARK}})}_{\mathbf{\Sigma}_{\text{IND}}}. \end{aligned} \tag{22}$$

This transformation improves tremendously the stability and convergence of any numerical work, as the induced effect is a deviation from the equilibrium dressed by disorder. This (the so-called dark polaron effect [14]) is duly taken into account in the second form of (22) for $\mathbf{G}$.

Correlations in the self-consistent Born approximation. The dark quantities are known beforehand and we solve numerically the Dyson equation in the second form together with the self-consistent Born approximation for the light induced part of the self-energy,

$$\begin{aligned}\mathbf{\Sigma}_{\text{IND}} &= \langle \mathcal{D}(\mathbf{G} - \mathbf{G}_{\text{DARK}})\mathcal{D} \rangle \\ &= \langle [\mathcal{D}_v + \mathcal{D}_c](\mathbf{G} - \mathbf{G}_{\text{DARK}})[\mathcal{D}_v + \mathcal{D}_c] \rangle \\ &= \mathbf{\Sigma}_{\text{IND};vv} + \mathbf{\Sigma}_{\text{IND};vc} + \mathbf{\Sigma}_{\text{IND};cv} + \mathbf{\Sigma}_{\text{IND};cc} \,. \end{aligned} \tag{23}$$

The off-diagonal parts of $\mathbf{\Sigma}_{\text{IND}}$ are the explicit source of the disorder induced correlation. It spreads, by self-consistence, also to the diagonal parts. We consider the case of a "diagonal disorder", when only the quasi-atomic levels in the Wannier representation fluctuate at random and independently of each other:

$$\mathcal{D} = \sum_{b=c,v} \sum_{\text{lattice}} |b\,i\rangle\, d_{b\,i}\, \langle b\,i| \,. \tag{24}$$

For a binary alloy, let the quasiatomic levels be $E_{cn} = E_c^A, E_c^B$; $E_{vn} = E_v^A, E_v^B$ at random and with average weights (concentrations) c^A, c^B. The fluctuating parts of the atomic potentials are $d_{b\,i} = E_{b\,i} - \langle E_{b\,i} \rangle$. The self-energy in the Born approximation is, according to (23) and (24), given by a 2×2 scalar matrix $\langle d_{b\,i} d_{b'\,i} \rangle \langle b\,i|\mathbf{G}|b'\,i\rangle$. The diagonal elements of $\langle d_{b\,i} d_{b'\,i} \rangle$ measure the intraband disorder strength, the off-diagonal elements pertain to the cross-correlation:

$$\begin{aligned}\langle d_{v\,i}^2 \rangle &= c^A c^B (E_v^A - E_v^B)^2 \equiv \lambda_v^2 \\ \langle d_{c\,i}^2 \rangle &= c^A c^B (E_c^A - E_c^B)^2 \equiv \lambda_c^2 \\ \langle d_{v\,i} d_{c\,i} \rangle &= c^A c^B (E_v^A - E_v^B)(E_c^A - E_c^B) \equiv \lambda_v \lambda_c \kappa \\ \langle d_{c\,i} d_{v\,i} \rangle &= c^A c^B (E_c^A - E_c^B)(E_v^A - E_v^B) \equiv \lambda_v \lambda_c \kappa \,. \end{aligned} \tag{25}$$

Depending on the sign interplay of the level differences, the correlation parameter κ is ± 1. We add in a formal manner the case $\kappa = 0$ and compare three cases:

$\kappa = -1$	0	1
anti-correlated	non-correlated	correlated

This terminology is the same as introduced in Sect. 2. As will be seen, the statistical correlation in the self-energy can be reformulated in terms of the "effective e–h interaction" also in the non-linear case, although the interpretation is less straightforward than in the linear response.

Vertex in the particle correlation function. The Green's functions for different correlation parameters will be distinguished by the left superscript:

${}^{\kappa}\mathbf{G}$ with $\kappa = 0, \pm 1$. To assess the effect of correlation, we compare ${}^{\pm 1}\mathbf{G}$ with ${}^{0}\mathbf{G}$. By (22), we have, putting $\bullet$ for ± 1,

$$\begin{aligned} {}^{\bullet}\mathbf{G}^{-1} &= {}^{0}\mathbf{G}^{-1} - {}^{\bullet}\mathbf{\Xi} \\ {}^{\bullet}\mathbf{\Xi} &= {}^{\bullet}\mathbf{\Sigma}_{\text{IND}} - {}^{0}\mathbf{\Sigma}_{\text{IND}} \,. \end{aligned} \tag{26}$$

More explicitly, the quantities ${}^{\bullet}\Xi^{R,A}_{bb'}(t,t')$, ${}^{\bullet}\Xi^{<}_{bb'}(t,t')$ specify the propagators ${}^{\bullet}G^{R,A}_{bb'}(k;t,t')$ and the particle function ${}^{\bullet}G^{<}_{bb'}(k;t,t')$. There are differences, however. The propagators have, in fact, only an auxiliary role; at the same time, they satisfy a simple Dyson equation ${}^{\bullet}G^{R,A} = {}^{0}G^{R,A} + 0G^{R,A\bullet}\Xi^{R,A\bullet}G^{R,A}$. The particle correlation function ${}^{\bullet}G^{<}$ plays a central role, by contrast, as it directly yields the averaged one-particle density matrix. The expression for it has a more involved structure, permitting to define the desired correlation vertex, which is not simply ${}^{\bullet}\Xi^{<}$. Skipping the straightforward, but lengthy derivation, we write directly the result:

$$\begin{aligned} {}^{\bullet}G^{<} &= {}^{0}G^{<} + {}^{0}G^{R}\,{}^{\bullet}\Theta^{<}\,{}^{0}G^{A} \\ {}^{0}G^{<} &= \mathrm{i}\hbar\,{}^{0}G^{R}\,P_{v}\,{}^{0}G^{A} + {}^{0}G^{R}\,{}^{0}\Sigma^{<}\,{}^{0}G^{A} \\ {}^{\bullet}\Theta^{<} &= {}^{\bullet}\Xi^{<} \\ &\quad +(1 + {}^{\bullet}\Xi^{R\bullet}G^{R}){}^{\bullet}\Xi^{<}(1 + {}^{\bullet}G^{A\bullet}\Xi^{A}) - {}^{\bullet}\Xi^{<} \\ &\quad +(1 + {}^{\bullet}\Xi^{R\bullet}G^{R}){}^{0}\Sigma^{<}(1 + {}^{\bullet}G^{A\bullet}\Xi^{A}) - {}^{0}\Sigma^{<} \,. \end{aligned} \tag{27}$$

The form of ${}^{0}G^{<}$, namely $G^{<} = G^{R}[\mathrm{i}\hbar\varrho_0 + \Sigma^{<}]G^{A}$, is the usual Dyson equation expressing the full $G^{<}$ as a decomposition into its coherent and incoherent parts. The newly introduced quantity ${}^{\bullet}\Theta^{<}$ plays clearly the role of a vertex transforming the uncorrelated particle function into the correlated one. It consists of several contributions. First, it is the corresponding change in the induced self-energy. The other two parts involve the correlated propagation. Particularly significant is the appearance of the last (third line) contribution, which only involves the correlation correction to the propagator, while the particle correlation proper, $\Xi^{<}$, does not enter at all. One last point should be made. This vertex correction Θ applies to a product of propagators, while the originally introduced Γ was correcting a product of spectral densities. This is not peculiar to our case, but rather a result already known from other physical problems.

5 Numerical Study: Onset and Evolution of Disorder Induced Correlations

To demonstrate the importance of the e–h disorder induced correlation, we present here a few numerical results obtained by solving the (22) and (23) using the method described in [4]. We have directly calculated full Green functions ${}^{\kappa}G^{R}_{ab}(k;t,t')$, ${}^{\kappa}G^{<}_{ab}(k;t,t')$ and self-consistent self-energies ${}^{\kappa}\Sigma^{R}_{\text{IND},ab}(t,t')$,

${}^{\kappa}\Sigma^{<}_{\mathrm{IND},ab}(t,t')$ for $a,b=c,v$ and $\kappa=0,\pm1$. We do not intend to analyze at this place the extensive results for the double time Green's functions and self-energies. The kinematical correlation is clearly revealed already in the behavior of much simpler single time observable quantities. All results are shown in the Galickii picture [4] suppressing the fast Ω-oscillations and extracting the envelopes characteristic of the pulse induced transients.

Calculations were performed for these parameters: $E_G=1.5\,\mathrm{eV}$, $\hbar\Omega=2.\,\mathrm{eV}$, $m_v^*=0.6$, $m_c^*=0.4$; *disorder*: c- and v- band lifetimes $\tau_c=160\,\mathrm{fs}$, $\tau_v=240\,\mathrm{fs}$, $c^A=0.05$; *pulse*: Gaussian, duration $\tau_m=120\,\mathrm{fs}$, strength – two pulses used, "weak" with $Q=0.0025\,\mathrm{eV}$ and "strong" with $Q=0.01\,\mathrm{eV}$. By this, their respective Rabi phase $3.3Q\tau_m/\hbar$ are adjusted to 0.48π, and 1.93π, in other words slightly below $\pi/2$, and 2π.

The central quantity is the averaged one-electron density matrix $\rho(t)=\langle\varrho(t)\rangle$. It is given by the time diagonal of $G^{<}$ as

$$ {}^{\kappa}\rho(t)=-\mathrm{i}\hbar\,{}^{\kappa}G^{<}(t,t)\,. \tag{28}$$

It is k-, but not band-, diagonal. To further condense the information, we trace over the k-vectors and obtain the total excitation per unit cell,

$$ {}^{\kappa}\rho^{\mathrm{TOT}}_{ab}(t)=N^{-1}\sum_{k\in\mathrm{BZ}}{}^{\kappa}\rho_{ab}(\mathrm{k},t),\quad a,b=c,v\,. \tag{29}$$

The time dependent deviation from equilibrium ${}^{\kappa}\rho^{\mathrm{TOT}}-P_v$ for the weaker pulse is in Fig. 4. The diagonal elements are related to total photoexcited populations, namely $n_e=\rho^{\mathrm{TOT}}_{cc}$, $n_h=1-\rho^{\mathrm{TOT}}_{vv}$. Their sum should be zero (particle number conservation). This is reasonably satisfied considering that computationally the way to each of the elements is rather remote [4]. The effect of correlation is distinct, but not striking. Two features can be noticed. First, the correlation develops only gradually; at the early times the populations appear as κ independent. This is in agreement with the notion of "final state interactions" discussed in Sect. 2.

Physically more relevant are the off-diagonal elements linked with the electric polarization by $M_x(t)=ex_{cv}\cdot 2\,\mathrm{Re}\,[\rho^{\mathrm{TOT}}_{cv}\mathrm{e}^{-i\Omega t}]$ (the dipole moment of one cell) and $P_x=\Upsilon_0^{-1}M_x$. The populations and the polarization have in common that they are correlation insensitive early after the pulse arrival, while the differences are marked at later times, again in a qualitative agreement with the notion of final state e–h interactions. This points to the important link between the correlation effects and the incoherent back-scattering which sets on gradually, but persists for long times. Formally, this is described by (27) which basically expresses the correlation vertex in terms of $\Sigma^{<}$.

The tendency is different, however: the diagonal elements increase with a stronger correlation, while the off-diagonal ones seem to decrease. In fact, this apparent discrepancy may be reconciled, if we notice that the late time overshoot in $\mathrm{Im}\rho^{\mathrm{TOT}}_{cv}$ is due to the coherent Rabi oscillation, which is then suppressed more, or less, depending on κ, and with the "correct" ordering.

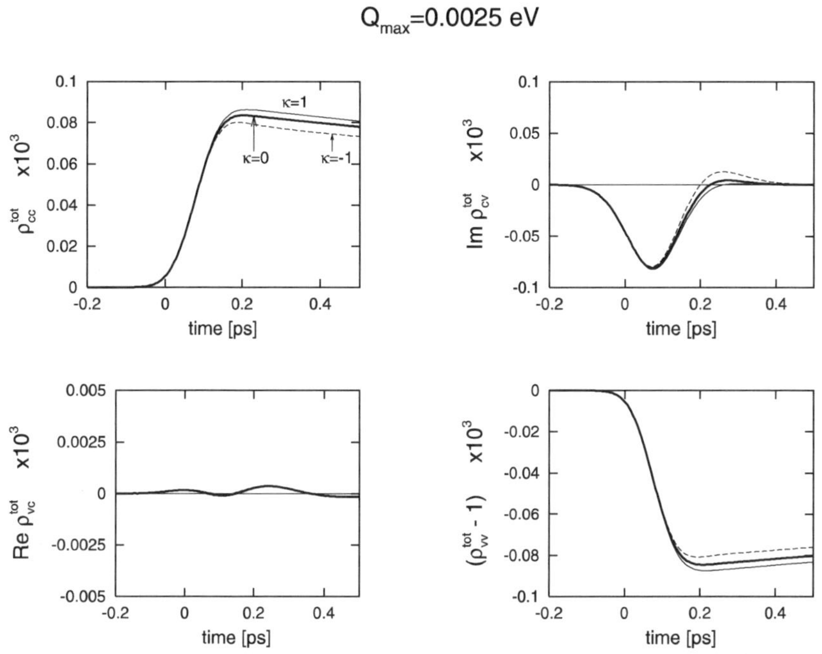

Fig. 4. Weak pulse, $Q = 0.0025\,\mathrm{eV}$: total one-electron density matrix (29) as a function of time. The diagonals are real, the off-diagonals are complex conjugate, and we show the real and the imaginary parts separately, as labeled. Three different cases of the e–h correlation, $\kappa = 0, +1, -1$ (uncorrelated, correlated, and anticorrelated disorder), have a line code defined in the ρ_{cc}^{TOT} panel

Finally, the tentatively suggested interpretation of this behavior is that *positive/negative* correlation *enhances/reduces* the incoherent backscattering, and, hence, the disorder effect on the optical response.

We may look at the effects of correlated disorder from a different angle by considering the energy balance. The "Joule heat" rate (actually, a non-dissipative energy transfer [4]) is

$$w = \boldsymbol{E} \cdot \boldsymbol{J} = \boldsymbol{E} \cdot \frac{\partial}{\partial t} \boldsymbol{P} \,. \tag{30}$$

As seen in Fig. 5, the gain is the least, and even becomes negative at the trailing side of the pulse, for $\kappa = -1$, while it is enhanced somewhat for $\kappa = 1$. By

$$W(t) = \int_{-\infty}^{t} \mathrm{d}t' \, w(t') \,, \tag{31}$$

this is transferred also on the integral energy transfer. Again, for $\kappa = \pm 1$ the incoherent backscattering is *enhanced/reduced*, which leads to the same tendency for the irreversible abduction of energy from the pulse into the bath.

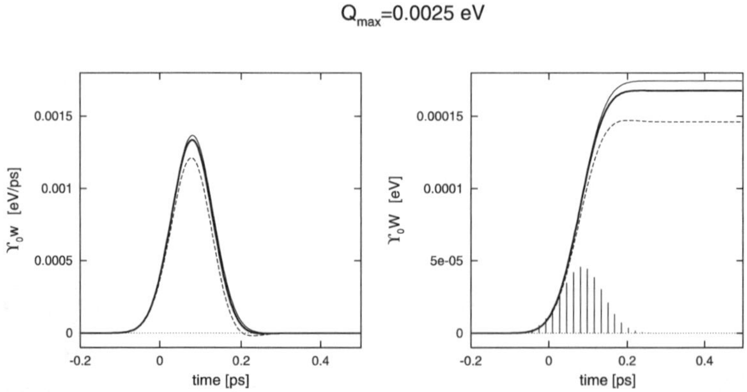

Fig. 5. Weak pulse, $Q = 0.0025\,\mathrm{eV}$: the absorbed power (*left panel*) and the integral energy gain (*right panel*) obtained as "Joule heat" from (31). Line code for different cases of the e–h correlation, $\kappa = 0, +1, -1$, as in Fig. 1. Hatched profile: pulse *intensity*, arbitrary scale

Let us briefly look on the strong pulse. In this case, the coherent action of light overcomes better the disorder (kept at the same level), so that the off-diagonal of the density matrix, shown in Fig. 6, nicely manifests the single Rabi oscillation. The populations are first coherently excited and then fall

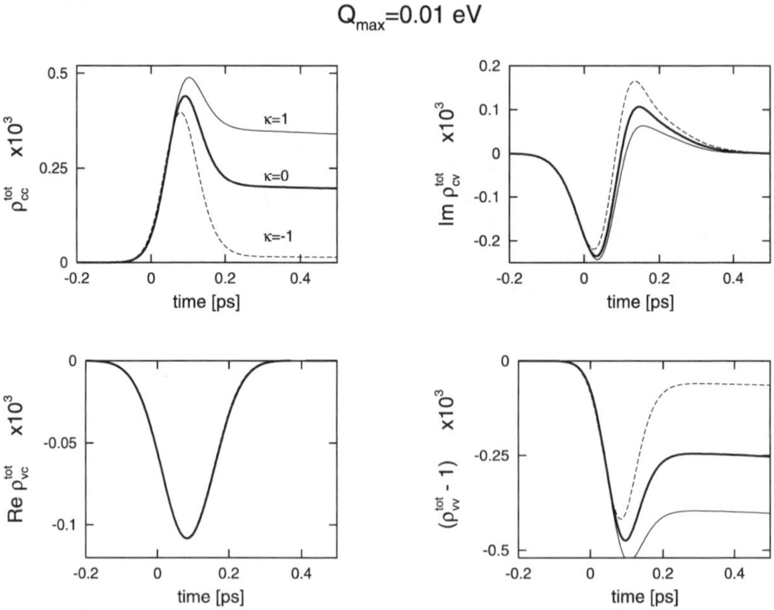

Fig. 6. Strong pulse, $Q = 0.01\,\mathrm{eV}$: total one-electron density matrix (29) as a function of time. The layout and coding of the figure is the same as in Fig. 4

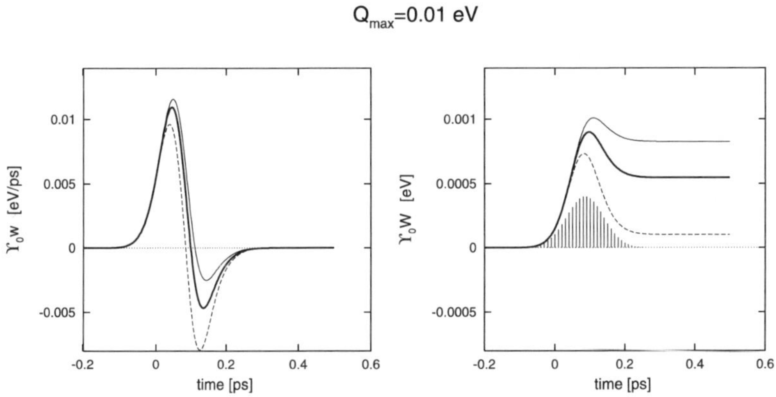

Fig. 7. Strong pulse, $Q = 0.01\,\mathrm{eV}$: the absorbed power (*left panel*) and the integral energy gain (*right panel*) obtained as "Joule heat" from (31). The layout and coding of the figure is the same as in Fig. 5

to their saturation values. Clearly, all of the discussion for the weak pulse fully applies here, too, only the effects are more pronounced. In particular, we may observe that for the anticorrelated disorder the final population is rather small. Without disorder, we would expect that at the end of one Rabi oscillation, all of the excitation would precisely return back to zero. This indicates that the anti-correlated case reduces the disorder effect on the e–h pair truly effectively. Similarly, the power gained from the pulse, Fig. 7, depends on κ; while the positive gain is at first comparable for all three cases, the negative gain, when energy is returned to the electromagnetic field during the second half-cycle, is strongly correlation dependent, and the anti-correlation leads virtually to a full compensation. We may conclude that the 2π pulses are particularly suited for measuring the kinematical correlations of e–h pairs in disordered systems.

6 Conclusion

To summarize our discussion, we may say first that indeed the statistical correlation of the disorder in a valence and a conduction band leads to a kinematical correlation between the particles in an e–h pair, which is fully observable and qualitatively important for systems with parameters resembling real semiconductor alloys.

A second conclusion is that in order to develop an efficient theoretical description of the photoexcitation process is such semiconductor alloys, we found as the most efficient way the use of Nonequilibrium Green's Function, which permit to define the correlation vertex and to evaluate the physical characteristics of such photoexcitation process in a straightforward numerical procedure.

In the course of the theoretical analysis and of the numerical work, we found a number of general conclusions concerning the behavior of the photoexcited electrons. The most interesting are as follows:

1. The stochastic correlation in motion of the photoexcited electron–hole pairs is predicted to have a strong effect on the absorption of light from a pulse
2. The theory developed presently extends the previous work for the linear response [8,11] to non-linear transients [4]
3. Main effect of the e–h kinematic correlation is to modify the late time asymptotics corresponding to the final state interaction in the linear theory and to the incoherent backscattering in the general non-linear case
4. correlated e–h disorder enhances the effect ("attraction")
   anticorrelated e–h disorder reduces the effect ("repulsion")

There is a number of open questions for future work. An obvious one is to analyze in detail the structure and properties of the correlation vertex ${}^{\kappa}\Theta$ and to obtain qualitative rules linking the behavior of this vertex with properties of the observable quantities.

More complicated is the question of improving the dynamical approximation for the self-energy beyond the Born approximation, and aiming at something like the coherent potential approximation.

A truly satisfactory theory leading to a picture permitting a direct comparison with experiments should, by necessity, incorporate also the real correlations due to the $e - p$ and the $e - e$ interactions. For this purpose, there is no theoretical framework available at present, however.

Acknowledgments

The research of this problem was conducted within the project AVOZ-010-914 of the Academy of Sciences of the Czech Republic. It was supported by the Grant Agency of the Czech republic under the project number 202/00/0643.

References

1. H. Haug, and A.P. Jauho, *Quantum Kinetics in Transport and Optics of Semiconductors*, (Springer, Berlin Heidelberg 1997), Vol. 123
2. M. Bonitz, *Quantum Kinetic Theory*, (Teubner, Stuttgart Leipzig 1998)
3. T. Schmielau, G. Manzke, D. Tamme, K. Hennenberger, *Phys. Stat. Sol. B* **221**, (1), 215–219
4. A. Kalvová and B. Velický, *Phys. Rev. B* **65**, 155329 (2002)
5. A.B. Chen and A. Sher, *Semiconductor Alloys*, (Plenum, New York 1995)
6. B. Velický and A. Pieczonková, *Physica scripta T* **19**, 558 (1987)
7. B. Seraphin, Editor, *Optical properties of solids: New developments*, (North Holland American Elsevier, Amsterdam 1976)

8. S. Abe and Y. Toyozawa, *J. Phys. Soc. Japan* **50**, 2185 (1981)
9. P.Y. Yu, and M. Cardona, *Fundamentals of Semiconductors*, (Springer Verlag, Berlin Heidelberg 1999)
10. R. J. Elliott, *Phys. Rev.* **108**, 1384 (1957)
11. K. C. Hass, *Topics in the Electronic Theory of Disordered Semiconductors*, Thesis, Ch. VIII., (Harvard University, Cambridge 1984)
12. N. F. Schwabe and R. J. Elliott, *Phys. Rev. B* **53**, 5301 (1996)
    N. F. Schwabe and R. J. Elliott, *Phys. Rev. B* **53**, 5318 (1996)
13. D. Langreth, and J.W. Wilkins, *Phys. Rev. B* **6**, 3189 (1972)
14. P. Gartner, L. Bányai, and H. Haug, *Phys. Rev. B* **60**, 14234 (1999)

Excitonic Correlations in the Nonlinear Optical Response of Two-dimensional Semiconductor Microstructures: A Nonequilibrium Green's Function Approach

Nai-Hang Kwong, Ryu Takayama, Ilia Rumyantsev, Zhen-Shan Yang and Rolf Binder

Abstract. Important Coulomb correlations among excitons have been extensively demonstrated in weakly-nonlinear optical measurements on quasi-2D semiconductor structures in the coherent regime. The Dynamics Controlled Truncation (DCT) scheme, a perturbation (in the applied field amplitude) theory within the density matrix formalism, has scored much success in elucidating the correlation structures revealed by these experiments. Practically, however, DCT is ill-suited to treat the incoherent aspects of the electronic or excitonic dynamics, and therefore its range of efficient applications is limited to short times and/or low carrier densities. Traditionally, the most powerful approach to study incoherent effects and correlations in highly excited semiconductors is that of nonequilibrium Green's functions (NGF). A combination of the insights and technical advantages of DCT and NGF can lead to more comprehensive theories for the nonlinear response of semiconductors. This contribution reviews some of our efforts in this direction.

We work within the NGF formalism. Our first step was a derivation within the NGF formalism of the most studied DCT equations of motion for electrons and holes – that to third order in the applied field. We summarize here the arguments that identify the set of NGF diagrams that constitute this equation and show the vanishing of the remaining diagrams (of the same order). This diagram set is to be included in the construction of higher-order kinetic equations for electrons and holes.

For processes involving predominantly excitons, we have formulated, via standard NGF techniques, a kinetic theory treating the excitons as bosons. The exchange effects due to the excitons' fermionic constituents are included through a judicious choice of exciton–exciton and exciton-photon couplings so that our effective exciton theory, when reduced to the third order coherent limit, yields formally identical optical responses as the third order electron-hole DCT theory. This effective theory can be applied to study higher order excitonic correlations and dephasing due to exciton–exciton scattering. We give an overview of this theory here.

Fundamental to a theoretical description of excitonic correlations is the exciton–exciton scattering amplitude (T-matrix), which can be (partially) characterized by coherent third order optical measurements. This T-matrix is usually calculated within a truncated exciton basis. We provide some justification of this approximation by comparing the resulting $\chi^{(3)}$ signals to a frequency-domain four wave mixing experiment on a quantum well microcavity.

1 Introduction

Besides its relevance to practical applications, nonlinear semiconductor optics offers a rare opportunity to study in detail the correlated motions of many particles out of equilibrium. For an overview, see e.g. [1,2]. The wealth of high-quality measurements from a variety of experimental configurations has spurred efforts to develop unifying theories that aim at achieving a comprehensive understanding of the microscopic physical processes involved.

In the excitation regime which is weakly nonlinear in the light field intensity and spectrally close to the lowest exciton, theories based on the Dynamics Controlled Truncation (DCT) scheme [3] or other formalisms [4] have been very successful in explaining the role of Coulomb correlations in the coherent exciton dynamics in quantum well structures (see e.g. [5–10]. DCT's efficiency arises from its demonstration that, in each order of the applied field, only a finite number of correlation functions need to be considered in the calculation of the coherent response. In this way it isolates for treatment the most important correlation functions in each order of the light field intensity. Practically, however, DCT is ill-suited to treat the incoherent aspects of the electronic or excitonic dynamics, and therefore its range of efficient applications is limited to short times and/or low carrier densities. Traditionally, the most powerful approach to study incoherent effects and correlations in highly excited semiconductors is that of nonequilibrium Green's functions (NGF). A combination of the insights and technical advantages of DCT and NGF can lead to more comprehensive theories for the nonlinear response of semiconductors. This contribution reviews some of our efforts in this direction.

We work within the NGF formalism. Our first step was a derivation within the NGF formalism of the most studied DCT equations of motion for electrons and holes – that to third order in the applied field. We summarize in Sect. 2 the arguments that identify the set of NGF diagrams that constitute this equation and show the vanishing of the remaining diagrams (of the same order). This diagram set is to be included in the construction of higher-order kinetic equations for electrons and holes. This set of diagrams were also obtained in [11] where they were combined with another set that are suitable for relatively high density electron-hole plasmas.

For processes involving predominantly excitons, the diagrammatic analysis of a theory in terms of electrons and holes is quite complicated. For efficient applications to higher order exciton processes and incoherent processes due to exciton–exciton scattering, a theory using excitons as the basic degrees of freedom would have the advantages of simplicity and physical transparency. The problem with following this approach is that the typical excitons in quantum wells made of say GaAs are not very small. Effects due to the exchange of the fermionic components among excitons are often not negligible. As quantum particles, excitons are almost bosons but not quite. In order to use established many-body theroy techniques, which are mostly formulated for fermions or bosons, theories based on 'elementary' excitons commonly take the excitons as strictly bosons and treat the fermionic corrections as additional interactions between the bosons (see e.g. [12–17]. We have formulated, via standard NGF techniques, a kinetic theory treating the excitons as bosons. The exchange effects due to the excitons' fermionic constituents are included through a judicious choice of exciton–exciton and exciton-photon couplings so that our effective exciton theory, when reduced to the third order coherent limit, yields formally identical optical responses as the third order electron-hole DCT theory. This effective theory can be applied to study higher order excitonic correlations and dephasing due to exciton–exciton scattering. One can also consider this bosonic exciton theory as a guide to select physically relevant diagrams in a full fermionic theory for theses processes. We give an overview of the boson theory in Sect. 3.

Fundamental to a theoretical description of excitonic correlations is the exciton–exciton scattering amplitude (T-matrix), which can be (partially) characterized by coherent third order optical measurements. In applications, this T-matrix is usually calculated within a truncated exciton basis. In fact a bosonic theory would not be efficient if one must include all internal exciton states in the calculations. In Sect. 4 we compare third order signals calculated with the intermediate states in the exciton–exciton T-matrix restricted to the 1s subspace with a frequency-domain four wave mixing experiment on a quantum well microcavity. The good agreement between theory and experiment provides some justification for using the 1s-approximation.

2 Green's Function Diagrammatic Analysis of the Third Order Optical Response of Semiconductors

In this section, we review some of the main features of a diagrammatic NGF approach to the third-order (or $\chi^{(3)}$) nonlinear response of a semiconductor system initially in its ground state. More details can be found in [18]. See e.g. [19,20] for general treatments of nonequilibrium Green's functions.

The model Hamiltonian for the electrons and holes consists of a single-particle energy term, the carrier-carrier basic (unscreened) Coulomb interaction, and a dipole coupling to a (classical) light field in the rotating-wave-approximation that acts to annihilate or create an electron-hole pair. The Green's functions relevant to $\chi^{(3)}$ order are two-point density-type and polarization-type functions, e.g.

$$\begin{aligned} G_{ee}(i,\bar{t},j,\bar{t}') &= -i\langle T_C[a_e(i,\bar{t})a_e^\dagger(j,\bar{t}')]\rangle \\ G_{eh}(i,\bar{t},j,\bar{t}') &= -i\langle T_C[a_e(i,\bar{t})a_h(j,\bar{t}')]\rangle \end{aligned} \tag{1}$$

and coherent four-point (biexcitonic) functions:

$$g_{eehh}(i,\bar{t},j,\bar{t'},k,\bar{t''},l,\bar{t'''}) = (-i)^2\langle T_C[a_e(i,\bar{t})a_e(j,\bar{t'})a_h(k,\bar{t''})a_h(l,\bar{t'''})]\rangle$$
$$-G_{eh}(i,\bar{t},l,\bar{t'''})G_{eh}(j,\bar{t'},k,\bar{t''})$$
$$+G_{eh}(i,\bar{t},k,\bar{t''})G_{eh}(j,\bar{t'},l,\bar{t'''})\,. \qquad (2)$$

Here $a_e(i,\bar{t})$, for example, denotes the annihilation operator for an electron orbital (band, spin) labeled i at time $\bar{t}$) on the Keldysh double-time contour. T_C denotes time ordering along the Keldysh contour C, and $\langle ... \rangle$ denotes averaging by the initial density operator. The density matrices that are the dynamical variables in the DCT formalism are equal-time limits of these Green's functions. The Green's functions can be expanded along standard lines in a diagrammatic perturbation series. Figure 1 shows the basic graphical elements: the free electron or hole propagators, the Coulomb interaction vertices, and the vertices corresponding to the creation and annihilation of an electron-hole pair through the external field.

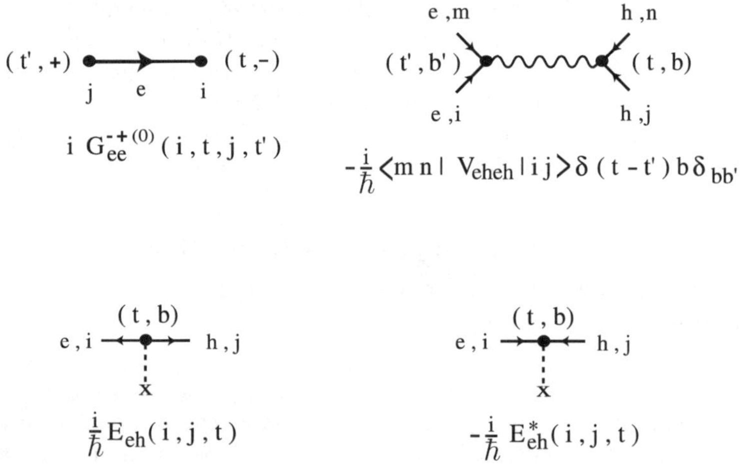

Fig. 1. Graphical elements in the perturbation theory for the fermion nonequilibrium Green's functions. For details, see [18]

Two assumptions are basic to the derivation of the DCT equations:
(1) the Coulomb vertices that effect a transition between a conduction band and a valence band are omitted,
(2) the system is initially in the electron-hole vacuum.
Assumption (2) immediately implies that the noninteracting one-particle Green's function only goes forward along the Keldysh contour C: $G^{(0)}_{eh}(i,\bar{t},j,\bar{t'}) = 0$ if $\bar{t}$ comes 'earlier' than $\bar{t'}$ on C. This in turn leads to

the vanishing of a vast class of diagrams [18]. For instance, a diagram equals zero if it contains a closed loop, traced along the arrow directions of free propagators and along either direction in an interaction line. This formal feature of our theory is similar to one in the equilibrium perturbation theory of dilute Bose gases at zero temperature [21]. With the Coulomb vertices allowed by assumption (1), we have classified in [18] the surviving diagrams in increasing order of the external field up to third order. It was shown that those diagrams (to all orders in the Coulomb interaction) contributing to the equal-time Green's functions – the density, the polarization, and the biexcitonic correlation – can be resummed to yield closed equations of motion that have been derived in $\chi^{(3)}$ DCT. For illustration, we show here the graphical series contributing to the first order (in E) polarization and some typical graphs contributing to the third order polarization. Figure 2 shows the surviving first-order graphs under the above two assumptions, the resummation of which yields the well-known linearized semiconductor Bloch equation (see e.g. [22]). The classification of the third-order polarization graphs first proceeds by the number (either zero or one) of fermion loops they contain. Guided by developments in DCT, we further group the graphs according to their factorizability properties. Figure 3 shows a representative diagram that contains no closed fermion loops and can be factorized into three first-order polarization graphs and a Coulomb line. This diagram is a part of the Hartree–Fock term in the $\chi^{(3)}$ equation of motion for the polarization. Figure 4 shows a representative graph that contains no fermion loop and can at most be factorized into parts containing a connected four-point function which is the equal-time limit of biexcitonic correlation Eq. (2) above. This graph contributes to the so-called correlation term in the DCT equation. A similar diagrammatic analysis of the second-order contributions to the coherent four-point function [18] yields the $\chi^{(2)}$ equation for the biexcitonic correlation function derived in DCT.

The foregoing is of course a rather laborious derivation of the $\chi^{(3)}$ equations which have already been derived quite straightforwardly within DCT. This perturbative diagrammatic analysis is expected to be useful in theories beyond $\chi^{(3)}$ where this particular diagram set, representing coherent scattering, can be combined with other sets representing other correlations, e.g. screening, incoherent scatterings.

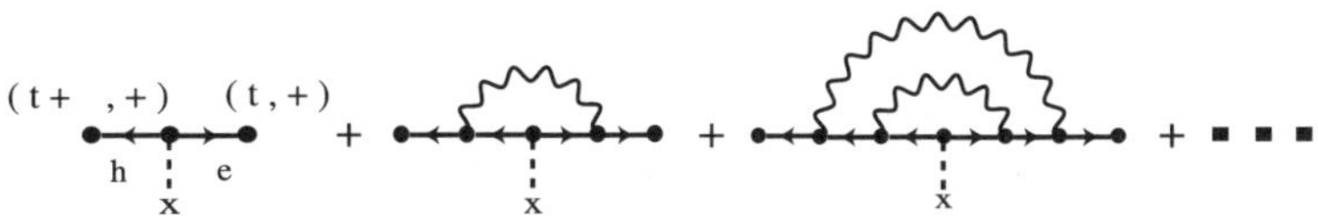

Fig. 2. Graphs contributing to the first order polarization function

3 Effective Boson Theory for Interacting Excitons

In this section, we give an overview of a low-energy effective boson theory for interacting excitons. Details can be found in [23]

As remarked in the Introduction, the motivation for using a bosonic theory is to have a more direct description of situations where the excitations are predominantly bound excitons and to take advantage of the established many-body formalisms, e.g. Green's function perturbation theory for boson systems. The main complication in following this approach comes from the fact that excitons in typical quantum wells, e.g. in GaAs, are not very small and not very strongly bound. A bosonic theory for these excitons should then be augmented by corrections due to exchange effects of their fermionic components. In our theory, these exchange corrections are included by choosing our boson Hamiltonian in such a way that this theory formally agrees with fermionic DCT up to $\chi^{(3)}$ order. In applying the theory to higher intensities and/or the incoherent regime, the assumption is that two-particle correlations, dynamical or exchange, are still the most important correlations. This assumption defines the limits of applicability of this bosonic theory.

The boson Hamiltonian is taken to be of a general form: $\hat{H} = \sum_{n=1}^{\infty} \hat{H}_n$, where $\hat{H}_n$ is a sum of terms each containing n creation/annihilation operators. The even-n terms are restricted to being number-conserving, while each odd-n term adds a particle to or subtracts one from the system and is linearly coupled to (multiplied by) an external classical radiation field. In a perturbation theory with $\hat{H}_2$ as the unperturbed term, each $\hat{H}_n, n > 2$ represents

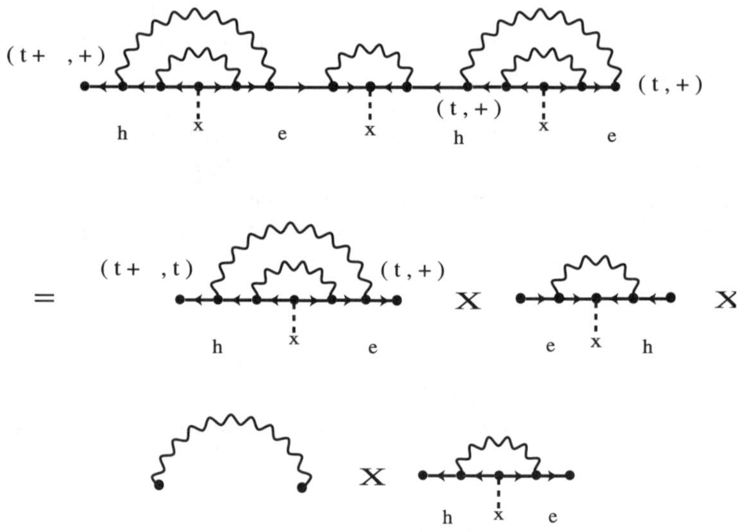

Fig. 3. Example of a graph contributing to the third order polarization function. This example shows a Hartree–Fock contribution

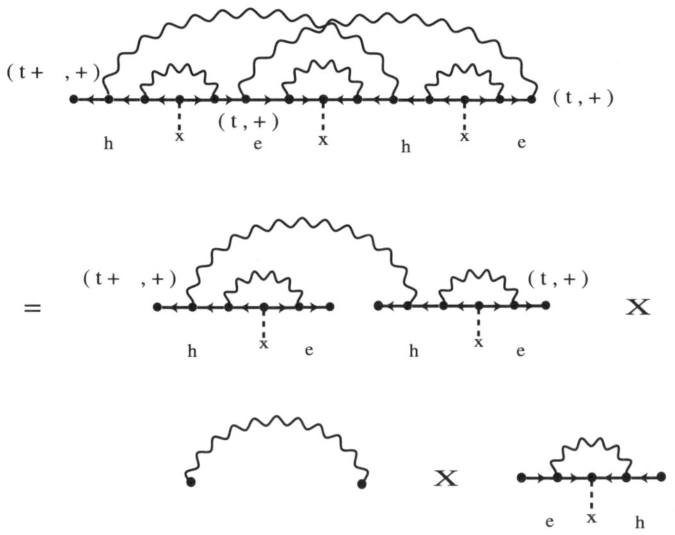

Fig. 4. Example of a graph contributing to the third order polarization function. This example shows a correlation contribution

an elementary scattering involving $n/2$ (for n even) or $(n+1)/2$ (for n odd) particles. Thus in a density regime where only binary collisions and correlations are important, we may truncate the series in the Hamiltonian to $n \leq 4$. These terms may fairly generally be written (with notations explained below) as:

$$\begin{aligned}
\hat{H} &= \hat{H}_1 + \hat{H}_2 + \hat{H}_3 + \hat{H}_4 \\
\hat{H}_1 &= \sqrt{\Omega} \sum_{sjn\mathbf{k}} \mathcal{E}_n^{sj}(\mathbf{k},t) {b_{n\mathbf{k}}^{sj}}^\dagger + h.c. \\
\hat{H}_2 &= \sum_{sjn\mathbf{k}} \varepsilon_n^{sj}(\mathbf{k}) {b_{n\mathbf{k}}^{sj}}^\dagger b_{n\mathbf{k}}^{sj} \\
\hat{H}_3 &= \frac{1}{\sqrt{\Omega}} \sum_{\{\alpha\}} \Big[F_{n_1 n_2 n_1' n_2'}^{s_1 j_1 s_2 j_2 (d)}(\mathbf{K}+\mathbf{q},\mathbf{K}) {b_{n_1\mathbf{k}_1+\mathbf{q}}^{s_1 j_1}}^\dagger {b_{n_2\mathbf{k}_2-\mathbf{q}}^{s_2 j_2}}^\dagger \\
&\quad + F_{n_1 n_2 n_1' n_2'}^{s_1 j_1 s_2 j_2 (e)}(\mathbf{K}+\mathbf{q},\mathbf{K}) {b_{n_1\mathbf{k}_1+\mathbf{q}}^{s_2 j_1}}^\dagger {b_{n_2\mathbf{k}_2-\mathbf{q}}^{s_1 j_2}}^\dagger \Big] \cdot \mathcal{E}_{n_2'}^{s_2 j_2}(\mathbf{k}_2,t) b_{n_1'\mathbf{k}_1}^{s_1 j_1} + h.c. \\
\hat{H}_4 &= \frac{1}{2\Omega} \sum_{\{\alpha\}} \Big[V_{n_1 n_2 n_1' n_2'}^{s_1 j_1 s_2 j_2 (d)}(\mathbf{K}+\mathbf{q},\mathbf{K}) {b_{n_1\mathbf{k}_1+\mathbf{q}}^{s_1 j_1}}^\dagger {b_{n_2\mathbf{k}_2-\mathbf{q}}^{s_2 j_2}}^\dagger \\
&\quad + V_{n_1 n_2 n_1' n_2'}^{s_1 j_1 s_2 j_2 (e)}(\mathbf{K}+\mathbf{q},\mathbf{K}) {b_{n_1\mathbf{k}_1+\mathbf{q}}^{s_2 j_1}}^\dagger {b_{n_2\mathbf{k}_2-\mathbf{q}}^{s_1 j_2}}^\dagger \Big] \cdot b_{n_1'\mathbf{k}_1}^{s_1 j_1} b_{n_2'\mathbf{k}_2}^{s_2 j_2} \,.
\end{aligned} \tag{3}$$

Here $b^{sj}_{n\mathbf{k}}$ denotes a boson annihilation operator for the orbital labeled by the momentum $\mathbf{k}$ and internal state indices s, the electron band, j, the hole band, and n, the relative-motion state of the electron and the hole around each other. $\{\alpha\}$ is a shorthand for $(s_1 j_1 s_2 j_2 n_1 n_2 n'_1 n'_2 \mathbf{k}_1 \mathbf{k}_2 \mathbf{q})$, Ω is the volume (area for 2D systems) of the normalization box, and $\mathbf{K}$ is the initial relative momentum between two interacting bosons. $\mathcal{E}^{sj}_n(\mathbf{k},t)$ is the matrix element for the creation of a boson in the $(sjn\mathbf{k})$ orbital by the external field. V is an effective two-body interaction, and F carries the fermionic exchange correction to the exciton-photon coupling $\hat{H}_1$ when another exciton is present. Both F and V are assumed to be diagonal (translational invariance) in and independent of the total momentum of the two bosons. We also assume F and V to be diagonal in the 'band' indices $(s_1 j_1 s_2 j_2)$. The superscripts (d) and (e) label direct and rearrangement (exchange of electrons or holes between the two excitons) scatterings respectively. More explicitly, $V^{(e)}$ and $F^{(e)}$ are matrix elements for processes which exchange the electron spins s_1 and s_2 of the two excitons. The effective-mass approximation is taken for the single-particle energy $\varepsilon^{sj}_n(\mathbf{k}) = \varepsilon^{sj}_n(\mathbf{0}) + \frac{\hbar^2 k^2}{2M_{sj}}$, where M_{sj} is the boson mass, which gives the relative momentum in the above definition as $\mathbf{K} = \mu_{j_2 j_1}\mathbf{k}_1 - \mu_{j_1 j_2}\mathbf{k}_2$, $\mu_{j_1 j_2} = M_{s_1 j_1}/(M_{s_1 j_1} + M_{s_2 j_2})$ etc. As stated above, the matrix elements $\mathcal{E}$, $\varepsilon, V^{(d)/(e)}, F^{(d)/(e)}$ will be determined within the $\chi^{(3)}$ response theory.

Just as in the fermionic theory, we formulate a nonlinear response theory for this boson-exciton system within the nonequilibrium Green's function formalism. The system's dynamics is again expressed in terms of multi-point Green's functions, the time variables of which take values on the Keldysh double-time contour [19,20]. The relevant fucntions are the one- and two-point functions and one particular three-point function, defined as follows:

$$
\begin{aligned}
&B^{sj}_n(\mathbf{k},\bar{t}) = \langle b^{sj}_{n\mathbf{k}}(\bar{t})\rangle \\
&g^{sjs'j'}_{nn'}(\mathbf{k},\bar{t},\mathbf{k}',\bar{t}') = -i\left[\langle T_C[b^{sj}_{n\mathbf{k}}(\bar{t}) b^{s'j'\,\dagger}_{n'\mathbf{k}'}(\bar{t}')]\rangle - B^{sj}_n(\mathbf{k},\bar{t}) B^{s'j'*}_{n'}(\mathbf{k}',\bar{t}')\right] \\
&g^{(a)sjs'j'}_{nn'}(\mathbf{k},\bar{t},\mathbf{k}',\bar{t}') = -i\left[\langle T_C[b^{sj}_{n\mathbf{k}}(\bar{t}) b^{s'j'}_{n'\mathbf{k}'}(\bar{t}')]\rangle - B^{sj}_n(\mathbf{k},\bar{t}) B^{s'j'}_{n'}(\mathbf{k}',\bar{t}')\right] \\
&Z^{sjs'j's''j''}_{nn'n''}(\mathbf{k},\bar{t},\mathbf{k}',\bar{t}',\mathbf{k}'',\bar{t}'') = \langle T_C[b^{sj}_{n\mathbf{k}}(\bar{t}) b^{s'j'}_{n'\mathbf{k}'}(\bar{t}') b^{s''j''\,\dagger}_{n''\mathbf{k}''}(\bar{t}'')]\rangle\,. \qquad (4)
\end{aligned}
$$

The initial state is assumed to be 'normal', i.e., all 'anomalous' or non-number-conserving functions, e.g. $B, g^{(a)}, \tilde{g}^{(a)}$ and Z in Eq. (4) above, are zero initially (when all time arguments are set to t_0). The external field $\mathcal{E}$, which is switched on after t_0, generates these anomalous functions.

The basic observable, the interband polarization, is defined in our theory through the Hamiltonian Equation (3) as:

$$P^{(b)}(\mathbf{k},t) = \sum_{sjn} p^{sj}_{n(b)}(\mathbf{k},t)$$

$$p^{sj}_{n(b)}(\mathbf{k},t) = B^{sj}_n(\mathbf{k},t) + \frac{1}{\Omega} \sum_{\substack{s_1 j_1 \mathbf{q}\mathbf{q}' \\ n_1 n_1' n_2'}}$$

$$\Big[F^{s_1 j_1 sj(d)^*}_{n_1' n_2' n_1 n}(\mathbf{K}+\mathbf{q},\mathbf{K}) Z^{s_1 j_1 sj s_1 j_1 \; +++}_{n_1' n_2' n_1}(\mathbf{q}'+\mathbf{q},t,\mathbf{k}-\mathbf{q},t+\epsilon,\mathbf{q}',t+2\epsilon)$$

$$+F^{s_1 j_1 sj(e)^*}_{n_1' n_2' n_1 n}(\mathbf{K}+\mathbf{q},\mathbf{K}) Z^{sj_1 s_1 j s_1 j_1 \; +++}_{n_1' n_2' n_1}(\mathbf{q}'+\mathbf{q},t,\mathbf{k}-\mathbf{q},t+\epsilon,\mathbf{q}',t+2\epsilon)\Big] \quad (5)$$

with $\mathbf{K} = \mu_{jj_1}\mathbf{q}' - \mu_{j_1 j}\mathbf{k}$ and $\epsilon \searrow 0$. Including the second term is crucial for getting the fermionic exchange correct. Other physical quantities are expressed in terms of equal-time limits of the 2-point functions.

Equations of motion of the Green's functions in Eq. (4) are derived through standard diagrammatic perturbation theory. The single-particle Hamiltonian $\hat{H}_2$ is taken to be the unperturbed Hamiltonian, and the remainder of the Hamiltonian Equation (3) are the perturbing terms. For simplicity, we specialize to a thin (zero-width) quantum well setting and assume normal incidence for the external field. The momentum space is then two-dimensional along the quantum well's plane. The photon field $\mathcal{E}^{sj}_n(\mathbf{k},t) = \delta_{\mathbf{k},\mathbf{0}}\mathcal{E}^{sj}_n(t)$ imparts no in-plane momentum, leading to momentum conservation in the system: $B^{sj}_n(\mathbf{k},\bar{t}) = \delta_{\mathbf{k},\mathbf{0}} B^{sj}_n(\bar{t})$, $g^{sjs'j'}_{nn'}(\mathbf{k},\bar{t},\mathbf{k}',\bar{t}') = \delta_{\mathbf{k},\mathbf{k}'} g^{sjs'j'}_{nn'}(\mathbf{k},\bar{t},\bar{t}')$ etc.

The assumption of an initial boson vacuum again eliminates large classes of diagrams. The surviving diagrams can again be grouped in ascending order of the external field. Identifying the $\chi^{(1)}$ expression for the interband polarization, $p^{sj(1)}_{n(b)}(t)$, with that from the standard electron-hole theory fixes the coefficient in $\hat{H}_1$ as the Rabi frequency for the exciton transition:

$$\mathcal{E}^{sj}_n(t) = \tilde{\phi}^{j*}_n(\mathbf{0})\mathbf{d}_{sj} \cdot \mathbf{E}(t)\,, \qquad (6)$$

where $\tilde{\phi}^j_n(\mathbf{0})$ is the configuration space exciton eigenfunction of quantum numbers (j,n) at relative electron-hole coordinate $\mathbf{r} = \mathbf{0}$, $\mathbf{d}_{sj}$ is the electron dipole moment, and $\mathbf{E}(t)$ is the applied field. To $\chi^{(3)}$ order, both terms on the right hand side of Eq. (5) contribute to $p^{sj}_{n(b)}(t)$. We list *all* graphs contributing to $B^{sj(3)}_n(t)$ in Fig. 5a and the contribution from the second term of Eq. (5) in Fig. 5b. The corresponding expression for $p^{sj(3)}_n(t)$ is compared with that from fermionic DCT. In this comparison, for simplicity, we further specialize to the case where only one conduction (electron) band is included, so that the electron band index s only runs through two spin projections $(\pm\frac{1}{2})$ along the normal to the plane, and where the exciton wavefunctions do not depend on the electron spin. V and F are then block diagonal in the electron spin triplet and singlet channels. They can then be written as matrices in the basis $(n, m, \mathbf{q})$ and labeled by the hole band indices (j, j') and the total-electron-

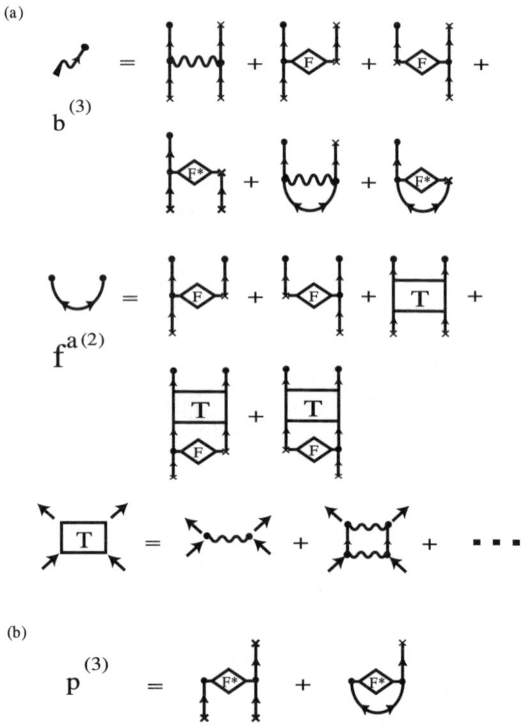

Fig. 5. All graphs in the boson theory contributing to the third order polarization function. (**a**) One point function and the equal-time two-point anomalous function. (**b**) Contributions from the seond term in Eq. (5)

spin of the two excitons:

$$\hat{V}^{jj'(d)/(e)} = \frac{1}{2}(\hat{V}^{jj'(+)} \pm \hat{V}^{jj'(+)}) \;\; ; \; \hat{F}^{jj'(d)/(e)} = \frac{1}{2}(\hat{F}^{jj'(+)} \pm \hat{F}^{jj'(+)}) , \quad (7)$$

where the hat denotes a matrix in the $(n, m, \mathbf{q})$ basis and $+/-$ here refers to the electron spin triplet/singlet channels. It is shown in [23] that the expressions from the $\chi^{(3)}$ boson and $\chi^{(3)}$ fermion theories for the interband polarization will be identical if one makes the following choice for V and F:

$$\begin{aligned} \hat{V}^{jj'\pm} &= [1 + \hat{F}^{jj'\pm}]^{-1} \cdot \hat{W}^{jj'\pm} \cdot [1 + \hat{F}^{jj'\pm}]^{-1} \\ &\quad + [1 + \hat{F}^{jj'\pm}]\varepsilon^{jj'}[1 + \hat{F}^{jj'\pm}]^{-1} - \varepsilon^{jj'} \\ \hat{F}^{jj'\pm} &= [1 \mp \hat{S}^{jj'}]^{\frac{1}{2}} - 1 , \end{aligned} \quad (8)$$

where $\hat{W}^{jj'\pm}$ is the Coulomb transition matrix between an initial pair and a final pair of exciton orbitals, and $\hat{S}^{jj'}$ is the overlap matrix of (nonorthogonal) two-exciton wavefunctions. The expressions of these two matrices in terms of electron-hole orbital wavefunctions of the excitons and the Coulomb potential can be found in e.g. [1,24]. $\varepsilon^{jj'}$ is a diagonal matrix with elements equal to the sum of the kinetic energies of the two bosons.

The expressions in Eq. (8) are physically quite reasonable. From them one can say that the fermionic corrections up to $\chi^{(3)}$ come basically from the fact that the (antisymmetrized) product states for two excitons do not form an orthonormal basis for the two-exciton subspace. Indeed, an equivalent expression for V was obtained in [17] by analyzing the nonorthogonality of these states.

The Hamiltonian equation (3) thus defined, one can now proceed, through a choice of self energy graphs, to write down general equations of motion for the Green's functions: Kadanoff–Baym (Dyson) equations for the two-point functions and generalized Gross–Pitaevskii (with collision terms) equations for the one-point functions. Some recent examples of such equations and their applications can be found in e.g. [25,26]. Formally the difference in our case is the additional graphs coming from the vertex $\hat{H}_3$. Use of these equations to study exciton dynamics and kinetics beyond $\chi^{(3)}$ is now in progress.

4 Exciton–Exciton Scattering in $\chi^{(3)}$ Four Wave Mixing Processes

In applying the theories in the previous two sections to excitonic processes, the equations are often solved in a truncated exciton basis, including the contributions only from a few exciton states to the sum over intermediate states in the scattering processes. In fact, the entire rationale for adopting a bosonic theory is the assumption that this truncation is reasonable. While we have given some numerical evidence that truncation to the lowest (1s) exciton subspace is not too unrealistic [24], it is not easy to determine accurately the magnitude of the error involved. It is important then to seek a certain degree of validation of this apparently drastic approximation by comparing with experiments. In this section we briefly review a comparison of our theory in the $\chi^{(3)}$ regime with a four wave mixing experiment on a quantum well microcavity [27]. The details can be found in [28].

The basic geometry is shown in Fig. 6. It consists of a semiconductor quantum well inside a microcavity, which is made of two distributed Bragg reflectors. The optical pump pulse enters the cavity at normal incidence, and an optical probe pulse induces a four wave mixing signal (FWM) in the direction opposite to the probe pulse propagaion. Both pump and probe pulses are relatively long in time and have the same frequency. The exciton frequency is set close to the cavity resonance frequency, resulting in the concentration of the four wave mixing signal in the spectral positions of the two polaritons. The relative strengths of this signals at the two peaks for various pump-probe-signal polarization configurations are sensitive to the energy and polarization dependencies of the exciton–exciton (off-shell) forward scattering amplitude

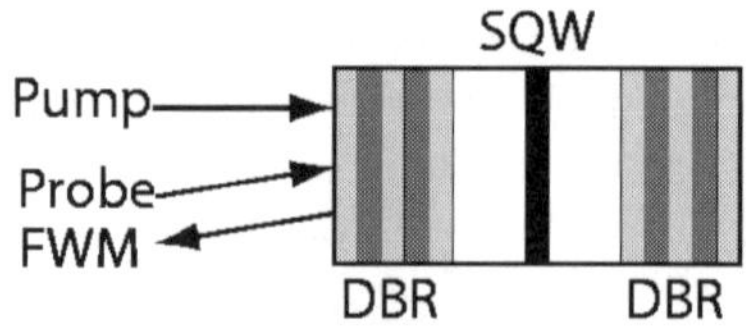

Fig. 6. Schematic of the microcavity four wave mixing configuration (DBR = distributed Bragg reflector, SQW = single quantum well)

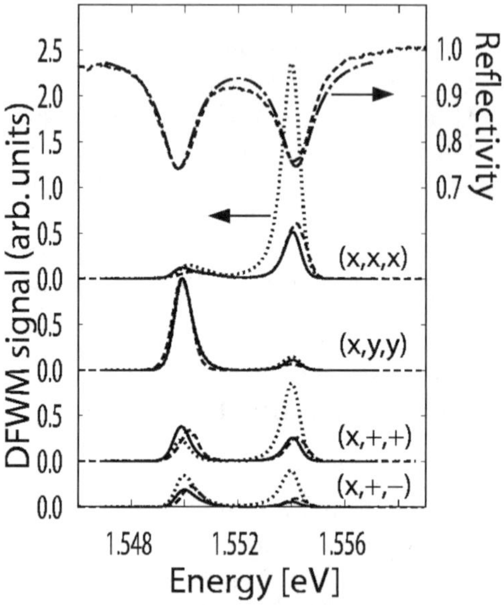

Fig. 7. Frequency-degenerate four wave mixing signals from a quantum-well microcavity: calculated full T-matrix solution (*solid*), calculated signal in 2nd Born approximation (*dotted*), experimental data taken from [27] (*dashed*). The polarization configurations are indicated in the order (pump,probe,signal). Also shown is the linear reflectivity (*upper panel*; theory: *dashed-dotted*, experiment: *dashed*). The cavity is tuned to the exciton resonance at 1.552 eV. For details, see [29]

(or T-matrix), and in some cases also to Pauli blocking correlations. We have solved the $\chi^{(3)}$ equation for the interband polarization, restricting to the 1s exciton subspace. The comparison with the experimental results is shown in Fig. 7. We show FWM signals for two cases: (i) the T-matrix evaluated in the 2nd Born approximation (BA) and (ii) the full, nonperturbative (but still restricted to 1s) solution for the T-matrix. The theoretical results are normalized to the experimental results at the lower polariton peak in the configuration (pump, probe, signal) $= (x, y, y)$. We then use the same normalization for the signals in all polarization configurations. As one can see, the full T-matrix solution gives excellent agreement with the experiment, while the 2nd BA yields significant discrepancies, mainly at the upper polariton peak. The connection of this significant failure of the 2nd BA with the dimensionality of the system was discussed in [29].

This analysis shows that microcavity FWM signals are quite sensitive to the nature of exciton–exciton scattering. It also confers confidence on using the 1s approximation, provided one then solves the resulting exciton scattering problem properly (nonperturbatively). In using the boson theory, this amounts to using the T-matrix as an effective interaction, which is a common practice in the general theory of dilute gases made of strongly interacting particles.

Acknowledgements

This work is supported by the Cooperative Excitation Project (ERATO-JST), NSF(DMR-9972195), JSOP, COEDIP(University of Arizona). We thank M. Kuwata-Gonokami and Y. Svirko for stimulating discussions and W. Schäfer for sending us part of his book ([1]) prior to publication.

References

1. W. Schäfer, M. Wegener: *Semiconductor Optics and Transport Phenomena*, (Springer, Berlin Heidelberg 2002)
2. J. Shah: *Ultrafast Spectroscopy of Semiconductors and Semiconductor Nanostructures*, (Springer, Berlin Heidelberg New York 1999)
3. V.M. Axt, A. Stahl: Z. Phys. B: Condens. Matter **93**, 195 (1994); **93**, 205 (1994)
4. T. Östreich, K. Schönhammer, L. Sham: Phys. Rev. Lett. **74**, 4698 (1995); Phys. Rev. B **58**, 12920 (1998)
5. W. Schäfer, D. Kim, J. Shah, T. Damen, J. Cunningham, K. Goosen, L. Pfeiffer, K. Köhler: Phys. Rev. B **53**, 16429 (1996)
6. P. Kner, S. Bar-Ad, M.V. Marquezini, D.S. Chemla, W. Schäfer: Phys. Rev. Lett. **78**, 1319 (1997)
7. G. Bartels, A. Stahl, V. Axt, B. Hasse, U. Neukirch, J. Gutowski: Phys. Rev. Lett. **81**, 5880 (1998)
8. C. Sieh, T. Meier, F. Jahnke, A. Knorr, S.W. Koch, P. Brick, M. Hübner, C. Ell, J. Prineas, G. Khitrova, H. Gibbs: Phys. Rev. Lett. **82**, 3112 (1999)
9. U. Neukirch, S.R. Bolton, N.A. Fromer, L.J. Sham, D.S. Chemla: Phys. Rev. Lett. **84**, 2215 (2000)
10. V. Axt, B. Hasse, U. Neukirch: Phys. Rev. Lett. **86**, 4620 (2001)
11. W. Schäfer, R. Lövenich, N. Fromer, D. Chemla: Phys. Rev. Lett. **86**, 344 (2001)
12. E. Hanamura: J. Phys. Soc. Jpn. **37**, 1545 (1974)
13. M. Combescot: Phys. Rep. **221**, 167 (1992)
14. G. Rochat, C. Ciuti, V. Savona, C. Piermarocchi, A. Quattropani, P. Schwendimann: Phys. Rev. B **61**, 13856 (2000)
15. S. Ben-Tabou de-Leon, B. Laiktman: Phys. Rev. B **63**, 125306 (2001)
16. M. Combescot, O. Betbeder-Matibet: Europhys. Lett. **58**, 87 (2002)
17. S. Okumura, T. Ogawa: Phys. Rev. B **65**, 035105 (2002)

18. N.H. Kwong, R. Binder: Phys. Rev. B **61**, 8341 (2000)
19. H. Haug and A.-P. Jauho: *Quantum Kinetics in Transport and Optics of Semiconductors*, (Springer, Berlin/Heidelberg, 1996)
20. P. Danielewicz: Ann. Phys. (N.Y.) **55**, 239 (1984)
21. A.L. Fetter, J.D. Walecka: *Quantum Theory of Many-Particle Systems*, (McGraw-Hill, New York 1971)
22. H. Haug, S.W. Koch: *Quantum Theory of the Optical and Electronic Properties of Semiconductors*, (World Scientific, Singapore 1993)
23. N.H. Kwong, Z.S. Yang, R. Takayama, M. Kuwata-Gonokami, R. Binder: (2003) in preparation
24. R. Takayama, N.H. Kwong, I. Rumyantsev, M. Kuwata-Gonokami, R. Binder: Eur. Phys. J. B **25**, 445 (2002)
25. O.M. Schmitt, L. Banyai, H. Haug: Phys. Rev. B **60**, 16506 (1999)
26. M. Imamovic-Tomasovic, A. Griffin: J. Low Temp. Phys. **122**, 617 (2001)
27. M. Kuwata-Gonokami, S. Inoue, H. Suzuura, M. Shirane, R. Shimano, T. Someya, H. Sakaki: Phys. Rev. Lett. **79**, 1341 (1997)
28. N.H. Kwong, R. Takayama, I. Rumyantsev, M. Kuwata-Gonokami, R. Binder: Phys. Rev. B **64**, 045316 (2001)
29. N.H. Kwong, R. Takayama, I. Rumyantsev, M. Kuwata-Gonokami, R. Binder: Phys. Rev. Lett. **87**, 027402 (2001)

Cluster Expansion in Semiconductor Quantum Optics

Walter Hoyer, Mackillo Kira, and Stephan W. Koch

1 Introduction

During recent decades, semiconductor research gradually progressed towards the quantum-optical regime. For example, the special quantum nature of light is apparent when semiconductor quantum dots emit well defined photons [1,2] or when light-matter entanglement influences optical experiments in microcavity structures [3,4].

Advanced theories of quantum electrodynamics are now very successful in the description of the interaction of atomic systems with a quantized light field where the coupling between the atoms plays only a minor role. These theories can mostly be applied to describe dilute and only weakly interacting atomic gases since relatively simple models like few-level systems are used to describe the material. On the other hand, semiconductor physics is usually dominated by the Coulomb interaction between the carriers such that the majority of theories have focused on the correct description of the fermionic many-body effects while the electromagnetic interaction has been limited mostly to classical fields. In this context, the coherent regime of classical optics has been investigated in detail. For excitations with ultrashort pulses, a diversity of experiments [5–10] and computations [11–15] have been performed. Since electrons and holes (missing electrons in the valence band) have opposite charges, they experience Coulomb attraction which may lead to the formation of atom-like bound electron-hole pairs (excitons) that often dominate both optical and transport properties [16–21].

In this context, photoluminescence spectra have been the focus of extensive research. In particular, photoluminescence experiments have been used as a measure of incoherent excitons. Several experimental publications directly conclude an exciton formation time from time-resolved photoluminescence spectra [22–26]. One typical experimental situation is to excite carriers resonantly high in the band and to follow the subsequent dynamics. The observed formation times in the literature vary between 20 ps and several hundred picoseconds. However, the deduction of such an exciton-formation time from luminescence experiments is not unambiguous. The formation of excitons, even without coupling to the vacuum fluctuations of light, takes place on a relatively long time scale of at least several hundreds of picoseconds [27] and it is not a priori clear in how far the exciton formation is altered if spontaneous emission processes of these excitons are considered. Especially since the radiative decay time of coherent excitons in two-dimensional quantum-well structures is known to be of the order of several picoseconds only [10], one can expect also incoherent excitons to decay equally fast. In that case, spontaneous emission should deplete excitonic populations much faster than they can build up.

This paper serves a two-fold purpose; after a brief review of the basic contributions to the Hamiltonian in Sect. 2, we present the cluster expansion and how it can be used to classify different N-particle correlations and to truncate the hierarchy of equations emerging due to both the many-body Coulomb interaction and the quantized light-matter interaction. In Sect. 3, we derive restrictions which must be fulfilled by the correlations as a consequence of the relation between density matrices of different order. After presenting the equations of motion for the photon correlations in Sect. 4, we finally investigate the properties of semiconductor photoluminescence in Sect. 5.

2 Carrier–Photon System

A semiconductor heterostructure is a complicated many-body system where electrons and holes are strongly interacting via their mutual Coulomb interaction, where interband or intraband transitions can be driven by optical or far-infrared fields, respectively, and where the carrier system is coupled to lattice vibrations such that energy can be directed out of or into the electronic system. The theoretical description of the many-body system can generally be formulated by introducing the system Hamiltonian $H_{\rm tot}$ with different interaction terms. These contributions have been discussed in literature [17,28,29] such that we only briefly state the result.

2.1 Bloch Electrons

The microscopic properties of the fermionic carriers can be described with the field operator

$$\Psi(\mathbf{r}) = \sum_{\lambda,k} \phi_{\lambda,k}(\mathbf{r}) a_{\lambda,k} \,, \tag{1}$$

where $a_{\lambda,k}$ is the annihilation operator for an electron in band λ with wave vector k along the heterostructure obeying fermionic anti-commutation relations. The Bloch function of such an electron is denoted by $\phi_{\lambda,k}$ where the band index λ may include different bands, subbands, and spins. For sufficiently narrow confinement, the carrier dynamics can be restricted to the lowest subband. In this paper, we consider a two-band model with one valence and one conduction band and use the operators $a_{\lambda=c,k}$ and $a_{\lambda=v,k}$ and their hermitian conjugates. The generalization to a multi-band system or the inclusion of spin is straight forward [17,30].

The Bloch functions are the solutions of the eigenvalue equation of one electron moving in an effective lattice periodic potential U. Consequently, this choice leads to a simple diagonal structure of the single-particle part of the general many-body Hamiltonian,

$$H_{\mathrm{kin}} = \sum_{\lambda,k} \varepsilon_k^{\lambda} a_{\lambda,k}^{\dagger} a_{\lambda,k} \,. \tag{2}$$

In general, the eigenenergies ε_k^{λ} must be obtained from precise band structure calculations. For the investigation of near-bandgap optical features, however, it is often a good approximation to assume a quadratic dispersion relation. In this case, the respective energies are given by

$$\varepsilon_k^c \equiv \varepsilon_k^e + E_{\mathrm{G}} = \frac{\hbar^2 k^2}{2m_e} + E_{\mathrm{G}}, \quad \varepsilon_k^v \equiv -\varepsilon_k^h = -\frac{\hbar^2 k^2}{2m_h} \,, \tag{3}$$

where m_e and m_h are the effective electron and hole masses. The unrenormalized bandgap is denoted with E_{G}.

In order to include the Coulomb interaction between the electrons in different bands beyond the effective single-particle level, we use the Hamiltonian [17,31]

$$H_{\mathrm{C}} = \frac{1}{2} \sum_{\substack{\lambda,\lambda' \\ k,k',q\neq 0}} V_q \, a_{\lambda,k}^{\dagger} a_{\lambda',k'}^{\dagger} a_{\lambda',k'+q} a_{\lambda,k-q} \,, \tag{4}$$

where V_q is the quantum-well or quantum-wire Coulomb matrix element.

2.2 Quantized Light Field

A standard approach for the quantization of the electromagnetic field starts from the vector potential $\mathbf{A}$ which is replaced by an operator within the canonical quantization scheme [35]. Using a complete mode basis, this operator $\mathbf{A}$ can be expressed in terms of creation and annihilation operators of photons in various modes via the relation

$$\mathbf{A}(\mathbf{r},t) = \sum_{\mathbf{q},\sigma} \frac{\mathcal{E}_{\mathbf{q}}}{\omega_{\mathbf{q}}} \left[\mathbf{u}_{\mathbf{q},\sigma}(r_\perp) \, \frac{e^{i\, q_\parallel \cdot r_\parallel}}{\sqrt{\mathcal{L}^d}} \, b_{\mathbf{q},\sigma}(t) + \mathrm{h.c.} \right] , \tag{5}$$

where the operators $b_{\mathbf{q},\sigma}$ and $b_{\mathbf{q},\sigma}^{\dagger}$ destroy or create a photon in mode $\mathbf{q}$ with polarization σ and energy $\hbar\omega_{\mathbf{q}}$. The mode function $\mathbf{u}_{\mathbf{q},\sigma}$ is the solution to the Helmholtz equation for a planar structure, possibly consisting of layers of different materials with different refractive indeces but still without the active semiconductor material. The quantization procedure [35] fixes the value of the expansion coefficient

$$\mathcal{E}_{\mathbf{q}} = \sqrt{\frac{\hbar\omega_{|\mathbf{q}|}}{2\epsilon_0}} \,, \tag{6}$$

which is often referred to as vacuum field amplitude, and the commutation relations for the photon operators

$$[b_{\mathbf{q}\sigma}, b_{\mathbf{q}'\sigma'}] = \left[b^\dagger_{\mathbf{q}\sigma}, b^\dagger_{\mathbf{q}'\sigma'}\right] = 0; \quad \left[b_{\mathbf{q}\sigma}, b^\dagger_{\mathbf{q}'\sigma'}\right] = \delta_{\mathbf{q},\mathbf{q}'}\delta_{\sigma,\sigma'}. \tag{7}$$

From the energy of the transverse field, we obtain the corresponding free-field Hamiltonian [36]

$$H_{\mathrm{em}} = \sum_{\mathbf{q},\sigma} \hbar\omega_{\mathbf{q}} \left(b^\dagger_{\mathbf{q}\sigma} b_{\mathbf{q}\sigma} + \frac{1}{2} \right). \tag{8}$$

This Hamiltonian formally corresponds to a Hamiltonian of a set of independent harmonic oscillators, one for each field mode.

The typical starting point for the description of light matter interaction is the minimal-substitution Hamiltonian containing the quantized vector potential. This interaction Hamiltonian describes the interaction between charge carriers and the quantized light field by the operator product $\mathbf{p} \cdot \mathbf{A}$. At the same time, it involves a term proportional to $\mathbf{A}^2$. By applying a gauge transformation, which for classical electromagnetic fields is known as Göppert–Mayer transformation [35], one can express the light matter interaction via

$$H_{\mathrm{D}} = - \sum_{\substack{q_\perp, q \\ \lambda, \lambda' \neq \lambda, k}} \left[i\mathcal{E}_{\mathbf{q}}\, \bar{u}_{\mathbf{q}}\, d_{\lambda,\lambda'}(\mathbf{q})\, a^\dagger_{\lambda,k+q} a_{\lambda',k} b_{q_\perp,q} \right] + \text{h.c.} \tag{9}$$

In this equation, three-dimensional vectors $\mathbf{r} = (r, r_\perp)$ are separated into components along and perpendicular to the heterostructure. Note that we do not explicitly label the parallel component. The overlap integral between the confinement wave functions and the mode function $u_{\mathbf{q}}(r_\perp)$ determines the effective interaction strength $\bar{u}_{\mathbf{q}}$ for the photon mode $\mathbf{q}$. The projection of the electric field onto the heterostructure is included in the effective dipole matrix element $d_{\lambda,\lambda'}(\mathbf{q})$.

Instead of the original A^2-term before the gauge transformation, one obtains a new dipole self-energy which, in principle, ensures the consistent coupling of the semiconductor polarization to the total electric field [36]. In practice, however, it often leads only to an energetic shift of the optical spectra. Since the difference between the displacement field $\mathbf{D}/\epsilon_0$ and the total electric field $\mathbf{E}$ is given by $\mathbf{P}/n^2$, this shift is particularly small for GaAs-like materials with a relatively large background refractive index of $n \approx 3.6$. For that reason, we have not included the dipole self-energy in the numerical calculations.

2.3 Total Hamiltonian

In this section, we have presented the relevant contributions to the total Hamiltonian for the description of photoluminescence or, more generally, quantum optics in semiconductors. These contributions include the non-interacting Bloch electrons and photons as well as the Coulomb interaction between carriers and the light-matter interaction in form of the dipole interaction. The starting point of all further investigations is the total Hamiltonian of Eqs. (2), (4), (8) and (9),

$$H_{\mathrm{tot}} = H_{\mathrm{kin}} + H_{\mathrm{em}} + H_{\mathrm{C}} + H_{\mathrm{D}}. \tag{10}$$

Starting from this Hamiltonian, one can compute the expectation value $\langle O \rangle$ for different operators O, corresponding to a variety of physical observables. In this process, by setting up the Heisenberg equation of motion

$$i\hbar \frac{\partial}{\partial t} O = [O, H_{\mathrm{tot}}], \tag{11}$$

single-particle (two-point) quantities as, e.g., the electron number operator $a^{\dagger}_{\lambda,k} a_{\lambda,k}$ are coupled to two-particle (four-point) quantities of the form $a^{\dagger} a^{\dagger} a a$ or to mixed carrier-photon operators of the form $b^{\dagger} a^{\dagger} a$. Since the equation of motion for a single photon operator,

$$i\hbar \frac{\partial}{\partial t} b_{q,q_{\perp}} = \hbar\omega_{q,q_{\perp}} b_{q,q_{\perp}} + \sum_{\lambda,\lambda' \neq \lambda,k} i\mathcal{E}_{\mathbf{q}}\, \bar{u}^{*}_{\mathbf{q}}\, d_{\lambda,\lambda'}(\mathbf{q})\, a^{\dagger}_{\lambda,k} a_{\lambda',k+q}, \tag{12}$$

shows that a photon operator is formally equivalent to a product of two carrier operators, both the many-body interaction, Eq. (4), and the quantized light-matter interaction, Eq. (9), lead to a similar hierarchy problem where an N-particle expectation value $\langle a^{\dagger}_1 \dots a^{\dagger}_N a_{N'} \dots a_{1'} \rangle \equiv \langle N \rangle$ is coupled to $(N+1)$-particle expectation values in the form

$$i\hbar \frac{\partial}{\partial t} \langle N \rangle = T_N[\langle N \rangle] + V_N[\langle N+1 \rangle]. \tag{13}$$

In this equation, T_N describes the coupling to other N-point expectation values due to single-particle contributions of the Hamiltonian and V_N denotes the coupling to the higher order correlations.

3 Cluster Expansion

The hierarchy problem defined by Eq. (13) seems to have a simple form; nevertheless, it is basically responsible for all the complications in many-body and semiconductor quantum-optical investigations. In fact, current theoretical and numerical approaches can solve the hierarchy problem exactly only in very limited cases [37]. Thus, much of the current research effort is devoted to developing consistent approximation schemes to deal with this hierarchy problem. Many-body techniques, such as non-equilibrium Green functions are available [31,38], but become rather tedious once one wants to deal with quantum-optical problems at the same level of sophistication as the Coulomb problem. Hence, we use a method known from quantum chemistry, where the truncation problem has successfully been approached with the so-called cluster expansion [39–41]. There, the electronic wave functions are divided into classes where electrons in an atom or molecule are: i) independent single particles (singlets), ii) coupled in pairs (doublets) iii) coupled in triplets, and iv) coupled in higher order clusters. The N-particle wave function is constructed from a suitable amount of coupled clusters including the correct antisymmetry of fermions. In other words, an approximative solution can be found by limiting the wave function to a certain level of coupled clusters. Typically, the cluster expansion method leads to rapidly converging results such that clusters up to doublets or triplets describe the system properties sufficiently accurately; beyond this, computer resources are usually exceeded.

3.1 Classification of Correlations

In semiconductors, the number of particles largely exceeds the electron number in atoms or molecules such that a direct solution of the wave function is not convenient. Thus, we rather utilize a density-matrix approach and evaluate the relevant expectation values from the corresponding Heisenberg equations of motion. In general, the system properties can be evaluated from $2N$-point expectation values

$$\langle N \rangle \equiv \langle a_1^\dagger \ldots a_N^\dagger a_{N'} \ldots a_{1'} \rangle, \tag{14}$$

where for simplicity the collective index $i \equiv (\lambda_i, k_i)$ has been introduced. If the system contains exactly $\mathcal{N}$ particles, $\langle \mathcal{N} \rangle$ fully describes the system properties. In this case, $2N$-point expectation values exist for all $N \leq \mathcal{N}$.

To truncate the hierarchy problem, one has to find a consistent way of approximating $\langle N \rangle$. Perhaps the simplest scheme is provided by the Hartree–Fock approximation where any $2N$-point expectation value is expressed in terms of two-point expectation values via

$$\langle N \rangle_{\mathrm{S}} = \langle N \rangle_{\mathrm{HF}} = \sum_\sigma (-1)^\sigma \prod_{i=1}^{N} \langle a_i^\dagger a_{\sigma(i')} \rangle. \tag{15}$$

The antisymmetry of the approximated $2N$-point expectation value is guaranteed by the permutation σ acting on the index set $\{1' \ldots N'\}$ such that even and odd permutations lead to $(-1)^\sigma = +1$ or -1, respectively. This factorization is exact in the case that the many-body system is described by a so-called Slater determinant of $\mathcal{N}$ independent single-particle wave functions. For a general many-body system, this factorized solution is only the first step in a controlled expansion in terms of N-particle correlation functions.

As a next step, we want to include also correlated pairs. The most simple N-particle wave function which may include a correlated pair is given by

$$|\psi\rangle_{\mathrm{D}} = \frac{1}{2}\sum_{1,2} c_{1,2}\, a_1^\dagger a_2^\dagger a_3^\dagger \ldots a_N^\dagger |0\rangle, \tag{16}$$

where formally the first two electrons are correlated. Of course, the use of the second-quantized operator product creates a correctly anti-symmetric wavefunction out of the vacuum state $|0\rangle$. We assume that $c_{1,2} = -c_{2,1}$ is already antisymmetric, and the prefactor compensates double-counting of identical physical states. Furthermore, we can assume that $c_{i,j} = 0$ if i or $j \in \{3 \ldots N\}$. The correct normalization of the state implies

$$\|\psi_{\mathrm{D}}\|^2 = \langle\psi|\psi\rangle = \frac{1}{2}\sum_{1,2} |c_{1,2}|^2 = 1. \tag{17}$$

Since in general $|\psi\rangle_{\mathrm{D}}$ cannot be expressed as a Slater determinant, we expect that a two-particle expectation value differs from its singlet approximation. And indeed, while the singlet approximation is given by

$$\langle a_i^\dagger a_m\rangle = \begin{cases} 1 & \text{if } i = j \in \{3 \ldots N\}, \\ \sum_{1'} c_{1',i}^* c_{1',m} & \text{if } i, m \notin \{3 \ldots N\}, \\ 0 & \text{otherwise}, \end{cases} \tag{18}$$

a general four-point expectation value is obtained via

$$\langle a_i^\dagger a_j^\dagger a_m a_n\rangle = \begin{cases} 1 & \text{if } i = n, j = m \in \{3 \ldots N\}, \\ -1 & \text{if } i = m, j = n \in \{3 \ldots N\}, \\ \sum_{1'} c_{1',i}^* c_{1',n} & \text{if } j = m \in \{3 \ldots N\} \text{ and } i, n \notin \{3 \ldots N\}, \\ -\sum_{1'} c_{1',i}^* c_{1',m} & \text{if } j = n \in \{3 \ldots N\} \text{ and } i, m \notin \{3 \ldots N\}, \\ -\sum_{1'} c_{1',j}^* c_{1',n} & \text{if } i = m \in \{3 \ldots N\} \text{ and } j, n \notin \{3 \ldots N\}, \\ \sum_{1'} c_{1',j}^* c_{1',m} & \text{if } i = n \in \{3 \ldots N\} \text{ and } j, m \notin \{3 \ldots N\}, \\ c_{i,j}^* c_{n,m} & \text{if } i, j, m, n \notin \{3 \ldots N\}, \\ 0 & \text{otherwise}. \end{cases} \tag{19}$$

Only the second to last row is truly related to the correlated pair. This motivates us to introduce a pair-correlation via

$$\Delta\langle a_i^\dagger a_j^\dagger a_m a_n\rangle = \langle a_i^\dagger a_j^\dagger a_m a_n\rangle - \Big(\langle a_i^\dagger a_n\rangle\langle a_j^\dagger a_m\rangle - \langle a_i^\dagger a_m\rangle\langle a_j^\dagger a_n\rangle\Big), \tag{20}$$

where the singlet (i.e. Hartree–Fock) part is subtracted. Computing this correlation for the ansatz wave-function from Eq. (16) gives

$$\Delta\langle a_i^\dagger a_j^\dagger a_m a_n\rangle = c_{i,j}^* c_{n,m} - \left(\sum_{1'} c_{1',i}^* c_{1',n}\right)\left(\sum_{2'} c_{2',j}^* c_{2',m}\right) + \left(\sum_{1'} c_{1',i}^* c_{1',m}\right)\left(\sum_{2'} c_{2',j}^* c_{2',n}\right), \quad (21)$$

if $i, j, m, n \notin \{3 \dots N\}$ and vanishes otherwise such that the only non-vanishing contributions are obtained if i and j as well as m and n belong to the correlated pair.[1]

In the same way, we can recursively define higher order correlations. For example, the three and four-particle correlations are defined by

$$\Delta\langle 3\rangle = \langle 3\rangle - \langle 1\rangle\langle 1\rangle\langle 1\rangle - \langle 1\rangle\Delta\langle 2\rangle, \quad (22)$$

$$\Delta\langle 4\rangle = \langle 4\rangle - \langle 1\rangle\langle 1\rangle\langle 1\rangle\langle 1\rangle - \langle 1\rangle\Delta\langle 3\rangle - \Delta\langle 2\rangle\Delta\langle 2\rangle, \quad (23)$$

where each term stands for all possible combinations of indeces with the proper sign depending on the number of permutations. In a similar manner, all higher order N-particle correlations can be defined [42].

3.2 Analytic Conservation Laws

The cluster expansion recursively classifies single-particle (singlet) contributions, pair (doublet) contributions and higher order clusters. We have motivated the definition of the two-particle correlation but for higher order correlations such a motivation becomes more and more cumbersome. Therefore, we check the consistency of the expansion by investigating the relation between reduced density matrices of different order. In particular, any approximation to describe, e.g., an N-particle expectation value only in term of singlets and doublets must automatically ensure that also all lower-order expectation values can be accurately described on the same level of approximation. This is not self-evident since N and $(N+1)$-particle expectation values are connected via the relation

$$\sum_\beta \langle a_1^\dagger \dots a_N^\dagger a_\beta^\dagger a_\beta a_{N'} \dots a_{1'}\rangle = (\mathcal{N} - N)\langle a_1^\dagger \dots a_N^\dagger a_{N'} \dots a_{1'}\rangle, \quad (24)$$

where we assumed a state with a fixed total number of $\mathcal{N}$ electrons. By expressing both sides of Eq. (24) in terms of the correlation functions, new conditions can be derived which must be fulfilled for the correlations.

[1] For the special case that $c_{1,2}$ can be expressed as a product of wave-functions, i.e., as Slater determinant, Eq. (21) correctly vanishes.

For example, the first two examples of additional restrictions are

$$\sum_{\beta} \langle a_{\beta}^{\dagger} a_{\beta} \rangle = \mathcal{N}, \tag{25}$$

$$\sum_{\beta} \Delta\langle a_1^{\dagger} a_{\beta}^{\dagger} a_{\beta} a_{1'} \rangle = \sum_{\beta} \langle a_1^{\dagger} a_{\beta} \rangle \langle a_{\beta}^{\dagger} a_{1'} \rangle - \langle a_1^{\dagger} a_{1'} \rangle. \tag{26}$$

The first equation relates the single-particle expectation values to the total number of particles. The second equation shows how the build-up of correlations is related to the single-particle properties. Furthermore, for vanishing correlations, it implies that the many-body state can be described by a Slater determinant of orthogonal wave functions.

In principle, similar restrictions are obtained by applying the cluster expansion to Eq. (24) for all N-particle expectation values. The consistency of the approach implies that *no new restrictions* relating, e.g., single and two-particle expectation values should emerge from applying Eq. (24) for larger N. And indeed, after some algebra, one can show that the general relation for $(N+1)$-particle correlations is given by

$$\sum_{\beta} \Delta\langle a_1^{\dagger} \ldots a_N^{\dagger} a_{\beta}^{\dagger} a_{\beta} a_{N'} \ldots a_{1'} \rangle + \sum_{\beta} \sum_{M,\sigma} (-1)^{\sigma} \Delta\langle M \rangle \Delta\langle N+1-M \rangle$$
$$= -N\, \Delta\langle a_1^{\dagger} \ldots a_N^{\dagger} a_{N'} \ldots a_{1'} \rangle. \tag{27}$$

In this equation, the sum over M and σ has to be understood such that one must take all possible ways of combining the $2N+2$ coordinates of the initial correlation into products of two correlation functions in which $a_{\beta}^{\dagger}$ and a_{β} do not occur within the same correlation. Furthermore, $\Delta\langle 1 \rangle = \langle 1 \rangle$ is a plain singlet expectation value. All other contributions which contain either more than two correlation functions or where $a_{\beta}^{\dagger} a_{\beta}$ occurs within the same correlation can be shown to cancel by applying the lower order relations.

By applying Eq. (27) under the assumption that the many-body state can be described by only singlets and doublets, i.e., by requiring $\Delta\langle N \rangle = 0$ for all $N \geq 3$, we obtain two more conditions in addition to Eqs. (25) and (26), namely

$$\sum_{\beta} \Big[\langle a_1^{\dagger} a_{\beta} \rangle \Delta\langle a_{\beta}^{\dagger} a_2^{\dagger} a_{2'} a_{1'} \rangle + \langle a_2^{\dagger} a_{\beta} \rangle \Delta\langle a_1^{\dagger} a_{\beta}^{\dagger} a_{2'} a_{1'} \rangle$$
$$+ \langle a_{\beta}^{\dagger} a_{2'} \rangle \Delta\langle a_1^{\dagger} a_2^{\dagger} a_{\beta} a_{1'} \rangle + \langle a_{\beta}^{\dagger} a_{1'} \rangle \Delta\langle a_1^{\dagger} a_2^{\dagger} a_{2'} a_{\beta} \rangle \Big] = 2\, \Delta\langle a_1^{\dagger} a_2^{\dagger} a_{2'} a_{1'} \rangle, \tag{28}$$

$$\sum_\beta \Big[\Delta\langle a_1^\dagger a_2^\dagger a_{2'} a_\beta\rangle \Delta\langle a_\beta^\dagger a_3^\dagger a_{3'} a_{1'}\rangle - \Delta\langle a_1^\dagger a_3^\dagger a_{2'} a_\beta\rangle \Delta\langle a_\beta^\dagger a_2^\dagger a_{3'} a_{1'}\rangle$$
$$+\Delta\langle a_2^\dagger a_3^\dagger a_{2'} a_\beta\rangle \Delta\langle a_\beta^\dagger a_1^\dagger a_{3'} a_{1'}\rangle - \Delta\langle a_2^\dagger a_3^\dagger a_{1'} a_\beta\rangle \Delta\langle a_\beta^\dagger a_1^\dagger a_{3'} a_{2'}\rangle$$
$$+\Delta\langle a_1^\dagger a_3^\dagger a_{1'} a_\beta\rangle \Delta\langle a_\beta^\dagger a_2^\dagger a_{3'} a_{2'}\rangle - \Delta\langle a_1^\dagger a_2^\dagger a_{1'} a_\beta\rangle \Delta\langle a_\beta^\dagger a_3^\dagger a_{3'} a_{2'}\rangle$$
$$+\Delta\langle a_1^\dagger a_3^\dagger a_{3'} a_\beta\rangle \Delta\langle a_\beta^\dagger a_2^\dagger a_{2'} a_{1'}\rangle - \Delta\langle a_1^\dagger a_2^\dagger a_{3'} a_\beta\rangle \Delta\langle a_\beta^\dagger a_3^\dagger a_{2'} a_{1'}\rangle$$
$$- \Delta\langle a_2^\dagger a_3^\dagger a_{3'} a_\beta\rangle \Delta\langle a_\beta^\dagger a_1^\dagger a_{2'} a_{1'}\rangle \Big] = 0. \tag{29}$$

All higher order restrictions vanish identically.

It is instructive to express Eqs. (25) and (26) for a semiconductor within a two-band model. If we assume vanishing two-particle correlations, we obtain

$$\mathcal{N} = \sum_{\tilde\lambda,\tilde k} \langle a^\dagger_{\tilde\lambda,\tilde k} a_{\tilde\lambda,\tilde k}\rangle, \tag{30}$$

$$0 = \sum_{\tilde\lambda,\tilde k} \langle a^\dagger_{\lambda,k} a_{\tilde\lambda,\tilde k}\rangle \langle a^\dagger_{\tilde\lambda,\tilde k} a_{\lambda',k'}\rangle - \langle a^\dagger_{\lambda,k} a_{\lambda',k'}\rangle. \tag{31}$$

Under the assumption of translational symmetry, i.e., momentum conservation along the heterostructure, only expectation values with a vanishing net momentum can occur. In this case, we can rewrite the last two equations as

$$\sum_k f_k^e = \sum_k f_k^h, \tag{32}$$

$$|P_k|^2 = f_k^e(1 - f_k^e), \tag{33}$$

$$0 = P_k^*(f_k^e - f_k^h), \tag{34}$$

with

$$f_k^e = \langle a^\dagger_{c,k} a_{c,k}\rangle, \tag{35}$$

$$f_k^h = 1 - \langle a^\dagger_{v,k} a_{v,k}\rangle, \tag{36}$$

$$P_k = \langle a^\dagger_{v,k} a_{c,k}\rangle. \tag{37}$$

These equations show the well-known fact that solving the semiconductor Bloch equations in Hartree–Fock approximation results in identical electron and hole distributions which are furthermore related to the microscopic polarization P_k via Eq. (33).

4 Equations of Motion

The general equation from the previous section can be used to derive the Coulomb contributions to the equations of motion for electron and hole distributions as well as for the different four-particle correlations. For the completely incoherent regime, after the coherences of an exciting pulse have decayed, only carrier distributions, Eqs. (35)–(36), and carrier–carrier as well as exciton correlations

$$c_{\lambda}^{q,k',k} = \Delta\langle a_{\lambda,k}^{\dagger} a_{\lambda,k'}^{\dagger} a_{\lambda,k'+q} a_{\lambda,k-q}\rangle\,, \tag{38}$$

$$c_{X}^{q,k',k} = \Delta\langle a_{c,k}^{\dagger} a_{v,k'}^{\dagger} a_{c,k'+q} a_{v,k-q}\rangle \tag{39}$$

are non-vanishing contributions involving only carrier operators.

The dominant consequence of the quantum nature of light for the incoherent regime investigated in this paper is the possibility of radiative decay of excited carriers via spontaneous emission of a photon. This emission is possible due to the broken translational symmetry perpendicular to the heterostructure [43–46]. Even without non-radiative decay channels, the coupling to the vacuum fluctuations of light leads to an intrinsic life time of excited carriers, see e.g. [10] and references therein.

In general, the computation of time resolved spectra is a subtle affair since the detection scheme itself must be included in the analysis [36,47]. A certain energy resolution requires a corresponding integration time of the detector. Consequently, early time spectra might exhibit transient narrowing of a sharp resonance which does not reflect the material properties but rather the finite detection time [48]. However, when we restrict ourselves to the stationary photoluminescence spectrum for an ideal detector with infinitely fine energy resolution, the spectrum is proportional to the rate of emitted photons [36,49]

$$I_{\mathrm{PL}}(\omega_q) = \frac{\partial}{\partial t}\Delta\langle b_{\mathbf{q}}^{\dagger} b_{\mathbf{q}}\rangle\,, \tag{40}$$

where the full three-dimensional wave number determines both energy and propagation direction of the photons. Under incoherent conditions with vanishing classical electric field, the correlated part $\Delta\langle b_{\mathbf{q}}^{\dagger} b_{\mathbf{q}}\rangle$ provides the only contribution to the photon number. The photoluminescence spectrum (40) can be directly obtained from the light-matter interaction Hamiltonian (9) as

$$\frac{\partial}{\partial t}\Delta\langle b_{\mathbf{q}}^{\dagger} b_{\mathbf{q}}\rangle = \frac{2}{\hbar}\mathrm{Re}\left[\sum_{k} \mathcal{E}_{\mathbf{q}} \bar{u}_{\mathbf{q}}^{*} d_{\mathbf{q}}^{*} \Delta\langle b_{q,q_\perp}^{\dagger} a_{v,k}^{\dagger} a_{c,k+q}\rangle\right]\,, \tag{41}$$

where we have used the rotating-wave approximation to neglect non-resonant contributions with a time dependent phase rotating with twice the optical frequency. According to Eq. (41), the spontaneous emission is driven by photon-assisted processes with probability amplitudes $\Delta\langle b_{q,q_\perp}^{\dagger} a_{v,k}^{\dagger} a_{c,k+q}\rangle$ where a photon is emitted under simultaneous transition of a conduction-band electron into the valence band [36,50]. The same photon-assisted transition amplitudes occur in the equation of motion of the carrier densities

$$\left[\frac{\partial}{\partial t} f_k^{\mathrm{e}}\right]_{H_{\mathrm{D}}} = \left[\frac{\partial}{\partial t} f_k^{\mathrm{h}}\right]_{H_{\mathrm{D}}} = \frac{2}{\hbar}\mathrm{Re}\left[\sum_{q,q_\perp} \mathcal{E}_{\mathbf{q}} \bar{u}_{\mathbf{q}}^{*} d_{\mathbf{q}}^{*} \Delta\langle b_{q,q_\perp}^{\dagger} a_{v,k-q}^{\dagger} a_{c,k}\rangle\right]. \tag{42}$$

In fact, we note that the change of the number of carriers due to recombination $\frac{\partial}{\partial t}\sum_k f_k^{\mathrm{e/h}}$ is equal to the number of emitted photons $\frac{\partial}{\partial t}\sum_{\mathbf{q}} \Delta\langle b_{\mathbf{q}}^\dagger b_{\mathbf{q}}\rangle$.

Due to the translational symmetry of the semiconductor heterostructure, the momentum along the structure has to be conserved, i.e., the photon momentum $\hbar q$ along the heterostructure is compensated by a small momentum change in the carrier transition $a_{v,k-q}^\dagger a_{c,k}$. The component $q_\perp$ perpendicular to the semiconductor structure is not restricted. Equation (42) directly shows that spontaneous emission depletes carriers with various momenta. Only the center-of-mass momentum of the recombining electron-hole pair is restricted by momentum conservation. This conservation of the center-of-mass momentum has direct consequences for the excitonic correlations and their coupling to the quantized light field. In order to see how they are influenced by spontaneous emission, we set up the equation of motion for c_{X} and get

$$\left[i\hbar\frac{\partial}{\partial t}c_{\mathrm{X}}^{q,k',k}\right]_{H_{\mathrm{D}}} = -i(1-f_{k'+q}^{\mathrm{e}}-f_{k'}^{\mathrm{h}})\sum_{q_\perp} d_{\mathbf{q}}\mathcal{E}_{\mathbf{q}}\bar{u}_{\mathbf{q}}\Delta\langle b_{q,q_\perp}^\dagger a_{v,k-q}^\dagger a_{c,k}\rangle^* \\ -i(1-f_k^{\mathrm{e}}-f_{k-q}^{\mathrm{h}})\sum_{q_\perp} d_{\mathbf{q}}^*\mathcal{E}_{\mathbf{q}}\bar{u}_{\mathbf{q}}^*\Delta\langle b_{q,q_\perp}^\dagger a_{v,k'}^\dagger a_{c,k'+q}\rangle, \qquad (43)$$

which shows that only exciton correlations with a center-of-mass momentum $\hbar q$ smaller than the photon momentum can be depleted by spontaneous emission because the wave number component q along the heterostructure is the same for both exciton correlation and photon [36,51].

In order to compute photoluminescence spectra and to include the effects of spontaneous emission into our analysis, we finally have to solve the dynamics for the photon-assisted polarizations. Computing the Heisenberg equation of motion for the carrier-photon subsystem, we obtain the semiconductor luminescence equations [36,50]

$$i\hbar\frac{\partial}{\partial t}\Delta\langle b_{q,q_\perp}^\dagger a_{v,k}^\dagger a_{c,k+q}\rangle = \left(\tilde{\varepsilon}_{k+q}^{\mathrm{e}}+\tilde{\varepsilon}_k^{\mathrm{h}}+E_{\mathrm{G}}-\hbar\omega_{q,q_\perp}\right)\Delta\langle b_{q,q_\perp}^\dagger a_{v,k}^\dagger a_{c,k+q}\rangle \\ -(1-f_{k+q}^{\mathrm{e}}-f_k^{\mathrm{h}})\Omega_{\mathrm{ST}}(k,\mathbf{q})+\Omega_{\mathrm{SE}}(k,\mathbf{q}) \qquad (44)$$

with the renormalized energies $\tilde{\varepsilon}$. Here, Ω_{ST} is a generalized photon-assisted Rabi frequency [36]

$$\Omega_{\mathrm{ST}}(k,\mathbf{q}) = id_{\mathbf{q}}\sum_{q'_\perp}\mathcal{E}_{q,q'_\perp}\bar{u}_{q,q'_\perp}\Delta\langle b_{q,q_\perp}^\dagger b_{q,q'_\perp}\rangle + \sum_{k'} V_{k-k'}\Delta\langle b_{q,q_\perp}^\dagger a_{v,k'}^\dagger a_{c,k'+q}\rangle. \qquad (45)$$

In addition, we have defined the source term

$$\Omega_{\mathrm{SE}}(k,\mathbf{q}) = i\mathcal{E}_{\mathbf{q}}\bar{u}_{\mathbf{q}}d_{\mathbf{q}}\left(f_{k+q}^{\mathrm{e}}f_k^{\mathrm{h}}+\sum_{k'}c_{\mathrm{X}}^{q,k,k'}\right) \qquad (46)$$

in Eq. (44) which initializes the build-up of photon assisted polarizations as soon as excited carriers are present in the system.

5 Photoluminescence

In order to demonstrate the capabilities of our theory, we numerically evaluate the equations for the coupled semiconductor-photon system by including electron and hole distributions as well as carrier–carrier and excitonic correlations. We show that most of the excitons are in dark states which do not couple to the light field. Using these results, we derive an analytical expression to compute photoluminescence spectra in direct analogy to the Elliott formula for absorption. This formula presents the fundamental features of semiconductor luminescence in a transparent manner such that one is able to understand and explain plasma vs. population features in emission spectra. While the full dynamic solution is currently feasible only for a one-dimensional model system, our analytic expression can be used to avoid part of the numerical complexity. It thus allows us to compute luminescence spectra also for two dimensional quantum-well systems.

5.1 Numerical Solution

In a previous publication [27], we have studied the build up of correlations out of an incoherent electron-hole plasma for a semiconductor structure with vanishing coupling to the light field. Since a strong suppression of spontaneous emission can be achieved by inserting a semiconductor structure inside a photonic bandgap material [52–54], these computations can be viewed as modeling a realistic setup. Even though in practice the detection of the weak photoluminescence signal might be difficult just because of the strong suppression of the spontaneous emission and the corresponding slow emission rate of photons, it is in principle possible. Also in the present publication we compare computed photoluminescence spectra for the case with and without suppression of the dipole coupling. The description of the quantized light field is obtained by including the dynamics according to Eqs. (41)–(44). All photoluminescence spectra are computed via Eq. (40), assuming a perfect energetic detector resolution. The suppression of spontaneous emission is obtained by switching off the photon-assisted contributions, Eqs. (42) and (43).

A first set of computed photoluminescence spectra is shown in Fig. 1. For these computations, spontaneous emission has been switched off. In all figures, spectra at three different times after the beginning of the computations for various lattice temperatures and carrier densities are displayed. All computations are initialized with vanishing correlations and quasi-equilibrium Fermi–Dirac distributions at the lattice temperature for carriers. Right from the beginning, the spectra are peaked at the excitonic resonance. This fact has been explained in terms of the Coulomb sum in Eq. (44) which introduces the excitonic resonance independently of the fact whether or not exciton correlations are formed [50]. Whereas for low densities and temperatures a growth of the photoluminescence peak by almost a factor of three is observed, this

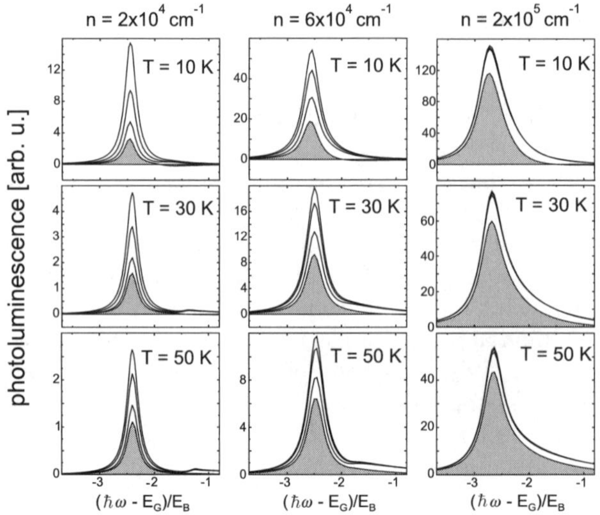

Fig. 1. Set of computed photoluminescence spectra for different carrier densities and lattice temperatures with suppressed spontaneous emission. In each figure the spectrum is shown 120 ps (*dotted line*), 600 ps (*dashed line*) and 1200 ps (*solid line*) after the start of the computation. The Hartree–Fock spectrum is shown for comparison as a *shaded area*

growth becomes negligible for high carrier densities and appreciably weaker for elevated temperatures.

In general, the source term, Eq. (46), shows that dynamic changes of the photoluminescence spectrum can equally well be due to changes in the exciton correlations $c_{\mathrm{X}}^{q,k',k}$ or to changing carrier distributions $f_k^{\mathrm{e/h}}$. Since the carrier distributions do not change appreciably during the computations, the growth of the peak height in Fig. 1 is directly related to the formation of exciton population correlations. For comparison, we also show the Hartree–Fock result of the steady-state spectrum obtained by including only the factorized source term $f_k^{\mathrm{e}} f_k^{\mathrm{h}}$ in Eq. (46). Whereas this approximation is in principle inconsistent according to the concept of the cluster expansion, it nevertheless gives reasonable results for the highest densities and temperatures shown in Fig. 1. In particular, it gives the main contribution at the lowest exciton resonance and underestimates mainly the luminescence at the higher bound states and in the continuum. For very low temperatures, this can even result in a negative luminescence signal around the renormalized band-edge. This possibility of a negative photoluminescence signal is unphysical and has been interpreted as a major limitation of our approach [55,56]. Therefore, we want to point out that the inclusion of the full source term (46) leads to a positive signal for any carrier distribution. This is nicely demonstrated from the numerical results in Fig. 2 where the luminescence spectra for different carrier densities at a lattice temperature of $T = 30\,\mathrm{K}$ are shown. The continuum lu-

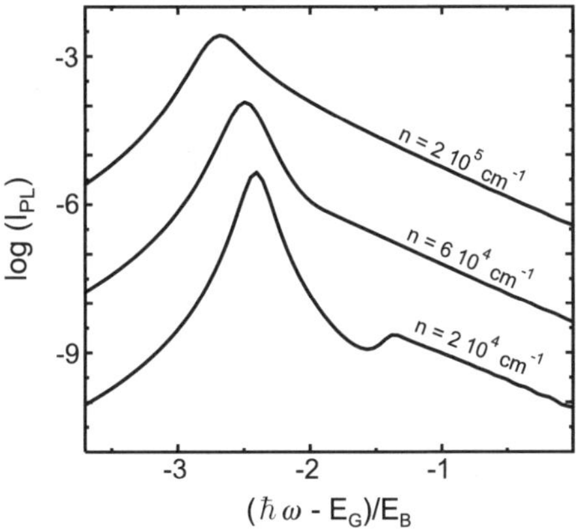

Fig. 2. Computed photoluminescence spectrum for different carrier densities at a lattice temperature of 30 K in semi-logarithmic scale

minescence shows a perfect exponential decay. The temperature fitted from this tail is approximately 40 K which is close to the initial carrier temperature. The deviation can be understood because the carriers are subject to a slow heating process in the course of their dynamic evolution.

When the computation is repeated without the suppression of spontaneous emission, the result obtained in Fig. 1 drastically changes. Figure 3 shows the result for a computation with an initial carrier density $n = 2 \times 10^4\,\mathrm{cm}^{-1}$ and a lattice temperature of 10 K. Now the excitonic peak of the photoluminescence spectrum even drops within a nanosecond. This drop is

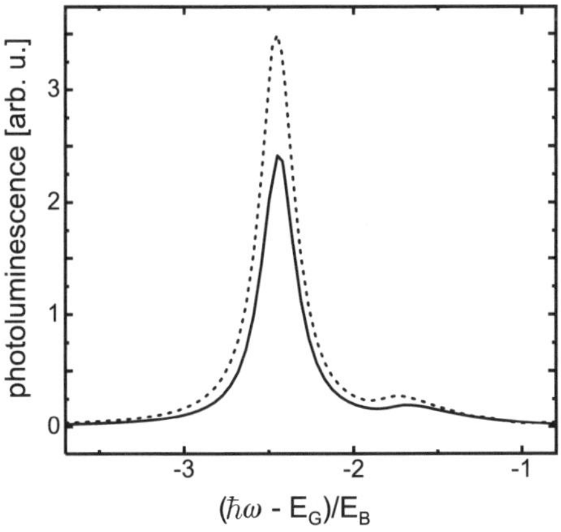

Fig. 3. Computed photoluminescence spectra for the case of full dipole coupling. The spectra are taken directly after the start of the computation (*dotted line*) and after 1 ns (*solid line*). Instead of a growing signal, one gets a reduction of the spectrum in time

mainly due to heating of the carrier distributions. In order to see which role the correlated part of the source term, Eq. (46), plays and to understand how the spontaneous emission influences the excitonic correlations, we compute the center-of-mass distribution of the lowest exciton population correlation. A comparison between the two computations with and without suppressed spontaneous emission is shown in Fig. 4.

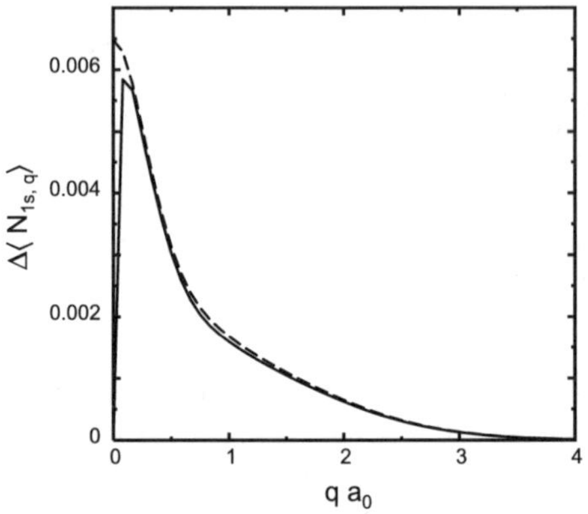

Fig. 4. Center-of-mass distribution of the lowest exciton population correlation $\Delta\langle N_{1s,q}\rangle$ with full spontaneous emission (*solid line*) or with strongly suppressed spontaneous emission (*dashed line*) for the same conditions as in Fig. 3. The coupling results in a strong hole burning at momenta within the radiative cone

Compared to the case without spontaneous emission, the momentum dependent exciton distribution exhibits a very strong hole burning at center-of-mass momenta within the radiative cone, i.e., for all wave vectors

$$q < q_{\text{photon}} \approx \frac{E_G}{\hbar c_0}, \tag{47}$$

when the spontaneous emission is not suppressed. Since the parallel component of the wave vector is conserved according to Eq. (43), only excitons which fulfill Eq. (47) can emit photons which can propagate outside the substrate to the experimental detection. Since the photon momentum is at most $q_{\text{photon}} \approx 0.1a_0^{-1}$, the radiative cone is very small compared to the full extent of the exciton distribution. Thus, we conclude that the dominant fraction of excitons is in dark states with a momentum too large to be transferred to a photon. Only after additional scattering processes into the radiative cone these excitons can be emitted. In general, the exact strength of the hole-burning effect depends on the balance between the spontaneous emission rate and the rate with which the hole is refilled via phonon or Coulomb scattering processes. It is known that the radiative life time of coherent excitons is on a picosecond time scale [11,43,57]. In principle, incoherent excitons should

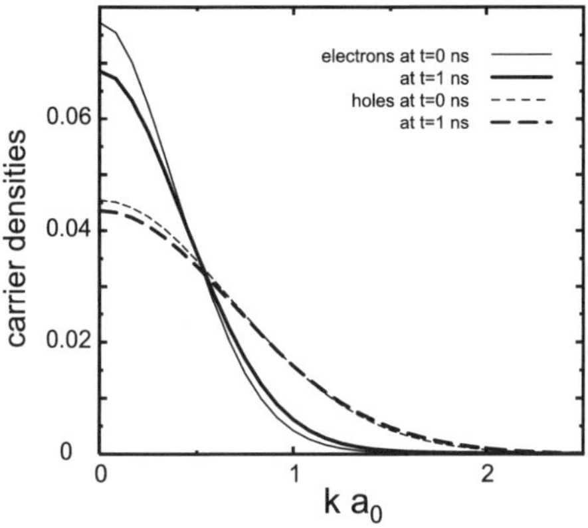

Fig. 5. Carrier distributions initially and after one nanosecond of computed time for identical parameters as in Fig. 3. The main effect is a slow heating, especially for the lighter electrons, whereas strong effects due to recombination are not visible

decay on the same picosecond time scale. Thus, the hole burning is always a dominant process in GaAs-like materials.

The electron and hole distributions do not exhibit a similar depletion and their recombination does not have any preferred k value. Instead, they are depleted continuously according to the k dependence of the excitonic wave functions. This is confirmed by our numerical calculations as can be seen from Fig. 5. The distributions only exhibit a weak heating but no traces of hole burning. Thus, even under conditions favorable for the build-up of excitonic correlations, the dominant part of all excitons occupies dark states and does not contribute to the measured photoluminescence spectrum.

5.2 Analytic Formula for Photoluminescence

In order to understand the nature of semiconductor luminescence more clearly and to derive an analytical expression for the computation of photoluminescence spectra, we start by transforming Eqs. (41) and (44) into the generalized exciton basis [27]. The photoluminescence spectrum is thus given by

$$I_{\mathrm{PL}}(\omega_{\mathbf{q}}) = \frac{2}{\hbar}\mathrm{Re}\left[\mathcal{F}^*_{\mathbf{q}} \sum_{\nu} \phi^{\mathrm{r}}_{\nu,q}(r=0) \Delta\langle b^{\dagger}_{q_\perp,q} X_{\nu,q}\rangle\right] , \tag{48}$$

where $X_{\nu,q}$ is the exciton annihilation operator and the effective matrix element $\mathcal{F}_{\mathbf{q}} = \mathcal{E}_{\mathbf{q}} \bar{u}_{\mathbf{q}} d_{\mathbf{q}}$ has been introduced. In Eq. (48), the vector $\mathbf{q}$ includes not only the information about the energy of the emitted photon $\hbar|\mathbf{q}|/c_0$, but also about the emission angle depending on the ratio between $q_\perp$ and q along the semiconductor structure. In Eq. (44), the stimulated term of photon–photon correlations can be neglected for systems without a cavity [36]. As a result, one obtains

$$i\hbar\frac{\partial}{\partial t}\Delta\langle b^{\dagger}_{q_\perp,q}X_{\nu,q}\rangle = (E_{\nu,q}+E_{\mathrm{G}}-\hbar\omega_{\mathbf{q}}-i\gamma)\Delta\langle b^{\dagger}_{q_\perp,q}X_{\nu,q}\rangle$$
$$+i\mathcal{F}_{\mathbf{q}}\sum_k \phi^{\mathrm{l}}_{\nu,q}(k)\left[f^{\mathrm{e}}_{k+q^{\mathrm{e}}}f^{\mathrm{h}}_{k-q^{\mathrm{h}}}+\sum_{k'}\Delta\langle c^{\dagger}_{k'+q^{\mathrm{e}}}v^{\dagger}_{k-q^{\mathrm{h}}}c_{k+q^{\mathrm{e}}}v_{k'-q^{\mathrm{h}}}\rangle\right]$$
$$=(E_{\nu,q}+E_{\mathrm{G}}-\hbar\omega_{\mathbf{q}}-i\gamma)\Delta\langle b^{\dagger}_{q_\perp,q}X_{\nu,q}\rangle$$
$$+i\mathcal{F}_{\mathbf{q}}\sum_{\nu'}\phi^{\mathrm{r}}_{\nu',q}(r=0)\left[\langle X^{\dagger}_{\nu',q}X_{\nu,q}\rangle_{\mathrm{S}}+\Delta\langle X^{\dagger}_{\nu',q}X_{\nu,q}\rangle\right], \quad (49)$$

where the singlet contribution to the source term

$$\langle X^{\dagger}_{\nu',q}X_{\nu,q}\rangle_{\mathrm{S}} = \sum_k \left(\phi^{\mathrm{r}}_{\nu',q}(k)\right)^{*}\phi^{\mathrm{r}}_{\nu,q}(k)f^{e}_{k+q^{e}}f^{h}_{k-q^{h}} \quad (50)$$

has been introduced. In Eq. (49), the total exciton energy $E_{\nu,q}$ includes the center-of-mass kinetic energy. The form of Eq. (49) furthermore assumes that the exciton wave functions change slowly in time. This adiabatic approximation is well valid since the carrier distributions typically do change slowly.

We proceed by solving Eq. (49) in Markov approximation,

$$\Delta\langle b^{\dagger}_{q_\perp,q}X_{\nu,q}\rangle = \frac{i\mathcal{F}_{\mathbf{q}}\sum_{\nu'}\phi^{\mathrm{r}}_{\nu',q}(r=0)\left[\langle X^{\dagger}_{\nu',q}X_{\nu,q}\rangle_{\mathrm{S}}+\Delta\langle X^{\dagger}_{\nu',q}X_{\nu,q}\rangle\right]}{\hbar\omega_{\mathbf{q}}-E_{\mathrm{G}}-E_{\nu,q}+i\gamma}. \quad (51)$$

Inserting this solution into Eq. (48) results in

$$I_{\mathrm{PL}}(\omega_{\mathbf{q}}) = -\frac{2}{\hbar}\mathrm{Im}\left[|\mathcal{F}_{\mathbf{q}}|^2\sum_{\nu,\nu'}\phi^{\mathrm{r}}_{\nu,q}(0)\phi^{\mathrm{r}}_{\nu',q}(0)\frac{\langle X^{\dagger}_{\nu',q}X_{\nu,q}\rangle_{\mathrm{S}}+\Delta\langle X^{\dagger}_{\nu',q}X_{\nu,q}\rangle}{\hbar\omega_{\mathbf{q}}-E_{\mathrm{G}}-E_{\nu,q}+i\gamma}\right], \quad (52)$$

which nicely shows that in general luminescence at a certain excitonic energy depends not only on the corresponding exciton population, but also on the correlated carrier plasma via the term $\langle X^{\dagger}_{\nu',q}X_{\nu,q}\rangle_{\mathrm{S}}$ and on all off-diagonal transition correlations $\Delta\langle X^{\dagger}_{\nu',q}X_{\nu,q}\rangle$. From this fundamental derivation one can already conclude that a photoluminescence experiment never detects exclusively exciton populations. A careful analysis and a comparison to a microscopic theory is thus very important.

From previous investigations we know that Coulomb processes typically lead to a very fast build-up of the off-diagonal transition correlations until they exactly cancel the singlet contribution [27]. Thus, it is justified to assume that this process is fast compared to all other time scales of interest and Eq. (52) can be further simplified to

$$I_{\mathrm{PL}}(\omega_{\mathbf{q}}) = -\frac{2}{\hbar}\mathrm{Im}\left[|\mathcal{F}_{\mathbf{q}}|^2\sum_{\nu}|\phi^{\mathrm{r}}_{\nu,q}(0)|^2\frac{\langle X^{\dagger}_{\nu,q}X_{\nu,q}\rangle_{\mathrm{S}}+\Delta\langle X^{\dagger}_{\nu,q}X_{\nu,q}\rangle}{\hbar\omega_{\mathbf{q}}-E_{\mathrm{G}}-E_{\nu,q}+i\gamma}\right]. \quad (53)$$

This equation summarizes all important aspects of semiconductor luminescence: i) the total luminescence spectrum has resonances given by the solution of the generalized Wannier equation due to the energy denominator, ii) these resonances are independent of the source term and a standard photoluminescence experiment cannot distinguish between contributions from exciton correlations and Coulomb correlated plasma; even with vanishing exciton correlations, the pure singlet source results in a strong peak at the 1s exciton resonance due to the large oscillator strength $|\phi^{\mathrm{r}}_{1\mathrm{s},q}(r=0)|^2$, iii) a typical photoluminescence spectrum is highly nonthermal in the sense that the Kubo–Martin–Schwinger relation [58], which relates photoluminescence to absorption measurements by a Bose–Einstein distribution via

$$I_{\mathrm{PL}}(\omega) = \alpha(\omega)\, g(\hbar\omega)\,, \tag{54}$$

cannot be applied in general.

In Fig. 6 we investigate how important the deviations are compared to the result expected from Eq. (54). For the same computation as in Fig. 3, absorption and photoluminescence spectra in a semi-logarithmic plot are shown in the upper picture, and the ratio between both in the lower picture. According to Eq. (54), this ratio would have to follow the dashed line which

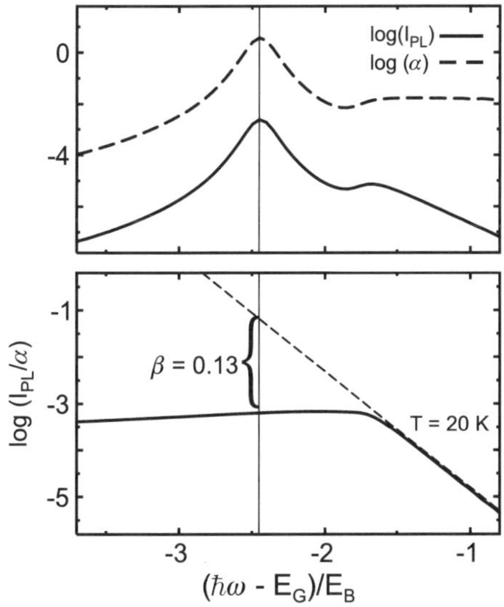

Fig. 6. *Top*: Absorption and photoluminescence spectra for a computation with an initial carrier distribution of $n = 2 \times 10^4\,\mathrm{cm}^{-1}$ at 10 K in semi-logarithmic scale. *Bottom*: ratio I_{PL}/α between photoluminescence spectrum and absorption spectrum. The *dashed line* indicates a Bose–Einstein distribution of $T = 20\,\mathrm{K}$ corresponding to the temperature obtained from the high energy tail. For bosons in thermal equilibrium, the ratio should coincide with this line

was obtained by fitting a temperature from the high-energy tail. In contrast, we observe an emission which lies below the thermal result by a factor of $\beta = 0.13$. This value is in agreement with experiments [59] where such an attenuation factor β had been introduced as a fitting parameter. Our theory explains β to be a fundamental physical quantity, related to the strong hole burning of excitons exposed to spontaneous emission. Due to this hole burning which is shown in Fig. 4, the emission according to Eq. (53) is typically dominated by the plasma contribution even under good formation conditions of low temperatures and low carrier densities. The nonthermal nature of the plasma contribution leads to the deviation from the naively expected thermal emission properties.

Equation (53) also reveals why the Hartree–Fock approximation of the source term works well for elevated temperatures [50]. Compared to Eq. (52), the main effect of the population correlations is to reduce the double sum over exciton states to a single sum where only diagonal exciton correlations enter. After that, it is obvious from Eq. (53) that the calculated photoluminescence including the effect of exciton correlations is positive for all frequencies.

If we assume for a moment identical electron and hole temperatures, we can rewrite the singlet source, Eq. (50), as

$$\begin{aligned}\langle X^\dagger_{\nu,q} X_{\nu',q}\rangle_{\mathrm{S}} &= \sum_k \phi^{\mathrm{l}}_{\nu,q}(k)\left(\phi^{\mathrm{l}}_{\nu',q}(k)\right)^* f^{\mathrm{e}}_{k+q^{\mathrm{e}}} f^{\mathrm{h}}_{k-q^{\mathrm{h}}} \\ &= \sum_k \phi^{\mathrm{r}}_{\nu,q}(k)\left(\phi^{\mathrm{l}}_{\nu',q}(k)\right)^* \frac{f^{\mathrm{e}}_{k+q^{\mathrm{e}}} f^{\mathrm{h}}_{k-q^{\mathrm{h}}}}{1 - f^{\mathrm{e}}_{k+q^{\mathrm{e}}} - f^{\mathrm{h}}_{k-q^{\mathrm{h}}}} \\ &= \sum_k \phi^{\mathrm{r}}_{\nu,q}(k)\left(\phi^{\mathrm{l}}_{\nu',q}(k)\right)^* g(E_{k,q})\end{aligned} \tag{55}$$

with a Bose–Einstein distribution

$$g(E) = \frac{1}{e^{\frac{E-\mu}{kT}} - 1}, \tag{56}$$

evaluated at the total kinetic energy $E_{k,q} = \frac{\hbar^2 k^2}{2\mu} + \frac{\hbar^2 q^2}{2M}$ of the electron-hole pair with the sum of the chemical potentials $\mu = \mu^{\mathrm{e}} + \mu^{\mathrm{h}}$. In the second line of Eq. (55), we have used a property of the exciton functions. For low temperatures, this expression confirms the nonthermal nature of the source term. For elevated temperatures, however, the Bose–Einstein distribution is sufficiently broad such that already the singlet source term can be approximated as

$$\langle X^\dagger_{\nu,q} X_{\nu',q}\rangle_{\mathrm{S}} \approx g(E_{k=0,q}) \sum_k \phi^{\mathrm{r}}_{\nu,q}(k)\left(\phi^{\mathrm{l}}_{\nu',q}(k)\right)^* = g(E_{k=0,q})\delta_{\nu,\nu'}\,. \tag{57}$$

Thus, also the pure singlet term is diagonal in the high temperature limit. As one can observe from Fig. 6, the Kubo–Martin–Schwinger relation is fully valid for the continuum emission.

5.3 Quantum-Well Luminescence

As we have seen in the previous section, the photoluminescence signal can very well be calculated from Eq. (53) once the source term is known. In order to simplify the further analysis and to extend our treatment to the computation of quantum-well photoluminescence spectra, we make use of the basic results obtained from the coupling of the semiconductor to the quantized light field: Since the dominant feature is the strong hole burning in the exciton distributions such that practically no excitons within the radiative cone are present, we assume from now on that the contribution from the exciton population correlations to the numerator of Eq. (53) are identical to zero. If we furthermore assume that the carrier densities are in thermal equilibrium with the lattice, we can compute photoluminescence spectra for various carrier temperatures and densities.

First, we show that these approximations are justified. In Fig. 7, the spectrum obtained from Eq. (53) is compared with the result of the full computation. And indeed, the result from Eq. (53) is in very close agreement with the case of full coupling to the quantized light field. It even overestimates the emission at the 1s resonance. But this small difference can be explained by the fact that the carrier distributions in the dynamical computation have heated up slightly such that the final carrier distributions are not exactly the ones used in Eq. (53).

With the same assumptions, the computation of luminescence spectra for two-dimensional quantum-well systems is now straight forward. We solve the generalized Wannier equation including the phase-space filling factor and

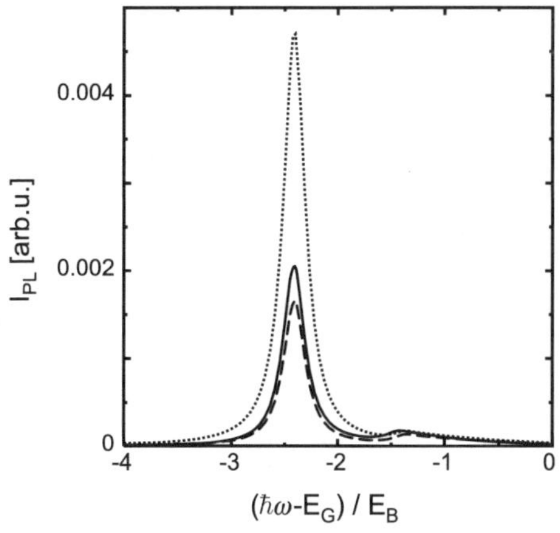

Fig. 7. Comparison between the photoluminescence spectrum obtained from Eq. (53) (*solid line*) and from the full dynamical computation with spontaneous emission (*dashed line*) or with a suppressed spontaneous emission (*dotted line*). The dynamic spectra are taken after 1 ns of evolution. All calculations are performed for a carrier density of $n = 2 \times 10^4\,\mathrm{cm}^{-1}$ and a lattice temperature of 30 K

a screened Coulomb potential in two dimensions [17,27], assume Fermi–Dirac distributions for electrons and holes and vanishing exciton population correlations, and apply Eq. (53) to compute the spectrum. The result is shown in Fig. 8(a). As in the one-dimensional case, Eq. (53) leads to a strong resonance at the 1s exciton, and also higher order excitons are still resolved with the small constant broadening γ which is used in the denominator of Eq. (53). A more realistic computation also includes microscopic Coulomb scattering

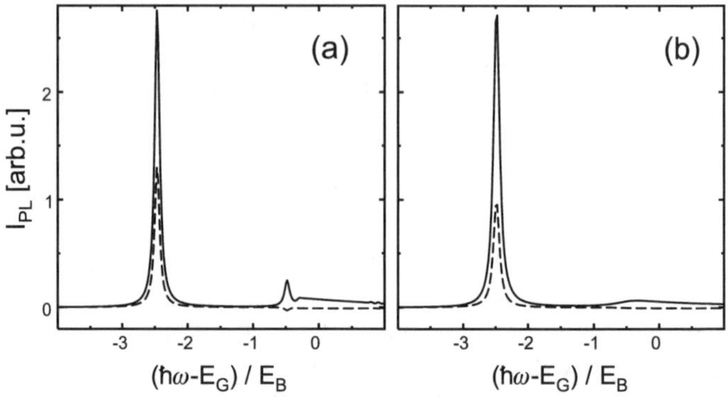

Fig. 8. (**a**) Photoluminescence spectrum obtained from Eq. (53) for a quantum-well at a temperature of $T = 70\,\mathrm{K}$ and a carrier density of $n = 1 \times 10^9\,\mathrm{cm}^{-2}$. The full result (*solid line*) is compared with the computation where the total factorized part of the source term (46) is included (*dashed line*). (**b**) Same as (**a**) but with microscopic scattering contributions due to higher order Coulomb scattering

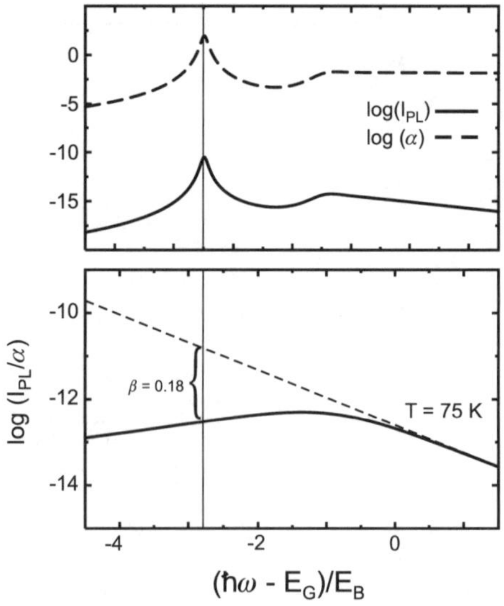

Fig. 9. Analogous figure to Fig. 6, but now computed for a quantum-well system computed for the same parameters as in Fig. 8

of the photon assisted polarizations. In this case, no constant γ is needed and the microscopic mechanism of Coulomb scattering leads to the spectrum which is shown in Fig. 8(b). Whereas the 1s peak does hardly change at all, we want to point out the strong broadening of the higher order excitonic resonances. Therefore, we have to conclude that even in the so-called "linear regime" higher order excitons can already exhibit typical nonlinear behavior.

Finally, we investigate the validity of the Kubo–Martin–Schwinger relation in two dimensions. Figure 9 shows the result of a computation with the same parameters as in Fig. 8. As in the one-dimensional case (see Fig. 6), one obtains a very strong suppression of the 1s luminescence peak compared to the result expected from simple thermodynamic arguments. Therefore, we must conclude that the nonequilibrium character of photoluminescence in the sense of a violation of the Kubo–Martin–Schwinger relation also holds in two dimensions. It is a challenge of the near future to measure time resolved absorption and emission spectra very precisely, using the same setup, in order to confirm these predictions experimentally.

6 Summary

We have presented the cluster expansion for semiconductor quantum optics which allows to consistently truncate the hierarchy of equations resulting from both the many-body Coulomb interaction and the interaction with a quantized light field. We have applied the truncation scheme for a detailed investigation of semiconductor photoluminescence. We have shown how the photoluminescence is an intrinsic non-equilibrium process and how it is dominated by the emission of a correlated plasma because even under good exciton formation conditions of low lattice temperature and moderate densities most of the excitons are in dark states. Only excitons with an extremely small center-of-mass momentum can be coupled to the light field.

Acknowledgements

This work was supported by the Deutsche Forschungsgemeinschaft through the Quantum Optics in Semiconductors Research Group, by the Humboldt Foundation and the Max-Planck Society through the Max-Planck Research prize, and by the Optodynamics Center of the Philipps-Universität Marburg. M. K. acknowledges funding from the Swedish Natural Science Research Council (NFR) and the Göran Gustafssons Stiftelse and thanks the Center for Parallel Computers (PDC) to make their computer resources available for this project.

References

1. E. Moreau, I. Robert, L. Manin, V. Thierry-Mieg, J.M. Gérard, and I. Abram. Quantum cascade of photons in semiconductor quantum dots. Phys. Rev. Lett., 87:183601, 2001.
2. O. Benson, C. Santori, M. Pelton, and Y. Yamamoto. Regulated and entangled photons from a single quantum dot. Phys. Rev. Lett., 84:2513, 2000.
3. Y.-S. Lee, T.B. Norris, M. Kira, F. Jahnke, S.W. Koch, G. Khitrova, and H.M. Gibbs. Quantum correlations and intraband coherences in semiconductor cavity QED. Phys. Rev. Lett., 83:5338, 1999.
4. C. Ell, P. Brick, M. Hübner, E.S. Lee, O. Lyngnes, J.P. Prineas, G. Khitrova, H.M. Gibbs, M. Kira, F. Jahnke, S.W. Koch, D.G. Deppe, and D.L. Huffaker. Quantum correlations in the nonperturbative regime of semiconductor microcavities. Phys. Rev. Lett., 85:5392–5395, 2000.
5. D.R. Wake, H.W. Yoon, J.P. Wolfe, and H. Morkoc. Response of excitonic absorption spectra to photoexcited carriers in GaAs quantum wells. Phys. Rev. B, 46:13452, 1992.
6. C. Weisbuch, M. Nishioka, A. Ishikawa, and Y. Arakawa. Observation of the coupled exciton-photon mode splitting in a semiconductor quantum microcavity. Phys. Rev. Lett., 69:3314, 1992.
7. T.B. Norris, J.-K. Rhee, C.-Y. Sung, Y. Arakawa, M. Nishioka, and C. Weisbuch. Time-resolved vacuum Rabi oscillations in a semiconductor quantum microcavity. Phys. Rev. B, 50:14663, 1994.
8. T. Rappen, U.G. Peter, M. Wegener, and W. Schäfer. Polarization dependence of dephasing processes: A probe for many-body effects. Phys. Rev. B, 49:10774, 1994.
9. P. Kner, W. Schäfer, R. Lövenich, and D.S. Chemla. Coherence of four-particle correlations in semiconductors. Phys. Rev. Lett., 81:5386, 1998.
10. G. Khitrova, H.M. Gibbs, F. Jahnke, M. Kira, and S.W. Koch. Nonlinear optics of normal-mode-coupling semiconductor microcavities. Rev. Mod. Phys., 71:1591, 1999.
11. F. Tassone, F. Bassani, and L.C. Andreani. Quantum-well reflectivity and exciton-polariton dispersion. Phys. Rev. B, 45:6023, 1992.
12. K. El Sayed, L. Bányai, and H. Haug. Coulomb quantum kinetics and optical dephasing on the femtosecond time scale. Phys. Rev. B, 50:1541, 1994.
13. T. Stroucken, A. Knorr, C. Anthony, A. Schulze, P. Thomas, S.W. Koch, M. Koch, S.T. Cundiff, J. Feldmann, and E.O. Göbel. Light propagation and disorder effects in semiconductor multiple quantum wells. Phys. Rev. Lett., 74:2391, 1995.
14. F. Jahnke, M. Kira, and S.W. Koch. Linear and nonlinear optical properties of quantum confined excitons in semiconductor microcavities. Z. Physik B, 104:559, 1997.
15. C. Sieh, T. Meier, F. Jahnke, A. Knorr, S.W. Koch, P. Brick, M. Hübner, C. Ell, J.P. Prineas, G. Khitrova, and H.M. Gibbs. Coulomb memory signatures in the excitonic optical stark effect. Phys. Rev. Lett., 82(15):3112, 1999.
16. H. Haug and E. Hanamura. Condensation effects of excitons. Phys. Rep., 33:209–284, 1977.
17. H. Haug and S.W. Koch. *Quantum Theory of the Optical and Electronic Properties of Semiconductors*. World Scientific Publ., Singapore, 3. edition, 1994.

18. D. Snoke. Coherent exciton waves. Science, 273:1351–1352, 1996.
19. D.G. Lidzey, D.D.C. Bradley, M.S. Skolnick, T. Virgili, S. Walker, and D.M. Whittaker. Strong exciton-photon coupling in an organic semiconductor microcavity. Nature, 395:53–55, 1998.
20. T. Lundstrom, W. Schoenfeld, H. Lee, and P.M. Petroff. Exciton storage in semiconductor self-assembled quantum dots. Science, 286:2312–2314, 1999.
21. M. Bayer, O. Stern, P. Hawrylak, S. Fafard, and A. Forchel. Hidden symmetries in the energy levels of excitonic 'artificial atoms'. Nature, 405:923–926, 2000.
22. R. Kumar, A.S. Vengurlekar, A.V. Gopal, T. Melin, F. Laruelle, B. Etienne, and J. Shah. Exciton formation and relaxation dynamics in quantum wires. Phys. Rev. Lett., 81:2578, 1998.
23. X. Marie, J. Barrau, P. Le Jeune, T. Amand, and M. Brosseau. Exciton formation in quantum wells. Phys. stat. sol. (a), 164:359, 1997.
24. A. Vinattieri, J. Shah, T.C. Damen, D.S. Kim, L.N. Pfeiffer, M.Z. Maialle, and L.J. Sham. Exciton dynamics in GaAs quantum-wells under resonant excitation. Phys. Rev. B, 50:10868, 1994.
25. P.W.M. Blom, P.J. Vanhall, C. Smit, J.P. Cuypers, and J.H. Wolter. Selective exciton formation in thin GaAs/Al_xGa_{1-x}As quantum wells. Phys. Rev. Lett., 71:3878, 1993.
26. B. Deveaud, F. Clérot, N. Roy, K. Satzke, B. Sermage, and D.S. Katzer. Enhanced radiative recombination of free excitons in GaAs quantum wells. Phys. Rev. Lett., 67:2355, 1991.
27. W. Hoyer, M. Kira, and S.W. Koch. Influence of coulomb and phonon interaction on the exciton formation dynamics in semiconductor heterostructures. Phys. Rev. B, 67:155113, 2003.
28. M. Bonitz. *Quantum Kinetic Theory.* Teubner, Stuttgart, 1998.
29. G.D. Mahan. *Many-Particle Physics.* Plenum, New York, 2. edition, 1990.
30. J. Hader, P. Thomas, and S.W. Koch. Optoelectronics of semiconductor superlattices. Prog. in Quant. Electr., 22:123, 1998.
31. H. Haug and A.-P. Jauho. *Quantum Kinetics in Transport & Optics of Semiconductors.* Springer-Verlag, Berlin, 1. edition, 1996.
32. C. Piermarocchi, F. Tassone, V. Savona, A. Quattropani, and P. Schwendimann. Nonequilibrium dynamics of free quantum-well excitons in time-resolved photoluminescence. Phys. Rev. B, 53:15834, 1996.
33. K. Siantidis, V.M. Axt, and T. Kuhn. Dynamics of exciton formation for near band-gap excitations. Phys. Rev. B, 65:035303, 2001.
34. S.R. Bolton, U. Neukirch, L.J. Sham, D.S. Chemla, and V.M. Axt. Demonstration of sixth-order coulomb correlations in a semiconductor single quantum well. Phys. Rev. Lett., 85:2002, 2000.
35. C. Cohen-Tannoudji, J. Dupont-Roc, and G. Grynberg. *Photons & Atoms.* Wiley, New York, 3. edition, 1989.
36. M. Kira, F. Jahnke, W. Hoyer, and S.W. Koch. Quantum theory of spontaneous emission and coherent effects in semiconductor microstructures. Prog. in Quant. Electr., 23:189, 1999.
37. see e.g. articles in *The Hubbard Model*, ed. by M. Rasetti, Series on Advances in Statistical Mechanics, World Scientific Publ. (1991).
38. W. Schäfer and M. Wegener. *Semiconductor Optics and Transport Phenomena.* Springer-Verlag, Berlin, 1. edition, 2002.

39. J. Cízek. On correlation problem in atomic and molecular systems. calculation of wavefunction components in ursell-type expansion using quantum-field theoretical methods. J. Chem. Phys., 45:4256, 1966.
40. G.D. Purvis and R.J. Bartlett. A full coupled-cluster singles and doubles model: The inclusion of disconnected triples. J. Chem. Phys., 76:1910, 1982.
41. F.E. Harris, H.J. Monkhorst, and D.L. Freeman. *Algebraic and Diagrammatic Methods in Many-Fermion Theory.* Oxford Press, New York, 1. edition, 1992.
42. J. Fricke. Transport equations including many-particle correlations for an arbitrary quantum system: A general formalism. Annals of Physics, 252(2):479, 1996.
43. V.M. Agranovich and O.A. Dubowskii. Effect of retarded interaction of exciton spectrum in 1-dimensional and 2-dimensional crystals. JETP Lett., 3:223, 1966.
44. E. Hanamura. Rapid radiative decay and enhanced optical nonlinearity of excitons in a quantum well. Phys. Rev. B, 38:1228, 1988.
45. L.C. Andreani, F. Tassone, and F. Bassani. Radiative lifetime of free excitons in quantum wells. Solid State Commun., 77:641, 1991.
46. D.S. Citrin. Comments Condens. Matter Phys., 16:263, 1993.
47. D.F. Walls and G.J. Milburn. *Quantum Optics.* Springer-Verlag, New York, 1. edition, 1994.
48. W. Hoyer, M. Kira, and S.W. Koch. Semiconductor Bloch Equations for Classical and Quantum Fields, pp. 15–62. In *Proceedings of the International School of Physics "Enrico Fermi", Course CL, Electron and Photon Confinement in Semiconductor Nanostructures*, eds. B. Deveaud, A. Quattropani and P. Schwendimann, IOS Press, Amsterdam (2003)
49. A. Thränhardt, S. Kuckenburg, A. Knorr, T. Meier, and S.W. Koch. Quantum theory of phonon-assisted exciton formation and luminescence in semicondutor quantum wells. Phys. Rev. B, 62(4):2706, 2000.
50. M. Kira, F. Jahnke, and S.W. Koch. Microscopic theory of excitonic signatures in semiconductor photoluminescence. Phys. Rev. Lett., 81:3263, 1998.
51. M. Kira, W. Hoyer, T. Stroucken, and S.W. Koch. Exciton formation in semiconductors and the influence of a photonic environment. Phys. Rev. Lett., 87:176401, 2001.
52. E. Yablonovitch. Inhibited spontaneous emission in solid-state physics and electronics. Phys. Rev. Lett., 58:2059–2062, 1987.
53. S. John. Strong localization of photons in certain disordered dielectric superlattices. Phys. Rev. Lett., 58:2086–2489, 1987.
54. D. Labilloy, H. Benisty, C. Weisbuch, T.F. Krauss, R.M. De La Rue, V. Bardinal, R. Houdré, U. Oesterle, D. Cassagne, and C. Jouanin. Quantitative measurement of transmission, reflection, and diffraction of two-dimensional photonic band gap structures at near-infrared wavelengths. Phys. Rev. Lett., 79:4147, 1997.
55. S. Glutsch, K. Hannewald, and F. Bechstedt. Green's function approach to photoluminescence in semiconductors. Phys. stat. sol. (b), 221:235, 2000.
56. K. Hannewald, S. Glutsch, and F. Bechstedt. Theory of photoluminescence in semiconductors. Phys. Rev. B, 62:4519, 2000.

57. G. Khitrova, D.V. Wick, J.D. Berger, C. Ell, J.P. Prineas, T.R. Nelson Jr., O. Lyngnes, H.M. Gibbs, M. Kira, F. Jahnke, S.W. Koch, W. Rühle, and S. Hallstein. Excitonic effects, luminescence, and lasing in semiconductor microcavities. Phys. stat. sol. (b), 206:3, 1998.
58. K. Hannewald, S. Glutsch, and F. Bechstedt. Quantum-kinetic theory of hot luminescence from pulse-excited semiconductors. Phys. Rev. Lett., 86:2451, 2001.
59. R.F. Schnabel, R. Zimmermann, D. Bimberg, H. Nickel, R. Lösch, and W. Schlapp. Influence of exciton localization on recombination line shapes: $In_xGa_{1-x}As/GaAs$ quantum wells as a model. Phys. Rev. B, 46:9873, 1992.

Part IV

Correlations and Dynamics Clusters and Nuclei

Analysis of Cluster Dynamics

Karsten Andrae, Mohamed Belkacem, Thi Phuong Mai Dinh, Eric Giglio, Ming Ma, Fabien Megi, Andreas Pohl, Paul-Gerhard Reinhard, and Eric Suraud

Abstract. Several different time scales compete in metal clusters, Mie plasmon oscillations, Landau damping, electronic collisions, and ionic motion. It is the aim of this contribution to disentangle these using a palette of different dynamical observables: optical absorption, photo-electron spectroscopy (in energy and angular distributions), and pump-and-probe analysis. Test cases are Na clusters. The theoretical description is based on time-dependent local-density approximation for the electrons coupled to simultaneous classical propagation of ionic dynamics. This provides a non-adiabatic treatment which is necessary in connection with dynamical processes induced by intense femtosecond lasers. The observables cover different dynamical regimes: Optical response and photo-electrons stay in the linear regime while pump-and-probe scenarios touch typically the semi-linear regime (i.e. non-linear but still non-destructive). Finally, we sketch laser-induced Coulomb explosion as an example from the non-linear and destructive regime.

1 Introduction

Cluster physics is an interesting field because it combines different aspects from atomic, molecular, and solid-state physics. Clusters are, on the one hand, finite systems seemingly "nothing but" large molecules. They can be viewed, on the other hand, as small pieces of bulk material because they are built, similar as solids, from arbitrary repetition of the same building block. For example, the onset of band-structure with increasing cluster size determines the electronic transport properties. But finite-size quantum-effects in the valence band lead to pronounced electronic shell effects in metal clusters, for reviews see e.g. [1–3]. The impact of different fields becomes even more apparent in cluster dynamics. Metal clusters, for example, display a pronounced collective excitation of the electron cloud, the Mie surface plasmon whose roots go back to the plasmon modes in bulk metals but whose actual appearance is closely linked to the finite cluster size, for an extensive discussion see e.g. [4]. It is obvious that this mix of aspects will lead also to a large variety of interesting scenarios in the realm of short-time dynamics and intense excitations. Correspondingly cluster dynamics in the non-linear regime is an active and growing field of research for which a comprehensive presentation can easily fill a book. This contribution concentrates on the special case of dynamics with metal clusters, particularly alkalines. The case is looked at from a theorists perspective employing various dynamical simulations. We aim at a guided tour through this sub-field picking a few instructive examples and we hope that the selected cases help to get an impression of the enormous potential of the field of cluster dynamics.

2 Time Scales in General

The basic electronic scales of alkaline clusters are summarized in Fig. 1. The right upper block shows the "natural units" of these clusters as they are given by: the Wigner–Seitz radius r_s as length scale, the Fermi energy ε_F as energy scale, and as time scale the interval in which an electron at the Fermi energy travels through one Wigner–Seitz radius (note that v_F is the Fermi velocity). There is, in fact, only one scale factor amongst these three quantities. Let us assume the length units, r_s, as given. Then the energy units scale as $\varepsilon_F \propto \hbar^2/(2mr_s^2)$ and the time units as $\tau_F = r_s/v_F \propto r_s^2 m/\hbar$. Expressing the relevant quantities in such natural units allows to compare all alkaline clusters on the same footing. This is done in the left part of Fig. 1 for

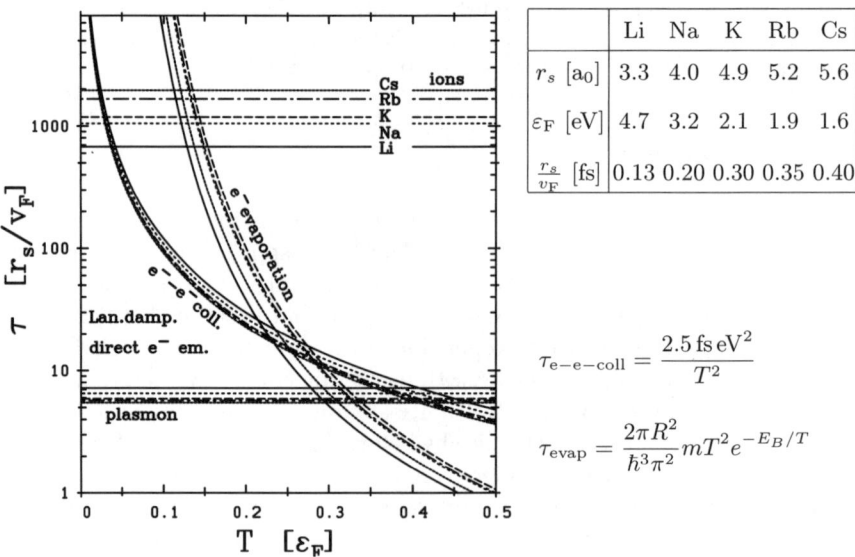

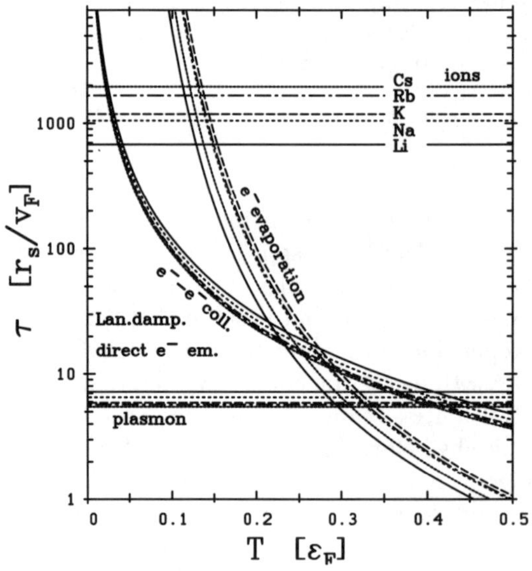

	Li	Na	K	Rb	Cs
r_s [a$_0$]	3.3	4.0	4.9	5.2	5.6
ε_F [eV]	4.7	3.2	2.1	1.9	1.6
$\frac{r_s}{v_F}$ [fs]	0.13	0.20	0.30	0.35	0.40

$$\tau_{\mathrm{e-e-coll}} = \frac{2.5\,\mathrm{fs\,eV}^2}{T^2}$$

$$\tau_{\mathrm{evap}} = \frac{2\pi R^2}{\hbar^3 \pi^2} m T^2 e^{-E_B/T}$$

Fig. 1. *Left part*: Time scales for alkaline clusters drawn versus internal excitation in terms of temperature T. Times and energies are expressed in natural units of each material according to the table on the right. Not drawn are the relaxation times for Landau damping and for direct electron emission. They reside at 20–40 r_s/v_F and are almost independent of T. *Right part*: The table on top shows gross properties of alkaline systems: Wigner–Seitz radius r_s, Fermi energy ϵ_F and microscopic time scale r_s/ϵ_F. The trends for collisional relaxation time (extracted from VUU calculations [5]) and thermal evaporation (according to the Weisskopf estimate [6]) are shown below

the time scales of the many different dynamical processes in a cluster. The dominant dynamical process is the collective oscillation of the electron cloud against the positive background, called the Mie plasmon. These oscillations quickly couple to single electron excitations which leads to Landau damping (damping of the collective mode due to its coupling to the mean field) and direct electron emission. The time scale for these two (related) processes is indicated by the position of the key words in the figure to avoid too many line crossing. It is typically a factor 3–5 longer than the Mie plasmon time. The Landau damping drives the single electrons in the cloud out of phase relative to each other. This gives rise to electron–electron collisions which add further damping and associated internal heating. The thermal energy of the electron cloud is transferred to ionic thermalization or released much later in terms of electron evaporation. The excited electron cloud also shakes the ions which react, of course, somewhat slower due to their comparatively huge mass. A typical time scale is here set by the cycle of ionic vibrations, keeping in mind that first effects of ionic motion can already be seen in the first quarter cycle. Subsequent ionic processes like fragmentation or monomer evaporation usually take much longer (several ionic cycles) but become faster with increasing violence of the excitation (as e.g. in a Coulomb explosion). Figure 1 demonstrates that all electronic time scales for the different alkaline materials are gathering close together when expressed in the natural units of the electron cloud. The ionic time scales, on the other hand, are dominated by the independent parameter of ionic mass and thus show a larger spread. Still, ionic time scales lie in a time range well separated from basic electronic scales. This is plausible in view of the large mass difference between ions and electrons. Figure 1 also shows the trends with internal excitation ($\equiv$ temperature T). The estimates for two most strongly varying processes, electron evaporation and electron–electron collisions, are printed at the right lower side of the figure. The different dynamical processes obviously depend in a different manner on the degree of internal excitation. This means that the balance between different reaction channels changes dramatically with the violence of the process. For example, electron–electron collisions are of minor importance in the linear regime of small excitations but compete with single electron times for temperatures around 2000 K.

The internal time scales of a cluster are to be complemented by the time scale of the excitation processes (not shown in the figure). Nanosecond lasers are beyond any time shown in Fig. 1. That is a regime where the frequency plays the dominant role. Collisions with highly charged and fast ions are below any time scale shown (typically in the sub-fs range). That covers the opposite regime where frequencies are unimportant and only forces count. Femtosecond lasers have pulse widths which interfere with the various time scales of metal clusters. It is obvious that this gives rise to a huge variety of accessible dynamical processes which can thus be triggered by laser experiments.

3 Computational Framework

The description is based on an independent single-particle picture for the electrons. They are represented by a set of single electron wavefunctions $\{\varphi_\alpha(\mathbf{r},t)\}$. The ions are treated classically in terms of conjugate coordinates $(\mathbf{R}_I,\mathbf{P}_I)$. The classical approximation is justified by their much larger mass. Density-functional theory at the level of the local-density approximation (LDA) is used for a most efficient and reliable handling of the electrons, see e.g. [7,8]. The starting point for LDA and time-dependent LDA (TDLDA) is an energy-density functional for the electrons. We aim at a coupled description together with ions. We thus write the total energy for electrons and ions

$$E_{\rm total} = E_{\rm kin}(\{\varphi_\alpha\}) + E_{\rm C}(\rho) + E_{\rm xc}^{\rm (LDA)}(\rho_\uparrow,\rho_\downarrow) + E_{\rm el,ion}(\rho,\{\mathbf{R}_I\}) + E_{\rm ion}(\{\mathbf{R}_I,\mathbf{P}_I\}) + E_{\rm ext}(\rho,t)\,, \tag{1a}$$

$$E_{\rm kin} = \sum_{\alpha=1}^{N_{\rm el}} \int d^3r \varphi_\alpha^+ \frac{\hat{p}^2}{2m}\varphi_\alpha\,, \tag{1b}$$

$$E_{\rm C} = \frac{e^2}{2}\int d^3r d^3r' \frac{\rho(\mathbf{r})\rho(\mathbf{r}')}{|\mathbf{r}'-\mathbf{r}|}\,, \tag{1c}$$

$$E_{\rm xc}^{\rm (LDA)} = \int d^3r \rho(\mathbf{r})\epsilon_{\rm xc}\left(\rho_\uparrow\mathbf{r},\rho_\downarrow\mathbf{r}\right)\,, \tag{1d}$$

$$E_{\rm el,ion} = \int d^3\,\rho(\mathbf{r})V_{\rm back}(\mathbf{r})\,, \tag{1e}$$

$$\rho(\mathbf{r}) = \rho_\uparrow(\mathbf{r}) + \rho_\downarrow(\mathbf{r})\,,\quad \rho_\sigma(\mathbf{r}) = \sum_\alpha \varphi_\alpha^+(\mathbf{r})\hat{\Pi}_\sigma\varphi_\alpha(\mathbf{r})\,,\quad \sigma\in\{\uparrow,\downarrow\}\,, \tag{1f}$$

$$E_{\rm ion} = \sum_I \frac{\mathbf{P}_I^2}{2M_I} + \sum_{I<J}\frac{e^2 Z_I Z_J}{|\mathbf{R}_I-\mathbf{R}_J|}\,, \tag{1g}$$

$$E_{\rm ext} = \int d^3\,\rho(\mathbf{r},t)\mathbf{E}_0\cdot e\mathbf{r}\exp\left(\imath\omega_{\rm las}t\right)f_{\rm pulse}(t)\,, \tag{1h}$$

where $\hat{\Pi}_\sigma$ is the projector on spin σ. The exchange-correlation energy-density $\epsilon_{\rm xc}$ is taken from one of the elsewhere provided functionals; we use actually [9]. Note that the matrix elements (in the kinetic energy) employ the hermitian conjugate rather than the complex conjugate because the wavefunction φ_α carries a Pauli spinor. The $V_{\rm back}$ in (1e) is the potential exerted from the ionic background. For most alkalines, it can be well approximated by a local pseudo-potential. A particularly soft profile (advantageous for numerics) is provided by $V_{\rm PsP}(\mathbf{r}) = \sum_{i=1}^2 c_i{\rm erf}(\sigma_i\mathbf{r})/|(\mathbf{r})|$. The appropriate parameter for Na are: $c_1 = 2.292$, $c_2 = -3.292$, $\sigma_1 = 0.681$, and $\sigma_2 = 1.163$ [10]. Attractive and repulsive parts are tuned such that the ground state reproduces approximately the energy of the $3s$ state. Note that the transition to the first excited states, the $3p$ state, is also well reproduced. This feature is crucial to guarantee transferability, i.e. the applicability of the potential also for molecules and clusters up to bulk matter. The performance of this pseudo-potential for electron dynamics was extensively tested in [10] and found to be very satisfying. The external excitation mechanism is parametrized in $E_{\rm ext}$. The form (1h) stands for a laser pulse with frequency $\omega_{\rm las}$, the field strength $\mathbf{E}_0$ determines intensity and polarization, and $f_{\rm pulse}$ describes the pulse profile. Note that the laser is parametrized as a classical, time-dependent source. That approximation complies with the mean-field description of the electrons and it is particularly suited for intense laser fields where the field amplitude is much larger than its uncertainty.

The static Kohn–Sham (KS) equations are derived by variation of the given energy (1a) with respect to the single-electron wavefunctions $\delta\varphi_\alpha^+$ and the time-dependent equations from variation of the action $\int dt\,[\langle \imath\partial_t\rangle - E_{\rm total}]$. Both versions contain formally the same Kohn–Sham potential

$$U_{{\rm KS},\sigma}(\mathbf{r}) = \frac{\delta\left(E_{\rm C} + E_{\rm xc}^{\rm (LDA)} + E_{\rm el,ion}\right)}{\delta\rho_\sigma(\mathbf{r})}\,, \tag{2}$$

where again $\sigma \in \{\uparrow,\downarrow\}$. The potential emerges by functional variation with respect to the local density (note that this simple form applies for local pseudo-potentials in $E_{\rm el,ion}$).

The final time-dependent Kohn–Sham equation is a single particle Schrödinger equation with the density dependent potential $U_{{\rm KS},\sigma}$. It propagates the electronic single particle wavefunctions $\varphi_\alpha(\mathbf{r},t)$. This is the TDLDA. Simultaneously, the ions are propagated by classical molecular dynamics (MD). Their equations of motion are derived by applying Hamilton's variation principle to the functional (1a). The coupled scheme is then the so called TDLDA-MD. The regime of energetic excitations also allows semi-classical approximations for the electronic dynamics, as the Vlasov-LDA [11,12]. This has the advantage that collisional correlations can be included through an Boltzmann–(Vlasov)–Ühling–Uhlenbeck collision term (BUU, VUU) [5] which becomes increasingly important just at high excitations, see Fig. 1. For details of the derivation of TDLDA-MD and its semi-classical variants, the emerging equations, and the numerical handling see the review article [13].

4 Attenuation of Laser Pulses

The basic laser parameter is its frequency $\omega_{\rm las}$. The basic dynamical mode of metal clusters is the Mie surface plasmon which is distinguished by one dominant peak at a frequency $\omega_{\rm Mie}$. The coupling between these two frequencies is most crucial for all further dynamical features. The attenuation of the laser field by the cluster is described by the optical absorption strength which is the most widely studied dynamical property, see e.g. [2,4,14–17]. The absorption strength is directly related to the proliferation of reaction products.

This is emphasized in Fig. 2 which shows the average number of emitted electrons (= ionization) as a function of frequency $\omega_{\rm las}$ (while all other laser

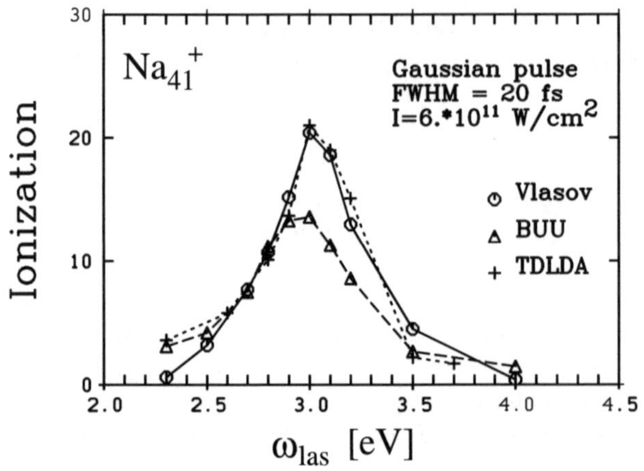

Fig. 2. The total ionization as a function of laser frequency for fixed laser intensity $I = 6 \times 10^{11}\,\mathrm{W/cm^2}$ and a Gaussian pulse profile with FWHM = 20 fs. We compare TDLDA-MD with Vlasov-LDA-MD and VUU-MD. The test case is the cluster Na_{41}^+ for which the Mie plasmon lies around 3 eV. From [18]

parameters were kept fixed). One sees clearly the Mie resonance peak. Mind the small frequency window of the plot. There is very little strength outside that window. In fact, the peak is already much broader than typically seen in the linear regime of small excitations. It was shown in [19] that the width of the peak grows with excitation. We have chosen here a strong excitation as can be seen from the dramatic values of ionization. The figure thus shows the enormous stability of the Mie plasmon even for strong excitations.

The strong excitation does also allow semi-classical approximations. Results from those are also shown in Fig. 2. The Vlasov-LDA agrees very well with its quantum mechanical ancestor TDLDA. This is one hint at the validity of semiclassical approximations in that regime. The VUU adds collisional relaxation. It is no surprise then that the peak becomes somewhat broader. This demonstrates the growing importance of collision effects with growing excitation. But one sees also that the collisional width is only one part of the total width. Although the excitation is high, we are still in a regime where mean-field effects dominate.

5 Effects of Laser Pulse Duration

After frequency, the next two important laser parameters are intensity I and pulse duration τ. The intensity regulates the strength of excitation in an obvious manner. The impact of the pulse length is more subtle. It plays a role only for short pulses in the range from fs to ps whose time scale interferes with typical cluster processes (see Sect. 2). Recent experiments with laser-induced Coulomb break-up of Pt clusters have shown marked dependence of the final ionization on the pulse length [20]. The laser frequency was below the Mie frequency. Intensities remained still in a regime where details of cluster dynamics are resolved ($10^{15}...10^{16}$ W/cm^2). The pulse length was varied while the fluence of the laser, $I * \tau$, was kept constant to arrange equal energy supply in each case. And yet, one finds a clear maximum of photo-ionization at a pulse length of about 600 fs [20]. Two much different mechanisms have been discussed as explanation: resonant enhancement due to a red-shifted Mie plasmon [21–23] and force-assisted enhanced tunneling rates [24]. It is not yet fully settled which mechanism applies. The high laser intensity hints that the second mechanism may be favorable. However, we present in the following the first mechanism of resonant enhanced ionization because it is interesting in itself and it is related to the subsequent discussion of a pump-and-probe scenario.

The basic idea of the resonant enhancement is that a small initial ionization leads to a slow radial expansion of the whole cluster due to Coulomb pressure. The growth of the radius R is known to yield a red-shift $\omega_{\mathrm{Mie}} \propto R^{-3/2}$ of the Mie plasmon [16]. This moves the plasmon into resonance with the laser which, in turn, leads to resonant response and subsequent strong ionization. The left panel of Fig. 3 illustrates the mechanism. The uppermost panel shows the ionization as a function of time. The initial ionization increases with intensity but maintains a moderate slope in any case. The initial ionization triggers a more or less slow Coulomb expansion depending on the intensity, see the lowest panel. The middle panel shows the mean Mie plasmon frequency estimated according to the actual ionic radius (lower panel) and charge state (upper panel). The fast initial ionization induces a blue-shift. The subsequent ionic expansion drives a red-shift $\propto R^{-3/2}$. Sooner or later, the actual Mie plasmon frequency crosses the laser frequency (horizontal line in the middle panel). The thus established resonant conditions blow up ionization as seen from the associated steep slopes in the uppermost panel. (Note that the time of steepest slopes does not perfectly match with the time of crossing frequencies; the estimate for the moving Mie plasmon was not refined enough.) Thus far the figure exemplifies the resonant enhancement mechanism. The pulse length was kept constant to keep things simple. Now imagine constant fluence, i.e. a pulse length inversely proportional to the intensity. The lowest intensity has a long sustained pulse. But the forces are very small such that initial ionization keeps low and resonant conditions are never reached. The highest intensity would then be associated with a pulse width of only $T = 100$ fs. This produces large initial ionization. But the time is too short to "see" the resonant conditions. Thus the final ionization is low in both limits and has a maximum inbetween, in qualitative agreement with the experimental findings.

A summary analysis is shown on the right panel of Fig. 3 for a variety of laser fluences. One sees a strong dependence of pulse length with a broad maximum at intermediate times. The effect is qualitatively the same for most

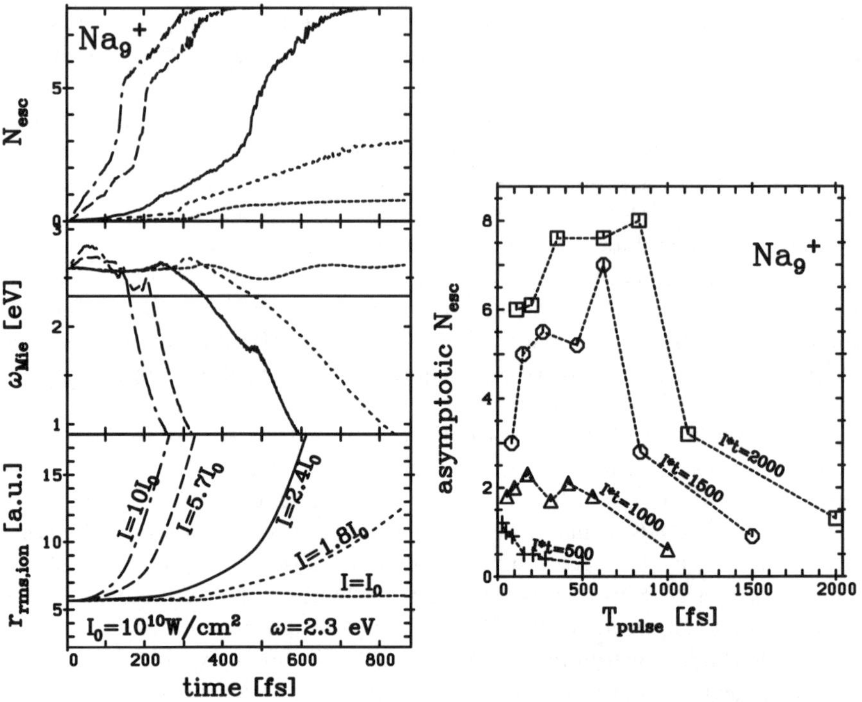

Fig. 3. *Left part*: Time evolution of key observables for $Na_9{}^+$ irradiated by laser pulses with varying intensity as indicated. All pulses have frequency $\omega = 2.3\,\mathrm{eV}$ and a pulse length of 1000 fs. *Left lowest panel*: Ionic r.m.s radius. *Left middle panel*: Frequency of the estimated average Mie plasmon peak; the *horizontal straight line* indicates the laser frequency. *Left uppermost panel*: Number of emitted electrons N_{esc}. *Right panel:* Summary of final ionization (= number of escaped electrons N_{esc}) versus pulse length for different laser fluences. Frequency was again the same $\omega = 2.3\,\mathrm{eV}$ for all cases. From [23]

of the fluences shown in the figure. Very low fluence, of course, never reaches sufficient ionization to trigger sufficient Coulomb pressure. Too large fluence, on the other hand, leads immediately to a nano-plasma which ranges in another regime of violent excitations where the actual cluster material becomes unimportant, see Sect. 8.

6 Pump and Probe Analysis of Ionic Collective Modes

The previous section has shown that already a simple variation of pulse length allows one to explore the cluster dynamics. More is possible with the much more flexible pump-and-probe analysis. This is meanwhile a heavily used standard tool in molecular physics and chemistry [25]. Very small clusters allow scenarios much like in simple molecules which can still be characterized in terms of molecular dynamics along Born–Oppenheimer surfaces, see e.g. the experiments on trimers [26,27] and associated theory [28]. Larger clusters hardly allow a detailed mapping of the multi-dimensional Born–Oppenheimer surfaces. Nonetheless, many different pump-and-probe scenarios are possible and have been realized experimentally. Most of them deal with deposited or embedded clusters for reasons of counting rates, see e.g. [29–31], although a few measurements with larger free clusters do also exist, e.g. [32] and measurements with free Na and Pt clusters are underway.

We exemplify here one possible scenario taking up the thoughts of the previous Sect. 5: the exploration of ionic breathing modes induced by Coulomb pressure. Figure 4 sketches the basic idea. The pump pulse is tuned to induce a moderate ionization which initiates cluster vibrations while leaving the cluster stable as a whole (leftmost parts of the figure). We assume an initially spherical cluster such that the Coulomb pressure excites preferably radial oscillations. The Mie frequency scales with the radius as $\omega_{\mathrm{Mie}} \propto R^{-3/2}$. Large radii are associated with low frequencies and vices versa. This is indicated in the figure. We now choose the laser frequency for the probe pulse safely below any Mie frequency occurring in the evolution. This is indicated in the figure. The response to a probe pulse then maps unambiguously the radius oscillations as indicated in the figure.

A quantitative example for such a scenario is shown in Fig. 5. The test case $\mathrm{Na}_{41}{}^{+}$ was chosen because it comes close to a spherical shape. The pump pulse ionizes the cluster into charge state $\mathrm{Na}_{41}{}^{++++}$ safely below the threshold for Coulomb explosion but large enough to produce sizeable Cou-

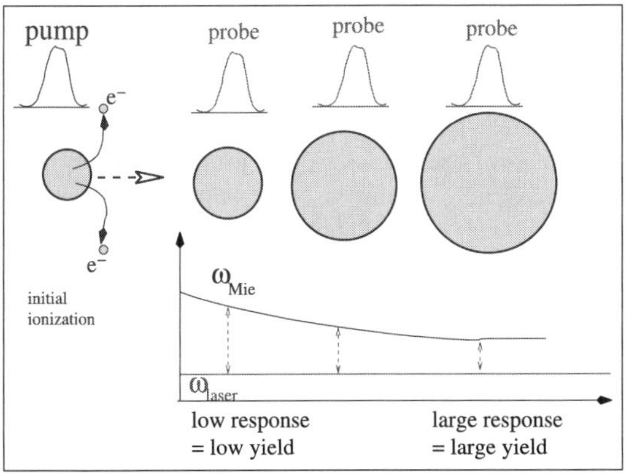

Fig. 4. Pump-and-probe scenario for measuring ionic breathing modes of a spherical cluster. The *upper line* sketches the envelopes of the laser pulses. Three different probe pulses are shown at different delay times. The *middle line* indicates the time evolution of the cluster radius after probe pulse. The *lowest line* shows the time-dependent Mie frequency in comparison to the laser frequency (which is chosen to stay always safely below)

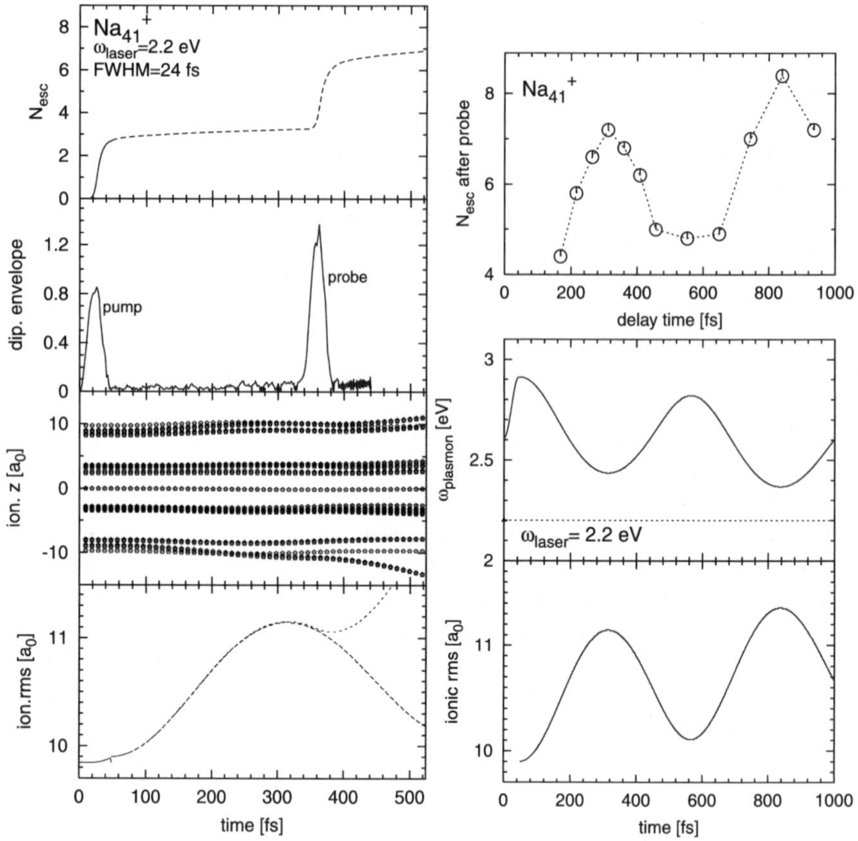

Fig. 5. Illustration of pump and probe spectroscopy of ionic breathing vibrations of ${Na_{41}}^+$ simulated with TDLDA-MD. The laser photon frequency is $\omega = 2.2\,\mathrm{eV}$, intensity $1.1 \times 10^{12}\,\mathrm{W/cm^2}$, and profile is a $\sin^2$ pulse with FWHM = 24 fs. *Left parts:* Time evolution of key observables for one pump and probe scenario for ${Na_{41}}^+$. The probe laser was active here around $t_{\mathrm{delay}} = 360\,\mathrm{fs}$. *Left lower panel:* ionic r.m.s. radius, $\sqrt{\sum_I \mathbf{R}_I^2}$ (*long dashed line:* without probe; *short dashed line:* after probe). *Left second panel* from below: detailed ionic z coordinates. *Left second panel* from above: envelope of the dipole oscillations $D(t)$. *Left upper panel:* number of emitted electrons (net ionization is $N_{\mathrm{esc}} + 1$). *Right parts*: Summary of final results from of a series of pump-and-probe shots with varied delay time. *Right upper panel:* Ionization after a probe pulse with varying delay time for the test case ${Na_{41}}^+$. The pump laser was tuned such that initially $N_{\mathrm{esc}} = 3$. The peak time of the probe laser is indicated by the *open circles. Right middle panel:* Time evolution of mean plasmon frequency after the initial pulse. The laser frequency is drawn (*dotted line*) for comparison. *Right lower panel:* The time evolution of the ionic r.m.s. radius after the initial pulse without probe. From [33]

lomb pressure for a well developed ionic oscillation. These oscillations of the cluster radius after the pump pulse are shown in the lowest panel of Fig. 5. Quadrupole components and other multipoles (not shown in the plot) were found to remain small. The Mie plasmon resonance depends on charge state and cluster radius. An estimate is shown in the middle panel. One sees the fast initial blue-shift which is due to the fast ionization accompanying the laser pulse (duration 50 fs). After that, one finds oscillations which perfectly map the radius oscillations with $\omega(t) \propto R(t)^{-3/2}$. The laser frequency for the probe pulse is indicated by an horizontal dashed line. It was safely chosen below the Mie resonance such that the actual Mie frequency always stays away from resonance with the laser. The electronic response to a probe pulse is small if the Mie frequency is far from the laser and larger if it comes close. The subsequent further ionization is proportional to the response. The net ionization (evaluated 200 fs after the probe pulse, and thus close to the asymptotic value) is shown in the uppermost panel. It behaves precisely as expected: large radii bring the Mie frequency closer to the laser and thus yield more ionization. This maps perfectly the radius oscillations. Even a detail as the slight drift to increasing average radii is well reproduced. This example shows that properly chosen laser conditions allow a direct mapping of global oscillations for metal clusters. Such scenarios are yet awaiting experimental tests.

The same setup applied to deformed clusters and exploiting different polarizations for the probe pulse allows to disentangle also possible ellipsoidal (quadrupole) oscillations of a metal cluster [34]. An extreme case of quadrupole motion is cluster fission. It is known that the optical absorption strength varies strongly along the fission path [35]. This invites a pump-and-probe scenario. An example for a laser-induced fission is shown in Fig. 6. Electronic shell effects drive the prolate ground state of Na_{14} to a large ellipsoidal deformation. The ionic structure (upper left insert) looks already like two preformed fragments. This makes it possible to find laser conditions under which a fission in two large fragments becomes possible (note that usually the dominating channel is extremely asymmetric fission into a trimer and a large fragment [36]). These are chosen here as demonstrated by the insert with the ionic configuration during the event. The lower part of the figure shows the dipole spectra along fission direction for the ${Na_{14}}^{+++}$ right after pump pulse and for both fragments. The spectrum of the ${Na_7}^{+}$ fragment coincides by chance with the initial peak. The high remaining electronic and ionic excitation stage leads to that red shift. The doubly charged second fragment has, naturally, a higher frequency. It develops monotonously from the initial peak of ${Na_{14}}^{+++}$. It is obvious that properly chosen probe pulses can map the fission path through the time evolution of the optical absorption strength. The setup has to be chosen here somewhat differently. The probe laser should have a frequency above the leading Mie frequency of the ${Na_{14}}^{+++}$. However, one cannot take the frequency too high because

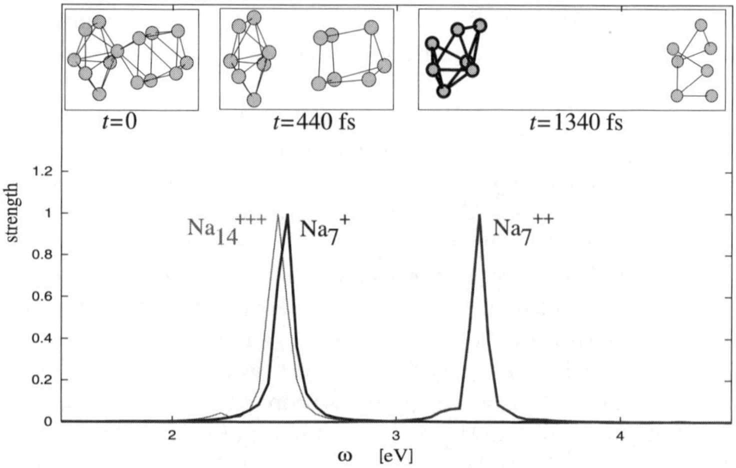

Fig. 6. Strength distributions (normalized to the maximum values) of dipole modes along z-axis for Na_{14} before and after laser-induced fission. The laser had the parameters: frequency = 2.3 eV, intensity 5×10^{10} W/cm^2, and pulse width 100 fs. It leads to an average ionization into a 3+ state. Three stages of fission are shown in the *upper inserts*: the initial stage to the *left*, the final stage to the *right*, and an intermediate stage in the *middle*. The initial Na_{14}^{+++} shows one Mie plasmon peak as indicated. The final state consist out of two different clusters, a Na_7^+ and a $Na_7{}^{++}$. It displays two separate Mie plasmon peaks which can be associated with each fragment as indicated

this may couple to other states which are more plentiful available for higher frequencies. The best would be to scan the probe frequency and to map the time of maximum response. Such a setup is, of course, much more elaborate than the simple breathing scenario shown above.

7 Angular Distributions

Higher excitation enhance the impact of electron–electron collisions (see Fig. 1). The collisional relaxation time can be measured in appropriate pump-and-probe scenarios [31]. We sketch here an alternative proposal for a time resolved measurement which could give access to collisional relaxation. The key idea is that the angular distribution of electrons emitted directly through the laser fields shows strong forward-backward emission while electrons released after thermalization radiate isotropically. Let us assume that one could measure angular distributions in sufficiently short time slices. For then one may deduce information on the collisional relaxation time.

The scenario is demonstrated in Fig. 7 which shows the time evolution of total ionization, dipole response, and sphericity for a laser excitation with a long pulse and frequency below resonance. The large intensity produces

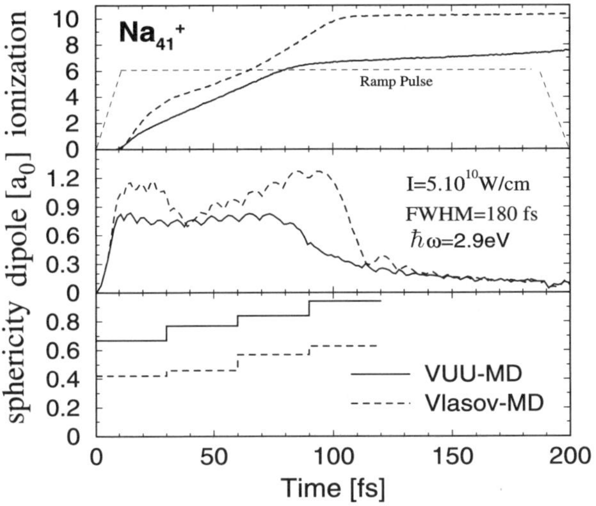

Fig. 7. Time evolution of total emission (*upper panel*), dipole amplitude (*middle panel*), and sphericity of the momentum distribution of the emitted electrons (*lower panel*) for excitation of ${Na_{41}}^+$ with a laser at frequency $\omega = 2.9\,\mathrm{eV}$ and intensity of $I = 5 \times 10^{10}\,\mathrm{W/cm^2}$. The pulse profile was a ramp of duration 200 fs and switching time of 20 fs. It is indicated by the *faint dashed line* in the *uppermost panel.* From [37]

high excitations and allows for a semi-classical treatment. Results from semi-classical Vlasov-LDA (*dashed*) calculations are compared with VUU (*solid line*) ones. The ionization (*upper panel*) for the VUU case is initially lower and shows substantial delayed emission (not fully shown in the plot). Electron-electron collisions deflect electrons from the direct emission path and feed them into internal thermalization. The thus larger internal excitation yields more thermal electrons. The dipole response (middle panel) corroborates that picture. VUU cuts somewhat the large response peaks which shows again that collisional damping is active. The lowest panel shows the "sphericity" of the angular distributions for the emitted electrons. To that end, an ellipsoid has been fitted to the velocity distribution of the emitted electrons. The ratio of the largest principle axis to the shortest one is defined to be a measure of sphericity. A value one stands for perfectly spherical emission and zero for perfectly directed emission. The sphericity (lowest panel) is, quite expectedly, generally larger for VUU. Moreover there is a clear increase of sphericity with time. It complies with an estimated dissipation time of about 25 fs. It is interesting to note that sphericity does also increase for mere Vlasov-LDA, although at a lower level. This shows that there is also some contribution from one-body dissipation (Landau damping). It requires complementary measurements to disentangle these two contributions.

8 Violent Processes

The situation changes dramatically for laser intensities above typically 10^{16} W/cm^2. The field strength is now larger than the binding forces for the valence electrons which leads to a large immediate ionization over the barrier. Moreover, the strong laser forces release many formerly well bound core electrons. Such pulses generate in a very short initial phase a hot and dense plasma with finite extension, a nano-plasma. This state is the same for all initial materials, may it have been rare gases, covalent atoms or metals. And indeed such experiments have been done with much different clusters with very similar results at the end (e.g. H [38], Kr [39], or Pb [40]). The final plasma state has some similarity to metal cluster dynamics, e.g., it is again mainly dominated by the plasma mode. The enormous ionization drives a violent Coulomb explosion emitting very energetic particles. Very large clusters are involved in those experiments to deliver a huge Coulomb pressure. A theoretical description of these processes thus requires macroscopic models as, e.g., the nano-plasma model [41] which combines rate equations for the various conversion processes (core ionization etc.) with hydrodynamical flow and classical electro-dynamics.

The result of a recent nano-plasma simulation [42] is presented in Fig. 8. It shows the temporal evolution of an exploding cluster of 5000 Xe atoms irradiated by a 100 fs, 780 nm laser pulse with a peak intensity of 10^{16} W/cm^2. Time $t = 0$ fs is the time at which the laser reaches the cluster with maximum intensity. We have represented in the upper part of the figure the time variation of the electron density inside the cluster normalized to the critical density n_c and in the lower part the electric field inside the cluster (*solid line*) together with the external one (*dashed line*). The critical density n_c is defined as that electron density at which the Mie plasma frequency comes into resonance with the laser. At the rising edge of the pulse, quasi-free electrons are produced inside the cluster from neutral atoms by tunnel ionization. The rapid increase of the number of quasi free electrons leads the system through a first modest resonance at $t \simeq -135$ fs when the density matches three times the critical density n_c. The electric field is then shielded due to the high electron densities reached. The tunnel ionization rate falls off, but electrons are still created through thermal collisions. From times $t \simeq -100$ fs onwards, the electron temperatures are high enough for some electrons to leave the cluster leading to a fall of the density inside the cluster until time $t \simeq 10$ fs when the second, more important resonance takes place. The inner field is then amplified to 1.5 times the external field, leading to a large heating of the cluster with electronic temperature reaching 1.2 keV. The rate of electron emission increases consequently giving a highly positively charged cluster which will explode due to the huge Coulomb repulsion between ions. After all, it is interesting to note that the time-evolution of resonant conditions plays again a dominating role similar to the previous examples at modest laser intensities but now in a totally different phase of matter.

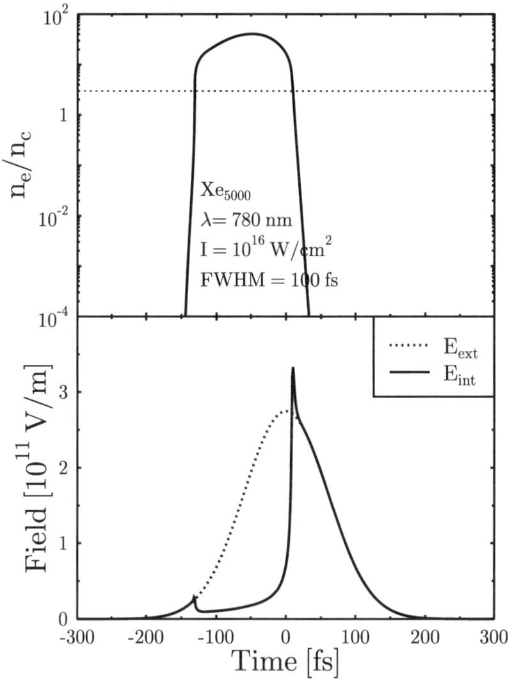

Fig. 8. Time evolution of the Coulomb explosion of a Xe cluster with 5000 atoms irradiated by a laser pulse with frequency of 1.6 eV (≡ 780 nm), pulse width of 100-fs and peak intensity 10^{16} W/cm^2. *Upper part*: electron density normalized to critical density n_c; *lower part*: amplitudes of internal (*solid line*) and external (*dashed line*) electric fields

9 Conclusions

We have discussed modern developments in cluster dynamics from a theoretical perspective with emphasis on metal clusters. The interplay of the various electronic and ionic time scales of the cluster with the close by time scales of modern laser setups gives rise to many interesting scenarios. The electronic response of metal clusters is dominated by the Mie surface plasmon which thus serves as the doorway to all mechanisms discussed here. The variation of laser pulse length, or more specifically of delay time in pump-and-probe setups, can give access to the time evolution of the basic ionic modes, breathing oscillations, quadrupole deformation, and optionally fission. Time resolved measurement of angular distributions would give complementing information on thermal relaxation times (although an experimental realization is uncertain). Finally, it is interesting to note that the very violent cluster excitations have in common the dominance of the Mie plasmon independent of the initial material.

Acknowledgments

This work has been supported by the French-German exchange program PROCOPE, contract number 99074, and by Institut Universitaire de France.

References

1. Walt A. de Heer. The physics of simple metal clusters: experimental aspects and simple models. *Rev. Mod. Phys.*, 65:611, 1993.
2. M. Brack. The physics of simple metal clusters: selfconsistent jellium and semiclassical approaches. *Rev. Mod. Phys.*, 65:677, 1993.
3. S. Bjornholm and J. Borggreen. Electronic shell structure in clusters as reflected in mass abundance spectra. *Phil. Mag.*, 79:1321, 1999.
4. U. Kreibig and M. Vollmer. *Optical properties of metal clusters*, volume 25. Springer Series in Materials Science, 1993.
5. A. Domps, P.-G. Reinhard, and E. Suraud. Two body collisions and relaxation in metal clusters. *Phys. Rev. Lett.*, 81:5524, 1998.
6. V. Weisskopf. *Phys. Rev.*, 52:295, 1937.
7. R.O. Jones and O. Gunnarsson. The density functional formalism, its applications and prospects. *Rev. Mod. Phys.*, 61:689, 1989.
8. R.M. Dreizler and E.K.U. Gross. *Density Functional Theory: An Approach to the Quantum Many-Body Problem.* Springer-Verlag, Berlin, 1990.
9. J.P. Perdew and Y. Wang. *Phys. Rev. B*, 45:13244, 1992.
10. S. Kümmel, M. Brack, and P.-G. Reinhard. Ionic geometries and electronic excitations of na_9^+ and na_{55}^+. *Euro. Phys. J. D*, 9:149, 1999.
11. P. L'Eplattenier, P.-G. Reinhard, and E. Suraud. *J. Phys. A*, 28:787, 1995.
12. M. Gross and C. Guet. *Z. f. Physik D*, 33:289, 1995.
13. F. Calvayrac, P.-G. Reinhard, E. Suraud, and C.A. Ullrich. Nonlinear electron dynamics in metal clusters. *Phys. Rep.*, 337:493, 2000.
14. W. Ekardt. Dynamical polarizability of small metal particles: Self-consistent spherical jellium background model. *Phys. Rev. Lett.*, 52:1925, 1984.
15. V. Bonacic-Koutecky, J. Pittner, C. Fuchs, P. Fantucci, M.F. Guest, and J. Koutecky. Ab initio predictions of structural and optical response properties of $Na_n{}^+$ clusters: Interpretation of depletion spectra at low temperature. *J. Chem. Phys.*, 104:1427, 1996.
16. P.-G. Reinhard, O. Genzken, and M. Brack. *Ann. Phys. (Leipzig)*, 5:1, 1996.
17. H. Haberland. *Optical and Thermal Properties of Sodium Clusters.* Wiley, New York, 1999.
18. E. Giglio, P.-G. Reinhard, and E. Suraud. "impact of two-body collisions on explosion dynamics of irradiated clusters". *J. Phys. B*, 34:1253, 2001.
19. C.A. Ullrich, P.-G. Reinhard, and E. Suraud. Metallic clusters in strong femtosecond laser pulses. *J. Phys. B*, 30:5043, 1997.
20. S. Teuber, T. Döppner, T. Fennel, J. Tiggesbäumker, and K.H. Meiwes-Broer. Ionic recoil energies in the coulomb explosion of metal clusters. *Euro. Phys. J. D*, 16:59, 2001.
21. P.-G. Reinhard, F. Calvayrac, C. Kohl, S. Kümmel, E. Suraud, C.A. Ullrich, and M. Brack. Frequencies, times, and forces in the dynamics of na clusters. *Euro. Phys. J. D*, 9:111, 1999.

22. E. Suraud and P.-G. Reinhard. Impact of ionic motion on ionization of metal clusters under intense laser pulses. *Phys. Rev. Lett.*, 85:2296, 2000.
23. P.-G. Reinhard and E. Suraud. Dynamics of na clusters in picosecond laser pulses. *Appl. Phys. B*, 73:401, 2001.
24. U. Saalmann and M. Rost. Ionization of clusters in strong X-ray laser pulses. *Phys. Rev. Lett.*, 89:133401, 2002.
25. A. H. Zewail. *Femtochemistry, Vol. I & II.* World Scientific, Singapore, 1994.
26. T. Leisner, S. Vajda, S. Wolf, and L. Wöste. The relaxation from linear to triangular ag3 probed by femtosecond resonant two-photon ionization. *J. Chem. Phys.*, 111:1017, 1999.
27. R. Heinicke and J. Grotemeyer. *Appl. Phys. B*, 71:419, 2000.
28. M. Hartmann, J. Pittner, V. Bonacic-Koutecky, A. Heidenreich, and J. Jortner. Theoretical exploration of femtosecond multi-state dynamics of metal clusters. *J. Chem. Phys.*, 108:3096, 1998.
29. J.-H. Klein-Wiele, P. Simon, and H.-G. Rubahn. Size-dependent plasmon lifetimes and electron-phonon coupling time constants for surface bound na clusters. *Phys. Rev. Lett.*, 80:45, 1997.
30. G. Seifert, M. Kaempfe, K.-J. Berg, and H. Graener. Femotsecond pump-probe investigation of ultrafast silver nanoparticle deformation in a glass matrix. *Appl. Phys. B*, 71:795, 2000.
31. C. Voisin, D. Christofilos, N. Del Fatti, F. Vallée, B. Prövel, E. Cottancin, J. Lermé, M. Pellarin, and M. Broyer. Size-dependent electron–electron interactions in metal nanoparticles. *Phys. Rev. Lett.*, 85:2200, 2000.
32. B. Bescos, B. Lang, J. Weiner, V. Weiss, E. Wiedemann, and G. Gerber. Real-time observation of ultrafast ionization and fragmentation of mercury clusters. *Euro. Phys. J. D*, 9:399, 1999.
33. K. Andrae, P.-G. Reinhard, and E. Suraud. Theoretical exploration of pump and probe in medium size na clusters. *J. Phys. B*, 35:1, 2002.
34. K. Andrae, P.-G. Reinhard, and E. Suraud. Pump and probe analysis of shape dynamics in metal clusters. Preprint, submitted to Phys. Rev. Lett., 2003.
35. P.-G. Reinhard, F. Calvayrac, and E. Suraud. Plasmons in fissioning metal clusters. *Z. f. Physik D*, 41:151, 1997.
36. U. Näher, S. Björnholm, S. Frauendorf, F. Garcias, and C. Guet. *Phys. Rep.*, 285:245, 1997.
37. E. Giglio, P.-G. Reinhard, and E. Suraud. Angular distribution of emitted electrons in sodium clusters: A semi-classical approach. *Phys. Rev. A*, 2003.
38. F. Gobet, B. Farizon, M. Farizon, M.J. Gaillard, J.P. Buchet, M. Carr, P. Scheier, and T.D. Märk. Direct experimental evidence for a negative heat capacity in the liquid-to-gas phase transition in hydrogen cluster ions: Backbending of the caloric curve. *Phys. Rev. Lett.*, 89:183403, 2002.
39. T. Ditmire, J. Zweiback, V.P. Yanovsky, T.E. Cowan, G. Hays, and K.B. Wharton. *Nature*, 398:489, 1999.
40. M.A. Lebeault, J. Viallon, J. Chevaleyre, C. Ellert, D. Normand, M. Schmidt, O. Sublemontier, C. Guet, and B. Huber. Resonant coupling of small size-controlled lead clusters with an intense laser field. *Euro. Phys. J. D*, 20:233, 2002.
41. T. Ditmire, T. Donnelly, A.M. Rubenchik, R.W. Falcone, and M.D. Perry. Interaction of intense laser pulses with atomic clusters. *Phys. Rev. A*, 53:3379, 1996.
42. F. Megi, M. Belkacem, M.A. Bouchene, E. Suraud, and G. Zwicknagel. *J. Phys. B*, 36:273, 2003.

Short Time-Scale Electron Kinetics in Bulk Metals and Metal Clusters

Arnaud Arbouet, Cyril Guillon, Dimitris Christofilos, Pierre Langot, Natalia Del Fatti, and Fabrice Vallée

Abstract. The first steps of nonequilibrium electron relaxation in noble metal films and clusters are discussed in the light of femtosecond time-resolved optical investigations. The presented results focus on the short time scale coherent electron-optical pulse coupling and electron scattering processes. The first effects of confinement are discussed by comparing results obtained in films and nanoparticles.

1 Introduction

With the advance of ultrashort femtosecond lasers and time-resolved spectroscopy, the properties of dense matter can now be investigated in strongly nonequilibrium situations that can be created on very short time scales. Direct time-domain monitoring of the transient material response offers the possibility to analyze both the properties of its elementary excitations in such a regime and the fundamental mechanisms at the origin of their phase and energy relaxation in different conditions, i.e., for instance, bulk or reduced dimensionality systems.

In the case of metallic materials, these femtosecond techniques are based on selective electron coupling with an optical pump pulse [1]. The electron kinetics is subsequently followed using a time-delayed probe pulse that monitors the time-evolution of an optical property of the material that depends on the electron and/or lattice properties, i.e., that is modified by electron excitation by the pump pulse. This approach discussed by Anisimov et al. [2] was first applied with picosecond resolution [3] and further extended to the femtosecond domain demonstrating the creation of a nonequilibrium electron distribution [4]. Different femtosecond techniques have then been developed to study the metal electron kinetics, monitoring for instance the transient optical response [5–13], transient harmonic generation [14–17] or induced photoelectron emission [18–22].

Depending on the probing condition and pulse duration, the different steps of the metal relaxation can be followed. They take place on a few femtoseconds to few hundred picoseconds as schematically shown in Fig. 1. Coherent electron-light coupling first takes place, the induced polarization decaying on a sub 10 fs time scale [16,17,23–25], much shorter than that

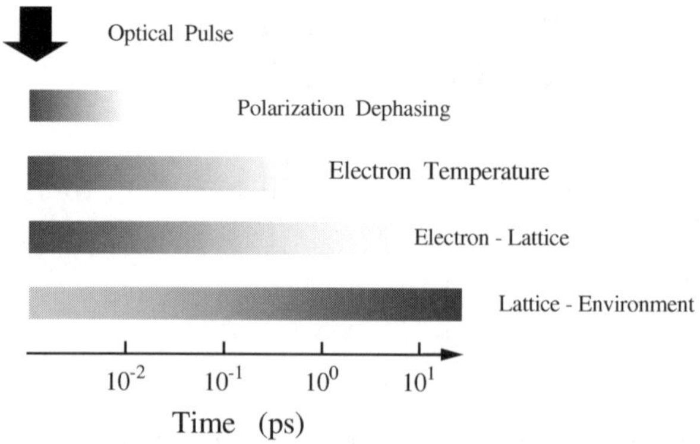

Fig. 1. Time scales of optically excited metal clusters

of the electron–electron or electron–lattice energy redistributions, creating a strongly nonequilibrium electron distribution [1,8,9,26]. The injected energy is subsequently redistributed among the electrons by electron–electron (e–e) scattering establishing a Fermi–Dirac distribution in few hundred femtoseconds [1,8,9,12,18–22,26], transferred to the lattice by electron–phonon (e–ph) interaction on a slightly longer time scale (typically one picosecond [7–15]) and eventually damped to the environment in few to few hundred picoseconds [27–29].

Time-resolved investigations have been recently extended to metal clusters [26,30]. They have permitted to study interactions of elementary excitations between themselves and with their environment in confined systems. New problems have thus been addressed: surface plasmon dephasing [16,17,23–25], ultrafast electron interactions (e–e and e–ph energy exchanges) in reduced dimensional systems [26,30–34,37,35,36,38–40,42–45], acoustic vibrations [46–51], and energy exchanges with the matrix [28,29]. In all of these processes, the quantum confinement effects (i.e. the presence of the surfaces) play key roles in the material response raising the question of their theoretical description.

In this paper we are discussing the first steps of nonequilibrium electron relaxation in bulk metals and metal clusters in the light of time-resolved optical pump-probe experiments. We will focus on noble metals because of their relatively simple band structure (Fig. 2) [52], and the simple connection between their optical properties and their electron energy distribution making them model systems for these studies [12,26]. Relatively large spherical nanoparticles will be considered (diameter $D \geq 2$ nm, i.e., more than few hundred atoms). They can be described using a small solid approach, rather than a molecular one for few atom clusters [53–55]. Their responses can thus be directly compared to the bulk material ones, permitting to trace the first impacts of confinement and surface effects on the ultrafast nonequilibrium electron relaxation.

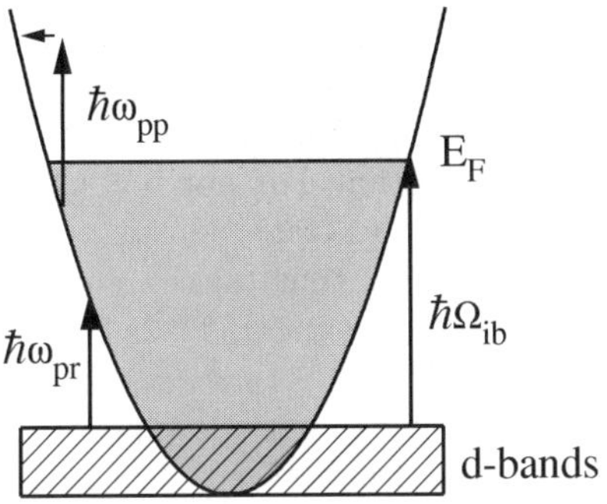

Fig. 2. Schematic electron band structure of noble metals. $\hbar\Omega_{\rm ib}$ is the interband transition threshold from the top of the d-bands to the Fermi surface. The *arrows* indicate intraband electron excitation at $\hbar\omega_{pp}$ and off resonant interband probing at $\hbar\omega_{pr}$

2 Optical Response

Ultrafast electron kinetics studies are based on optical excitation and detection processes. They rely on the close connection between the metal optical response and the electronic properties that we briefly recall in the following.

The optical dielectric constant $\epsilon^{\rm bulk}$ of a bulk metal at the frequency ω can be written as the sum of contributions of the conduction and bound electrons

$$\epsilon^{\rm bulk}(\omega) = \epsilon^{\rm f,bulk}(\omega) + \delta\epsilon^{\rm b}(\omega)\,. \tag{1}$$

In noble metals, the main contribution to the bound electron term $\delta\epsilon^{\rm b}(\omega) = \epsilon^{\rm b}(\omega) - 1$ is associated to the interband transitions from the fully occupied d-bands below the Fermi energy to the half filled s-p conduction band (Fig. 2). The conduction electrons follow a quasi-free electron behavior and their contribution $\epsilon^{\rm f}$ to ϵ is well described by a Drude formula [52,56]

$$\epsilon^{\rm f,bulk}(\omega) = 1 - \frac{\omega_p^2}{\omega[\omega + i/\tau_o^{\rm bulk}(\omega)]}\,, \tag{2}$$

where ω_p is the plasma frequency ($\omega_p^2 = n_e e^2/\epsilon_0 m$, n_e and m being the conduction electron density and effective mass, respectively). $\tau_o^{\rm bulk}(\omega)$ is the electron optical relaxation time which is determined by electron–phonon and electron–electron scattering with simultaneous exchange of a photon energy $\hbar\omega$ [57].

The quasi-free electron motion is modified in a nanocrystal due to electron interaction with the interface, and the concomitant breakdown of translational invariance. Introducing confinement in a classical or quantum mechanical way, the intraband contribution to the dielectric constant of a metal cluster has been shown to keep a Drude-type form (2) [25,58], with the scattering time now being

$$\frac{1}{\tau^{\rm nano}(\omega)} = \frac{1}{\tau_o^{\rm nano}(\omega)} + \frac{2v_F}{D} g_S(\omega)\,. \tag{3}$$

The first term, τ_o^{nano}, reflects the intrinsic bulk-like electron scattering processes in the particles. The second term, proportional to the Fermi velocity v_F divided by the diameter D of the assumed spherical nanoparticle, is a consequence of the confinement of the electron motion. For not too small nanoparticles ($D \geq 3\,\mathrm{nm}$), modification of the interband transitions is negligible [58]. The total crystallite dielectric constant ϵ^{nano} thus takes a similar form as the bulk one (1) using the bulk $\delta\epsilon^{\mathrm{b}}$ and replacing τ_o^{bulk} by τ^{nano} in (2).

In most experiments one is dealing with an ensemble of particles dispersed in a transparent dielectric matrix. For a low volume fraction $p \ll 1$ of small spheres ($R \ll \lambda$, where λ is the optical wavelength), the optical properties of this composite material can be described by introducing an effective dielectric constant $\widetilde{\epsilon}$ and its absorption coefficient can then be written [25,58]:

$$\widetilde{\alpha}(\omega) = \frac{9\,p\,\epsilon_d^{3/2}}{c} \frac{\omega\,\epsilon_2^{\mathrm{nano}}(\omega)}{\left[\epsilon_1^{\mathrm{nano}}(\omega) + 2\epsilon_d\right]^2 + \left[\epsilon_2^{\mathrm{nano}}(\omega)\right]^2}\,, \tag{4}$$

where $\epsilon^{\mathrm{nano}}(\omega) = \epsilon_1^{\mathrm{nano}}(\omega) + i\epsilon_2^{\mathrm{nano}}(\omega)$ and ϵ_d is the matrix dielectric constant. As compared to the bulk metal, the absorption is resonantly enhanced around the frequency, Ω_R, minimizing the denominator which is the condition for the surface plasmon resonance (SPR) [25,58].

It can be related to a resonance between the applied optical field and the collective electron oscillation. The applied electromagnetic field builds-up a surface charge distribution oscillating at the frequency ω_{pp} inside each nanoparticle. The surface charge generate a restoring force acting on the electron motion. When the optical field is in resonance with the collective electron oscillation induced by this force a large enhancement of the optical response as compared to the bulk material one is observed. This effect is concomitant with the enhancement of the electric field of the optical wave inside the particle due to the charge displacement and can be interpreted in terms of the local field effect [27,59]. This effect is illustrated in the inset of Fig. 4 that shows the absorption spectrum of silver nanoparticles. Conversely to other noble metals, the SPR is away from the interband transitions in silver ($\hbar\Omega_R \approx 3\,\mathrm{eV}$ as compared to $\hbar\Omega_{ib} \approx 4\,\mathrm{eV}$) and shows-up as a well defined resonance in the absorption spectrum.

In time-resolved pump-probe experiments, the time dependent changes of the sample optical properties (transmission, T, and/or reflectivity, R) induced by a pump pulse are monitored by a probe pulse of frequency ω_{pr} delayed by t_D. These changes reflect alterations of the material dielectric function, $\Delta\epsilon(\omega_{pr})$, which are usually sufficiently weak to allow the use of a perturbational approach:

$$\frac{\Delta T(t_D)}{T} = \frac{\partial \ln T}{\partial \epsilon_1}\Delta\epsilon_1(t_D) + \frac{\partial \ln T}{\partial \epsilon_2}\Delta\epsilon_2(t_D) \tag{5}$$

$$\frac{\Delta R(t_D)}{T} = \frac{\partial \ln R}{\partial \epsilon_1}\Delta\epsilon_1(t_D) + \frac{\partial \ln R}{\partial \epsilon_2}\Delta\epsilon_2(t_D)\,, \tag{6}$$

with $\epsilon = \epsilon^{\text{bulk}}$ in a metal film and $\epsilon = \tilde{\epsilon}$ in a composite material. $\Delta T(t_D)$ ($\Delta R(t_D)$) is defined as the difference between the sample transmission (reflection) at time t_D minus the one without perturbation. The coefficients can be calculated from the material equilibrium optical response [12,26]. In metal film, $\Delta T(t_D)$ and $\Delta R(t_D)$ exhibit similar amplitudes permitting experimental determination of $\Delta\epsilon_1^{\text{bulk}}(t_D)$ and $\Delta\epsilon_2^{\text{bulk}}(t_D)$.

This cannot be done in dilute composite materials because of their small reflectivity and induced reflectivity changes (since $\tilde{\epsilon}_1 \approx \epsilon_d$). In these systems, $\Delta T/T$ is only determined by $\Delta\tilde{\epsilon}_2$, or, equivalently, by the sample absorption change $\Delta\tilde{\alpha}$ [26,42]:

$$\frac{\Delta T}{T}(t_D) = -\Delta\tilde{\alpha}(t_D)L = a_1 \Delta\epsilon_1^{\text{nano}}(t_D) + a_2 \Delta\epsilon_2^{\text{nano}}(t_D) \, , \tag{7}$$

where L is the sample thickness. It is a linear combination of the changes of the real and imaginary parts of $\Delta\epsilon^{\text{nano}}(\omega_{pr})$. The coefficient a_1 and a_2 are related to the equilibrium $\epsilon^{\text{nano}}(\omega_{pr})$ (4) and are entirely determined by the linear absorption properties [26]. The response in a metal film, assumed to be identical to a bulk material, can thus be compared to the one in nanoparticles.

3 Experimental Setup

All the experiments described in the following were performed using a high repetition rate femtosecond pump-probe technique. The pulses were created from the pulse train delivered by a home-made 25 fs Ti:sapphire oscillator tunable in the range 1.07–0.82 μm. Part of the output of the laser was used as the near-infrared pump (some measurements have also been performed by frequency doubling to directly excite the SPR in Ag). $\Delta T/T$ or $\Delta R/R$ was probed in the vicinity either of the silver or gold SPR (around 420 or 520 nm, respectively) or interband transition threshold (around 310 or 530 nm, respectively). The necessary probe pulses were created by either frequency doubling the remaining part of the pulse train, yielding 30 fs pulses in the 535–410 nm range, or frequency tripling it yielding 55 fs UV pulses [12].

The two beams were sent into a standard pump-probe setup, with mechanical chopping of the pump beam at 1.5 kHz and differential and lock-in detection of ΔT or ΔR. Taking advantage of the high stability and high repetition rate (76 MHz) of the oscillator, very high sensitivity measurements were performed with a noise level for $\Delta T/T$ in the 10^{-6} range. This sensitivity is essential to study metallic systems in the weak perturbation regime (pump fluences in the 5–200 μJ/cm^{-2} range). It is convenient to characterize the energy injected by the pump pulse by defining a maximum equivalent electron temperature rise ΔT_e^{me} as the temperature increase of a thermalized electron gas for the same injected energy. In the experiments described here ΔT_e^{me} is of the order of 100–200 K. The electron heat capacity being much smaller than the lattice one, the final temperature rise of the fully thermalized electron–lattice system is typically 1 K.

The optically thin gold and silver metal films used in these studies were deposited on a sapphire or glass substrate using a standard evaporation technique. In the case of nanoparticles, most of the experimental investigations were performed in spherical silver nanocrystals embedded either in a 50BaO-$50P_2O_5$ or in a Al_2O_3 matrix. The former samples were prepared by a fusion and heat treatment technique. The average particle diameter D ranges from 4 to 30 nm [60]. The metal volume fraction, p, was in the range 1×10^{-4}–5×10^{-4} with a sample thickness $L \approx 15\,\mu$m. The latter samples, with $3 \leq D \leq 4$ nm, were grown using low energy cluster beam deposition (LECBD) with codeposition of alumina [61]. The fraction p was typically a few percent with $L \approx 0.2\,\mu$m. Studies were also performed in silver and gold colloidal solutions prepared by chemical synthesis using a reverse micelle (Ag) [62] or a radiolysis (Au) [63] technique. The thiol stabilized nanoparticles were either left in solution or embedded in PMMA or dispersed on a glass substrate. All samples have a narrow size dispersion with standard size deviations smaller than 10 % of diameter (measured by TEM).

4 Coherent Electron–Light Coupling

The first step in the electron interaction with the optical pulses involves their coherent coupling with the applied electromagnetic field. In noble metals, optical absorption is associated to intra-conduction band transitions (quasi-free electron absorption) and, above a threshold Ω_{ib}, to interband transitions from the full d-bands to empty states above the Fermi energy E_F (Fig. 2). We will consider here the case of a pulse of frequency ω_{pp} below the interband transitions threshold ($\omega_{pp} < \Omega_{ib}$). In a bulk metal, the optical electric field couples with the electrons inducing their forced oscillation at the optical frequency ω_{pp}. The electron motion generates a polarization (or equivalently a current) that radiates an electromagnetic field. This simple picture leads to the Drude expression for the quasi-free electron dielectric function (2). Light absorption takes place with the polarization decay due to electron scattering, i.e., electron–phonon and electron–electron scattering in bulk metal. It can be globally interpreted by the single particle excitation model describing the free electron absorption: one photon is absorbed by one conduction band electron with assistance of a third particle to conserve energy and momentum.

In the case of metal nanoparticles dispersed in a dielectric matrix, a similar approach can be introduced, taking into account electron localization at a nanoscale. As discussed in Sect. 2 the optical field induces an electron density oscillation at the optical frequency ω_{pp} in the nanoparticles. This electron movement generates an oscillating dipole in each particle that radiates at the same frequency modifying the applied electromagnetic field. Such coherent superposition of material polarization and electromagnetic field has been described in bulk dielectric material using mixed material – photon excitations introducing the polariton concept [64]. Light absorption takes place with polariton damping which usually reflects damping of its material part, i.e., decay of the induced material polarization. The same approach can be used here, the polarizable entities being the nanoparticles. Electron optical excitation in a nanoparticle can be understood in terms of electron polarization decay. As discussed by Kawabata and Kubo, it takes place with single electron excitation, which is similar to Landau damping of the collective plasmon mode in a plasma [25]. As in bulk metal, decay is induced by electron scattering in each nanoparticle, i.e. by electron–phonon, electron–electron and, using a classical approach, the additional contribution of electron-surface scattering (3).

In both bulk and confined metallic materials, optical absorption eventually leads to excitation of single electrons, each of them increasing its energy by $\hbar\omega_{pp}$. In this intraband process, conduction electrons with an energy E between $E_F - \hbar\omega_{pp}$ and E_F are excited above the Fermi energy with a final energy between E_F and $E_F + \hbar\omega_{pp}$ (Fig. 3). Describing the electron distribution by a one-particle function f and assuming an isotropic parabolic conduction band, the induced distribution change $\Delta f(E) = f(E) - f_0(E)$ is given by

$$\Delta f^{exc}(E) = B\{\sqrt{E - \hbar\omega_{pp}}\, f_0(E - \hbar\omega_{pp})[1 - f_0(E)] - \sqrt{E + \hbar\omega_{pp}}\, f_0(E)[1 - f_0(E + \hbar\omega_{pp})]\}, \quad (8)$$

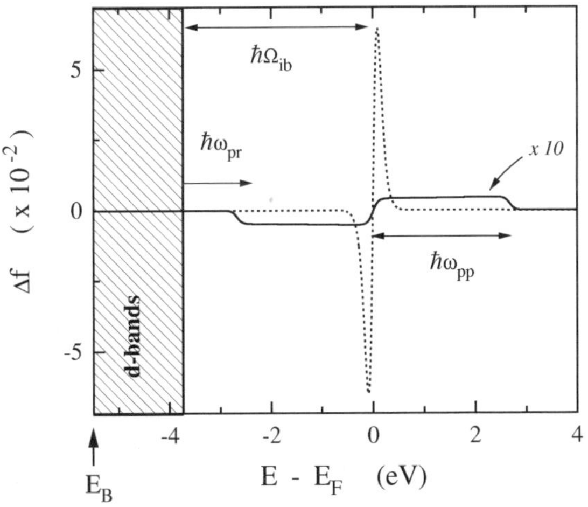

Fig. 3. Initial electron distribution change assuming ultrafast electron excitation, (8) (*full line*) and after thermalization (*dotted line*). The electron excess energy is the same in the two distributions. $\hbar\Omega_{ib}$ is the interband transition threshold between the top of the d-bands and the Fermi surface ($\hbar\Omega_{ib} \approx 3.9$ eV in silver). The *arrows* indicate intraband electron excitation at $\hbar\omega_{pp} \approx 3$ eV and off resonant probing at $\hbar\omega_{pr} \approx 1.5$ eV. E_B indicates where would be the bottom of the conduction band if one could assume a perfectly parabolic isotropic band shape

where f_0 is the electron distribution before optical excitation, and B a pump intensity dependent parameter. It exhibits a step like shape as shown in Fig. 3.

An important aspect here is the time scale of the single electron excitation, i.e., of that of the coherent polarization decay. For an optical pulse in resonance with the surface plasmon resonance, the induced polarization is usually described in terms of collective electron oscillation in a particle (i.e., of coherent superposition of single electron modes) and its decay discussed in terms of surface plasmon resonance dephasing. The polarization decay in large disk-shape noble-metal particles has been monitored in the time domain using time-resolved second and third harmonic generation [16,17,23]. Sub-10 fs decay times have been deduced, consistent with the estimated electron scattering times τ_0^{bulk} in noble metals. In the spectral domain hole burning measurements have been recently performed to estimate the homogeneous surface plasmon resonance linewidth as a function of the size, shape and environment of oblate silver particles. Dephasing times in the same range were inferred by these spectral results [24].

The Landau type of decay mechanism can be confirmed by directly monitoring the reduction of the occupation of electron states well below the Fermi energy (i.e. for $E_F - E \gg k_B T_e$), that are fully occupied in thermal equilibrium (Figs. 2 and 3). This occupation reduction reflects in induced absorption for a probe pulse with a frequency ω_{pr} such that d-band electrons can be excited to these open conduction band states. This can be done provided that ω_{pr} satisfies:

$$\hbar\omega_{pp} > \hbar\Omega_{ib} - \hbar\omega_{pr} \gg k_B T_e . \tag{9}$$

This transient interband absorption thus reflects in a transient increase of the imaginary part of the metal dielectric function.

In the case of a bulk metal, the temporal evolutions of the changes of the real and imaginary parts of the metal dielectric constant can be deduced from simultaneous measurements of the transmission and reflectivity changes in an optically thick film (using (5) and (6)). The results are shown in Fig. 4 for a 23 nm thick silver film using $\hbar\omega_{pr} = 1.5\,\mathrm{eV}$ and $\hbar\omega_{pp} = 3\,\mathrm{eV}$. The measured $\Delta\epsilon_2^{\mathrm{bulk}}$ is influenced by both the interband and intraband contributions to the metal dielectric function. The latter is responsible for the long delay ($t_D > 50\,\mathrm{fs}$) signal. This part of the response is consistent with the one observed in off-resonant conditions (with $\hbar\omega_{pp} = \hbar\omega_{pr} \approx 1.5\,\mathrm{eV}$) for which no interband absorption is induced. Assuming as a crude approximation that the Drude model is still applicable on this very short time-scale (2), the observed slow rise of $\Delta\epsilon_2^{\mathrm{bulk}}$ up to $t_D \approx 100\,\mathrm{fs}$ (Fig. 4) has been ascribed to a delayed modification of the electron scattering time τ_o^{bulk} [11]. However, its interpretation is still an open question and requires modeling of the properties of a nonequilibrium electron gas and of correlation effects on this time scale.

On a shorter time scale, $\Delta\epsilon_2^{\mathrm{bulk}}$ exhibits a fast rise and fall that can be ascribed to the interband term and is a signature of the large energy extension of the created nonequilibrium distribution. Interband absorption at ω_{pr} involves

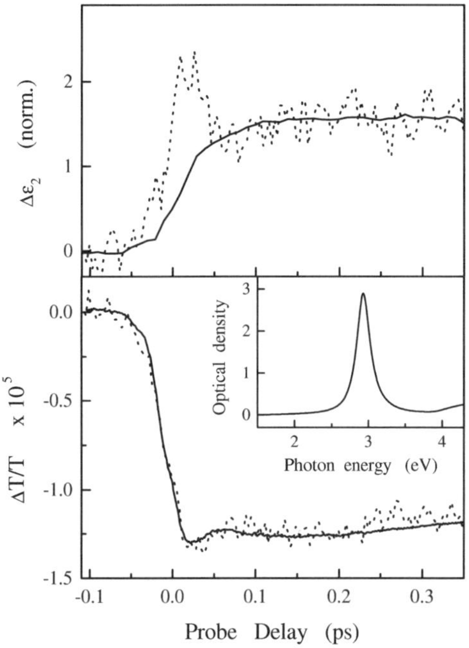

Fig. 4. (**a**) Change of the imaginary part of the dielectric function $\Delta\epsilon_2^{\text{bulk}}(\omega_{pr})$ measured in a 23 nm thick silver film for $\hbar\omega_{pr} = 1.5$ eV and a near infrared or blue pump pulse $\hbar\omega_{pp} = 1.5$ or 3 eV (*full* and *dotted line* respectively). (**b**) Time dependence of the transmission change $\Delta T/T$ for the same probing condition and resonant excitation of the surface plasmon resonance $\hbar\omega_{pp} \approx \hbar\Omega_R \approx 3$ eV in D = 26 nm Ag nanoparticles in a $50BaO\text{-}50P_2O_5$ glass matrix (*full line*). The *dashed line* is the normalized $\Delta T/T$ calculated from the measured $\Delta\epsilon_1^{bulk}(\omega_{pr})$ and $\Delta\epsilon_2^{\text{bulk}}(\omega_{pr})$ using (7). The *inset* shows the measured sample absorption spectrum

final conduction band states around $E_e = E_F - \hbar(\Omega_{ib} - \omega_{pr}) \approx E_F - 2.5$ eV, in the energy region where the electron occupation number is reduced by single electron excitation (from $E_F - \hbar\omega_{pp}$ to E_F, Fig. 3). Conduction band states are thus opened for interband absorption at ω_{pr} leading to an absorption increase that relaxes with population relaxation of these states, i.e., in a few femtoseconds ($\sim$ 3 fs for electrons 2.5 eV above E_F [20], the "holes" below E_F having a similar dynamics as the corresponding electrons above E_F [12]). This leads to a transient increase of $\Delta\varepsilon_2^{\text{bulk}}$, whose temporal shape is essentially limited by the pump-probe cross-correlation. This interpretation is confirmed by the absence of a transient peak when exciting in the near infrared for the same probe frequency (Fig. 4). No interband absorption is then induced since the final states (with energy $\approx E_F - 2.5$ eV) are unperturbed as they are 1 eV below the minimum energy of the perturbed states ($E_{\min} = E_F - \hbar\omega_{pp} \approx E_F - 1.5$ eV).

A similar behavior is observed in clusters as shown in Fig. 4 for $\hbar\omega_{pr} = 1.5\,\mathrm{eV}$ and $\hbar\omega_{pp} = 3\,\mathrm{eV}$, corresponding to resonant excitation of the nanoparticle surface plasmon resonance (inset of Fig. 4). For this probe frequency, the imaginary part of the metal dielectric constant dominates the $\Delta T/T$ signal ($a_2/a_1 \approx 2.7$ in (7)). The observed time behavior is consistent with the one measured for $\Delta\epsilon_2^{\mathrm{bulk}}$. In particular, a short time delay peak is observed followed by an increase over a 100 fs scale and a picosecond decay due to electron energy transfer to the lattice.

For the investigated nanoparticle size ($D = 26\,\mathrm{nm}$) the electron dynamics is almost identical to the bulk metal one [30,26]. The measured $\Delta T/T$ can be quantitatively compared to the bulk response using (7) and $\Delta\epsilon^{\mathrm{bulk}}$ measured for the same pump and probe frequencies. A very good agreement between the experimental and calculated $\Delta T/T$ responses is obtained (Fig. 4), showing that the bulk and confined materials behave in a very similar way. As for $\Delta\epsilon_2^{\mathrm{bulk}}$, the transient interband induced absorption is actually at the origin of the observed $\Delta T/T$ transient peak.

5 Nonequilibrium Electron Kinetics

To further analyze the nonequilibrium electron excitation process, we have compared the measured response with the theoretical one assuming direct single electron excitation by the pump pulse (i.e., neglecting the coherent step). Even for very short femtosecond pulses, electron–electron (e–e) scattering and, to a lesser extent, electron–phonon (e–ph) interaction modify the electron distribution. In bulk metal, electron kinetics has been shown to be well described by the electron Boltzmann equation [8,12]:

$$\frac{df(E,t)}{dt} = \left.\frac{df(E,t)}{dt}\right|_{\mathrm{e-e}} + \left.\frac{df(E,t)}{dt}\right|_{\mathrm{e-ph}} + \frac{df^{exc}}{dt}(E,t)\ . \tag{10}$$

The first term describes e–e scattering via a statically screened Coulomb potential. Assuming a Fermi–Dirac distribution at zero temperature, this term yields the usual expression for the electron lifetime in the vicinity of the Fermi surface due to inelastic e–e collisions [21,65]:

$$\frac{1}{\tau_{\mathrm{e-e}}(E)} = \frac{me^4\,(E-E_F)^2}{64\pi^3\hbar^3\varepsilon_0^2(\varepsilon_b^0)^2 E_S^{3/2}\sqrt{E_F}}\left[\frac{2\sqrt{E_F E_S}}{4E_F+E_S} + \arctan\sqrt{\frac{4E_F}{E_S}}\right]\ . \tag{11}$$

with $E_S = \hbar^2 k_S^2/2m$ where k_S is the screening wave vector. This expression yields the well known $(E-E_F)^{-2}$ variation of $\tau_{\mathrm{e-e}}$ with the electron energy E, due to phase space filling (Pauli exclusion effect).

The second term on the right hand side is the e–ph scattering rate. Its expression is given in [12] for a bulk metal for deformation potential e–ph coupling. Though it is a rough approximation in metals, it has been shown that the exact nature of the e–ph coupling does not influence the computed electron dynamics. This is a consequence of the fact that the lattice temperature T_L being larger than the Debye temperature Θ_D, electron distribution changes on the energy scale of a phonon have a minor influence on the overall dynamics.

The last term describes perturbation by the finite duration pump pulse. It is identical to Δf^{exc} (8), with f_0 being replaced by $f(t)$ and B proportional to the pump pulse intensity $I_p(t)$.

The time dependent electron distribution function computed using (10) are shown in Fig. 5 in the case of silver for a 25 fs pump pulse with $\hbar\omega_{pp} = 1.5$ eV. Fast relaxation of the high energy electrons leads to a build up of the induced distribution change amplitude $\Delta f(E,t) = f(E,t) - f_0(E)$ around E_F that takes place concurrently with excitation by the pump pulse. This fast energy redistribution is at the origin of the $\Delta f(E)$ shape distortion as compared to instantaneous excitation (Fig. 3).

The interband term, $\Delta\epsilon_2^b(\omega_{pr})$, can be calculated from the time-dependent electron distribution using a band structure model for the d-bands to conduction band transitions [66,67]. As ω_{pr} is far from the interband transition threshold, the simple model of a parabolic conduction band and undispersed d-bands can be used [12]. In agreement with the qualitative discussion of Sect. 4, $\Delta\epsilon_2^b(\omega_{pr})$ is found to be nonzero only for very short time delays. The

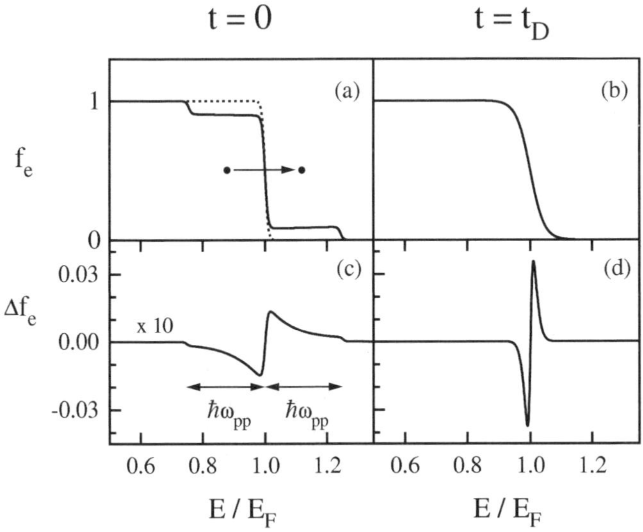

Fig. 5. Energy dependent electron occupation number f assuming instantaneous intraband excitation by a pump pulse with $\hbar\omega_{pp} = 0.24E_F$ at $t = 0$ (the initial distribution f_0 is shown by the *dotted line*) (**a**) and after establishment of an electron temperature after $t = t_D > 0$ (**b**). Computed distribution changes $\Delta f = f - f_0$ for $\Delta T_e^{me} = 100$ K and a pump-pulse duration of 25 fs, (**c**) and (**d**)

second contribution to $\Delta\epsilon_2^{\rm bulk}$, $\Delta\epsilon_2^{\rm f,bulk}$ (1) cannot be simply deduced from Δf. As a first approximation, it is assumed to be identical to the one measured for infrared pumping, ϵ_2^b being then unchanged (Fig. 4). Summing-up these two terms, a very good description of the film data is obtained confirming the interpretation of Sect. 4. This good agreement also shows that coherent effects negligibly influence the measured response. It is also the case in metal nanoparticles, for resonant or nonresonant excitation of the surface plasmon resonance, the composite material response closely following the one deduced from the film study (Fig. 4).

The absence of any detectable influence of coherent effects with the 25–30 fs pulses used here, is consistent with a sub 10 fs polarization decay as deduced from time or spectral domain surface plasmon resonance dephasing time measurements [16,17,23,24] and optical electron scattering time. Note that it is even faster in nanoparticles because of the additional surface scattering contribution (3) [25,24]. On the time scale of the experiments discussed here, electron excitation by a femtosecond pulse can thus be simply described in term of incoherent single electron excitation (i.e., quasi-free electron absorption).

6 Electron Energy Exchanges: Surface Effects

The first step in the relaxation of the nonequilibrium electrons is the establishment of an electron temperature T_e by electron–electron scattering. Electron–electron scattering is strongly reduced close to E_F by Pauli exclusion effect and the electron distribution stays athermal for few hundred femtoseconds as shown in Fig. 5. This has been first experimentally demonstrated in bulk noble metals by Bokor and coworkers [18] using time-resolved two-photon photoemission and confirmed by transient optical property modulation [8,12], surface plasmon polariton kinetics [9,68] and picosecond ultrasonics [69].

The optical property modulation technique has the main advantage of yielding a precise absolute determination of the characteristic thermalization times and of being easily extendable to metal clusters embedded in different matrices. The results obtained in various samples can thus be quantitatively compared and confronted to the prediction of theoretical modeling. This technique is based on the sensitivity of the absorption around the interband transition threshold Ω_{ib} on the occupancy of the electron states around E_F (Fig. 2), i.e., on the thermal or athermal character of the distribution [8,12].

In optically thin gold and silver films characteristic internal thermalization times τ_{th} have been extracted assuming a monoexponential rise of the measured response and fitting it using [8,12]:

$$u(t) = H(t)\,[1 - \exp(-t/\tau_{th})]\,\exp(-t/\tau^0_{\rm e\text{-}ph})\ , \qquad (12)$$

where $H(t)$ is the Heaviside function (Fig. 6). The exponential decay with the time $\tau^0_{\text{e-ph}}$ stands for electron energy transfer to the lattice [12]. Characteristic times $\tau_{th} \approx 350$ fs and 500 fs were obtained in bulk silver and gold, respectively [8,12].

As stressed above, the long thermalization time in metal is a consequence of the large screening of the electron–electron Coulomb interaction and of Pauli exclusion effect close to the Fermi energy. In these experiments, only the distribution change around E_F is detected and although τ_{th} a priori reflects electron redistribution over the full excited region, it is actually essentially determined by e–e collisions around the Fermi level, whose probability is largely reduced due to Pauli exclusion. The faster thermalization in silver than in gold, though the conduction electron properties are similar in the two metals, has been ascribed to the weaker screening by the bound electrons in Ag. Assuming static screening, the dependence of the scattering time of an electron out of its state on the density n_e and core electron screening amplitude (related to $\epsilon_{sc} = \epsilon_1^b(0)$) in the bulk metal is given by [65]:

$$\tau_e \propto n_e^{5/6} \epsilon_{sc}^{1/2} \,. \tag{13}$$

For a weak perturbation, τ_{th} is proportional to τ_e [12]. It is thus expected to scale as $\epsilon_{sc}^{1/2}$ from silver to gold (n_e being almost unchanged). It leads to an estimated time about 35% larger in gold than in silver ($\tau_{th}^{\text{Au}}/\tau_{th}^{\text{Ag}} \approx 1.35$) in close agreement with the experimental result $\tau_{th}^{\text{Au}}/\tau_{th}^{\text{Ag}} \approx 1.4$.

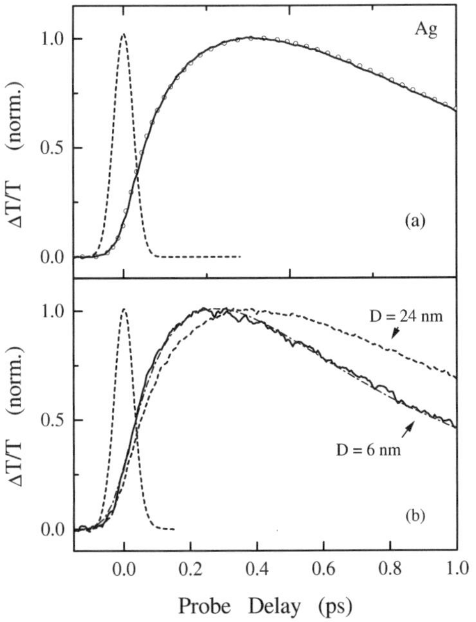

Fig. 6. Differential transmissions $\Delta T/T$ measured (*full lines*) and computed (*open dots*) for interband probing $\hbar\omega_{pr} = 4$ eV in optically thin silver (**a**). Same in silver nanoparticle with diameter $D = 24$, and 6 nm, the *dash-dotted line* is a fit for the $D = 6$ nm case using (12) (**b**). The *dotted lines* are the pump-probe cross-correlations

A similar approach can be used in noble metal nanoparticles where the interband properties are almost unaffected by confinement (for $D \geq 3\,\mathrm{nm}$). As in films, the interband term dominates the response around Ω_{ib} permitting to analyze the impact of confinement on electron–electron energy exchanges. As previously observed in films, the transmission changes measured in gold or silver nanoparticle composite materials exhibit a delayed rise demonstrating the existence of a non-Fermi distribution in metal nanoparticles on a few hundred femtosecond time scale. The deduced τ_{th} were however found to strongly decrease for sizes smaller than 10 nm [31]. Comparable results were obtained for silver nanoparticles embedded either in a $50BaO\text{-}50P_2O_5$ or in a Al_2O_3 matrix (Fig. 7) or deposited on a substrate, showing that the results are independent of the environment and sample preparation method (Fig. 7). Similar size dependencies were observed in gold nanoparticles, with however larger thermalization times consistent with previous measurements in gold film. A similar approach has been used in gold colloids with $D = 9$ and 48 nm by El-Sayed et al. [26,30,70]. A delayed signal rise is observed around Ω_{ib} with τ_{th} values almost identical to the bulk one [70]. This absence of variation in this size range is consistent with our results. The τ_{th} decrease with D thus demonstrates an intrinsic confinement induced fastening of the electron–electron energy exchanges [31].

The reduction of τ_{th} with size has been ascribed to surface induced modification of the electron correlation effect close to a surface. It has been

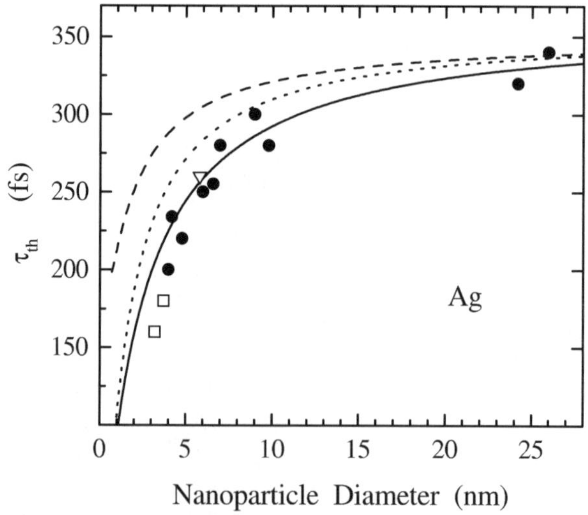

Fig. 7. Size dependence of the electron thermalization time τ_{th} for Ag nanoparticles in a $BaO\text{-}P_2O_5$ (*dots*) and Al_2O_3 (*open squares*) matrix or dispersed on a glass substrate (*open triangle*). The *full line* shows the computed τ_{th} taking into account both the spillout and d-electron localization effects. The *dotted* and *dashed lines* include only the spill-out and d-electron effect, respectively

shown that the wave functions of the conduction electrons extend beyond the particle radius defined by the ionic lattice (electron spillout [71]). This leads to a reduction of their density n_e and thus of ω_p that manifests itself by a red shift of the SPR with size reduction in alkali metal clusters [72]. Conversely, the d-electron wave-functions are localized in the inner region of the particle leading to an incomplete embedding of the conduction electrons in the core electron background [73]. Both effects lead to less efficient screening of the e–e Coulomb interactions close to a surface in noble metals and are thus expected to increase the effective e–e scattering rate in small nanoparticles.

As a first approximation, their contributions can be estimated assuming a bulk like behavior and averaging τ_e^{-1} over the particle volume using position dependent n_e and ϵ_{sc} [31]. Although this local approach is a crude approximation, it is partly justified by the short screening length of the Coulomb interaction (of the order of the inverse of the Thomas–Fermi wave vector, ~ 1 Å in metals). The spatial variations of ϵ_{sc} and n_e in silver clusters were previously quantitatively modeled to analyze the size dependence of the surface plasmon resonance frequency [74]. The d-electron surface exclusion effect was described using a phenomenological two-region dielectric model, where screening by the d-electrons is effective only in a core sphere of radius $R_c = R - c$ with $c = 3.5$ a.u. (i.e., $\epsilon_{sc} = \epsilon_{sc}^{\text{bulk}}$ inside the core sphere and $\epsilon_{sc} = 1$ outside) [74,73]. n_e was calculated using the jellium dielectric sphere model [74,71]. Using this model, a good reproduction of the experimental τ_{th} size dependence has been obtained with similar contributions from the two screening reduction effects (Fig. 7). It is actually related to the increasing percentage of electrons in the perturbed surface layer (about 25% of the electrons are out of the R_c radius core sphere for $R = 2\,\text{nm}$), i.e., the increasing influence of the surface in small clusters. This enhanced e–e scattering close to the surface is consistent with the larger dephasing times measured for surface than for bulk electrons using a space resolved technique [75]. It actually reflects the impact of the surface on electron correlation effects.

Although a good reproduction of the data is obtained, a systematic deviation is observed for small sizes ($D \leq 4\,\text{nm}$). The above model, based on a simple extension of the bulk calculations, overlooks specific features of the confined materials. In particular, the e–e scattering rate has been derived using the bulk electron wave functions and includes momentum conservation. This is relaxed in confined systems, due to, classically, electron scattering off the surfaces, leading to opening of additional e–e scattering channels and thus further size dependent modifications of the e–e interactions (this effect is comparable to the increase of the electron optical scattering rate due to surface effects, (3)). A more correct description requires nonlocal calculations of the e–e scattering rate and many body effects in a nanoparticle using the confined electron wave functions.

7 Electron–Lattice Energy Exchanges: Size Effect

The transient signals measured in metallic systems decay in a few picoseconds with the electron gas excess energy and thus contain information on the energy exchanges of the electrons with their environment. In the case of films, this is essentially limited to the metal lattice permitting direct measurement of the electron–phonon interaction time. Assuming that both the electron and lattice are thermalized at different temperatures T_e and T_L, this energy transfer can be described using the bulk metal two-temperature model [76]. The electron gas is then entirely described by its temperature and its cooling dynamics can be simply modeled using the rate equation system

$$C_e(T_e)(\partial T_e/\partial t) = -G(T_e - T_L),$$
$$C_L(\partial T_L/\partial t) = G(T_e - T_L) \;, \tag{14}$$

where G is the effective electron–phonon coupling constant. In time-resolved femtosecond measurements, this model can only be used after internal electron thermalization, i.e., few hundred femtoseconds to one picosecond.

For a weak perturbation, $\Delta T_e^{me} \ll T_0$ where T_0 is the initial temperature, $C_e(T_e)$ can be identified with $C_e(T_0)$ leading to an exponential decay of the electron temperature rise ΔT_e and of the electron excess energy Δu_e with the same time constant $\tau^0_{\text{e–ph}} \approx C_e(T_0)/G$ [12].

The energy losses of the electrons to the lattice can be selectively followed by measuring the sample transmission change $\Delta T/T$ of a femtosecond probe pulse in off-resonant conditions (i.e. for a probe photon energy $\hbar\omega_{pr}$ well below the interband transition threshold $\hbar\Omega_{ib}$). In these conditions and for a not strongly out of equilibrium situation (i.e., for Δf extending over an energy range much smaller than $\hbar(\Omega_{ib} - \omega_{pr})$ below E_F), $\Delta T/T$ is almost insensitive to the details of the electron distribution and proportional to the electron gas excess energy density Δu_e [9,12]:

$$\Delta T/T(t_D) \propto \Delta u_e(t_D) \propto \int E\Delta f(E, t_D)dE \;. \tag{15}$$

For nanoparticles the electron energy can also be damped to the surrounding solvent or matrix either directly or via the metal lattice, possibly modifying the observed relaxation. This coupling is frequently assumed to be sufficiently slow to be neglected on the scale of the metal electron–lattice energy exchange. However, it strongly increases with size reduction and may play a role in the observed electron cooling, especially for large electron excitation [34,38]. As long as only energy exchanges inside a nanoparticle are considered, the two-temperature model can also be used.

In silver, the non-resonant conditions are realized for a near-infrared or blue probe pulse after few tens of femtoseconds. As expected on a picosecond time scale, $t_D \geq 1\,\text{ps}$, the measured transmission changes $\Delta T/T$ decay exponentially (Fig. 8), with a decay time clearly decreasing for small sizes, indicating increase of the electron–lattice coupling in small clusters.

Measurements performed for D ranging from 4.2 to 30 nm have shown that $\tau^0_{\text{e–ph}}$ is comparable to its bulk value ($\tau^0_{\text{e–ph}} \approx 0.85$ ps in Ag) for large nanoparticles ($D \geq 10$ nm, typically) but significantly decreases for smaller ones (with $\tau^0_{\text{e–ph}} \sim 0.6$ ps for $D = 4.2$ nm in Ag) [41] with similar results in gold. In contrast, using extrapolation of the strong perturbation measurements, no size dependence of $\tau^0_{\text{e–ph}}$ (or G) has been estimated in the case of gold colloidal nanoparticles with size between 2.4 and 100 nm by Hartland et al. [37,47]. A similar conclusion was drawn by El-Sayed et al. for sizes ranging from 4 to 100 nm [30]. A non monotonic size dependence has been observed by Zhang et al. in gold [77] with possible modification of the decay time due to strong excitation.

A decrease of the $\Delta T/T$ decay time with D has also been reported in tin [35] and gallium [36] nanoparticles for size ranging from 4 to 12 nm and 10 to 18 nm, respectively. The size dependence is however much larger than for silver, $\tau_{\text{e–ph}}$ being almost proportional to D over the investigated range. This variation has been interpreted in terms of electron coupling with the surface acoustic modes of the particles and quenching of the electron-bulk phonon interactions using the Belotskii and Tomchuk model [78,79]. However, this

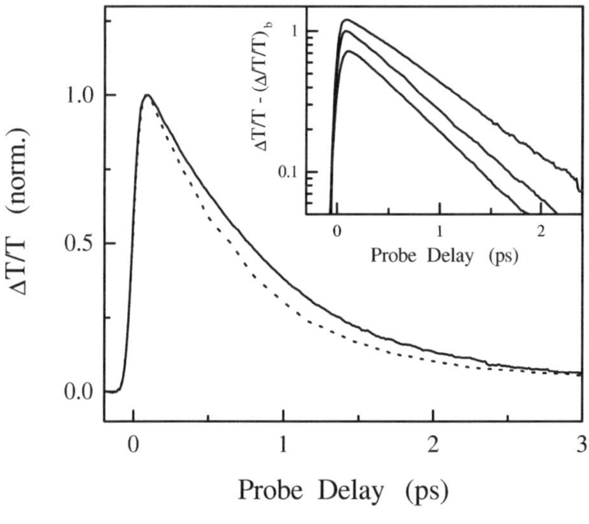

Fig. 8. Time behavior of the normalized transmission change $\Delta T/T$ measured off-resonance with the interband transitions ($\hbar\omega_{pp} \approx 3$ eV) for $\hbar\omega_{pr} \approx 1.5$ eV in $D = 26$ nm (*full line*) and $D = 6$ nm (*dotted line*) Ag nanoparticles in glass. The *inset* shows the same data on the logarithmic scale after subtraction of the long delay residual signal $(\Delta T/T)_b$ for the $D = 26$, 6 and 4 nm samples (from *top* to *bottom*, the curves being vertically shifted for clarity). The decay time is $\tau^0_{\text{e–ph}} = 820$, 680 and 600 fs, respectively

model, only applicable to small sizes, predicts a much slower electron–lattice energy transfer than in the bulk [79], in contrast to the observations in noble metals.

The origin of the different behaviors observed by the different groups in different experimental conditions is not clear. An important point here could be the influence of the environment [29,34,80], and of the interface layer and quality (as has been observed for the SPR width [55]). This can be particularly important for small sizes for which the particles are strongly coupled with their environment, the ratio of the surface to volume atom numbers strongly increasing as D decrease. Actually, the decay times measured in 18 nm gold particles have been shown to depend on the solvent (water or cyclohexane) [80]. Similarly, relaxation has been found to be faster in Ag particles embedded in aluminate than in silica glass [34]. An important effect could be local heating of the matrix around the nanoparticles with thus a strong influence of the thermal conductivity of the matrix material [29]. However, these matrix effects have been observed only in the strong excitation regime, while in contrary no environment influence has been recently demonstrated in weak perturbation studies [39]. In this regime, the observation of a $\tau^0_{\text{e–ph}}$ size dependence in silver similar to that of τ_{th} suggests a confinement induced acceleration of these exchanges due to increase of the electron–lattice mode coupling as in semiconductor nanoparticles [81]. It can take its origin both in increase of the electron-ion interaction and in modification of the phonon properties by confinement, as observed in surface experiments [82]. Additional theoretical investigations are clearly needed to understand electron–lattice coupling in confined metallic systems.

8 Nonequilibrium Electron–Lattice Energy Exchanges

On a short time scale, the electron distribution is out of equilibrium and the two-temperature model is no more applicable. In this regime, electron–phonon interaction in a nonequilibrium situation can be analyzed. For both nanoparticles and films, the short time scale $\Delta T/T$ decay clearly shows that the electron gas energy loss rate to the lattice is initially slow and increases over a time scale of a few hundred femtoseconds to reach its long delay value (Fig. 9). This behavior is a direct consequence of the existence of a long living athermal electron gas. During the excitation process, a very small number of electrons gains a large excess energy as compared to k_BT_0. As a crude approximation, separating the electron gas into unperturbed and nonequilibrium electrons, only the latter ones can lose energy by phonon emission. Their number increases with time as e–e scattering redistributes energy among the carriers: this leads to an overall increase of the energy loss rate to the lattice during the early stages of the internal electron gas thermalization [9,11]. As internal thermalization is approached, the above separation is no more valid and a constant energy loss rate is eventually reached, corresponding to col-

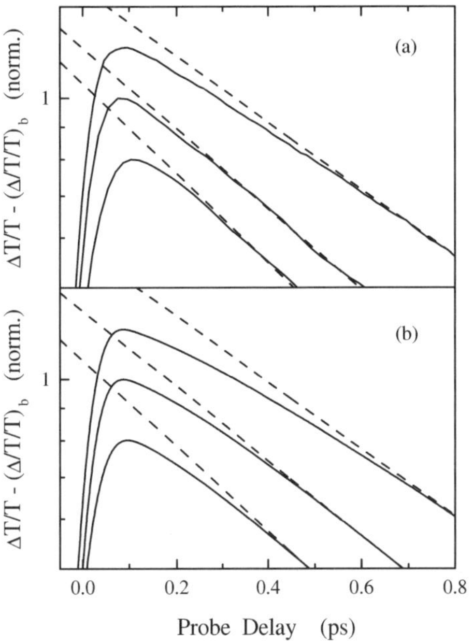

Fig. 9. (**a**) Measured and (**b**) calculated short time scale evolution of the transmission change $\Delta T/T - (\Delta T/T)_b$ in Ag nanoparticles with average diameter $D = 26\,\mathrm{nm}$ (*top*), 6 nm (*middle*) and 4 nm (*bottom*) on a logarithmic scale. The *dashed lines* correspond to long delay exponential decay (see Fig. 8)

lective electron gas-lattice interaction as described by the two-temperature model. This evolution from a quasi-individual to a collective electron behavior is responsible for the observed short-time non-exponential decay of the excess energy.

In small metal nanoparticles, the electron gas internal thermalization is faster (Sect. 6). The transient nonequilibrium regime for e–ph energy exchanges is thus expected to be observed only on a shorter time scale. This is in agreement with the experimental results, the transient regime being barely observable for the smallest nanoparticles (Fig. 9).

The above description of the interplay between electron thermalization and energy losses can be made more quantitative by computing the electron gas relaxation dynamics using the electron Boltzmann equation (Sect. 5, (10)). The time evolution of Δu_e calculated for the bulk silver conditions is shown in Fig. 9 and exhibits the same behavior as the experimental ones for the film or large nanoparticle cases. Similar calculations performed assuming instantaneous internal thermalization of the electron gas only show an exponential decay of Δu_e in agreement with the two temperature model.

For nanoparticles, as a first approximation, the simulations were repeated modifying the e–e and e–ph scattering rate to match their increase with cluster size reduction. The results are in very good agreement with the experimental ones showing the same transient evolution of the signal decay to an exponential behavior on a reduced time scale with size reduction. This confirms the strong interplay between e–e and e–ph scattering observed in bulk metals and yields further evidence of the acceleration of the electron energy exchanges in small clusters.

9 Conclusion

Time-resolved optical techniques have emerged as powerful tools for the selective investigation of electron kinetics in metallic materials, offering the unique possibility of separating the different interaction processes. The performed investigations have greatly improved our understanding of electron interaction processes in bulk metals and their modification in reduced dimensionality systems. They are very promising and should yield new insights into the impact of confinement on the electron kinetics and its evolution from a bulk-like to molecular type of behavior.

Experimental results have now been obtained on surface plasmon resonance relaxation, on electron–electron and electron–phonon energy redistribution and their interplay in noble metal particles over very large size ranges and for different types of environment. Electron–electron energy exchanges have been shown to be almost unaffected by confinement for sizes larger than 10 nm and to strongly increase for smaller sizes. Extension of these first measurements to other metals or type of confinement (1D or 2D systems) and the use of other techniques, such as time-resolved photoemission [43–45] would be particularly interesting here. The size dependence of electron–lattice energy exchanges is still controversial, probably because of the influence of the surrounding media and/or difficulty of comparing experiments performed in very different conditions. A very interesting extension of these types of measurement would be time-resolved investigation of spin dynamics in clusters, similarly to what has been done in bulk metals [83–85].

On the theoretical side, understanding of the electronic response in highly nonequilibrium situation is still partial. It requires a development of theoretical approach describing electron interactions and correlation effects in this regime. In nanoparticle electron dynamics, the modeling is mostly based on a small solid approach, i.e., modifications are introduced to the bulk material response. This is justified for not too small sizes, i.e., formed by more than few hundred atoms, for which quantum mechanical confinement only introduces correction to their properties. It clearly only constitutes a first approximation. Improved modeling of the electron scattering based on a quantum mechanical approach for both the electrons and phonons is necessary for understanding electron interactions in a confined system and properly taking into account surface effects.

Acknowledgements

The authors wish to thank C. Voisin and C. Flytzanis for their very important contributions to some of the described studies and for very helpful discussions. We are also indebted to A. Nakamura, Y. Hamanaka, S. Omi for helpful discussions and for providing some of the silver nanoparticle samples. We also thank B. Prével, M. Gaudry, E. Cottancin, J. Lermé, M. Pellarin, M. Broyer, M. Maillard and M.P. Pileni for their help in the theoretical and experimental parts of this work and for providing us with very good quality samples. Financial support by the Conseil Regional d'Aquitaine is also acknowledged.

References

1. For a review see F. Vallée, C. R. Acad. Sci. **2**, 1469 (2001).
2. S.I. Anisimov, B.L. Kapeliovitch, and T.L. Perelman, Sov. Phys. JETP **39**, 375 (1974).
3. G.L. Eesley, Phys. Rev. Lett. **51**, 2140 (1983); Phys. Rev. B **33**, 2144 (1986).
4. J.G. Fujimoto, J.M. Liu, E.P. Ippen, and N. Bloembergen, Phys. Rev. Lett. **53**, 1837 (1984).
5. H.E. Elsayed-Ali, T.B. Norris, M.A. Pessot, and G.A. Mourou, Phys. Rev. Lett. **58**, 1212 (1987).
6. R.W. Schoenlein, W.Z. Lin, J.G. Fujimoto, and G.L. Eesley, Phys. Rev. Lett. **58**, 1680 (1987).
7. S.D. Brorson, A. Kazeroonian, J.S. Modera, D.W. Face, T.K. Cheng, E.P. Ippen, M.S. Dresselhaus, and G. Dresselhaus, Phys. Rev. Lett. **64**, 2172 (1990).
8. C.K. Sun, F. Vallée, L.H. Acioli, E.P. Ippen, and J.G. Fujimoto, Phys. Rev. B **50**, 15337 (1994).
9. R. Groeneveld, R. Sprik, and A. Lagendijk, Phys. Rev. B **51**, 11433 (1995).
10. H.E. Elsayed-Ali, T. Juhasz, G.O. Smith, and W.E. Bron, Phys. Rev. B **43**, 4488 (1991).
11. N. Del Fatti, R. Bouffanais, F. Vallée, and C. Flytzanis, Phys. Rev. Lett. **81**, 922 (1998).
12. N. Del Fatti, C. Voisin, M. Achermann, S. Tzortzakis, D. Christofilos, and F. Vallée, Phys. Rev. B **61**, 16956 (2000).
13. T. Juhasz, H.E. Elsayed-Ali, X.H. Hu, and W.E. Bron, Phys. Rev. B **45**, 13819 (1992).
14. J. Hohlfeld, D. Grosenick, U. Conrad, and E. Matthias, Appl. Phys. A **60**, 137 (1995).
15. J. Hohlfeld, S.S. Wellershoff, J. Güdde, U. Conrad, V. Jähnke, and E. Matthias, Chem. Phys. **251**, 237 (2000).
16. B. Lamprecht, A. Leitner, and F.R. Aussenegg, Appl. Phys. B **68**, 419 (1999).
17. B. Lamprecht, J.R. Krenn, A. Leitner, and F.R. Aussenegg, Phys. Rev. Lett. **83**, 4421 (1999).
18. W.S. Fann, R. Storz, H.W.K. Tom, and J. Bokor, Phys. Rev. Lett. **68**, 2834 (1992), Phys. Rev. B **46**, 13592 (1992).
19. A. Knoesel, A. Hotzel, and M. Wolf, Phys. Rev. B **57**, 12812 (1998).
20. M. Aeschlimann, M. Bauer, S. Pawlik, W. Weber, R. Burgermeister, D. Oberli, and H.C. Siegmann, Phys. Rev. Lett. **79**, 5158 (1997).

21. For a review see: H. Petek and S. Ogawa, Progr. Surf. Science **56**, 239 (1997).
22. J. Cao, Y. Cao, H.E. Elsayed-Ali, R.J.D. Miller, and D.A. Mantell, Phys. Rev. B **58**, 10948 (1998).
23. T. Vartanyan, M. Simon, and F. Träger, Appl. Phys. B **68**, 425 (1999).
24. J. Bosbach, C. Hendrich, F. Stietz, T. Vartanyan, and F. Träger, Phys. Rev. Lett. **89**, 257404 (2002).
25. A. Kawabata and R. Kubo, J. Phys. Soc. Jap. **21**, 1765 (1966).
26. C. Voisin, N. Del Fatti, D. Christofilos, and F. Vallée, J. Phys. Chem. B **105**, 2264 (2001).
27. C. Flytzanis, F. Hache, M.C. Klein, D. Ricard, and P. Roussignol, In *Progress in Optics Vol.XXIX*; E. Wold, Ed. (North Holland: Amsterdam, 1991) p. 321.
28. M.J. Bloemer, J.W. Haus, and P.R. Ashley, J. Opt. Soc. Am. B **7**, 790 (1990).
29. R.F. Haglund, G. Lupke, D.H. Osborne, H. Chen, R.H. Magruder, and R.A. Zuhr, In *Ultrafast Phenomena XI*; T. Elsaesser, J.G. Fujimoto, D.A. Wiersma, and W. Zinth Eds. (Springer-Verlag: Berlin, 1998) p. 356.
30. S. Link and M.A. El-Sayed, J. Phys. Chem. B **103**, 8410 (1999).
31. C. Voisin, D. Christofilos, N. Del Fatti, F. Vallée, B. Prével, E. Cottancin, J. Lermé, M. Pellarin, and M. Broyer, Phys. Rev. Lett. **85**, 2200 (2000).
32. T. Tokizaki, A. Nakamura, S. Kavelo, K. Uchida, S. Omi, H. Tanji, and Y. Asahara, Appl. Phys. Lett. **65**, 941 (1994).
33. J.Y. Bigot, J.C. Merle, O. Cregut, and A. Daunois, Phys. Rev. Lett. **75**, 4702 (1995).
34. V. Halté, J.Y. Bigot, B. Palpant, M. Broyer, B. Prével, and A. Pérez, Appl. Phys. Lett. **75**, 3799 (1999).
35. A. Stella, M. Nisoli, S. De Silvestri, O. Svelto, G. Lanzani, P. Cheyssac, and R. Kofman, Phys. Rev. B **53**, 15497 (1996).
36. M. Nisoli, S. Stagira, S. De Silvestri, A. Stella, P. Tognini, P. Cheyssac, and R. Kofman, Phys. Rev. Lett. **78**, 3575 (1997).
37. J.H. Hodak, A. Henglein, and G.V. Hartland, J. Chem. Phys. **112**, 5942 (2000).
38. Y. Hamanaka, J. Kuwabata, I. Tanahashi, S. Omi, and A. Nakamura, Phys. Rev. B **63**, 104302 (2001).
39. A. Arbouet, C. Voisin, D. Christofilos, P. Langot, N. Del Fatti, F. Vallée, J. Lermé, G. Celep, E. Cottancin, M. Gaudry, M. Pellarin, M. Broyer, M. Maillard, M.P. Pileni, and M. Treguer, Phys. Rev. Lett. **90**, 177401 (2003).
40. M. Perner, P. Bost, U. Lemmer, G. von Plessen, J. Feldmann, U. Becker, M. Mennig, M. Schmitt, and H. Schmidt, Phys. Rev. Lett. **78**, 2192 (1997).
41. N. Del Fatti, C. Flytzanis, and F. Vallée, Appl. Phys. B **68**, 433 (1999).
42. N. Del Fatti, F. Vallée, C. Flytzanis, Y. Hamanaka, and A. Nakamura, Chem. Phys. **251**, 215 (2000).
43. M. Fierz, K. Siegmann, M. Scharte, and M. Aeschlimann, Appl. Phys. B **68**, 415 (1999).
44. J. Lehmann, M. Merschdorf, W. Pfeiffer, S. Voll, and G. Gerber, Phys. Rev. Lett. **85**, 2921 (2000).
45. M. Merschdorf, C. Kennerknecht, K. Willig, and W. Pfeiffer, New J. Phys. **4**, 95.1 (2002).
46. J.H. Hodak, I. Martini, and G.V. Hartland, J. Phys. Chem. B **102**, 6958 (1998).
47. J.H. Hodak, A. Henglein, and G.V. Hartland, J. Chem. Phys. **111**, 8613 (1999).
48. C. Voisin, N. Del Fatti, D. Christofilos, and F. Vallée, Appl. Surf. Science **164**, 131 (2000).

49. M. Nisoli, S. De Silvestri, A. Cavalleri, A.M. Malvezzi, A. Stella, G. Lanzani, P. Cheyssac, and R. Kofman, Phys. Rev. B **55**, R13424 (1997).
50. N. Del Fatti, C. Voisin, F. Chevy, F. Vallée, and C. Flytzanis, J. Chem. Phys. **110**, 11484 (1999).
51. N. Del Fatti, C. Voisin, D. Christofilos, F. Vallée, and C. Flytzanis, J. Phys. Chem. A **104**, 4321 (2000).
52. N.W. Ashcroft and N.D. Mermin, *Solid State Physics* (Holt-Saunders, Tokyo, 1981).
53. M. Kaveh and N. Wiser, Adv. in Phys. **33**, 257 (1984).
54. V.A. Gasparov and R. Huguenin, Adv. in Phys. **42**, 393 (1993).
55. H. Hovel, S. Fritz, A. Hilger, U. Kreibig, and M. Vollmer, Phys. Rev. B **48**, 18178 (1993).
56. P.B. Johnson, R.W. Christy, Phys. Rev. B **6**, 4370 (1972).
57. J.B. Smith and H. Ehrenreich, Phys. Rev. B **25** 923 (1982).
58. U. Kreibig and M. Vollmer, *Optical Properties of Metal Clusters* (Springer, Berlin 1995).
59. F. Vallée, N. Del Fatti, and C. Flytzanis, In *Nanostructured Materials*; V.M. Shalaev, and M. Moskovits Eds. (American Chemical Society: Washington, 1997) p. 70.
60. K. Uchida, S. Kaneko, S. Omi, C. Hata, H. Tanji, Y. Asahara, A.J. Ikushima, T. Tokisaki, and A. Nakamura, J. Opt. Soc. Am B **11**, 1236 (1994).
61. B. Palpant, B. Prével, J. Lermé, E. Cottancin, M. Pellarin, M. Treilleux, A. Perez, J.L. Vialle, and M. Broyer, Phys. Rev. B **57**, 1963 (1998).
62. A. Taleb, C. Petit, and M.P. Pileni, Chem. Mater. **9**, 950 (1997).
63. M. Tréguer, C. de Cointet, H. Remita, J. Khatouri, J. Amblard, J. Belloni, and R. de Keyzer, J. Phys. Chem. B, **102**, 4310 (1998).
64. D.L. Mills and E. Burstein, Rep. Prog. Phys. **37**, 817 (1974).
65. D. Pines and P. Nozières, *The Theory of Quantum Liquids* (Benjamin, New-York, 1966).
66. R. Rosei, Phys. Rev. B **10**, 474 (1974).
67. R. Rosei, F. Antonangeli, and U.M. Grassano, Surf. Sci. **37**, 689 (1973).
68. R.H.M. Groeneveld, R. Sprik, and Ad. Lagendijk, Phys. Rev. Lett. **64**, 784 (1990).
69. G. Tas and H.J. Maris, Phys. Rev. B **49**, 15046 (1994).
70. S. Link, C. Burda, Z.L. Wang, M.A. El-Sayed, J. Chem. Phys. **111**, 1255 (1999).
71. W. Ekardt, Phys. Rev. B **29**, 1558 (1984).
72. C. Bréchignac, P. Cahuzac, J. Leygnier, and A. Sarfati, Phys. Rev. Lett. **70**, 2036 (1993).
73. A. Liebsch, Phys. Rev. B **48**, 11317 (1993).
74. J. Lermé, B. Palpant, B. Prével, M. Pellarin, M. Treilleux, J.L. Vialle, A. Perez, and M. Broyer, Phys. Rev. Lett. **80**, 5105 (1998).
75. L. Bürgi, O. Jeandupeux, H. Brune, and K. Kern, Phys. Rev. Lett. **82**, 4516 (1999).
76. M.I. Kaganov, I.M. Lifshitz, and L.V. Tanatarov, Zh. Eksp. Teor. Fiz. **31**, 232 (1957) [Sov. Phys. JETP **4**, 173 (1957)].
77. B.A. Smith, J.Z. Zhang, U. Giebel, and G. Schmid, Chem. Phys. Lett. **270**, 139 (1997).
78. E.D. Belotskii and P.M. Tomchuk, Surf. Science **239**, 143 (1990).
79. E.D. Belotskii and P.M. Tomchuk, Int. J. Electr. **73**, 955 (1992).

80. J.Z. Zhang, Acc. Chem. Res. **30**, 423 (1997).
81. T. Takagahara, J. Lum. **70**, 129 (1996).
82. A. Eiguren, B. Hellsing, F. Reinert, G. Nicolay, E.V. Chulkov, V.M. Silkin, S. Hufner, and P.M. Echenique, Phys. Rev. Lett. **88**, 066805 (2002).
83. E. Beaurepaire, J.C. Merle, A. Daunois, and J.Y. Bigot, Phys. Rev. Lett. **76**, 4250 (1996).
84. J. Hohlfeld, E. Matthias, R. Knorren, and K.H. Bennemann, Phys. Rev. Lett. **78**, 4861 (1997).
85. G. Ju, A. Vertikov, A.V. Nurmikko, C. Canady, G. Xiao, R.F.C. Farrow, and A. Cebollada, Phys. Rev. B **57**, R700 (1998).

Origin of the Pseudogap in High Temperature Superconductors

Richard A. Klemm

Abstract. The physical properties of hole-doped high temperature superconductors (HTCS) are characterized by a 'normal' state (NS) for temperatures $T > T^*$, and a pseudogap (PG) in the electronic spectrum, for $T^* > T > T_c$. *Strikingly* similar behavior occurs in the charge-density wave (CDW) state, $T_0 > T > T_c$, in the transition metal dichalcogenides (TMD) $2H$-MX_2, where M = Ta, Nb, and X = S, Se, both in the NS ($T > T_0$) and in the incommensurate charge-density wave ($T_{ICDW} > T > T_c$) states. Such strikingly similar behavior has also been seen in the organic layered superconductors (OLS) κ-$(ET)_2X$, where ET is bis(ethylenedithio)tetrathiafulvalene, and X = $Cu[N(CN)_2]Cl$, $Cu[N(CN)_2]Br$, and $Cu(SCN)_2$, both in the NS, $T > T_{SDW} > T_c$, and in the spin-density wave regime, $T_{SDW} > T > T_c$. In all three materials classes, the anomalous transport and thermodynamic properties associated with the pseudogap or density-wave regime are completely independent of the applied magnetic field strength, whereas the same properties below T_c are all strongly field dependent. Hence, we propose that the pseudogap in the HTSC arises from charge- and/or spin-density waves, but not from either superconducting fluctuations or charged quasiparticle "preformed pairs".

1 Introduction

The mechanism for superconductivity in the high temperature superconductors (HTSC) is a continuing topic of great debate. Most workers consider the important keys to unlocking the secrets of the mechanism for superconductivity to be found either by determining the symmetry of the order parameter (OP), or by understanding the nature of the region in temperature T above the superconducting transition temperature T_c, which has usually been known as the 'normal', or non-superconducting state. Although many experiments have been interpreted as giving evidence for an OP of $d_{x^2-y^2}$ wave symmetry in the HTSC [1], a number of more recent experiments supported an OP with mixed s- and $d_{x^2-y^2}$-wave symmetry, and some supported an OP or either pure or at least predominant s-wave symmetry [2–4]. A discussion of most of the relevant experiments has been given recently [5]. It was even proposed that the HTSC might be s-wave in the bulk and $d_{x^2-y^2}$-wave on the surfaces [6]. Thus, this topic is highly controversial, and there is really no consensus on this question.

I therefore turn to the other proposed technique for probing the mechanism for superconductivity in the HTSC: the study of the 'normal' state. In this work I shall only consider the hole-doped HTSC. As we shall see, the region above T_c is actually characterized by two or three regions, and is by no means 'normal' in the usual sense of the word. It very likely involves both an unusual normal state (NS) at high T, and a second (and possibly a third) phase in the intermediate regime between the NS and T_c. The transport and thermodynamic behavior of the NS is rather different from what is usually found above T_c in conventional superconductors. In addition, depending somewhat upon the hole doping x, at least part of this intermediate region in T is characterized by a pseudogap (PG) in the quasiparticle energy spectrum. This PG region depends upon the doping x in the HTSC, but in most hole-doped systems appears at $T^* > T_c$ over most, if not all, of the hole-doped phase diagram for which $T_c > 0$. This PG is evidenced not only in angle-resolved photoemission spectroscopy (ARPES) experiments [7], but also in optical reflectivity [8], nuclear magnetic resonance (NMR) [9–11], transport [12–14], Knight shift [12], thermodynamic measurements [15,16], and neutron scattering experiments [17,18].

In most experiments, the PG features become prominent below one or two crossover PG temperatures T_0 and T^*, where generally $T_0 > T^* > T_c$. A qualitative $T(x)$ phase diagram of this HTSC characterization, based upon the latest experimental evidence, is shown in Fig. 1. In this diagram, the region of the PG is largest in the 'underdoped' part of the phase diagram [to the left of the maximum in $T_c(x)$]. It is still a matter of considerable debate as to whether this PG is a precursor to or is additional to the superconducting gap in the electronic spectrum below T_c.

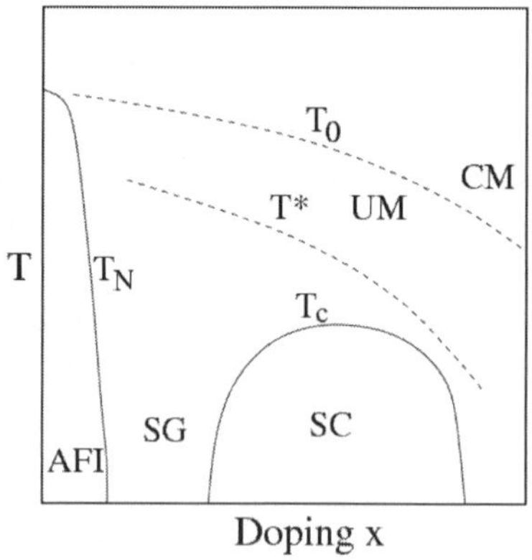

Fig. 1. Schematic T vs. doping concentration x phase diagram of the hole-doped HTSC, indicating the Neél temperature T_N, the superconducting transition T_c, and the pseudogap onset T_0 and secondary temperature T^*. The acronyms SC, SG, AFI, CM, and UM represent superconducting, spin glass, antiferromagnetic insulator, conventional metal, and unconventional metal, respectively

There is substantial experimental evidence, both direct and circumstantial, that this intermediate T regime, within part of which the PG appears, is characterized by some sort of charge- (CDW) or spin-density wave (SDW), each of which has an origin that is unrelated to superconductivity. Hence, I maintain that an understanding of the 'normal' state still does not settle the issue of the mechanism for superconductivity in the HTSC. In investigating this question, however, we shall see that there was a lot of very interesting work done long ago on two classes of compounds, both of which are most likely s-wave superconductors, which have very similar PG states, one of which was known to arise from a CDW, the other from a SDW.

2 Two Gaps Seen in Tunneling Experiments

Recently, a number of groups have presented break junction, point contact, and especially scanning tunneling microscopy (STM) measurements of the density of states for different dopings of pristine and $HgBr_2$-intercalated $Bi_2Sr_2CaCu_2O_{8+\delta}$ (Bi2212) and $(HgBr_2)$Bi2212, respectively [19–25]. These data indicate a PG well above T_c, which has a T dependence *distinct* from the superconducting gap (SG). Regardless of the stoichiometry, there are two different gaps, with different T and magnetic field **H** dependences. From Fig. 2, one can readily see that the T dependence of the gap in the electronic spectrum reveals strong evidence of two gaps. In particular, the right panel of Fig. 2 indicates that the SG vanishes at T_c in overdoped Bi2212, whereas the other gap, the PG, is rather T-independent. In Fig. 3, the V-shaped PG is shown to be field independent, but the quasiparticle peaks characteristic of the SG vanish at the upper critical field $H_{c2,\perp}(T)$ for $\mathbf{H}||\mathbf{c}$ at this T.

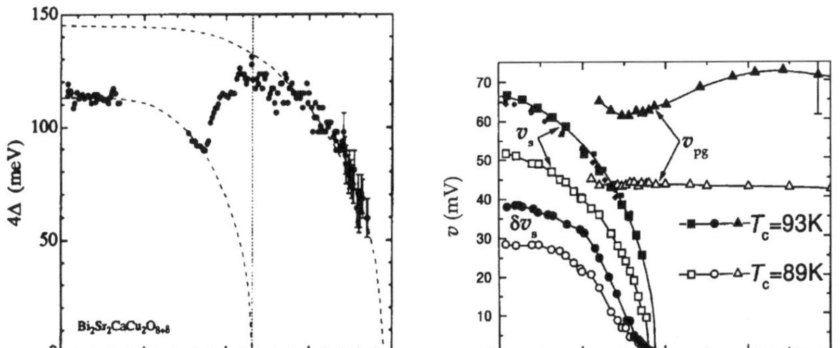

Fig. 2. *Left*: break junction determination of the total gap amplitude in Bi2212, from Ekino *et al.* [21] *Right*: Mesa tunneling determination of the temperature dependence of the two gaps in overdoped Bi2212, from Krasnov *et al.* [22]

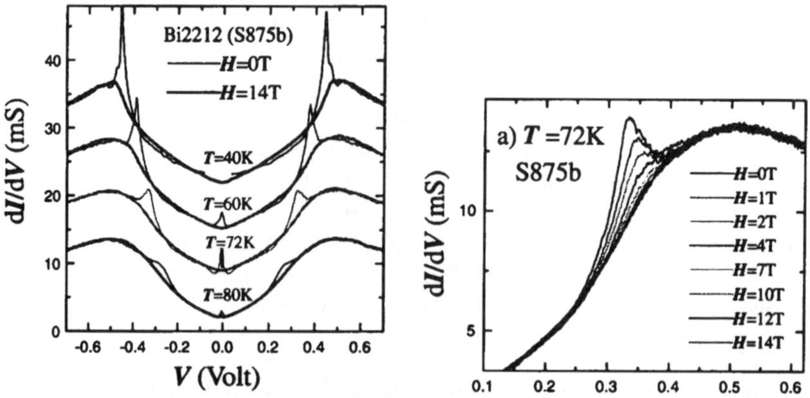

Fig. 3. *Left*: Mesa determination of the temperature and field dependences of the two gaps in overdoped Bi2212. *Right*: Field dependence of the density of states for one overdoped Bi2212 sample, from Krasnov *et al.* [22]

3 Similarities with the TMD and OLS

Our main point is that behaviors in *both* the NS and PG regimes of the HTSC are *strikingly* similar to those seen in the $2H$ polytypes of the transition metal dichalcogenides (TMD), $2H$-MX_2, where M = Nb, Ta and X = S, Se [26], and in the κ-$(ET)_2X$ organic layered superconductors (OLS), where ET is bis(ethylenedithio)tetrathiafulvalene, and X = $Cu(SCN)_2$, $Cu[N(CN)_2]Br$, and $Cu[N(CN)_2]Cl$ [27], the building blocks of which are pictured in Fig. 4.

In the OLS, both these NS and PG-like behaviors are observed above and below their respective SDW transition temperatures, T_{SDW}, indicated in the right panel of Fig. 5 [28]. In the TMD, these same behaviors are seen above and below their respective incommensurate CDW transition temperatures, T_{ICDW}, indicated in the left panel of Fig. 5. Although little is known from experiments about the details of the electronic structure of the OLS, both the TMD and the HTSC have an NS exhibiting *saddle bands*. These saddle bands

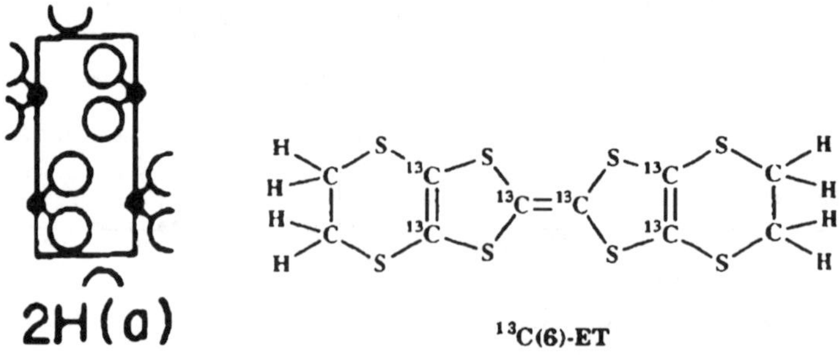

Fig. 4. Basic building blocks of the TMD (*left*) and of the OLS, ET (*right*)

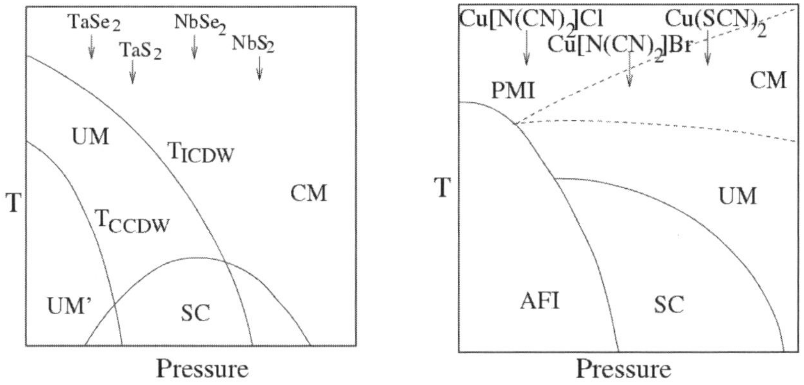

Fig. 5. Schematic T versus internal pressure phase diagrams of the TMD (*left*) and the OLS (*right*). The regions of different physical behavior are denoted conventional metal (CM), unconventional metal (UM, UM'), paramagnetic insulator (PMI), antiferromagnetic insulator (AFI), and superconductor (SC)

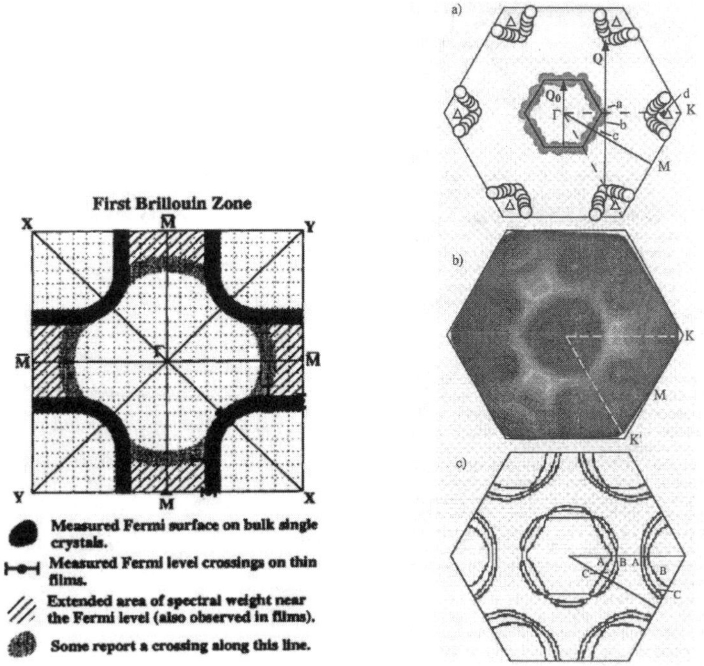

Fig. 6. *Left*: Measured Bi2212 Fermi surface obtained from thin films [30]. The *saddle bands* are the hatched regions near the $\overline{M}$ points. *Right*: Data taken on $2H$-$TaSe_2$ by Liu *et al.*[29] The *saddle bands* in the central figure are the extended regions of scattering intensity around the Fermi surfaces centered at the K points

are extended, flat regions of electronic states with saddle-like dispersions lying very near to the Fermi energy E_F, that dominate the physical behavior in the NS at high T. The CDW's in the TMD and the PG in the HTSC arise from instabilities on or near to these saddle bands, pictured in Fig. 6 [29,30]. The PG features in the quasiparticle density of states are clearly observed in both the HTSC and TMD compounds, as shown in Fig. 7.

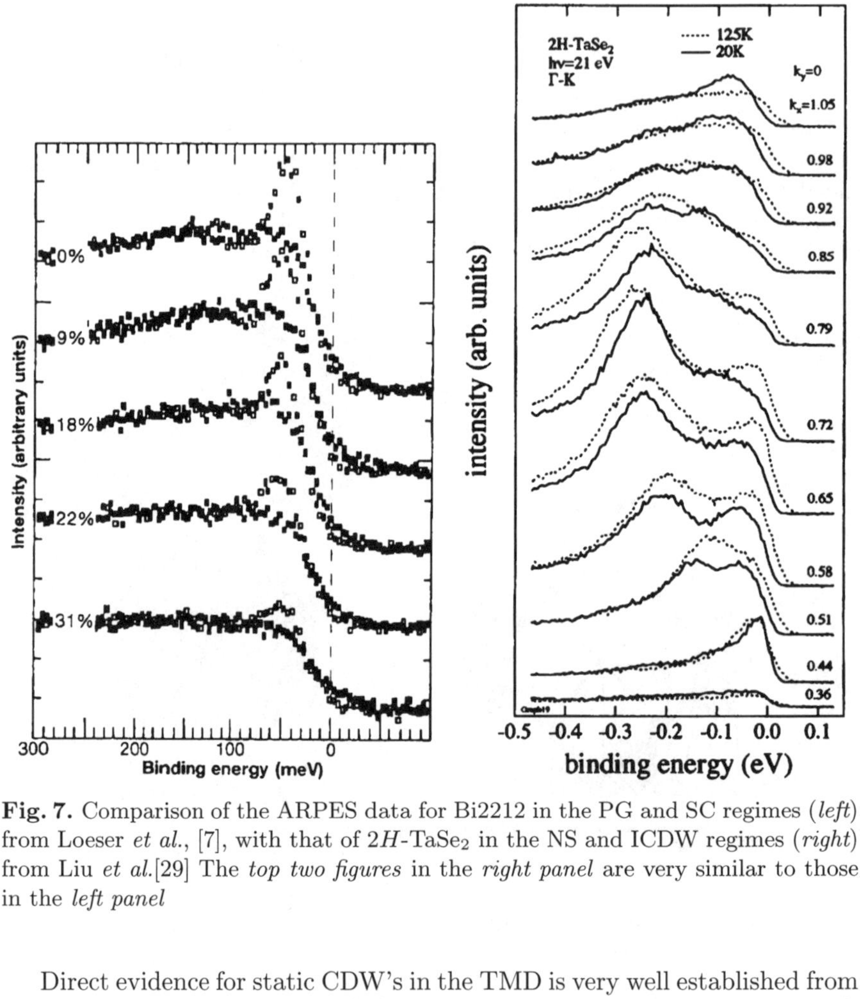

Fig. 7. Comparison of the ARPES data for Bi2212 in the PG and SC regimes (*left*) from Loeser *et al.*, [7], with that of 2*H*-$TaSe_2$ in the NS and ICDW regimes (*right*) from Liu *et al.*[29] The *top two figures* in the *right panel* are very similar to those in the *left panel*

Direct evidence for static CDW's in the TMD is very well established from x-ray and electron diffraction [26], neutron diffraction [31], and STM [32]. The neutron diffraction evidence is shown in Fig. 8. The second-order transitions at T_{ICDW} have large λ-like specific heat anomalies (except in 2*H*-$NbSe_2$, where it is BCS-like), with noticeable transport and PG features just below it, but the first-order transition at T_{CCDW} in 2*H*-$TaSe_2$ exhibits a small, but hysteretic specific heat anomaly, with very subtle transport

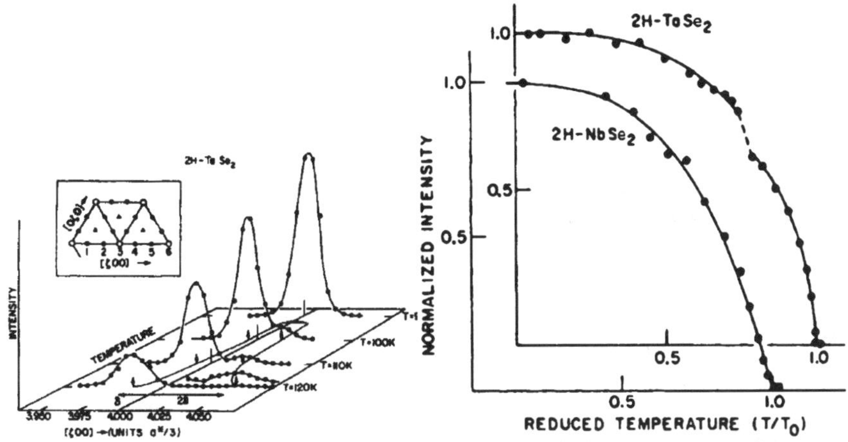

Fig. 8. *Left*: Neutron diffraction evidence for the commensurate CDW in $2H$-$TaSe_2$. *Right panel*: Intensity v. T data for $2H$-$TaSe_2$ and $2H$-$NbSe_2$ [31]

changes [33,34]. This is pictured in Fig. 9, along with similar data for the HTSC, $YBa_2Cu_3O_{7-\delta}$ (YBCO) [15].

Similar direct evidence for long-range antiferromagnetic order in the OLS κ-$(ET)_2Cu[N(CN)_2]Cl$ was seen in magnetization M experiments [35], and evidence for SDW order in the OLS κ-$(ET)_2Cu(SCN)_2$ and κ-$(ET)_2Cu[N(CN)_2]Br$ was observed in the NMR spin-lattice relaxation rate, $1/T_1$ [36], as shown in Fig. 10. Those authors also observed the SDW onset in the zero-field magnetic susceptibility $\chi_\perp(T)$ in the OLS [36].

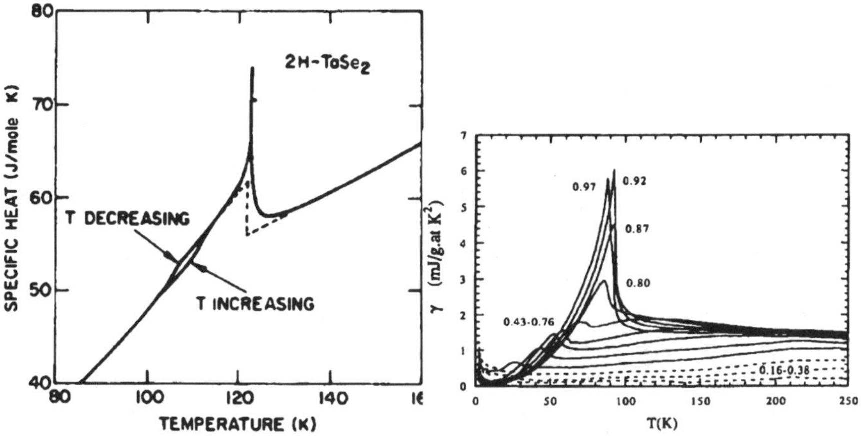

Fig. 9. *Left*: Specific heat data near the CDW transitions in $2H$-$TaSe_2$ [33]. *Right*: Thermodynamic data for $YBa_2Cu_3O_{7-\delta}$ [15]. The doping levels δ are indicated

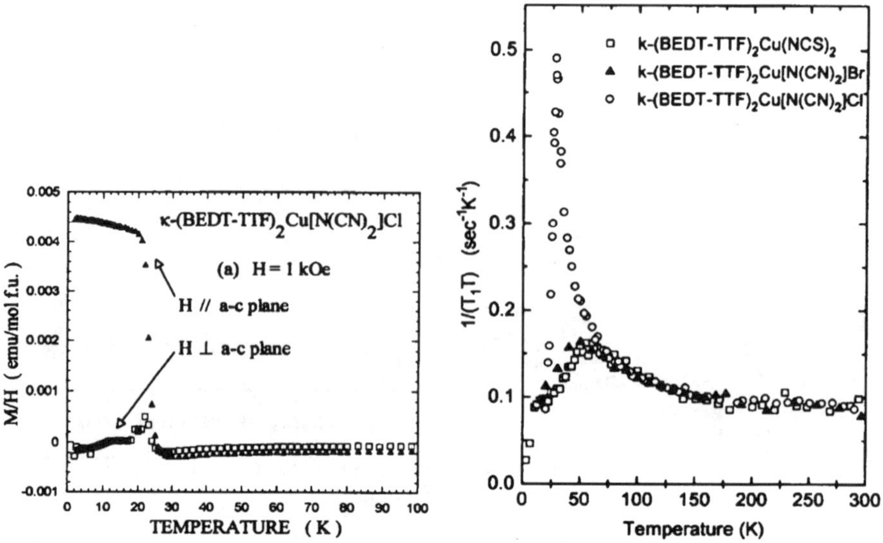

Fig. 10. *Left*: Magnetization evidence that κ-$(ET)_2Cu[N(CN)_2]Cl$ at ambient applied pressure is an antiferromagnet [36]. *Right*: ^{13}C NMR $1/(T_1T)$ versus T data for the OLS, demonstrating the SDW for $T_c < T < T^* \approx 20$–$50\,K$, and the rather constant, Korringa behavior of an ordinary metal in the NS for $T > T^*$ [36]

4 Zero-Field Resistivities

The resistivities of layered superconductors are generically given by an in-plane resistivity $\rho_{||}(T)$ that is rather linear in T in the NS, with a decrease in $\rho_{||}(T)$ below the onset of the PG (or the CDW/SDW), which has often been misinterpreted as evidence for 'ordinary metallic' behavior. In the perpendicular direction, the NS resistivity $\rho_\perp(T)$ is also rather linear in T, but in the PG or CDW/SDW regime, it generically exhibits behavior characteristic of an insulator. This behavior is pictured in Fig. 11. In the HTSC, one often sees a downturn in $\rho_{||}(T)$ at T_0, followed by a flattening of $\rho_{||}(T)$ at T^*, as sketched. In the OLS, it is often difficult to observe these features at $H = 0$, because of disorder effects in the intermediate T regime. However, the in-plane $\rho_{||}(T)$ at $H = 0$ is also linear in the regime $T_c < T < 2T_c$.

In the HTCS, the generic behavior is most easily seen in the more anisotropic materials, as pictured for Bi2212 in Fig. 12. $\rho_{||}(T)$ and $\rho_\perp(T)$ are shown, respectively, in the left and right panels. The effect of the PG on $\rho_{||}(T)$ is evident by the decrease below the points indicated as T^* in the figure, which corresponds to our T_0. $\rho_\perp(T)$ has a minimum at this T.

In the TMD, this behavior is very dramatic, as pictured in Fig. 13 for $2H$-TaS_2 [37]. Note the sharp behavior of the onset of the CDW at $T = 75\,K$. As T decreases below $T_{ICDW} = 75\,K$, $\rho_{||}(T)$ decreases below a small bump (that is often smeared by impurities), and $\rho_\perp(T)$ flattens out and

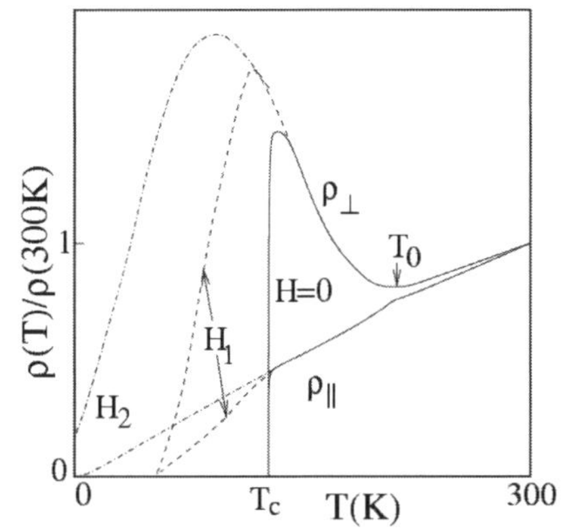

Fig. 11. Generic curves for the in-plane $\rho_{||}(T)$ and out-of-plane $\rho_{\perp}(T)$ resistivities for the HTSC, TMD, and OLS compounds, for $H = 0$ and for $H_2 > H_1 > 0$

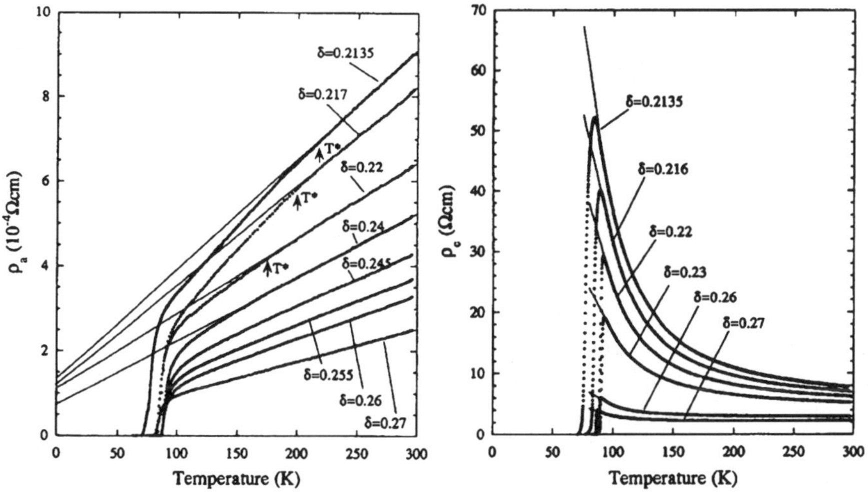

Fig. 12. In-plane (*left*) and *c*-axis (*right*) resistivities $\rho_{||}(T)$ and $\rho_{\perp}(T)$ of Bi2212, as a function of oxygen doping δ in zero field [14]

then rises for $T < T_{ICDW}$. Similar linear-in-T behavior for $\rho_{||}(T)$ has been seen out to 600 K in $2H$-$TaSe_2$, but $\rho_{\perp}(T)$ has not yet been measured in that material [38]. Thus, the behavior of $\rho_{||}(T)$ below T_{ICDW} is *not* that of a 'conventional' metal. Instead, it is an 'unconventional' metal, containing an incommensurate CDW with long-range order.

Upon intercalation with pyridine [39],$2H$-TaS_2 becomes the much more anisotropic $2H$-TaS_2(pyridine)$_{1/2}$, but the evidence for the CDW has disappeared, as demonstrated dramatically in the right panel of Fig. 13. In this figure, $\rho_{||}(T)/\rho_{||}$ (300) K and $\rho_{\perp}(T)/\rho_{\perp}(300)$ K are compared for typical quality, pristine $2H$-TaS_2 and intercalated $2H$-TaS_2(pyridine)$_{1/2}$ [40].

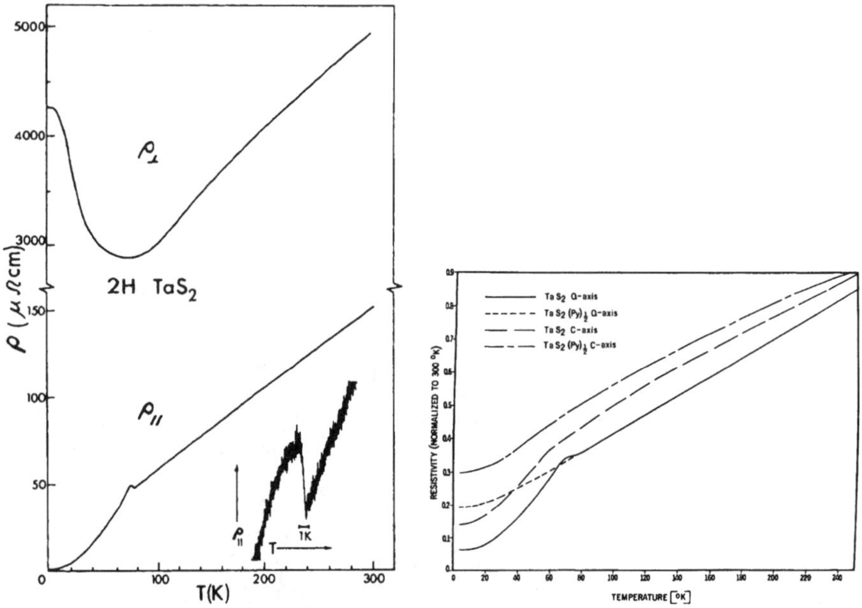

Fig. 13. *Left*: $\rho_{||}(T)$ and $\rho_{\perp}(T)$ for a high quality single crystal of $2H$-TaS_2, showing the sharp onset of the incommensurate CDW at $T_{ICDW} = 75\,K$. At the *lower right* is a blowup of $\rho_{||}(T)$ curve at T_{ICDW} [37]. *Right*: $\rho_{||}(T)$ and $\rho_{\perp}(T)$ of an ordinary quality $2H$-TaS_2 crystal before and after intercalation with pyridine [40]

However, intercalation of Bi2212 with $HgBr_2$ does not remove the PG, as shown in the right panel of Fig. 14 [23]. In this figure, the c-axis $\rho_{\perp}(T)$ in Bi2212 and two samples of $(HgBr_2)$Bi2212 are shown, and the results are very similar to those shown in the right panel of Fig. 12 [14,23]. We note that the resistivity anisotropy in $(HgBr_2)$Bi2212 is huge, as $\rho_{\perp}(175\,K) \approx 2\,k\Omega$ cm [23].

One material that appears at first sight not to follow these trends is optimally doped YBCO. In this case, the c-axis transport is coherent and metallic, and $\rho_{||}(T)$ only shows a slight rounding, rather than the generic features of the PG, as pictured in the left panel of Fig. 14. By comparing the data for $2H$-TaS_2 and optimally doped YBCO in Figs. 13 and 14, respectively, it is evident that $\rho_{||}(300\,K)$ and $\rho_{\perp}(300\,K)$ are nearly the same in these materials, except for the approximate factor of 2 in-plane anisotropy of $\rho_{||}(T)$ in YBCO. This is indicated in Table 1.

Thus, the TMD and the HTSC have very similar zero-field resistivities in the NS. Intercalation of the TMD removes the CDW, but intercalation of the HTS does not remove the PG. In both cases, intercalation greatly increases the resistivity anisotropy. Although the TMD have in-plane resistivities $\rho_{||}(T)$ in the NS, $T > T_{ICDW}$, that are at least as linear in T as are those of the HTSC for $T > T_0$, the $\rho_{||}(T)$ of these TMD in the CDW regime $T_{ICDW} > T > T_c$ are also strikingly similar to those of the HTSC in the PG regime

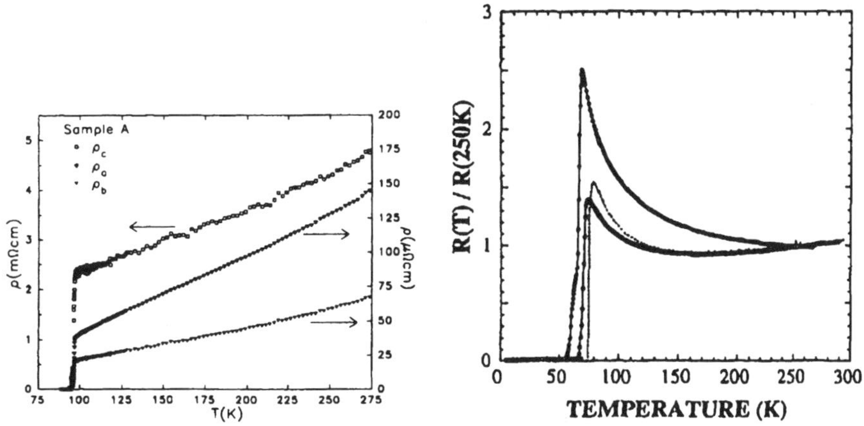

Fig. 14. *Left*: Resistivities in all three directions a, b, c for optimally doped YBCO [41]. *Right*: c-axis $\rho_\perp(T)$ of overdoped Bi2212 (*central curve*) and two samples of $(HgBr_2)$Bi2212 [23]

Table 1. Room temperature $\rho_{||}$ and $\rho_\perp$, magnetic susceptibility for $\mathbf{H}||\mathbf{c}$, and Hall constant R_H for the $\mathbf{I} \perp \mathbf{H}||\mathbf{c}$, in the NS of three layered superconductors. The references are: (a): [34], (b): [41], (c): [16], (d): [26], (e): [13], and (f): [37]

Compounds	$\rho_{\vert\vert}(300\,\mathrm{K})$	$\rho_\perp(300\,\mathrm{K})$	$\chi_\perp(300\,\mathrm{K})$	$R_{H,\perp}(300\,\mathrm{K})$
	$10^{-4}\,\Omega\,\mathrm{cm}$	$10^{-3}\,\Omega\,\mathrm{cm}$	$10^{-7}\,\mathrm{cm}^3/\mathrm{g}$	$10^{-4}\,\mathrm{cm}^3/\mathrm{C}$
$2H$-TaS_2	$1.2^{(a)}$, $1.5^{(f)}$	$4.8^{(f)}$	$5.7^{(d)}$	$2.2^{(a)}$
$2H$-$TaSe_2$	$1.3^{(a)}$		$4.1^{(d)}$	$1.8^{(a)}$
$YBa_2Cu_3O_7$	$1.46^{(b)}$, $0.68^{(b)}$	$4.8^{(b)}$	$4.2^{(c)}$	$4.0^{(e)}$

$T_0 > T > T_c$, appearing more 'metallic' there than in the NS. In pure samples, the c-axis resistivity $\rho_\perp(T)$ of the TMD, while quasi-linear above T_{ICDW}, *increases* with decreasing T for $T < T_{ICDW}$ [37], qualitatively similar to that of Bi2212 for all dopings that have been studied. Furthermore, the values of $\rho_{||}(300\,\mathrm{K})$ and $\rho_\perp(300\,\mathrm{K})$ are almost exactly the *same* as for the optimally doped HTSC, YBCO [37,41]. However, the PG in optimally doped YBCO is not obviously discernible from the resistivity data shown in Fig. 14. The PG is more evident in underdoped YBCO [42]. The similarities of the TMD and the HTSC YBCO are also present in the Hall constant $R_{H\perp}(T)$ for current $\mathbf{I} \perp \mathbf{c}||\mathbf{H}$, and in the Seebeck coefficient [13,34,36,43].

5 Field-Dependence of the Resistivities

We now turn to the magnetic field dependence of the resistivities of the three classes of layered superconductors under discussion. Here we are mainly interested in the most field-sensitive case of the field perpendicular to the layers. The generic behavior is shown in Fig. 11. In Fig. 15, we show the field dependence of the in-plane resistivity for both field directions in highly underdoped YBCO [42], and for both $\rho_{||}(T)$ and $\rho_{\perp}(T)$ in a field for a sample of Bi2212 [44], the behavior of which is very similar to that obtained for overdoped Bi2212 [22,23]. We note that $\rho_{||}(H,T)$ for the TMD 2H-$NbSe_2$ is qualitatively similar to that shown in the upper left panel of Fig. 15 for the $\rho_{||}(H,T)$ of underdoped YBCO [42].

In Fig. 16, the field dependences for $\mathbf{H}||\mathbf{c}$ of the resistivities $\rho_{\perp}(T)$ and $\rho_{||}(T)$ in the OLS κ-$(ET)_2Cu(SCN)_2$ are shown [45]. There are striking resemblances to the field dependences of the resistivities of Bi2212, shown in the right panel of Fig. 15.

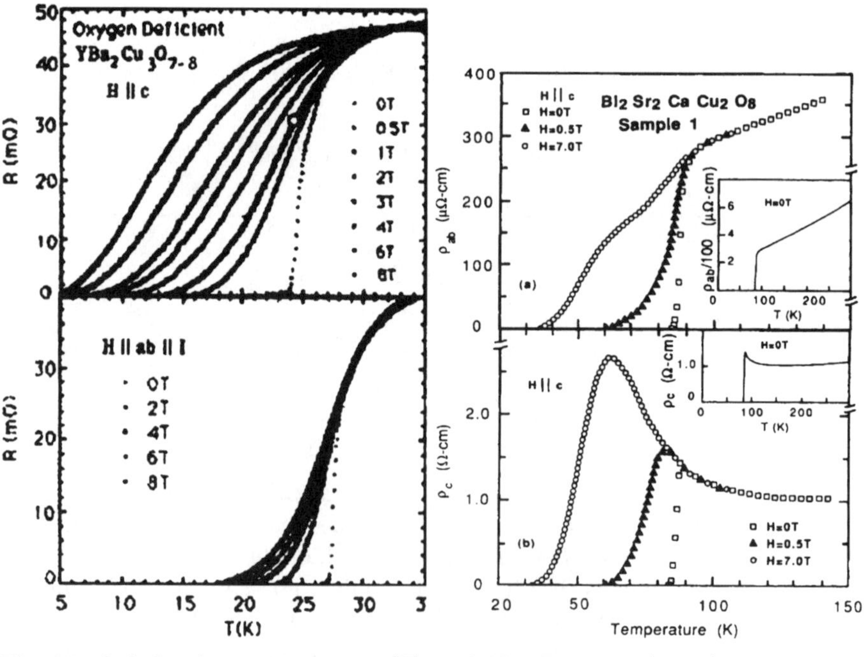

Fig. 15. *Left*: In-plane resistivity $\rho_{||}(T)$ in fields $\mathbf{H}$ normal (*upper*) and parallel (*lower*) to the layers of highly underdoped YBCO, with $T_{c0} \approx 28\,\mathrm{K}$ [42]. *Right*: Resistivities $\rho_{||}(T)$ (*upper*) and $\rho_{\perp}(T)$ (*lower*) in fields $\mathbf{H}||\mathbf{c}$ in Bi2212 [44]

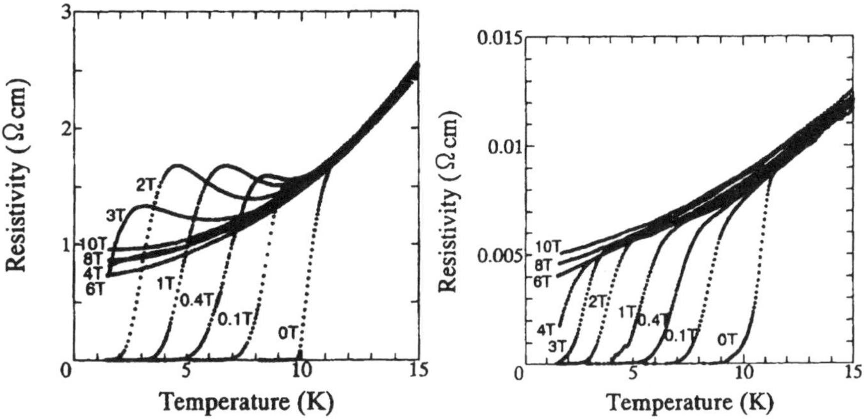

Fig. 16. Resistivities $\rho_\perp(T)$ (*left*) and $\rho_{||}(T)$ (*right*) in fields normal to the layers in the OLS κ-$(ET)_2Cu(SCN)_2$ [45]

Thus, for all three classes of layered superconductors under discussion, a field **H** normal to the layers has a dramatic effect on the resistivities. For $\rho_{||}(T)$, an increasing field strength increasingly spreads out the transition. As for $\rho_\perp(T)$, **H** normal to the layers also spreads out the transition, but the increase in $\rho_\perp(T)$ with decreasing T continues into the region below the zero-field transition T_{c0}. However, there is no discernible change in $\rho_\perp(T)$ with field *above* T_{c0}. Thus, no obvious field dependences of the resistivities were observed above T_c, *even* in the cases when the PG onset is far above T_c, as for the underdoped YBCO sample shown in the left panel of Fig. 15. This field independence of the resistivities above T_c is also found for the TMD

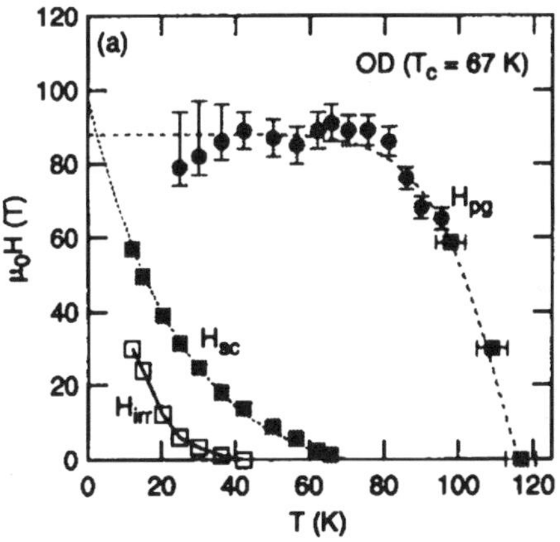

Fig. 17. Phase diagram in very strong d.c. and pulsed magnetic fields of overdoped Bi2212 [46]

and OLS, for which the region above T_c is characterized by a CDW or SDW, respectively. Thus generically, the field greatly affects the resistivity below T_{c0}, but not in the PG or CDW/SDW region above T_c.

However, in a recent experiment in *extremely* strong constant and pulsed fields, the field dependence of the PG regime for three differently-doped Bi2212 samples was measured [46]. The results for an overdoped sample are shown in Fig. 17. Unlike the SC state, which is destroyed by the orbital pair-breaking, the PG is destroyed by the Zeeman energy, or when $g\mu_B H = k_B T_c$, *precisely* what would be expected if the PG were constructed of chargeless, spinless, particle-hole pairs [47].

6 Magnetic Susceptibilities

We now turn to the magnetic susceptibilities $\chi_\perp(T)$ for **H** normal to the layers in the three classes of compounds. This is important, because it has generally been assumed that a decrease in $\chi_\perp(T)$ with decreasing T signals the onset of a 'spin gap'. However, such a decrease is generic to all three classes of materials, although the TMD are known to have only a CDW, not a SDW. The generic behavior for $\chi_\perp(T)$ at $H = 0$ and for $\mathbf{H}||\mathbf{c} \neq 0$ is shown in Fig. 18. Although the H dependence of $\chi_\perp(T)$ in many of these compounds has been measured, we only summarize the data in this generic figure, and concentrate on $\chi_\perp(T)$ for $H = 0$.

As seen in Fig. 18, $\chi_\perp(T)$ for $H = 0$ generically decreases at T_0, and then flattens out at T^*, similar to $\rho_{||}(T)$. In a field, the only discernible changes are for $T < T_{c0}$, so that the behaviors in the PG, CDW, or SDW regimes are unaffected. In Fig. 19, we show the HTSC $\chi_\perp(T)$ as a function of doping for $La_{2-x}Sr_xCuO_{4+y}$ and for YBCO, respectively. The generic zero-

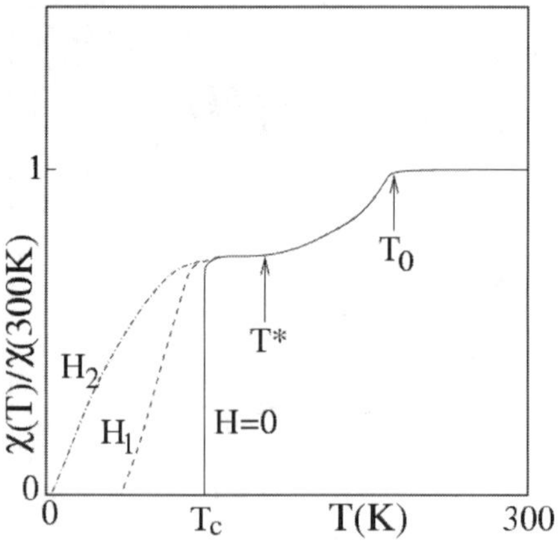

Fig. 18. Generic magnetic susceptibility $\chi_\perp(T)$ for $\mathbf{H}||\mathbf{c}$, both for $H = 0$ and for $H_2 > H_1 > 0$

field behavior is easy to see. As the HTSC become increasing underdoped, the inflection points T^* and T_0 increase, compatible with Fig. 1.

In Fig. 20, we show the $\chi_\perp(T)$ for two OLS and two TMD, respectively. The OLS show smooth behavior, similar to that of the HTSC, but the TMD pictured show sharp decreases at T_{ICDW}.

In Fig. 21, we show the effect of intercalation upon $\chi_\perp(T)$ for the TMD. In this figure, the curves for a typical $2H$-TaS_2 sample and for a sample of intercalated $2H$-TaS_2(pyridine)$_{1/2}$ are shown. Clearly, intercalation removes the effect of the CDW upon $\chi_\perp(T)$, as well as upon $\rho_{||}(T)$.

Thus, all three of these classes of layered superconductors have very similar $\chi_\perp(T)$, even though the TMD have not been shown to have any significant magnetic excitations. Hence, the behavior pictured in Fig. 18 can arise from *either* a 'spin gap' or a 'charge gap'.

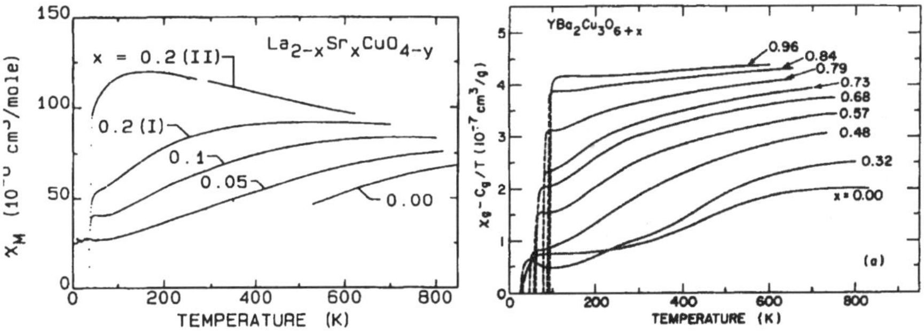

Fig. 19. Magnetic susceptibillities $\chi_\perp(T)$ at $H = 0$ as a function of doping for the HTSC $La_{2-x}Sr_xCuO_{4+y}$ (*left*) and YBCO (*right*) [16]

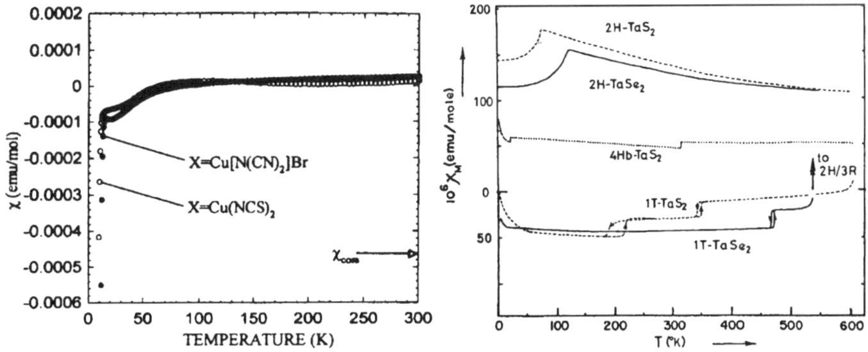

Fig. 20. Magnetic susceptibillities $\chi_\perp(T)$ at $H = 0$ for the OLS κ-$(ET)_2X$ for $X = Cu[N(CN)_2Br$ and $Cu(SCN)_2$ (*left*), and for the TMD $2H$-TaS_2 and $2H$-$TaSe_2$. Also pictured are $\chi_\perp(T)$ curves for three other dichalcogenide polytypes [26]

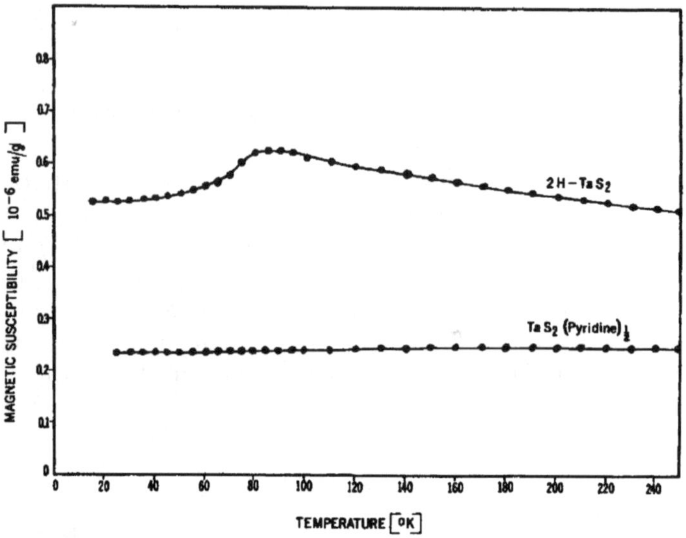

Fig. 21. $\chi_\perp(T)$ for unintercalated $2H$-TaS_2 (*upper curve*) and intercalated $2H$-$TaS_2(\text{pyridine})_{1/2}$ (*lower curve*) [40]

7 NMR Spin-Lattice Relaxation Rate

In Fig. 22, we show the generic behavior of the spin-lattice relaxation rate $1/T_1$ measured in NMR experiments on the HTSC and the OLS for **H** normal to the layers. There have been insufficient studies of the H dependences of $1/T_1$ in the TMD to date [48].

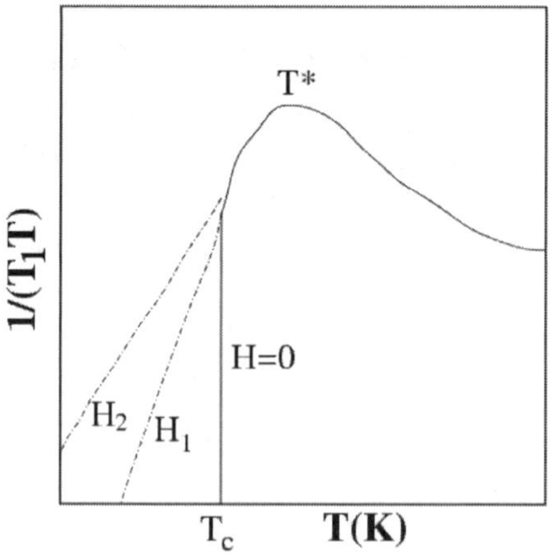

Fig. 22. Generic plot of the spin-lattice relaxation rate $1/T_1$, relative to T versus T of the HTSC and OLS, both for small fields ($H = 0$) and for $H_2 > H_1 > 0$

In the left panel of Fig. 22, we show the behavior of $1/(T_1T)$ versus T in fields up to 23.2 T in the stoichiometric HTSC $YBa_2Cu_4O_8$ [9]. As seen from this figure, there are strong changes below T_{c0} in increasing field strengths, but no discernible changes above T_{c0}. Zheng *et al.* have since increased the (static, not pulsed) field strength up to 42 T, and again show increased changes below T_{c0} but none above T_{c0} [9]. Similar behavior in fields up to 14.8 T has been seen in YBCO, as shown in the right panel of Fig. 23 [10].

In Fig. 24, we show the zero-field $1/(T_1T)$ for two OLS, and the normal **H** dependence of $1/(T_1T)$ in one of them, κ-$(ET)_2Cu[N(CN)_2]Br$. Although

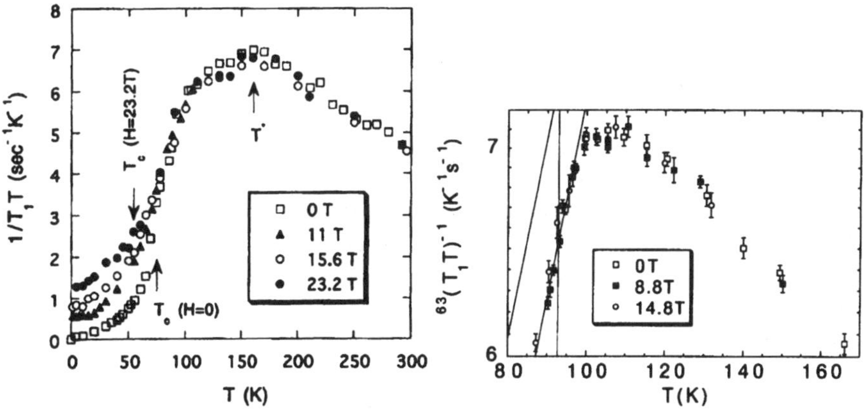

Fig. 23. NMR measurements of $1/(T_1T)$ versus T in HTSC as a function of $\mathbf{H}||\mathbf{c}$. *Left*: Behavior in $YBa_2Cu_4O_8$ in fields up to 23.2 T [9]. *Right*: Behavior in YBCO in fields up to 14.8 T [10]

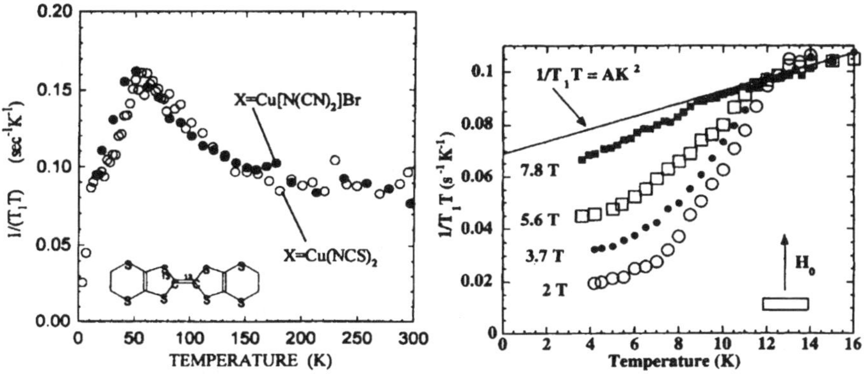

Fig. 24. NMR behavior of $1/(T_1T)$ versus T in OLS. *Left*: Zero-field behavior of κ-$(ET)_2X$ for X = $Cu[N(CN)_2]Br$ and $Cu(SCN)_2$, showing the onset of the SDW at $T_{SDW} \approx 50\,K$ [36]. *Right*: Field dependence of $1/(T_1T)$ versus T in κ-$(ET)_2Cu[N(CN)_2]Br$ from Mayaffre *et al.* [49]

there is a strong field dependence to $1/(T_1T)$ for $T < T_{c0}$, there do not appear to be any changes with fields up to 7.8 T above T_{c0} [49].

Thus, the field dependences of $1/(T_1T)$ versus T in the HTSC and in the OLS appear to be strikingly similar, with strong field effects below T_{c0}, and no discernible field effects above T_{c0}.

8 Discussion

Since superconducting fluctuations and/or 'preformed pairs' involve particles of charge q moving with velocity $\mathbf{v}$ [50], in a field they would classically experience the Lorentz force, $\mathbf{F} = q\mathbf{v} \times \mathbf{B}/c$, where $\mathbf{B}$ is the magnetic induction, and quantum-mechanically, they would localize within Landau orbits, leading to fluctuation diamagnetism. There would be a strong field dependence to the fluctuation diamagnetism and fluctuation conductivity well above T_c if the PG were due to superconducting fluctuations [50]. The fact that almost no field effects are measured in the resistivity, spin-lattice relaxation rate, and magnetization studies of a variety of HTSC comprises very strong evidence that the PG cannot be mainly due to either superconducting fluctuations or to 'preformed pairs'. But in extremely large fields [46], strong evidence has been presented that the quasiparticles are only destroyed by the Zeeman effect (or Pauli pairbreaking of chargeless pairs), implying that they have charge zero [47]. Most likely, the quasiparticles are particle-hole pairs that form into some sort of CDW (or $M_s = 0$ SDW) state, the details for which are not presently known.

For example, a mistake was made long ago, by falsely ascribing the T dependence of $\chi_\perp(T)$ in partially intercalated $2H$-TaS_2(pyridine)$_{1/2}$ to superconducting fluctuations [51]. However, when more systematic studies were

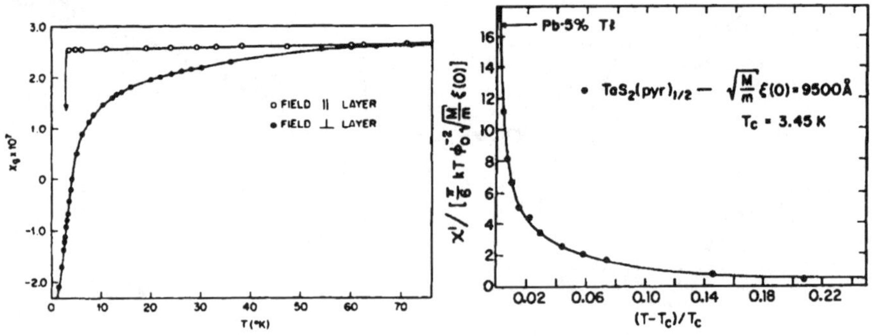

Fig. 25. *Left*: Plot of $\chi_\perp(T)$ of partially intercalated $2H$-TaS_2(pyridine)$_{1/2}$, for which precursor effects of superconductivity up to 35 K were claimed [51]. *Right*: Magnitude of the fluctuation diamagnetism, appropriately normalized, of fully intercalated $2H$-TaS_2(pyridine)$_{1/2}$, *solid circles*, along with the data from a Pb-5% Tl alloy (*solid curve*) [53]

done, taking careful account to intercalate all of the material (as in the lower curve of Fig. 20), the effect went away [52,53]. The curves showing these false signals, along with the correction, are shown in Fig. 25. Note that the true effect of fluctuations was only clearly observable up to a few percent above T_c, and $\chi_\perp(T)$ due to superconducting fluctuations was identical to that of dirty 3D superconductors.

In conclusion, we note that this topic is presently under considerable debate [54], and we hope that by showing the published data on related compounds, those with open minds will be able to make informed judgments.

References

1. J.R. Kirtley and C.C. Tsuei, Rev. Mod. Phys. **72**, 969 (2000).
2. Q. Li, Y.N. Tsay, M. Suenaga, R.A. Klemm, G.D. Gu, and N. Koshizuka, Phys. Rev. Lett. **83**, 4160 (1999).
3. A. Bhattacharya, I. Zutić, O.T. Valls, A.M. Goldman, U. Welp, and B. Veal, Phys. Rev. Lett. **82**, 3132 (1999).
4. Y. Takano, T. Hatano, A. Fukuyo, A. Ishii, M. Ohmori, S. Arisawa, K. Togano, and M. Tachiki, Phys. Rev. B **65**, 140513 (2002); R.A. Klemm, cond-mat/0211423.
5. A. Bille, R.A. Klemm, and K. Scharnberg, Phys. Rev. B **64**, 174507 (2001).
6. K.A. Müller, Phil. Mag. Lett. **82**, 279 (2002).
7. H. Ding *et al.*, Nature (London) **382**, 51 (1996); A.G. Loeser *et al.*, Science **273**, 325 (1996).
8. D.N. Basov *et al.*, Science **283**, 49 (1999); A.V. Puchkov *et al.*, Phys. Rev. Lett. **77**, 3212 (1996).
9. G.-Q. Zheng *et al.*, Phys. Rev. B **60**, R9947 (1999).
10. K. Gorny *et al.*, Phys. Rev. Lett. **82**, 177 (1999); K. Gorny, private communication (unpublished).
11. G.V. Williams *et al.*, Phys. Rev. Lett. **78**, 721 (1997).
12. B. Wuyts, V.V. Moschalkov, and Y. Bruynseraede, Phys. Rev. B **53**, 9418 (1996).
13. T. Tamegai and Y. Iye, Phys. Rev. B **44**, 10 167 (1991).
14. T. Watanabe, T. Fujii, and A. Matsuda, Phys. Rev. Lett. **79**, 2113 (1997).
15. J.W. Loram *et al.*, Phys. Rev. Lett. **71**, 1740 (1993).
16. D.C. Johnston, Phys. Rev. Lett. **62**, 957 (1989); J.M. Tranquada *et al.*, Phys. Rev. B **38**, 2477 (1988).
17. H.F. Fong *et al.*, Phys. Rev. Lett. **82**, 1939 (1999); Y. Sidis *et al.*, Phys. Rev. Lett. **84**, 5900 (2000).
18. P. Dai, H.A. Mook, and F. Dogan, Phys. Rev. Lett. **80**, 1738 (1998); H.A. Mook *et al.*, Nature (London) **395**, 580 (1998), *ibid.* **401**, 145 (1999); *ibid.* **404**, 729 (2000).
19. Ch. Renner *et al.*, Phys. Rev. Lett. **80**, 149 (1998).
20. M. Franz and A.J. Millis, Phys. Rev. B **58**, 14 572 (1998).
21. T. Ekino, Y. Sezaki, and H. Fujii, Phys. Rev. B **60**, 6916 (1999).
22. V.M. Krasnov *et al.*, Phys. Rev. Lett. **84**, 5860 (2000); V.M. Krasnov *et al.*, Phys. Rev. Lett. **86**, 2657 (2001); V.M. Krasnov, Phys. Rev. B **65**, 140504 (2002).

23. A. Yurgens, D. Winkler, T. Claeson, S.-J. Hwang, and J.H. Choy, Int. J. Mod. Phys. B **13**, 3758 (1999).
24. S. Heim, T. Nachtrab, M. Mößle, R. Kleiner, R. Koch, S. Rother, O. Waldmann, P. Müller, T. Kimura, and Y. Tokura, Physica C **367**, 348 (2002).
25. K.M. Lang, V. Madhavan, J.E. Hoffman, E.W. Hudson, H. Eisaki, S. Uchida, and J.C. Davis, Nature (London), **415**, 412 (2002).
26. J.A. Wilson, F.J. DiSalvo, and S. Mahajan, Adv. Phys. **24**, 117 (1974).
27. J.M. Williams *et al.*, *Organic Superconductors (Including Fullerenes)* (Prentice-Hall, Englewood Cliffs, NJ, 1992).
28. R. McKenzie, Science **278**, 820 (1997).
29. R. Liu *et al.*, Phys. Rev. Lett. **80**, 5762 (1998); Phys. Rev. B **61**, 5212 (2000).
30. D.S. Marshall *et al.*, Phys. Rev. B **52**, 12 548 (1995).
31. D.E. Moncton, J.D. Axe, and F.J. DiSalvo, Phys. Rev. B **16**, 801 (1977).
32. R.V. Coleman *et al.*, Adv. Phys. **37**, 559 (1988).
33. R.A. Craven and S.R. Meyer, Phys. Rev. B **16**, 4583 (1977).
34. M. Naito and S. Tanaka, J. Phys. Soc. Jpn. **51**, 219 (1982).
35. K. Miyagawa *et al.*, Phys. Rev. Lett. **75**, 1174 (1995).
36. A. Kawamoto, K. Miyagawa, Y. Nakazawa, and K. Kanoda, Phys. Rev. Lett. **74**, 3455 (1995); Phys. Rev. B **52**, 15522 (1995).
37. J.P. Tidman *et al.*. Phil. Mag. **30**, 1191 (1974).
38. V. Vescoli *et al.*, Phys. Rev. Lett. **81**, 453 (1998).
39. F.R. Gamble, F.J. DiSalvo, R.A. Klemm, and T.H. Geballe, Science **168**, 568 (1970).
40. A.H. Thompson, F.R. Gamble, and R.F. Koehler, Phys. Rev. B **5**, 2811 (1972); A.H. Thompson, Sol. State Commun. **13**, 1911 (1973).
41. T.A. Friedmann *et al.*, Phys. Rev. B **42**, 6217 (1990).
42. W.W. Kwok, Mod. Phys. Lett. B **5**, 547 (1991); W. Kwok, private communication (unpublished).
43. D.J. Huntley and R.F. Frindt, Can. J. Phys. **52**, 861 (1974).
44. G. Briceño, M.F. Crommie, and A. Zettl, Phys. Rev. Lett. **66**, 2164 (1991).
45. H. Ito *et al.*, Physica B **201**, 470 (1994).
46. T. Shibauchi, L. Krusin-Elbaum, M. Li, M.P. Maley, and P.H. Kes, Phys. Rev. Lett. **86**, 5763 (2001).
47. R.A. Klemm, M.R. Beasley, and A. Luther, Phys. Rev. B **12**, 877 (1975).
48. K. Ishida *et al.*, J. Phys. Soc. Jpn. **65**, 2341 (1996).
49. H. Mayaffre *et al.*, Phys. Rev. Lett. **75**, 4122 (1995).
50. R.A. Klemm, M.R. Beasley, and A. Luther, Phys. Rev. B **8**, 5072 (1973); R.A. Klemm, J. Low Temp. Phys. **16**, 381 (1974).
51. T.H. Geballe *et al.*, Phys. Rev. Lett. **27**, 314 (1971).
52. F.J. DiSalvo, in *Low Temperature Physics-LT13*, K.D. Timmerhaus *et al.*, Eds. (Plenum, New York, 1974) **3**, pp. 417–427.
53. D.E. Prober, M.R. Beasley, and R.E. Schwall, Phys. Rev. B **15**, 5245 (1977).
54. R.S. Markiewicz, Phys. Rev. Lett. **89**, 229703 (2002).

Electric Fields in Superconductors

Jan Koláček and Pavel Lipavský

Abstract. The electric field in the superconductor is discussed for the equilibrium and the quasi-stationary regime. For the equilibrium we present a theory that is microscopic on the level of the Ginzburg–Landau theory. For the quasi-stationary regime we discuss a phenomenologic picture for quantities averaged over the Abrikosov lattice of vortices.

1 Introduction

In any conductor there are various electric fields that can be crudely sorted into two families. The first family collects build-in fields which are present already in the equilibrium. Assume for example a junction of two metals with different work functions. In absence of electrostatic forces, the metal with the lower work function would drain nearly all electrons from the other. In reality, already a very small fraction of the total number of electrons transferred across the junction creates an electrostatic potential which balances the difference between work functions. Similar fields appear at surfaces and with any gradient, would it be of impurity concentration, strain, temperature or an applied magnetic field.

The second family of electric fields is enforced by external sources. In contrast to the above cases, these fields do not prevent but drive the transport of electrons. Such fields can enter the conductor via contacts, as an electromagnetic wave or they appear due to time changes of the magnetic flux encircled by the conductor. These fields cause dissipation and exist only under non-equilibrium conditions.

In most cases, one can focus on one of the fields neglecting the others. Two closely related extremes represent a magneto-conductivity in normal and superconducting metals. In normal metal the magneto-conductivity is studied for infinite homogeneous samples and the electrostatic field is treated as an enforced one opposed by dissipative processes. In the superconductor in the Meissner state, there is no dissipation so that the electrostatic field appears merely as the build-in one due to gradients of the magnetic field and condensate density. The later field bears a somehow confusing name of the Bernoulli potential according to its low temperature limit and the magneto-hydrodynamical picture of superconductivity in early theories.

Both families of the electric field coexist with comparable amplitudes in the type-II superconductors in the mixed state. While the motion of vortices creates the dissipative processes and the related enforced electromotoric force, the motion of surrounding condensate results in the Bernoulli potential with a gradient not aligned with the electromotoric force. The relative contributions of both mechanisms naturally depend on material parameters that change with the temperature. One of spectacular physical effects controlled by the competition of these two fields is the Hall voltage which can two or three times change its sign as the temperature decreases from the critical temperature to zero.

In this paper we want to discuss both families of the electric field in superconductors. The paper is organized as follows. In Sect. 2 we discuss the electrostatic potential in the equilibrium superconductor starting from a historical review and concluding with the recent treatment on the level of the Ginzburg–Landau theory. In Sect. 3 we discuss the type-II superconductor with the moving vortex lattice. A phenomenologic theory for mean values of the electric field and electric currents is presented and compared with experimental data on the FIR magneto-spectroscopy and the Hall voltage. We conclude this section with a possible modification of the Josephson relation. In Sect. 4 we summarize.

2 Bernoulli Potential

As it is with many properties of superconductors, the theory of the Bernoulli potential can be divided into the pre-BCS and the BCS-type approaches. Early activities in this field has been terminated by a disappointing experimental result that the electrostatic field cannot serve as a probe of the pairing mechanism. Moreover, there was a principal disagreement between the theory and experiment and this problem has been overcome only recently. We describe these studies in the historical order.

2.1 Pre-BCS theories

The pre-BCS predictions of the electrostatic potential date back to 1937 when Bopp applied the magneto-hydrodynamic approach to superconductors [1]. He found an expectable result – inertial and Lorentz forces which would drive the system far from the charge neutrality are balanced by the Coulomb force so that only a very small deviation from the neutrality appears. These forces can be represented by the electrostatic potential

$$e\varphi = -\frac{1}{2}m\mathbf{v}^2\,, \tag{1}$$

where e and m are electron charge and mass, and $\mathbf{v}$ is the velocity of the electron fluid. In the thermodynamic picture, the electrostatic potential provides one of the contributions to the internal pressure. In this sense, the potential (1) has been named the Bernoulli potential.

From next studies in this period, the most remarkable is the one due to Sorokin [2], who assumed the free energy of condensation. With the free energy of Gorter and Casimir [3,4], his approach corresponds to the generalized London theory put forward by Bardeen and Stephen [5] within their study of vortex motion. The limit to low magnetic field gives the so called quasiparticle screening of van Vijfeijken and Staas [6] and the contribution to the electric potential obtained two decades later by Rickayzen [7] and known as the thermodynamic correction.

London included the Bernoulli potential in his famous book [8] and showed that trajectories dictated by the vector potential are in conflict with the Newton equation, unless electric field supplies the missing forces. On the other hand, he disregarded the free energy part of these forces as unknown and likely unimportant, so that his derivation applies only to low temperatures and low fields, where the density of condensate equals to the total density of electrons. Paradoxically, London is mostly cited as the author of the Bernoulli potential in superconductors.

Since Sorokin's derivation was not sufficiently appreciated, the last important pre-BCS contribution is the quasiparticle screening of van Vijfeijken and Staas [6]. In analogy with the fountain effect, they introduced a force between the normal electrons and the condensate. This force balances the electric force acting on normal electrons to keep them at rest. According to the action-reaction law, the force on the condensate is increased so that a smaller Bernoulli potential

$$e\varphi = -\frac{n_s}{n}\frac{1}{2}m\mathbf{v}^2 \,, \tag{2}$$

is sufficient to balance forces on the condensate. Here n_s/n is the fraction of condensate in the total density of electrons.

Finally, there is a controversial theory of Jakeman and Pike [9]. From the stationary limit of the time dependent Ginzburg–Landau (GL) theory they have arrived at the Poisson equation with the screening on the Thomas–Fermi length and the potential (2) as the source term. Except for the misprint in factor of two, their never published derivation apparently includes two serious neglects. First, using the local limit they have neglected gradients of the GL function, so that their theory does not apply at vortex cores and close to the surface. In spite of it, they have derived the surface charge localized on the Thomas–Fermi length, while the later non-local treatment [10] provides the surface charge extending on the scale of the GL coherence length. Second, they have ignored the density dependence of GL parameters which yield the thermodynamic correction.

2.2 BCS Theories

The BCS studies of the electrostatic potential were started by Adkins and Waldram [11] who evaluated the deformation of the BCS gap by the supercurrent at the zero temperature. They found that for a fixed Fermi energy the deformation would result in changes of the density of electrons. Accordingly, a local shift of the Fermi energy is necessary in order to stay close to the charge neutrality. Since the sum of the Fermi energy and the electrostatic potential has to be constant in the equilibrium system, shifts of the Fermi energy are compensated by the electrostatic potential. They showed that for the isotropic band structure, the potential simplifies to the Bernoulli potential (1).

Rickayzen [7] recovered results of Adkins and Waldram using the thermodynamic approach and formulated the electrostatic potential in terms of thermodynamic functions. His formulation allows one to choose between microscopic BCS relations or their simpler phenomenologic counterparts. This makes it possible to study the systems with any band structure at finite temperatures. For the isotropic electron mass the Bernoulli potential reads

$$e\varphi = -\frac{\partial n_s}{\partial n}\frac{1}{2}m\mathbf{v}^2 \, . \tag{3}$$

The derivative can be evaluated from $n_s = n(1 - T^4/T_c^4)$,

$$e\varphi = -\left(1 - \frac{T^4}{T_c^4}\right)\frac{1}{2}m\mathbf{v}^2 - 4\frac{T^4}{T_c^4}\frac{\partial \ln T_c}{\partial \ln n}\frac{1}{2}m\mathbf{v}^2 \, , \tag{4}$$

where one has to take into account that the critical temperature T_c depends on the density of electrons, what results in the so called thermodynamic correction. The BCS estimate of the derivative, $(\partial \ln T_c/\partial \ln n) \approx (4/3)\ln(\theta_D/T_c)$, gives values between unity and ten, so that the thermodynamic correction is in fact the dominant term at temperatures above $(2/3)T_c$. Such clear prediction was a challenge for experimentalists.

2.3 Measurements of the Bernoulli Potential

The first attempt to measure the Bernoulli potential with the standard Hall voltage setup appeared already in 1914 by Kamerlingh Onnes and Hof [12]. They found no voltage. One of possible explanations was that in this pioneering experiment the system was in the intermediate state, which was believed to spoil the measurement. The later measurements by Lewis [13], however, excluded this complication while zero Hall voltage was confirmed.

Lewis also argued that theoretical explanations of the zero result are wanting [14]. The correct argument is attributed to de Gennes although commonly with no literature cited. He pointed out that by measurements with Ohmic contacts one observes difference of the Gibbs chemical potential, which is a sum of the electrostatic potential and the local Fermi energy. In equilibrium the Gibbs potential is always constant so that no voltage can be observed.

To isolate the electrostatic potential from the shifts of the Fermi energy, Hunt [15] proposed to use the Kelvin capacitive pickup. This approach was adopted by Bok and Klein [16] and Brown and Morris [17] who both observed the Bernoulli potential in agreement with formula (1). It should be noted that they made measurements at about half the critical temperature, where nearly all electrons are in the condensate and the thermodynamic corrections are very small. These measurements were not sufficiently accurate, nevertheless, Brown and Morris speculated that perhaps they have seen some traces of the thermodynamic corrections.

The next measurement of Morris and Brown [18] chilled such expectations. In a wide range of temperatures and with a high precision they showed that the observed Bernoulli potential corresponds to formula (2), having no traces of the thermodynamic corrections. The zero result reduced an effort in this direction. In particular, the absence of the thermodynamic corrections remained unexplained till recently. From the Budd–Vannimenus theorem it was proved that the thermodynamic correction is compensated by the current induced change of the surface dipole [19]. This compensation parallels the surface tension, which cancels the internal pressure of fluids.

2.4 Recent Theories

The most sophisticated treatment of the Bernoulli potential has been performed by Hong [20] who employed the Gorkov equations. His general results are too complicated for implementations while his simplified formulas include the dangerous local approximation. Nevertheless, Hong reformulated the thermodynamic correction in terms of the BCS gap.

A new interest in the shift of the Fermi energy and related charge transfer appeared with the high-T_c materials, where the shift is enhanced by the reservoir of holes with no or weak pairing forces [21,22]. These theories describe the mechanism on the level of Hong's approach using the local value of the gap and currents. They pointed out, however, that in the high-T_c materials the charge transfer is much larger and it encouraged a new interest in the subject and new experimental methods were offered [23].

The non-local modification of the Bernoulli potential on the basis of the GL theory has been proposed in [24]. It corresponds to a simple replacement of the kinetic energy by its quantum mechanical counterpart. The full GL theory with the non-local kinetic energy, the quasiparticle screening, and the thermodynamic corrections has been derived from the free energy in [25,26].

The first implementations of the non-local theory have been devoted to mesoscopic superconductors [27]. It has been shown that the GL boundary condition implies the total charge neutrality so that no surface charge on the scale of the Thomas–Fermi screening length is necessary. The surface charge has been studied for flat superconductors and shown that it extends over the region comparable with the GL coherence length [10].

The next effort was aimed at surfaces of conventional superconductors. Using the identity called the Budd–Vannimenus theorem, the missing thermodynamic correction in the Bernoulli potential observed in contactless measurements has been shown to follow from their cancellation by the surface dipole [19]. The disturbing disagreement between the theory and experiment has been thus lifted. The surface dipole also contributes to electric forces acting on the crystal lattice which properly sums to the Lorentz force.

The non-local theory was also applied to the high-T_c materials [28], in particular, to describe the charge transfer in the Abrikosov vortex lattice in the $YBa_2Cu_3O_7$ observed recently by Kumagai, Nozaki and Matsuda [29] via the nuclear magnetic resonance. The experiment and theory confirmed expectations of van der Marel [21] that charge reservoirs of the high-T_c materials strongly enhance the charge transfer.

2.5 Theory of Ginzburg–Landau Type

Following [26] we write down equations of GL type for the properties of the superconductor in equilibrium including the electrostatic potential. It is based on the variation of the free energy with respect to the GL wave function ψ, the vector potential $\mathbf{A}$, and the density of normal electron n_n.

The free energy is composed of four parts. The condensation is covered by the Gorter–Casimir free energy

$$\mathcal{F}_s = \int d\mathbf{r} \left(U - \varepsilon_{\text{con}} \varpi - \frac{1}{2} \gamma T^2 \sqrt{1-\varpi} \right) , \tag{5}$$

where U is the internal energy of the metal in the normal state at zero temperature, $\varepsilon_{\text{con}} = \gamma T_c^2/4$ is the condensation energy per unitary volume, γ is the linear coefficient of the specific heat, and $\varpi = 2|\psi|^2/n$ is the fraction of condensate in the density of pairable electrons $n = n_n + 2|\psi|^2$. The kinetic energy is of the GL type,

$$\mathcal{F}_{\text{kin}} = \int d\mathbf{r} \frac{1}{2m^*} \left| \left(-i\hbar\nabla - e^*\mathbf{A} \right) \psi \right|^2 , \tag{6}$$

where $m^* = 2m$ and $e^* = 2e$. The Helmholtz magnetic free energy is

$$\mathcal{F}_M = \int d\mathbf{r} \frac{1}{2\mu_0} (\mathbf{B} - \mathbf{B}_a)^2 , \tag{7}$$

where $\mathbf{B}_a$ is the applied magnetic field. Finally, the Coulombic energy reads

$$\mathcal{F}_C = \frac{1}{2} \iint d\mathbf{r} d\mathbf{r}' \frac{1}{4\pi\epsilon} \frac{1}{|\mathbf{r} - \mathbf{r}'|} \rho(\mathbf{r}) \rho(\mathbf{r}') , \tag{8}$$

where $\rho = e^*|\psi|^2 + e n_n + \rho_{\text{latt}}$ is the charge density. The Coulomb interaction also determines the electrostatic potential,

$$\varphi(\mathbf{r}) = \int d\mathbf{r}' \frac{1}{4\pi\epsilon} \frac{1}{|\mathbf{r} - \mathbf{r}'|} \rho(\mathbf{r}') . \tag{9}$$

Definition (9) is equivalent to the Poisson equation $-\epsilon\nabla^2\varphi = \rho$.

The equilibrium state is characterized by the minimum of free energy $\mathcal{F} = \mathcal{F}_s + \mathcal{F}_{\mathrm{kin}} + \mathcal{F}_M + \mathcal{F}_C$ and it can be found by the variation principle. Variation with respect to vector potential $\mathbf{A}$ gives the second GL equation

$$\nabla \times \nabla \times \mathbf{A} = \mu_0 \frac{e^*}{m^*} \mathrm{Re}\, \bar{\psi}(-i\hbar\nabla - e^*\mathbf{A})\psi \,, \tag{10}$$

where $\bar{\psi}$ denotes the complex conjugate of ψ. By variation with respect to $\bar{\psi}$ one finds the first GL equation,

$$\frac{1}{2m^*}(-i\hbar\nabla - e^*\mathbf{A})^2\psi - \left(2\frac{\varepsilon_{\mathrm{con}}}{n} - \frac{\gamma T^2}{2n} \frac{1}{\sqrt{1 - \frac{2|\psi|^2}{n}}} \right) \psi = 0 \,. \tag{11}$$

For $T \to T_c$ this equation achieves the customary GL form [30] with the potential linear in $|\psi|^2$. The present form is Bardeen's extension toward low temperatures [31,32] which has an additional advantage that its parameters have known dependence on the density of electrons. By variation with respect to n_n one finds the third GL equation

$$e\varphi - \lambda_{TF}^2 \nabla^2 e\varphi = -\bar{\psi}\frac{(-i\hbar\nabla - e^*\mathbf{A})^2}{2m^* n}\psi + \frac{\partial \varepsilon_{\mathrm{con}}}{\partial n}\frac{2|\psi|^2}{n} + \frac{T^2}{2}\frac{\partial \gamma}{\partial n}\sqrt{1 - \frac{2|\psi|^2}{n}} \,. \tag{12}$$

The first term on the rhs of (12) is the non-local Bernoulli potential with quasiparticle screening. The second and the third terms form the thermodynamic correction. The screening on the Thomas–Fermi length present on the lhs of (12) can be neglected in most cases, at least at the vortex cores and also at the surfaces with the vacuum.

We note that the charge transfer brings only a minor correction to the first and second GL equations via the density dependence of material parameters $\varepsilon_{\mathrm{con}}$ and γ. It is thus justified to use solutions of customary GL equations and to evaluate the electrostatic field from (12) without iterations. The distribution of charge density is obtained from the electrostatic potential via the Poisson equation.

The presented theory is microscopic only on the scale of the GL coherence length. The fully microscopic treatment formulated from the Bogoliubov-de Gennes equations shows additional modulation of the potential and the charge on the smaller scale [33].

2.6 Summary

To summarize this section, we remind that the non-local theory with the surface dipole accounted for is in agreement with experimental data. From properties of the electrostatic field we want to point out three morals that are important for understanding of the electric field in the mixed state.

- The electrostatic potential and the Gibbs chemical potential have to be distinguished. While the Gibbs potential is constant at equilibrium, the electrostatic potential reflects inertial and Lorentz forces and inhomogeneities of the condensate.
- The macroscopic Lorentz force on the crystal lattice is covered by the electrostatic potential.
- On the surface the thermodynamic corrections of the Bernoulli potential are cancelled by the surface dipole, consequently they are not visible from outside.

3 Josephson Relation

While the electrostatic potential in the equilibrium superconductor is by now well theoretically described and individual contributions to it experimentally tested, less is known about the electric field which accompany the motion of vortices. The topic attracted a lot of interest, however, the final answer is still missing. In particular, there is no widely accepted theory to explain the far infra-red (FIR) magneto-spectroscopy and the anomalous Hall voltage.

A common picture of the Hall voltage relies on the Josephson relation [34]

$$\mathbf{E}' = -\nabla(\varphi + \mu/e) - \partial\mathbf{A}/\partial t = -\mathbf{v_V} \times \mathbf{B}\,, \tag{13}$$

which links the 'electric' field with the velocity of vortices $\mathbf{v_V}$ and the mean magnetic field $\mathbf{B}$ generated by the vortex lattice. This relation has to be applied with a caution, however. The field $\mathbf{E}'$ is not the electric field treated in the Maxwell theory, but it is the effective electric field given by a gauge-invariant generalization of the gradient of the Gibbs electrochemical potential $\varphi + \mu/e$, with μ being the local Fermi energy.

At equilibrium in the absence of vortex motion, the Josephson relation naturally yields $\mathbf{E}' = 0$. This is in agreement with the fact that the Gibbs potential is constant and there is no transport of normal electrons confirmed by the zero Hall voltage measured by the Ohmic contacts. On the other hand, the electric field $\mathbf{E} = -\nabla\varphi - \partial\mathbf{A}/\partial t$ is non-zero and can be measured by contactless capacitive pickup (Kelvin method) or via the charge transfer effect on the NMR. It is thus essential to distinguish which of the two fields is relevant for a physical process in question.

In this paper we want to discuss the FIR magneto-spectroscopy in which the electric field of the laser light has to be matched with the electric field $\mathbf{E}$ inside the superconductor. In the FIR the Josephson relation thus does not apply and one has to evaluate the electric field in some alternative way. Of course, a contribution similar to the Josephson relation has to be expected as the electric field is linked with the time derivative of the vector potential associated with the moving vortices.

Here we present a simple phenomenologic theory which can qualitatively describe the FIR spectra and the anomalous Hall effect. It is rooted in the above mentioned assumption that all the forces acting inside the superconductor are balanced, in contrast to former theories in which the sum of forces equals to the Lorentz force. According to studies of the equilibrium state presented above, Lorentz force is covered by the surface currents. It corresponds to the fact, that the Lorentz force represents the interaction of the electric current with the external electric field and not with the vortex lattice.

3.1 Force Balance in the Mixed State

In equilibrium the type-II superconductor in the mixed state can be characterized by the density of condensate n_s, the density of normal electrons n_n, and the density of vortices n_{V}. The density of vortices is linked to the mean value of the magnetic field $\mathbf{B} = n_{\mathrm{V}}\Phi_0\mathbf{z}$, where Φ_0 is the elementary magnetic flux and $\mathbf{z}$ is a unitary vector in the direction of vortices that we associate with the axis z. Close to the surface the magnetic field also includes the contribution of the external field penetrating on the scale of the London depth, but we will not discuss this contribution for simplicity.

As the system is driven out of equilibrium, the vortices, superconducting fluid and the normal state fluid move with velocity $\mathbf{v}_{\mathrm{V}}$, $\mathbf{v}_s$ and $\mathbf{v}_n$, respectively. Our aim is to identify forces which control the motion of these components. Let us start from known forces acting on a vortex. When moving, the vortex dissipates energy what is described by the friction force

$$\mathbf{F}_{\mathrm{V}}^{\mathrm{dis}} = -\eta\mathbf{v}_{\mathrm{V}} = -\frac{m_{\mathbf{V}}}{\tau_{\mathrm{V}}}\mathbf{v}_{\mathrm{V}} \,. \tag{14}$$

The second form expresses the viscosity coefficient η in terms of the vortex mass $m_{\mathbf{V}}$ and the vortex relaxation rate $1/\tau_{\mathrm{V}}$. The simplest model of pinning keeps vortices by parabolic wells,

$$\mathbf{F}_{\mathrm{V}}^{\mathrm{pin}} = -\kappa\mathbf{r}_{\mathrm{V}} = -m_{\mathbf{V}}\alpha^2\mathbf{r}_{\mathrm{V}} \,, \tag{15}$$

where $\mathbf{r}_{\mathrm{V}}$ is a displacement of the vortex linked to the velocity, $\dot{\mathbf{r}}_{\mathrm{V}} = \mathbf{v}_{\mathrm{V}}$. The pinning constant κ is expressed via the mass and the pinning frequency α. The interaction between vortices and the condensate is mediated by the Magnus force [35]

$$\mathbf{F}_{\mathrm{V}}^{\mathrm{Mag}} = -\pi n_s\hbar(\mathbf{v}_s - \mathbf{v}_{\mathrm{V}}) \times \mathbf{z} = m_{\mathrm{V}}\varpi\Omega(\mathbf{v}_s - \mathbf{v}_{\mathrm{V}}) \times \mathbf{z} \,, \tag{16}$$

where $\varpi = n_s/(n_s+n_n)$ is the fraction of condensate, and Ω is the frequency of the cyclotron vortex motion.

According to the assumption that all forces are balanced [36], the Magnus force has its counterpart acting on the condensate

$$\mathbf{F}_{\mathrm{s}}^{\mathrm{Mag}} = -\frac{n_{\mathrm{V}}}{n_s}\mathbf{F}_{\mathrm{V}}^{\mathrm{Mag}} = -m\omega_c(\mathbf{v}_s - \mathbf{v}_{\mathrm{V}}) \times \mathbf{z} \,, \tag{17}$$

where $\omega_c = eB/m = \pi n_{\mathrm{V}}\hbar/m$ is the cyclotron frequency.

The motion of vortices also acts on normal electrons. From the Aharonov–Casher Lagrangian [37] one finds the Iordanskii force

$$\mathbf{F}_{\mathrm{n}}^{\mathrm{Ior}} = \pi n_{\mathrm{V}} \hbar (\mathbf{v}_n - \mathbf{v}_{\mathrm{V}}) \times \mathbf{z} = m\omega_c (\mathbf{v}_n - \mathbf{v}_{\mathrm{V}}) \times \mathbf{z} \,. \tag{18}$$

Its counterpart acting on vortices reads

$$\mathbf{F}_{\mathrm{V}}^{\mathrm{Ior}} = -\frac{n_n}{n_{\mathrm{V}}} \mathbf{F}_{\mathrm{n}}^{\mathrm{Ior}} = -m_{\mathrm{V}} (1 - \varpi) \Omega (\mathbf{v}_n - \mathbf{v}_{\mathrm{V}}) \times \mathbf{z} \,. \tag{19}$$

Finally, the normal electrons dissipate energy what we express via the relaxation rate,

$$\mathbf{F}_n^{\mathrm{dis}} = -\frac{m}{\tau_n} \mathbf{v}_n \,. \tag{20}$$

3.2 Equations of Motion

Having draw up all interaction forces we can now write down Newton's equations of motion. The condensate is accelerated by the electric force $e\mathbf{E}$ and by the Magnus force,

$$m\dot{\mathbf{v}}_s = e\mathbf{E} + \mathbf{F}_s^{\mathrm{Mag}} \,. \tag{21}$$

The normal electrons are driven by the electric and the Iordanskii force and they suffer the friction,

$$m\dot{\mathbf{v}}_n = e\mathbf{E} + \mathbf{F}_n^{\mathrm{Ior}} + \mathbf{F}_n^{\mathrm{dis}} \,. \tag{22}$$

The vortices are driven by the Magnus force, the Iordanskii force, the pinning and the friction,

$$m_{\mathrm{V}} \dot{\mathbf{v}}_{\mathrm{V}} = \mathbf{F}_{\mathrm{V}}^{\mathrm{Mag}} + \mathbf{F}_{\mathrm{V}}^{\mathrm{Ior}} + \mathbf{F}_{\mathrm{V}}^{\mathrm{pin}} + \mathbf{F}_{\mathrm{V}}^{\mathrm{dis}} \,. \tag{23}$$

For thin layers used in the FIR transmission measurement, the electric and magnetic fields are not essentially influenced by the motion in the superconductor. With the electric field perpendicular to the magnetic field, (21–23) accompanied by the relation $\dot{\mathbf{r}}_{\mathrm{V}} = \mathbf{v}_{\mathrm{V}}$ form a closed set of eight first order differential equations for x and y components of velocities and the vortex position. Its solution is easily obtained for the monochromatic light, $\mathbf{E} \propto \mathrm{e}^{i\omega t}$, when it reduces to the set of ordinary linear equations. This solution can be found in [36] where we refer to the reader for details.

3.3 FIR Magneto-Spectroscopy

The above set of equations of motion has been originally developed to describe the FIR magneto-spectroscopic measurements [38]. Let us test its validity on this experiment.

For a long time the only accessible data were for the low frequency limit in the microwave region and the rather high frequencies $\omega > 25\,\mathrm{cm}^{-1}$. With the circularly polarized light Lihn *et al.* covered in their measurement the missing frequency range going from 5.26 to 200 cm^{-1} [39]. They found that the vortex dynamics does not correspond to any of existing models but that it behaves as a mixture of the Gittleman-Rosenblum model of massless vortices [40,41] and the clean limit model of Hsu [42]. A particularly important feature is the pronounced peak near the zero frequency that is not properly covered by any of the models.

As shown in [36], the above set of equations yields the desirable peak of magneto-conductivity presented in Fig. 1. The non-symmetry in frequency reflects the circular polarization in the presence of the magnetic field. It is worthwhile to note that the resulting formula for conductivity does not have the delta function component, considered to be the hallmark of superconductivity. The reason lies in the fact that moving vortices dissipate energy and consequently the real part of resistivity is nonzero. The material is superconducting in the sense that the charge carriers are condensed, the superconducting fluid forms vortices, nevertheless due to vortex motion the d.c. resistivity is nonzero. Accordingly, the delta function in the formula for conductivity has to be absent.

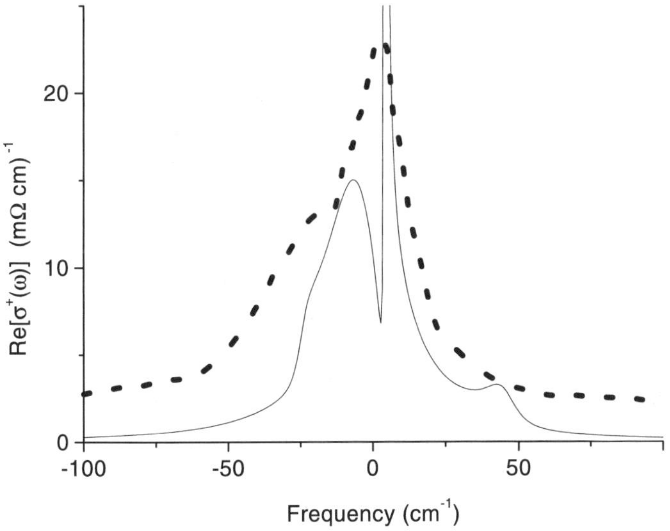

Fig. 1. Conductivity for the circularly polarized light. The experimental data from [39] are in the *dotted line*

3.4 Hall Voltage

In the normal metal, the Hall voltage reflects the Lorentz force on the current and it can thus be used to observe whether the conductivity is dominated by electrons or holes. As discussed in Sect. 2, in superconductors in the Meissner state, the Hall voltage measured by Ohmic contacts vanishes since the electrostatic potential created by the Lorentz force is cancelled by changes of the local Fermi energy.

The type-II superconductor in the mixed state has a finite resistance due to a motion of vortices so that the Hall voltage can be observed by the Ohmic contacts [43]. Vinen's remark [44] that the Magnus force will lead to a large Hall angle, was not in agreement with first measurements on samples of short mean free path [45–47]. Bardeen argues that the concept of the Magnus force is in error when applied to superconductors [48]. The semi-microscopic study by Bardeen and Stephen [5], in which they use a force on vortices $\mathbf{F}_{\mathrm{V}} = -\pi n_s \hbar \mathbf{v}_s \times \mathbf{z}$ that differs from the Magnus force being independent of the vortex velocity, predicts the Hall angle to be equal to the one in the normal state.

Later measurements on samples with a different mean free path have revealed the large Hall angle [49,50] including an anomalous feature – the sign reversal below T_c [51,52]. It has been found, and particularly well observed on the high-T_c materials, see e.g. [53–59], that the Hall voltage reverses its sign close below T_c, recovers its normal state sign at lower temperatures and eventually again reverses its sign at very low temperatures. The low temperature sign reversal has been so far observed only on HgBaCaCuO by Kang *et al.* [60].

These features have not been satisfactorily explained so far. In fact, there are many theories and speculations, see e.g. [61–67], but none of them is generally accepted. It should be noted that approaches based on the Josephson relation (13) provide the reversed sign only if vortices move upstream the electric current.

Here we briefly present one of simple theories [67] based on the above equations of motion developed for the FIR spectroscopy. Since the finite resistance requires to overcome the pinning forces, we set the pinning constant to zero, $\kappa = 0$. In the set of equations (21–23) we take the zero frequency limit, $\omega = 0$ and evaluate the current, $\mathbf{j} = en_s\mathbf{v}_s + en_n\mathbf{v}_n$ as a function of the electric field $\mathbf{E}$. For the x axis parallel to the current, the resulting Hall resistivity reads

$$\begin{aligned}\rho_{xy} &= \frac{E_y}{j_x} = \frac{\omega_c m}{ne^2}\\ &\times \frac{(2\varpi-1)(\varpi(1-\varpi)\tau_{\mathrm{V}}^2\Omega^2 - \tau_n^2\omega_c^2) - \varpi(1+4(1-\varpi)\tau_n\tau_{\mathrm{V}}\omega_c\Omega)}{(\tau_{\mathrm{V}}^2\Omega^2 + (2\varpi-1)^2)\,\tau_n^2\omega_c^2 + 8\varpi^2(1-\varpi)\tau_n\tau_{\mathrm{V}}\omega_c\Omega + \varpi^2(2\varpi-1)^2\tau_{\mathrm{V}}^2\Omega^2 + \varpi^2}\,.\end{aligned} \tag{24}$$

The temperature enters this lengthy but simple expression via the fraction of condensate, $\varpi = 1 - T^4/T_c^4$, and the angular frequency of vortex motion, $\Omega = \pi n\hbar/m_{\mathrm{V}}$. From the Hsu approximation of the vortex mass [42], $m_{\mathrm{V}} = (\pi^3/2)mn\xi$ with $\xi = \hbar v_F/\pi\Delta$, where v_F is the Fermi velocity and Δ is the gap, one finds $\Omega = \Delta^2/E_F$. While the Fermi energy E_F is a constant, the gap depends on the temperature, $\varDelta \approx \varDelta_0 \sqrt{\cos(\pi T^2/2T_c^2)}$, giving $\varOmega = \varOmega_0 \cos(\pi T^2/2T_c^2)$.

The plot of the temperature dependence of the Hall resistivity resulting from formula (24) is in Fig. 2. One can see that as the temperature decreases from the normal to the superconducting state, the Hall resistivity reverses its sign close below the critical temperature and regains its normal state sign again at a slightly lower temperature. The formula also provides the third sign reversal at very low temperatures.

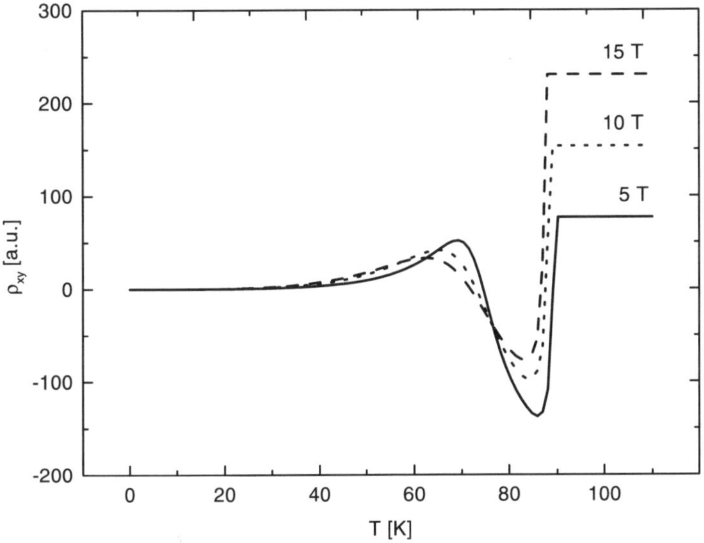

Fig. 2. The Hall resistivity as a function of temperature for different magnetic fields

In the competition of many advanced and sophisticated theories, we do not want to claim that the above simple theory is the right one. Our main objection was to demonstrate that equations developed for the FIR response describe the zero frequency limit in a reasonable agreement with experimental data.

3.5 Modified Josephson Relation

Perhaps it is not necessary to remind that in the above approach, the Josephson relation (13) has not been employed. In fact, a relation between the vortex velocity and the electric field has to result from the set (21–23) under a proper limit.

Following Josephson we assume stationary conditions. Since the condensate is not accelerated, $\dot{\mathbf{v}}_s = 0$, from the balance of forces on the condensate

(21) and the counterpart of the Magnus force (17) we obtain

$$\mathbf{E} = (\mathbf{v}_s - \mathbf{v}_\mathrm{V}) \times \mathbf{B} \,. \tag{25}$$

This equation reminds the Josephon relation so that we would like to call it the modified Josephson relation. There are two important differences, however. First, the left hand side of (25) is the mean value of the true electric field defined in the sense of the Maxwell theory, while in the Josephson relation (13) the effective field appears. Second, the value of electric field depends on the relative velocity of vortices with respect to the condensate while the Josephson relation includes the absolute velocity of vortices.

To our understanding, relations (13) and (25) are not in conflict since they apply to different physical quantities. We do not want to specify where and how to use the Josephson relation as it has been exploited already many times. From possible implementations of the modified Josephson relation we merely remind the FIR spectroscopy where the external electric field of the light has a nature of the Maxwell field.

4 Summary

The electric field in superconductors is still an open question which now attracts a growing attention as its understanding is an important tool for mastering vortex motion. The electrostatic fields in the equilibrium are now well understood, at least the long lasting controversy between theoretical predictions and experimental results has been resolved. The electric field in non-equilibrium superconductors still represent a problem which creates more questions than answers. It is nevertheless clear that unlike in normal metals, in superconductors one has to be careful about distinction between the Maxwell and the effective field.

Since this paper is a contribution to the interdisciplinary meeting, we feel encouraged to add a general remark. The shift of the Fermi energy which is responsible for the Bernoulli potential results from a correlation between electrons. It can be expressed via a local change of the Fermi momentum, i.e., it corresponds to the correlated density discussed in [68] for the nuclear matter above the critical temperature. Superconductors thus offer a unique possibility to observe this, otherwise very subtile, physical phenomenon.

References

1. F. Bopp, Z. Phys. **107**, 632 (1937).
2. V.S. Sorokin, JETP **19**, 553 (1949).
3. C.J. Gorter and H.B.G. Casimir, Phys. Z. **35**, 963 (1934).
4. C.J. Gorter and H.B.G. Casimir, Z. Tech. Phys. (Leipzig) **15**, 539 (1934).
5. J. Bardeen and M.J. Stephen, Phys. Rev. **140**, A1197 (1965).

6. A.G. van Vijfeijken and F.S. Staas, Phys. Lett. **12**, 175 (1964).
7. G. Rickayzen, J. Phys. C **2**, 1334 (1969).
8. F. London, *Superfluids* (Wiley, New York, 1950), Vol. I, Sect. 8.
9. E. Jakeman and E.R. Pike, Proc. Phys. Soc. London **91**, 422 (1967).
10. J. Koláček, P. Lipavský and E.H. Brandt, Physica C **369**, 55 (2002).
11. C.J. Adkins and J.R. Waldram, Phys. Rev. Lett. **21**, 76 (1968).
12. H. Kamerlingh Onnes and K. Hof, Leiden Comm. No. 142b (1914).
13. H.W. Lewis, Phys. Rev. **92**, 1149 (1953).
14. H.W. Lewis, Phys. Rev. **100**, 641 (1955).
15. T.K. Hunt, Phys. Lett. **22**, 42 (1966).
16. J. Bok and J. Klein, Phys. Rev. Lett. **20**, 660 (1968).
17. J.B. Brown and T.D. Morris, in *Proceedings of 11th International Conference on Low Temperature Physics*, edited by J.F. Allen, D.M. Finlayson and D.M. McCall (St. Andrews, Scotland, 1968), Vol. 2, p. 768.
18. T.D. Morris and J.B. Brown, Physica (Amsterdam) **55**, 760 (1971).
19. P. Lipavský, J. Koláček and J.J. Mareš, Phys. Rev. B **65**, 2507 (2002).
20. K.M. Hong, Phys. Rv. B **12**, 1766 (1975).
21. D. van der Marel, Physica C **165**, 35 (1990).
22. D.I. Khomskii and F.V. Kusmartsev, Phys. Rev. B **46**, 14245 (1992).
23. G. Blatter, M. Fiegal'man, V. Geshkenbein, A. Larkin and A. van Otterlo, Phys. Rev. Lett. **77**, 566 (1996).
24. J. Koláček, P. Lipavský and H.E. Brandt, Phys. Rev. Lett. **86**, 312 (2001).
25. J. Koláček and P. Lipavský, Physica C **364-365**, 138 (2001).
26. P. Lipavský, J. Koláček, K. Morawetz and H.E. Brandt, Phys. Rev. B **65**, 144511 (2002).
27. S.V. Yampolskii, B.J. Baelus, F.M. Peeters and J. Koláček, Phys. Rev. B **64**, 144511 (2001).
28. P. Lipavský, J. Koláček, K. Morawetz and H.E. Brandt, Phys. Rev. B **66**, 134525 (2002).
29. K. Kumagai, K. Nozaki and Y. Matsuda, Phys. Rev. B **63**, 144502 (2001).
30. W.L. Ginsburg and L.D. Landau, Ž. eksper. teor. Fiz. **20**,1064 (1950).
31. J. Bardeen, Phys. Rev. **94**, 554 (1954).
32. J. Bardeen, *Theory of Superconductivity* in Handbuch der Physik, Bd. XV. 274 (1956).
33. M. Machida and T. Koyama, Physica C **378-381**, 443 (2002).
34. B.D. Josephson, Phys. Lett. **16**, 242 (1965).
35. P. Ao and D.J. Thouless, Phys. Rev. Lett. **70**, 2158 (1993).
36. J. Koláček and E. Kawate, Phys. Lett. A **260**, 300 (1999).
37. Y. Aharonov and A. Casher, Phys. Rev. Lett. **53**, 319 (1984).
38. J. Koláček, R. Tesař and Z. Šimša, Physica B **284**, 773 (2000).
39. H.-T.S. Lihn, S. Wu, H.D. Drew, S. Kaplan, Qi Li and D.B. Fenner, Phys. Rev. Lett. **76**, 3810 (1996).
40. J.I. Gittleman and B. Rosenblum, Phys. Rev. Lett. **16**, 734 (1966).
41. M.W. Coffey and J.R. Clem, Phys. Rev. Lett. **67**, 386 (1991).
42. T.C. Hsu, Physica (Amsterdam) **213**C, 305 (1993).
43. P.G. de Gennes and J. Matricorn, Rev. Mod. Phys. **36**, 45 (1964).
44. W.F. Vinen, Rev. Mod. Phys. **36**, 48 (1964).
45. A.I. Schindler and D.J. Gillepse, Phys. Rev. **130**, 953 (1963).
46. A.R. Strnad, C.F. Hempstead and Y.B. Kim, Phys. Rev. Lett. **13**, 794 (1964).

47. C.F. Hempstead and Y.B. Kim, Bull. Am. Phys. Soc. **9**, 29 (1964).
48. J. Bardeen, Phys. Rev. Lett. **13**, 747 (1964).
49. A.K. Niessen and F.S. Staas, Phys. Lett. **15**, 26 (1965).
50. W.A. Reed, E. Fawcett and Y.B. Kim, Phys. Rev. Lett. **14**, 790 (1965).
51. N. Usui, T. Ogasawara and K. Yasukochi, Phys. Lett. **27**A, 139 (1968).
52. C.H. Weinsenfeld, Phys. Lett. **28**A, 362 (1968).
53. M. Galfy and E. Zirngiebl, Solid State Comm. **68**, 929 (1988).
54. L. Forró and A. Hamzić, Solid State Comm. **71**, 1099 (1989).
55. S.N. Artemenko, I.G. Gorlova and Yu.I. Latyshev, Phys. Lett. A **138**, 428 (1989).
56. S.J. Hagen, C.J. Lobb, R.L. Greene and M. Eddy, Phys. Rev. B **43**, 6246 (1991).
57. D.M. Ginsberg and J.T. Manson, Phys. Rev. B **51**, 515 (1995).
58. Y. Matsuda, T. Nagaoka, G. Suzuki, K. Kumagai, M. Suzuki, M. Machida, M. Sara, M. Hiroi and N. Kobayashi, Phys. Rev. B **52**, R15749 (1995).
59. C.C. Almasan, S.H. Han, K. Yoshira, D.A. Gajewski, L.M. Paulius, M.B. Maple, A.P. Paulikas, Chun Gu and B.W. Veal, Phys. Rev. B **51**, 3981 (1995).
60. W.N. Kang, B.W. Kang, Q.Y. Chen, J.Z. Wu, Y. Bai, W.K. Chu, D.K. Christen, R. Kerchner and S.-I. Lee, Phys. Rev. B **61**, 722 (2000).
61. A.G. Aronov and S. Hikami, Phys. Rev. B **41**, 9548 (1990).
62. D.I. Khomskii and A. Freimuth, Phys. Rev. Lett. **75**, 1384 (1995).
63. N.B. Kopnin and A.V. Lopatin, Phys. Rev. B **51**, 15291 (1995).
64. N.B. Kopnin, Phys. Rev. B **54**, 9475 (1996).
65. E.B. Sonin, Phys. Rev. B **55**, 485 (1997).
66. Y. Kato, J. Phys. Soc. Japan **68**, 3798 (1999).
67. J. Koláček and P. Vašek, Physica C **336**, 199 (2000).
68. K. Morawetz and P. Lipavský, in this book, page 114.

Cluster Emission in Complex Nuclear Reactions

W. Udo Schröder and Jan Tõke

Abstract. Possible mechanisms of the emission of nuclear clusters in heavy-ion induced nuclear reactions are discussed in the context of the overall dissipative reaction environment. Essentials of model approaches to the reaction dynamics are briefly reviewed. Analysis of the overall reaction dynamics provides an estimate of time scales for cluster emission and the buildup or preexistence of nucleonic correlations. Experimental results demonstrate the presence of at least two different mechanisms of cluster emission. Energetic nuclear clusters are likely emitted during early stages of a nuclear reaction, containing memory of entrance-channel conditions. In addition, sequential, statistical cluster emission is observed from the hot remnants of projectile and target nuclei. The difficulty encountered previously by statistical models to describe statistical emission of massive clusters from excited nuclei is resolved in the framework of a simple interacting Fermi-gas model of expanded nuclear matter. It is suggested that sequential cluster emission is entropy-driven and can be understood in terms of a rearrangement of the surface of hot nuclei produced in energetic nuclear reactions.

1 Introduction

The atomic nucleus is one of the simplest quantal A-body systems, for which the interactions of the free constituents are well known. Thus, to gain an understanding of the structure of the nuclear system, its self-organization, and its response to external perturbations constitutes a prominent and fundamental challenge to many-body theory (see [1] for a recent interdisciplinary review of many-body problems). Conventional theory of nuclear structure and reactions has emphasized mean-field and shape degrees of freedom of the nucleus, as well as mean-field interactions, one-body dissipation, and Markovian transport phenomena based on nucleon transfer and in-medium scattering. Phenomenology of the decay of highly excited, metastable nuclear systems produced in reactions is typically addressed in terms of equilibrium-statistical phase space models.

While relatively successful at low energies, conventional nuclear theory appears to be challenged already by the few available comprehensive sets of data on nuclear reactions at bombarding energies in the Fermi energy domain. Of notable interest are nucleonic cluster and correlation phenomena which have achieved ubiquity in reactions induced by heavy ions or relativistic nucleons, both leading to hot nuclear systems. The inability of conventional statistical approaches to account for the high multiplicities of nuclear cluster emission in a process termed "multi-fragmentation" has motivated (see [2]) for a recent review explorations of new nuclear decay modes. Proposals of hypothetical nuclear liquid-gas phase transitions, observable for finite nuclei, seemed to promise a rather direct access to properties of the equation of state of nuclear matter and an experimental characterization of its critical behavior. Subsequent studies have shown a more complex and interesting behavior of colliding nuclear systems than assumed by the early reaction theories.

In the following, a brief survey is given over the essentials of most frequently used theoretical approaches to analyze the dynamics of complex nuclear reactions induced by heavy ions at intermediate bombarding energies (E/A – 20–100 MeV). Following this discussion, a selection of experimental data on heavy-ion reactions is presented, in order to convey an idea of the reaction environment in which nuclear cluster emission is observed experimentally. In this context, one hopes to answer specific questions as to the dominantly statistical or dynamical origin of cluster emission. These questions pertain to whether clusters are produced in statistical processes, e.g., result from a spectacular spontaneous disassembly of a supercritical nuclear system or are emitted in a sequence of individual binary statistical decays of hot primary reaction products. Conversely, dynamical cluster emission could be a consequence of a fracture-like breakup of projectile and/or target on impact or signal the snapping of a matter bridge ("neck") between massive remnants re-separating after a violent collision. Progress will be reported on a new understanding of a sequential, statistical cluster emission process driven by surface phenomena, casting new light on nuclear multi-fragmentation.

2 Heavy-Ion Reaction Dynamics

2.1 General

A great many experimental studies and theoretical analyses have crystallized [3] to a generally accepted picture of nuclear heavy-ion reactions at bombarding energies E/A of a few MeV per nucleon, corresponding to a few times the respective interaction barriers. As illustrated in Fig. 1 (*top*), a projectile nucleus approaching the target nucleus at a high angular momentum (l) or impact parameter (b) is repelled by strong Coulomb and centrifugal interactions.

For more central collisions, projectile and target form a di-nuclear system that rotates ("orbits") about its center of gravity, under the influence of attractive conservative and dissipative forces. The frictional loss of kinetic energy of relative motion is dominantly transformed into excitation energy of the interacting nuclei forming a transient di-nuclear complex. The heated constituents of the di-nuclear complex expand, deform, and mostly re-separate after some finite interaction time (a few times 10^{-22} to 10^{-21} s). For heavy

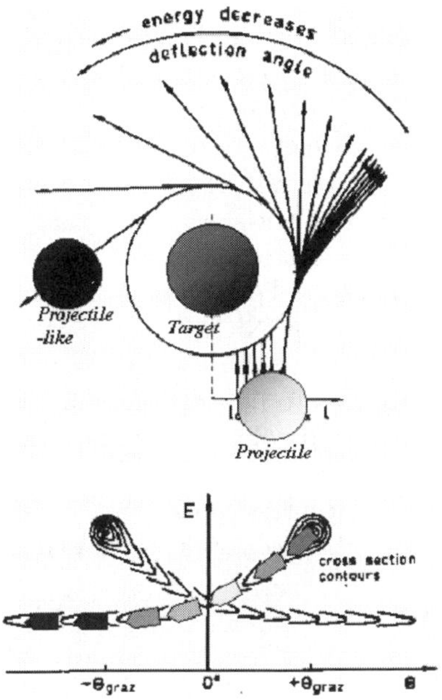

Fig. 1. Di-nuclear reaction scenario, dissipative trajectories (*top*) leading to characteristic angle-energy correlations (*bottom*)

systems and bombarding energies of up to E/A = 60 MeV per nucleon, complete nuclear fusion occurs, if at all, with only relatively small probabilities of the order of not more than a few % of the total reaction cross section (see, e.g., [4,2]). In the overwhelming majority of cases, projectile-like (PLF) and target-like (TLF) remnants of the collision partners are observed in the exit channel.

This dynamical collision scenario leads to characteristic, "tell-tale", correlations between the emission angle of the projectile (Θ) and its final kinetic energy (E) or the amount of dissipated energy (E_{loss}). In Fig. 1 (*bottom*), this dissipative orbiting process is illustrated by the schematic cross section contour lines of the "Wilczyński plot" [5]. Here, a forward-going ridge of reaction yield, starting from the Coulomb rainbow "grazing angle" (Θ_{graz}), is associated with increasing amounts of dissipated energy, as the reaction angle decreases, and as the di-nuclear system eventually rotates beyond the beam axis ($\Theta = 0°$) to the region of negative angles. Since experiments typically do not distinguish between trajectories originating from different sides of the beam, one measures a superposition of cross section features from both collision geometries. In the experimental distributions, a yield ridge of partially damped (positive angles) and a ridge of strongly damped events (negative angles) appear superimposed for the measured range $0 \leq \Theta \leq \Theta_{graz}$.

The above dissipative-orbiting reaction type is distinguished from a schematic fragmentation-type or "participant-spectator" reaction scenario [2,6] often invoked in the description of heavy-ion reactions at intermediate and higher bombarding energies. In this latter scenario, the interaction region, defined by the geometrical overlap of projectile and target on a collision trajectory, decouples from the projectile and target "spectators" and forms an independent third nuclear system, a source of a variety of decay products. While spectator remnants of projectile and target remain relatively cold, the intermediate participant region has, in this participant-spectator scenario, all predicates of a hot, equilibrated intermediate system that can disintegrate spontaneously into light particles and clusters. For central collisions, in particular for lighter reaction systems such as Ar+Ni [2,7], the participant region can presumably encompass the entire projectile-target system. As will be shown further below, the reality appears to lie somewhere in between these two reaction scenarios: Most heavy-ion collisions produce hot and massive projectile and target remnants in the exit channel. In addition, a third entity appears in peripheral, as well as in central collisions with different degrees of excitations. How heavy-ion reaction scenarios are viewed in theoretical approaches is discussed next.

2.2 Microscopic Reaction Models

In most reaction models, the time dependent mean-field (one-body) nucleus-nucleus interaction is thought to be responsible for the conservative forces influencing the trajectories of relative motion of projectile and target nuclei. In addition, as has been discussed extensively in the earlier literature [3], one-body nuclear interactions produce particle-hole (p.h.) excitations in the nuclear reaction partners and effect or allow the exchange of nucleons between them. Nucleon exchange (and ph excitation) processes give rise to characteristic transport mechanisms driven by and dissipating the energy of relative motion. Consequently, macroscopic properties of the interaction partners, such as A_i, Z_i, E_i^* (thermal excitation), and I_i (spin), vary in a stochastic fashion for each nucleus i. With certain probabilities, transferred nucleons have enough energy in the recipient nucleus to exceed its binding energy and may be emitted directly, or after a few in-medium collisions, as "Fermi jets" [8,9] or prompt particles [10]. Similar prompt-particle emission is predicted by all major theoretical approaches.

The one-body nuclear exchange model NEM developed by Randrup and collaborators [11–16,70] is one of the simplest mean-field reaction models. It predicts joint probability distributions for all variables of the relative motion and the intrinsic nuclear systems, including statistical correlations. The model implicitly assumes a Markov transport process of nucleons between the interacting nuclei. Here memory of the collision history is lost on a fast time scale, typically of the order of 10^{-22} s (40–50 fm/c). Such a fast memory loss is implicitly postulated and attributed to higher-order interactions. It assures that the intrinsic system of nucleons in each nucleus is at equilibrium virtually at all times during a nuclear collision and gives rise to dissipative features.

Well-known consequences of the one-body model are the "window and wall" friction forces [17]. In this reaction model, a family of mono- or di-nuclear potential wells enclosing Fermi gases of neutrons and protons at normal density represents the interacting nuclei. A time-dependent joint probability distribution $P(\mathbf{x}, t)$ is defined for the assumed set of macroscopic reaction variables, $\mathbf{x} = \{x_i\}$, following a Fokker–Planck equation,

$$\frac{\partial}{\partial t}P(\mathbf{x},t) = -\Sigma_i \frac{\partial}{\partial x_i}\{v_i P(\mathbf{x},t)\} + \Sigma_{i,j}\frac{\partial^2}{\partial x_i \partial x_j}\{D_{i,j}P(\mathbf{x},t)\} \, . \tag{1}$$

Here, the quantities v_i and $D_{i,j}$ denote the effective drift and diffusion coefficients, respectively, for the corresponding macroscopic observables. The time dependence of the mean trajectories $\{\mathbf{x}_i, \dot{\mathbf{x}}_i\}$ is determined by the Lagrange–Rayleigh equations of motion

$$\left\{\frac{d}{dt}\frac{\partial}{\partial \dot{\mathbf{x}}_i} - \frac{\partial}{\partial \mathbf{x}_i}\right\}\mathbf{L} = -\frac{\partial}{\partial \dot{\mathbf{x}}_i}\mathbf{F} \, . \tag{2}$$

The Lagrangian $\mathbf{L} = \mathbf{T} - \mathbf{V}$, with kinetic energy ($\mathbf{T}$) and mean-field interaction potential ($\mathbf{V}$), depends on the degrees of freedom explicitly considered. The Rayleigh dissipation function $\mathbf{F}$ in Eq. (2) describes the combined action of one-body "wall and window" frictional forces [17]. Through Eqs. (1) and (2) the NEM links macroscopic nuclear shapes and the trajectories of the relative nucleus-nucleus motion to the transmutation of projectile and target nuclei. These relations are dynamically dependent on bombarding energy, impact parameter, and the energy loss in a collision. In an approximate fashion, dynamical processes, such as the emission of fast, nonequilibrium nucleons and clusters initiated by the exchange of nucleons between projectile and target, can be modeled [9,18] by integrating corresponding kinetic rate equations along the trajectories defined by Eq. (2). The above approach has been surprisingly successful in the description of experimental fragment mass, charge, and energy distribution, even though it is difficult to justify its physical foundation for other than very peripheral interactions amenable to perturbative treatment.

Some of the reaction models most often used for the interpretation of heavy-ion reactions represent variations of the Boltzmann–Uehling–Uhlenbeck approach. Such transport models consider the total average s.p. probability distribution function f, whose time evolution is governed by the classical Boltzmann equation

$$\frac{df}{dt} = \frac{\partial f}{\partial t} + \mathbf{v}\cdot\nabla_r f - \nabla_r U \cdot \nabla_p f = \left(\frac{df}{dt}\right)_{coll} \, . \tag{3}$$

The function f is a sum of functions f_i for individual nucleons, and U is the combined mean field of the system of two interacting nuclei. This "BUU" model represents individual nucleons by sets of point-like "test" particles [19,20] and propagates them in the combined, density-dependent (average) mean field $U = U_{projectile} + U_{target}$ according to classical equations of motion. NN collisions due to residual interactions are modeled by the collision term on the r.h.s. of Eq. (3). In its simplest form, the collision term $(1+2 \Rightarrow 1'+2')$ for a given nucleon ($i=1$) is written as

$$\left(\frac{df_1}{dt}\right)_{coll} = \int \frac{d\mathbf{p_2}\mathbf{p_1}'\mathbf{p_2}'}{(2\pi)^5}\sigma_{sc}v_{12}[f_1'f_2'\bar{f}_1\bar{f}_2 - f_1 f_2 \bar{f}_1'\bar{f}_2'] \times \delta_{\mathbf{p_1}+\mathbf{p_2}-\mathbf{p_1}'-\mathbf{p_2}'}\delta_{\epsilon_1+\epsilon_2-\epsilon_1'-\epsilon_2'} \quad (4)$$

in terms of the in-medium scattering cross section, σ_{sc}, the relative NN velocity, v_{12}, and the Pauli blocking factors, which are of the form $f_i\bar{f}_j = f_i(1-f_j)$. The collisions lead to gains and losses in the probability $f_{i=1}$ and are, in this model, responsible for any dissipative features of the reaction mechanism. The BUU model incorporates most of the simplifications mentioned above for the general mean-field approach and is therefore valid in the weak-coupling limit, i.e., for dilute gases [21]. Collisions between two given nucleons 1 and 2 are assumed to be instantaneous, point-like, and time-reversal invariant. Each collision process occurs only once and leads from a well-defined initial two-particle state to a well-defined final state. It is independent of all other particles and of the reaction history (stochastic Markov process). Correlations that occur in dense gases and liquids, notably those that form nuclear clusters are not considered. In further developments, the BUU approach has been extended [15,22–28] to include certain fluctuations along particle trajectories, such that the r.h.s of the BUU equation contains an additional Langevin fluctuation functional of f. If the collision dynamics are stable, these fluctuations merely generate a narrow distribution about the average system trajectories. Fluctuations become more dramatic, if the dynamics lead to bifurcation or to unstable nuclear systems, where fluctuations are amplified without bounds. Specific correlation [29–32] and cluster effects have been added to the BUU model, which go somewhat beyond its basically mean-field approach.

The BUU model has then also been applied to describe the spinodal decomposition of expanding nuclei [2,28,33] and cluster formation associated with fluctuations. Coherent nuclear response can express itself in terms of resonant excitations of instabilities of neck-like structures or in the coherent emission of fast particles in diabatic "squeezing-mode" processes [34]. Figure 2 illustrates the formation of a cluster from the stretched matter bridge ("neck") between projectile and target in the exit channel of a Sn+Ca collision at 35 MeV per nucleon. According to the model, substantial mixing of projectile and target nucleons has taken place, up until the time of separation, such that the neck matter should have no memory of an entrance-channel asymmetry.

Such cluster configurations appear in the model [22] only for a small band of impact parameters. Even if such non-compact nuclear geometries are quantum-mechanically not likely to form [35], their existence could result in a characteristic angular correlation of the emitted cluster(s) with the PLF-TLF separation axis.

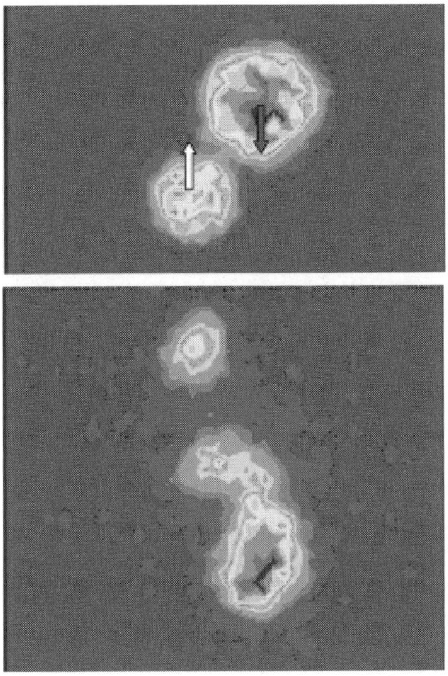

Fig. 2. BUU simulation of a Sn+Ca collision at = 35 MeV in configuration space. Density contours (log) (x,y) of projectile and target nuclei are shown for entrance- (*top*) and exit-(*bottom*) channel configurations

Since BUU is a microscopic model, it should automatically predict an evolution of the macroscopic collision mechanism with projectile-target consistencies, bombarding energy, and impact parameter. For example, it should describe some dissipative effects, chaotic excitations, as well as predict an isospin-dependent collective response to compression and/or the direct ejection of nucleons in the process. In practice however, the spectrum of actual responses may be reduced, due to the details of the model realization in terms of potential, scattering cross section, number of test particles, etc. Furthermore, it is often difficult and ambiguous [36] to extract from theoretical calculations the expectation values of fundamentally interesting observables, such as product excitation energies or promptly emitted particle spectra and angular distributions. Clearly, the BUU approach deserves further technical improvements (see e.q. page 114 this book), which would allow more meaningful tests of the basic approach to be performed. Experiments can help test the basic assumptions underlying BUU theory and resolve the associated uncertainties.

A competing theoretical approach to the dynamics of heavy-ion collisions is represented by (Quantum) molecular dynamics (QMD, MD) models [37–41,66,67,69,70]. Again, a mean field is calculated which governs the evolution of independent particles. This theory describes individual nucleons by Gaussian (minimum) test wave packets of the form

$$\varphi(\mathbf{x},t) = \left(\frac{2}{\pi L}\right)^{\frac{3}{4}} \mathrm{e}^{\frac{(\mathbf{x}-\langle\mathbf{x}(t)\rangle)^2}{L}} \mathrm{e}^{ip(x-\langle x(t)\rangle)} \mathrm{e}^{-i\frac{p^2(t)}{2m}} , \tag{5}$$

where $\mathbf{x}$ and $\mathbf{p}$ are position and momentum vectors, and L is a model parameter. The total nuclear A-body wave function is a direct product of such wave packets, neglecting quantum requirements of an anti-symmetrized wave function. Free model parameters are determined using a variational method minimizing the energy with the above trial wave functions.

Preparation of nuclei in their ground-state configurations requires the adoption of some operational procedure or definition, just like in BUU. In some models, sharp spheres or spherical Saxon–Woods-type density distributions containing the centroids of the wave packets represent the initial nuclei. Others minimize the Hamiltonian and reproduce the liquid-drop binding energies. In time, the individual particles, i.e., the centers of the corresponding wave packets, are evolved classically according to Hamilton's equation of motion, an expression of Ehrenfest's theorem for expectation values,

$$\dot{\mathbf{r}}_\mathbf{i} = \frac{\mathbf{p}_\mathbf{i}}{m} + \nabla_{\mathbf{p}_\mathbf{i}} \Sigma_j \langle V_{ij} \rangle \, , \qquad \bar{\mathbf{p}}_\mathbf{i} = -\nabla_{\mathbf{r}_\mathbf{i}} \Sigma_{j \neq i} \langle V_{ij} \rangle \, . \tag{6}$$

Here, $\Sigma_{j \neq i} \langle V_{ij} \rangle$ is the mean field calculated with the above trial wave functions.

For the effective NN interaction V_{ij} a combination of Skyrme and finite-range Yukawa forms is chosen in some MD approaches, with parameters adjusted such that at equilibrium it is equivalent to an equation of state of specified stiffness for nuclear matter. The Yukawa range is adjusted to describe the diffuse nuclear surface. The quantity L in Eq. (5) remains a model parameter that can be adjusted with some freedom. In some MD models, no explicit treatment of nucleonic isospin is performed, and all nucleons carry an average effective charge. In others, neutrons are distinguished from protons in their interactions. Typically, binary collisions of two independent, point-like nucleons are considered within these models, even though several of these collisions may take place at the same time. The free NN cross section is used as a geometrical criterion to determine the occurrence of a collision. Models differ in their assumptions of isotropic or anisotropic NN scattering. Quantum statistics for NN collisions is considered schematically in terms of Pauli blocking factors (cf. Eq. (4)). Recent implementations of the MD approach account for some average quantal antisymmetrization effects by introducing a "Pauli potential" [42].

The emission of particles from all stages of the intra-nuclear equilibration cascades initiated by a heavy-ion collision occurs naturally in MD models. In a study [41] of the reaction ^{40}Ca+^{40}Ca at 35 and 55 MeV, the crucial role of the in-medium cross section σ_{sc} (Eq. (4)) was demonstrated for the number of NN collisions and the damping of the projectile motion. Such a connection between dissipation of the relative motion and the emission of nonequilibrium particles is intuitive. Also as expected, the simulations for typical cases of heavy asymmetric systems, e.g., E/A = 45-MeV ^{159}Tb+^{84}Kr [41], show that high-energy protons are emitted at early times (≤ 100 fm/c) in the collision process. In addition, emission of nuclear clusters is predicted, based on criteria of the vicinity of nucleons in phase space (coalescence model). Figure 3 illustrates the initial configuration and the result of a central Sn+Xe collision at $\epsilon = 50\,\mathrm{MeV}$ in a QMD simulation by Aichelin et al. [43].

The bottom panel clearly shows that projectile and target break up individually into nucleons and small clusters. Such a result is reminiscent of the breakup/shattering of non-viscous materials. As indicated by the respective shades, to a large extent the nuclear clusters emerging in the exit channel retain memory of their origin in projectile or target nucleus. Interestingly, according to the simulations, the nucleons of projectile and target do not mix very thoroughly, even in a central collision. It appears that correlations between nucleons resulting in real cluster formation are not created in a collision but pre-exist within projectile and target nuclei. Unfortunately and like in BUU, so too in QMD, it is not straightforward to evaluate and predict from the model observables such as intrinsic excitation energy, temperature, and spin of reaction products. Furthermore, from the consideration of only average antisymmetrization effects and the neglect of quantal correlations and fluctuations, cluster formation and emission are typically not modeled realistically in present-day QMD. Clusters do typically not appear in bound states but are designated ensembles of unbound nucleons in some proximity in phase space, disregarding the strong quantal anticorrelations that should make the phase space around each nucleon inaccessible to others. The final decay and cooling of the primary QMD reaction products is not treated in the QMD model directly but require additional statistical model calculations.

Even though the QMD approach in its current realizations exhibits deficiencies and has not been as widely used and promoted as BUU-type simu-

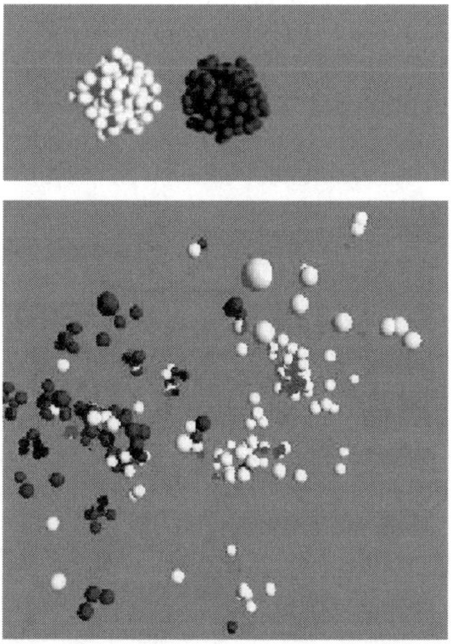

Fig. 3. QMD simulation of a central collision between a Sn and a Xe nucleus at 50 MeV per nucleon. initial (*top*), final (*bottom*), [43]

lations, QMD shows interesting potential for further sophistication and improvement, in particular, as computing power becomes more readily available. The potential of the QMD approach has been demonstrated by the AMD and FMD quantal molecular dynamics theories by Ono [44] and Feldmeier [45], respectively. While most dynamical (BUU and QMD) models are essentially classical approaches, Feldmeier et al. [45] have developed a quantal model of Fermionic Molecular Dynamics (FMD), where correlations are treated on the microscopic level of nucleonic wave functions. This model has potentially much predictive power and should be able to describe static properties such as ground-state bound nuclei and α-particle correlations, as well as nuclear dynamics such as dissipative collisions, fusion, and multi-fragmentation.

2.3 Nuclear Thermodynamics

The large number of small and intermediate-mass fragments (IMFs or clusters) emitted in various types of collision events constitutes one of the most intriguing features of heavy-ion reactions at intermediate energies. The emission of massive clusters in statistical nuclear decay processes has not been expected to contribute significantly, based on predictions by traditional statistical nuclear decay models. The barriers for cluster emission from a heavy nucleus are typically one order of magnitude higher than typical temperatures reached in heavy-ion collisions at the bombarding energies of interest. The new multi-cluster process was termed spontaneous "multi-fragmentation" (MF) and likened to a nuclear liquid-gas phase transition [19,46,47], the transition from the ordinary nuclear liquid (the nucleus at normal density) to a gas phase with embedded clusters. The phenomenon seemed to offer an opportunity to study the nuclear equation of state (EOS) in the low-density, spinodal region. To isolate equilibrated experimental "participant" systems at maximum excitations has proven rather difficult. Tentative experimental information on associated nuclear thermodynamic properties are typically strongly model-dependent (see, e.g., analyses by [48,49] and often rely on disputable prerequisites [50,51] or idealizations [52,47].

Theoretical approaches to a statistical interpretation of MF, such as MMMC [52], SMM [47], or percolation [53] models, have focused on approximate statistical final state populations, but ignored the reaction path leading to such states and the nuclear decay proper. Rigorously, these scenarios appear physically unattainable or are not directly applicable to the fragmentation decay of actual nuclei. For example, equilibrium models of nuclear phase coexistence rely implicitly on the existence of a stable, equilibrated gas phase beyond the liquid (nuclear) interface sustained by a container with ideally reflective walls. Several such models postulate a statistical population of arbitrarily large "freeze-out" volumes with clusters of different sizes and monomers, imposed by a strict spatial confinement. Recent, more phenomenological interpretations [53,54] consider the inverse process, that of an equilibrium condensation [55] of dilute interacting monomeric gases.

In other cases, particularly convenient recipes are adopted [47,52,56,57] to approximate (and boost) the phase space available for fragmentation. Unfortunately, effect and justification of disputable [58] assumptions underlying several popular statistical MF models have rarely been discussed.

In any case, the above statistical-model descriptions address only one class of cluster emission processes, in which nuclear systems at thermodynamic equilibrium are formed which disassemble spontaneously into clusters. This corresponds at most to only a small fraction of all cluster processes observed in heavy-ion collisions. Moreover, the experimental evidence gathered to date for such a transformation has remained incomplete and controversial. A new approach seems to be required to make further progress in the pursuit of an explanation of the different aspects of the multi-fragmentation phenomenon. Further below, a new approach of surface cluster emission from nuclear systems is discussed, which addresses one of two frequent mechanisms of cluster emission in heavy-ion reactions. To explain the competition between this and other processes presents a scientific challenge reaching beyond issues of equilibrium thermodynamics.

3 Experimental Characterization of Heavy-Ion Reactions

3.1 The Dissipative Reaction Environment

As illustrated previously, there are different potential mechanisms of cluster production in heavy-ion collisions. For an exploration of their respective properties and their mutual competition, it is desirable to distinguish collisions with respect to their impact parameter and violence. Thereby, one could transiently produce nuclear systems in a range of excitations, from low temperatures perhaps to those exceeding the critical temperature for finite nuclei [59,60] ($T_c \approx 10\,\mathrm{MeV}$). The ability to select experimentally equilibrated systems with different temperatures is a prerequisite to the exploration of formation and decay pathways of metastable finite nuclei. Since collision energies of interest here are sufficiently high to lead to complete disintegration of the involved projectile and target nuclei, a large range of particles and reaction fragments is typically produced in heavy-ion reactions and has to be measured.

The experiments to be reported below have used combinations of 4π detection devices for neutrons and light charged particles and clusters, augmented by special detectors sampling the massive remnants of projectile and target, PLF and TLF, respectively. Figure 4 illustrates the size of the neutron detector [61] used in these experiments.

It provides event-by-event measurements of the multiplicity (m_n) and total kinetic energy of neutrons, as well as providing coarse angular information for the emitted neutrons. In the figure, the segmented SuperBall detector

Fig. 4. The Rochester Super-Ball neutron calorimeter at its beam line at the National Superconducting Cyclotron Laboratory (MSU) in Lansing/MI

is opened to expose the scattering chamber containing the set of charged-particle detectors, including the Washington University Dwarf Array [62]. This latter detector covers a large fraction of the total solid angle and provides particle identification (mostly atomic number Z), a measurement of their energies, as well as the multiplicities of light charged particles (m_{lcp}) and intermediate-mass clusters (m_{IMF}). Additional, position-sensitive solid-state detector telescopes are used to measure atomic number, energy, and emission angle of projectile-like fragments (PLF).

As a typical result, Fig. 5 exhibits a logarithmic contour diagram of the joint multiplicity distribution $P(m_n, m_{lcp})$ of neutrons and light charged particles (LCP) measured for the reaction $^{209}\mathrm{Bi} + {}^{136}\mathrm{Xe}$ at E/A = 62 MeV. Different contours are separated by a factor of 2. Experimental results of average ridge lines of the probability distributions for this and other bombarding energies are indicated by dotted curves.

One observes in Fig. 5 the characteristic shape of the joint multiplicity distribution, which has now been observed for approximately a dozen intermediate-energy heavy-ion reactions. Since neutrons and LCPs are mostly statistically evaporated from hot reaction products, the distribution in Fig. 5 represents, non-linear and convoluted, information on the combined thermal excitation energy (E^*) distribution of the primary reaction products,

$$P(E^*) = \tilde{P}[P(m_n, m_{lcp})] \ . \tag{7}$$

Nonequilibrium particle emission processes do not contribute significantly to the energy balance. Simulation calculations performed with statistical-model codes for heavy, neutron-rich reaction systems have shown that the distribution does not depend significantly on the split of the intermediate system

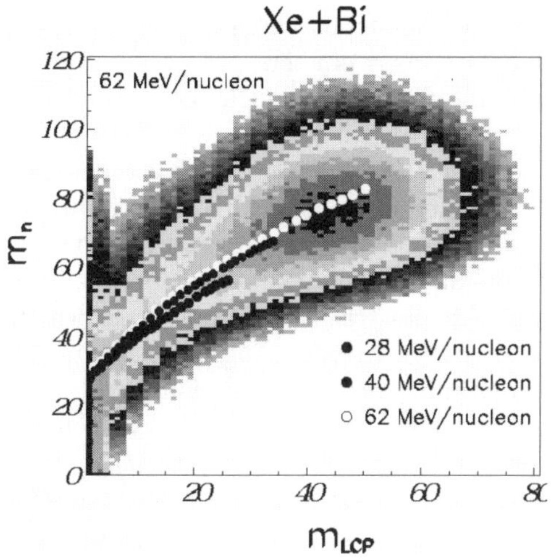

Fig. 5. Joint probability distribution of neutrons and LCPs from the reaction $^{209}Bi + ^{136}Xe$ at E/A=62 MeV (log contour diagram). The *dotted lines* represent the ridge line of these distributions for the same reaction at three bombarding energies

into the primary reaction fragments or on the relative distribution of the total excitation energy among these fragments. Then, selecting slices of the multiplicity distribution aligned approximately perpendicular to the ridge line of maximum probability selects events with approximately constant average excitation energy $\langle E^* \rangle$. This quantity is related to the dissipated energy (E_{loss}) and collision impact parameter (b). However, it is important to note that there is no one-to-one relation between excitation energy (or b) and multiplicity. A fixed excitation energy E^* corresponds to a distribution in both m_n and m_{lcp}, and vice versa. The widths of the respective distributions represent the statistical fluctuations of the evaporation process. A more complete picture can be produced by viewing a multi-dimensional joint distribution that is differential in the identities of $LCPs$ and clusters.

As seen from Fig. 5, the joint multiplicity distribution exhibits a ridge line of yield extending from the origin of the plot, first parallel to the m_n axis at $m_{lcp} \approx 0$. This is the range of relatively low dissipated energies and excitations of the primary reaction products that are sufficient only for the emission of neutrons. LCP emission has to overcome the respective Coulomb barriers. At larger neutron multiplicities, i.e., for higher excitation energies of the products, LCP and neutron multiplicities increase in a correlated fashion. The width of the distribution represents the statistical fluctuations of the evaporation process. At very high excitations, corresponding to a region around $(m_n, m_{lcp}) \approx (75, 40)$, a broad bump appears in the joint distribution, which indicates a broadly defined extremum in excitation energy appearing in the region of central collisions. It is significant to note that this excitation energy is far lower than the energy available in the entrance channel. Only about some (50–60)% of this energy reappears in evaporated light particles, for the reactions in question ("incomplete damping"). Potentially higher excitation energies are apparently not discernible in these reactions, based on the multiplicities of evaporated light particles, neutrons and LCPs. Additional dissipated energy appears in intermediate-mass clusters, which apparently displace the normal evaporation process by which hot nuclei are thought to decay.

The relation between average multiplicity and fragment excitation energy can be verified by measuring the kinetic energies of associated reaction fragments, e.g., of the PLFs. One finds there is a characteristic relation between intrinsic excitation of reaction fragments and their kinematic properties such as kinetic energy and emission angle. An illustrative example is shown in Fig. 6, for three heavy-ion reactions. In these contour diagrams, events are displayed according to final PLF energy and laboratory reaction angle.

A clear correlation is observed in these plots (and many others), which are reminiscent of the orbiting picture discussed in the context of the schematics in Fig. 1. In fact, dynamical calculations with the simple NEM reaction model (cf. Eqs. (1) and (2)), represented by the curves in Fig. 6 give a qualitative account of the observed PLF reaction features and support a dissipative-orbiting mechanism [3].

This mechanism virtually exhausts the reaction cross section, albeit within sizeable experimental uncertainties (10%). Hence, in heavy-ion reactions in the intermediate-energy domain, there is very little room for complete fu-

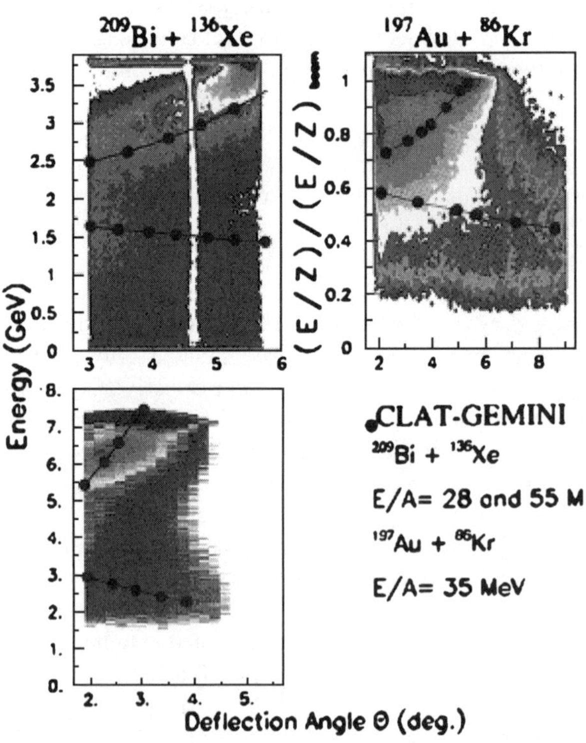

Fig. 6. Contour plots of energy-angle correlations (Wilczynski Plots) for three reactions. The *circles* and *solid curves* indicate pre-dictions by the NEM reaction model, accounting for sequential de-excitation of the fragments

sion leading to a thermally equilibrated nuclear system. One is then led to assume a reaction environment, in which two or at most a few massive fragments emerge in the primary reaction step. These primary fragments, which are typically far from mutual thermal equilibrium, are hot and equilibrate intrinsically on a fast time scale (-10^{-22} s) after the breakup of the intermediate di-nuclear complex. The primary fragments decay subsequently, mainly via evaporation of particles. Fluctuations in the decay stage are responsible for much of the dispersion of the cross section features observed in Fig. 6.

The distributions shown in Fig. 6 have been measured with detectors sampling only a representative portion of the total solid angle. Such data have been used to calibrate the multiplicity observables discussed in the context of Fig. 5. Figure 7 illustrates a procedure in which 5 coarse conditions are set on the total excitation energy of fragments produced in the ^{209}Bi + ^{136}Xe reaction at E/A = 62 MeV. On the left, these bins are indicated as diagonal lines cutting through the (unconditional) experimental joint multiplicity distribution.

The middle panel corresponds to events, in which no IMF-clusters ($Z > 2$) have actually been detected by the ($0.9 \times 4\pi$) charged-particle array. Obviously, such $m_{IMF} \approx 0$ events are concentrated at low total excitation energies. In other words, the actual cluster emission probability is relatively low for peripheral ^{209}Bi+^{136}Xe collisions. However, as discussed below, the cluster emission probability for such collisions, though small, is actually non-zero.

On the other hand, the right panel of Fig. 7 demonstrates that most clusters are emitted in more central collisions. The condition that at least one cluster be detected illuminates a region in m_n, m_{lcp} space associated with high excitations, while peripheral collisions leading to low excitations of the fragments do not contribute much to the cluster emission process. This is a general feature of many heavy-ion reactions in the intermediate range of bombarding energies: Clusters are emitted dominantly in central collisions, while peripheral collisions lead to such emission less frequently.

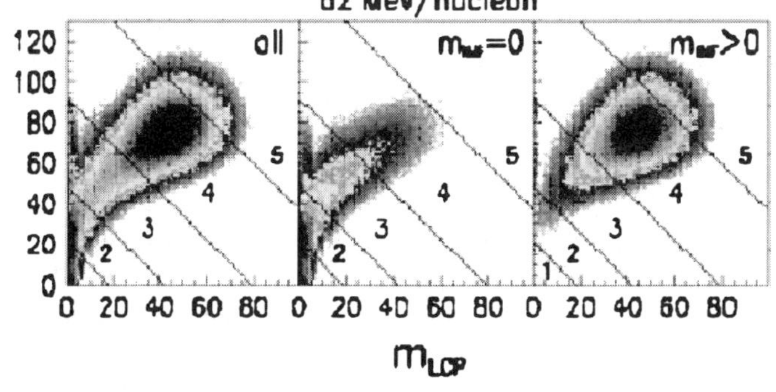

Fig. 7. Joint multiplicity distribution of Fig. 5 (*left*), measured absence of clusters (*middle*), and with at least one additional cluster per event (*right*)

3.2 Nuclear cluster emission in dissipative heavy-ion reactions

As pointed out in previous sections, heavy-ion reactions at E/A = (20–100) MeV are characterized by the appearance of two massive remnants of projectile and target (PLF and TLF, resp.) in the exit channel. In addition, intermediate-mass and $-Z$ clusters are produced, mainly in mid-central to central collisions. In Fig. 8, samples are shown for integrated charge and (transverse) energy distributions of clusters emitted in the ^{209}Bi + ^{136}Xe reaction at E/A = 28 MeV, as functions of cluster multiplicity m_{IMF}.

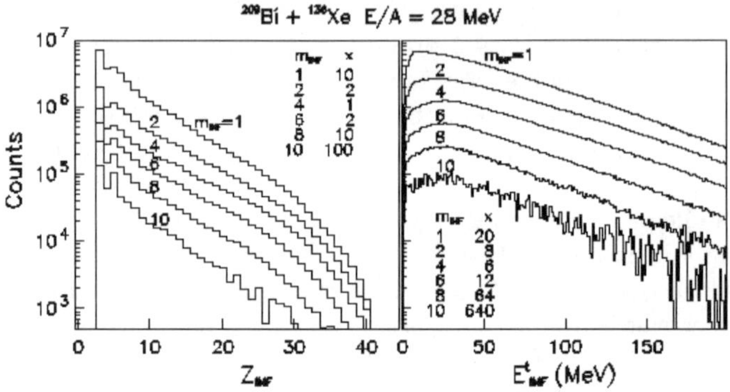

Fig. 8. Atomic-number (Z, *left*) and transverse-energy distributions (*right*) for clusters from the ^{209}Bi+ ^{136}Xe reaction, as functions of the cluster multiplicity, m_{IMF}. Individual spectra are scaled by factors x

Already at such a relatively low bombarding energy, the average number of clusters per event is larger than $\langle m_{IMF} \rangle = 1$. The IMF cluster distributions illustrated in Fig. 8 are averaged over all measured events but stem dominantly from central collisions. This one concludes from the associated excitation energy distributions represented by the conditional joint multiplicity distribution (compare Fig. 7). Approximately 25% of the cross section is associated with at least 2 clusters with $Z_{IMF} > 2$, and a maximum of 13 clusters have been observed per event. From Fig. 8 (*left*), one gathers that the average atomic number Z_{IMF} of a cluster emitted in this reaction is about $\langle Z_{IMF} \rangle = 8$ (oxygen). The Z_{IMF}-distributions are relatively independent of the cluster multiplicity, which suggests certain independence [48,63] in the production process. Like the Z_{IMF} distributions, the transverse-energy distributions shown on the right of Fig. 8 are of a generally exponential character and rather independent of the number (m_{IMF}) of clusters emitted in an event. In fact, the shapes of the energy distributions are reminiscent of statistical Maxwell–Boltzmann energy spectra.

Closer inspection of these IMF distributions, for ^{209}Bi + ^{136}Xe as well as for other reactions, reveals a more complex picture and demonstrates in particular the non-statistical nature of the energy spectra. While light-particle (p, d, t, ...) spectra indicate emitter temperatures of not higher than $T = 5.5\,\text{MeV}$, the spectral slope parameters characteristic of the IMF energy distributions are of the order of $E_0 = 30\,\text{MeV}$. It is not obvious, how these two types of spectra with very different slopes could be reconciled within an equilibrium-statistical decay model. Most likely, they are produced in different processes. This expectation is supported by other experimental data. Galilei-invariant velocity plots of light particles and IMF clusters are shown in Fig. 8 for the same reaction, measured in coincidence with a projectile remnant at forward angles. Here, the invariant cross section,

$$\frac{d\sigma}{d^3\mathbf{v}} = \frac{d\sigma}{v_\perp \cdot dv_\perp dv_\parallel} \tag{8}$$

is plotted vs. the particle velocity components parallel and perpendicular to the beam.

In such a picture, statistical emission in random directions from a moving emitter produces a (semi-) circular pattern in velocity space. The radius of this distribution corresponding to the Coulomb barrier for emission and its center is located at the tip of the velocity vector of the emitter. In Fig. 8, a superposition of two such ring shapes is observed to overlap, due to sequential evaporation of hydrogen (p, t) from the moving PLF and TLF, respectively. In addition, a fast non-statistical component of protons or tritons is visible at sideways angles, producing the bulge seen in the corresponding velocity plots. This additional component is attributed to a hypothetical, not necessarily physically well defined, third or "intermediate-velocity" source (IVS). For the IMF clusters (Li, C), on the other hand, shown in Fig. 8 for the same events, the velocity pattern is concentrated just at intermediate velocities,

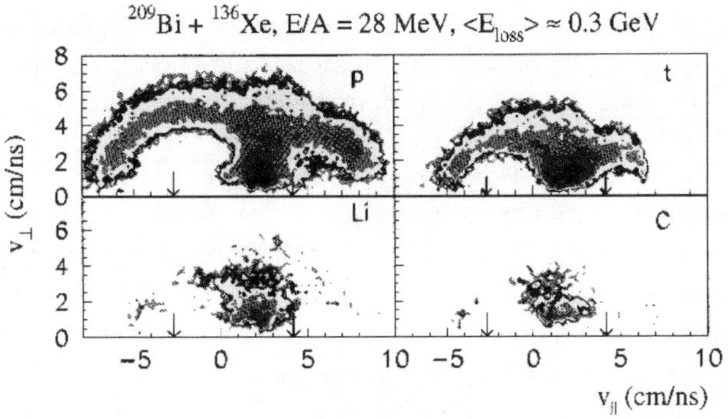

Fig. 9. Galilei-invariant cross section for charged products in peripheral collisions, plotted vs. two velocity components. The parallel is defined with respect to the beam. *Arrows* indicate the velocities of PLF and TLF, resp.

i.e., in the velocity range between the two arrows indicating the velocities measured of PLF and TLF, respectively. Emission characteristic of this hypothetical intermediate-velocity source (IVS) accounts for most of the IMF clusters emitted in highly dissipative collisions. However, it is significant to note that such clusters are already produced with low probabilities ($\sim 10^{-1}$) in relatively peripheral collisions where interaction times are of the order of a few times 10^{-22} s. Emission of IMFs from the relatively cold PLF and TLF does apparently not take place, for the relatively low energy loss of $\langle E_{loss} \rangle = 0.3\,\mathrm{GeV}$ and the corresponding low fragment temperatures represented by the data in Fig. 8.

A likely scenario for the above non-statistical, dynamical cluster emission involves the interactions of the surfaces of the approaching collision partners. As illustrated in Fig. 10, nucleons in the overlap ("participant") zone could coalesce into a cluster, or participant matter could be "squeezed" out.

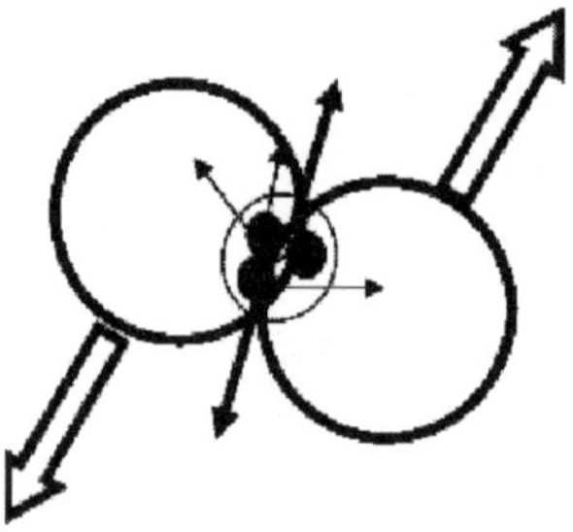

Fig. 10. Cluster formation through coalescence in the projectile-target overlap zone

In either case, clusters would appear with low velocities in the nucleon-nucleon center-of-mass frame of reference, kinematically between PLF and TLF. The emission probabilities associated with such a cluster mechanism are expected to depend on the collision dynamics, as well as on the composition of nuclear surfaces and the viscosity of nuclear matter. The experimental ability to control the collision trajectories via bombarding energy, dissipated kinetic energy of relative motion, and reaction angle (cf. Fig. 6) points to interesting opportunities for future detailed studies of cluster correlations in peripheral heavy-ion reactions in the time range $10^{-22}\,\mathrm{s} \leq t \leq 10^{-21}\,\mathrm{s}$. Below, some preliminary studies employing a hybrid coalescence model will be discussed.

In addition to the fast, dynamical cluster emission process discussed above, cluster components are observed experimentally that are associated with the sequential statistical decay of hot primary reaction fragments. The two top panels in Fig. 11 indicate the overall Galilei-invariant velocity plots of IMF clusters ($Z_{IMF} > 2$) in the $^{209}\mathrm{Bi}+^{136}\mathrm{Xe}$ reaction at two bombarding energies.

In each plot, representations of sequential emission from a fast-moving and a slow-moving source are clearly identified by semi-circular cross section ridges. As expected, the associated features are stretched farther apart at the higher bombarding energy. In addition, an IVS component is visible in the intermediate velocity range. Here, the larger spacing of the logarithmic contours indicates that the velocity spectrum of IVS clusters is much shallower (on average more energetic) than that of the spectra of clusters evaporated from PLF or TLF. For the following qualitative analysis, the IVS component is defined by two lines in velocity space, as indicated in the plots shown in the top row of Fig. 11. PLF cluster appear forward, TLF clusters backward of the IVS region.

In order to assess the impact parameter dependence of the two distinct cluster components, the associated joint $m_n - m_{lcp}$ multiplicity distribution is subdivided into three domains (cf. Fig. 7). Here, the domain of "peripheral" collisions is less strictly defined than in the previous discussion of Fig. 8. Consequently, there is some admixture of dynamical IVS clusters and sequentially emitted clusters in all three domains. The associated IMF cluster multiplicity distributions $P(m_{IMF})$ are shown in Fig. 11 in the three rows below the top.

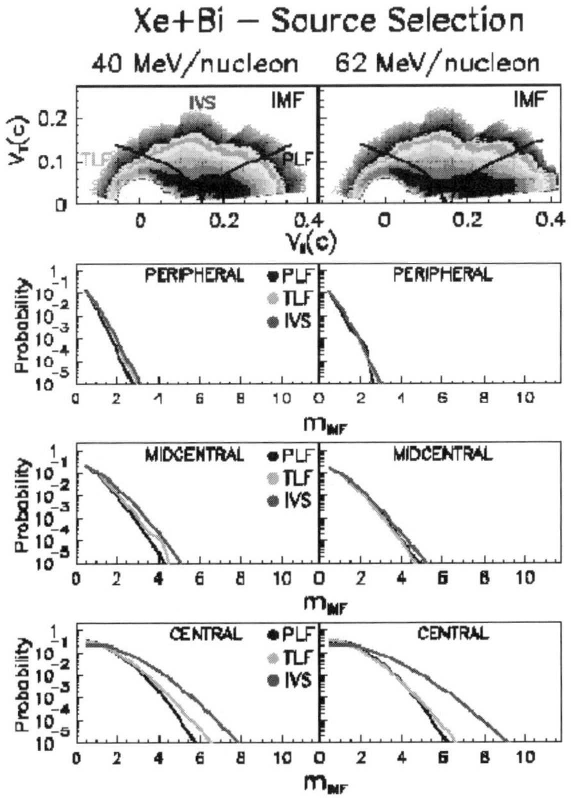

Fig. 11. Impact parameter dependence of cluster emission in the reaction ^{209}Bi + ^{136}Xe at two bombarding energies. An IVS was defined by the lines shown on the invariant velocity plots in the *top row*

It so happens that the multiplicity distributions of clusters emitted sequentially from PLF or TLF almost coincide in all three domains. For a range of peripheral collisions, the IVS cluster component distribution is very similar to those associated with either PLF or TLF. As the impact parameter decreases and the collisions become more violent, all cluster multiplicity distributions broaden. Cluster emission generally becomes more important for central collisions, where PLF and/or TLF may emit several IMF clusters. Curiously, such events are not associated with larger light-particle multiplicities. The IVS component is seen to gain the most intensity, such that in central collisions, the process appears to be dominated by the dynamical component. In addition to the multiplicity distributions, the two cluster mechanisms also appear to produce different energy spectra. Preliminary data for the reaction ^{209}Bi$+$^{136}Xe (E/A = (28–62) MeV) show effective spectral slope parameters of $T_{eff} \approx 10\,\mathrm{MeV}$ for statistical cluster spectra but $T_{eff} \approx (23\text{–}26)\,\mathrm{MeV}$ for dynamically emitted clusters. For the same events, light-particle $(p-, \alpha-)$ spectra indicate average temperatures of the order of $T_{eff} \approx (5\text{–}6)\,\mathrm{MeV}$. In comparison to the evaporation of light particles, spectra for heavier particles are affected by spin-off contributions, which can increase the effective temperature to values well above the temperatures of the respective emitters.

The differences between the two types of emission processes become more obvious at the higher bombarding energies. However, for most event selection criteria and all systems studied, both statistical and non-statistical cluster components appear intertwined. This requires a differential consideration of experimental data, rather than the customary average analysis.

3.3 Model Interpretation of Dynamical Cluster Emission

As suggested in Fig. 10, the interaction (participant) region of a colliding heavy-ion system is subjected to strong mechanical stresses and matter pile-up. A simple microscopic picture of such occurrence is provided by the nucleon exchange model, assuming a dominant role of the exchange of individual nucleons between projectile and target in relative motion. Sosin et al. [18] have adopted a simplified version of the NEM model to the problem of random cluster coalescence at the distance of closest approach in a heavy-ion collision. Formation and decay are assumed to be governed by the respective level densities of all entities present at that time. Trajectory dynamics have been treated only in a cursory fashion. For light-ion induced fusion reactions, where such a neglect is somewhat justified, resulting cluster distributions can be accounted for relatively well [64]. Predictions of invariant IMF velocity plots by this model are depicted in Fig. 12, for the ^{209}Bi + ^{136}Xe reaction, using the same detection and analysis conditions ("filter") as for the experimental data.

The calculation (2^+) shown in this figure assumes cluster decay only for PLF and TLF, while the last row of theoretical plots (3^+) considers cluster coalescence. From a comparison of calculations to the data shown in the

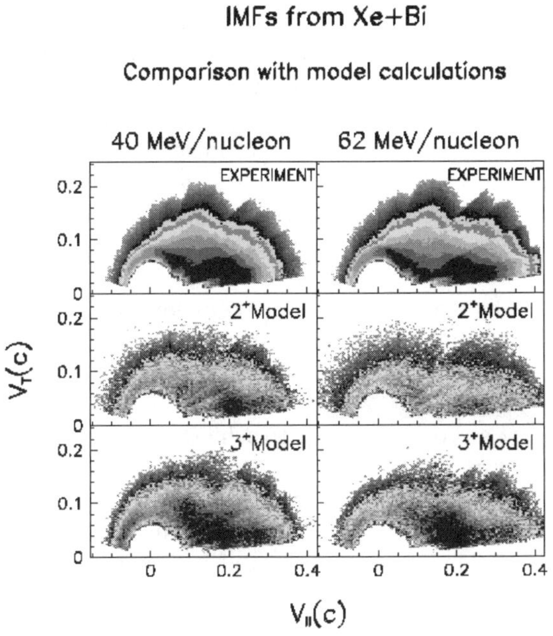

Fig. 12. Experimental (*top row*) and theoretical invariant cluster velocity plots. The calculations are based on a stochastic coalescence model without (2+) and with (3+) cluster coalescence

top row, it is clear that clusters emerging from the participant zone acquire kinematical properties similar to the experimental observations. However, the discrepancies between model calculations (bottom row in Fig. 12) and data are obvious. The calculations cannot account for the high velocities of the IVS clusters suggested by the data. Further comparison exhibits more discrepancies: The predicted multiplicity distributions are too narrow, and the Z_{IMF} distribution falls off too rapidly. This shortcoming of the model comes not unexpected, since the probability to produce a substantial cluster by random aggregation of individual nucleons decreases exponentially with the number of nucleons involved. Similar difficulties are encountered with QMD calculations, which also describe the random collisions of PLF and TLF nucleons and treat their coalescence in a schematic fashion, based on the proximity of nucleons in phase space. The phenomena appear to suggest a breakup cluster production mechanism, similar to light-particle pre-equilibrium emission, rather than an inverse. Theory is challenged to explain this interesting dynamical cluster production mechanism.

3.4 New Ideas about Statistical Cluster Emission from Hot Nuclei

Traditional statistical-model descriptions of nuclear decay, which have been successful to the extent that they are now considered reliable tools for the interpretation of experimental data, break down when the emission of nuclear clusters from nuclei with temperatures of up to $T = 6\,\mathrm{MeV}$ is concerned. This failure is attributed to the fact that the barrier heights for emission of such clusters from heavy nuclei are of the order of $B = (50–60)\,\mathrm{MeV}$. Yet this type of cluster emission is clearly associated with statistical emission patterns. Nuclear fission, another well-known nuclear decay mechanism with statistical and dynamical aspects, is not prevalent for medium-weight nuclei such as ^{136}Xe, since also here barriers are very high ($\approx 40\,\mathrm{MeV}$). Furthermore, the cluster emission mechanism does not appear to compete with light-particle emission in a statistical fashion but the former overwhelms the latter, in contrast to fission of heavy nuclei. Attempts to link cluster emission to a nuclear phase transition face the same inherent problems of high emission barriers and therefore do not address such issues. Clearly, new approaches have to be found to explain cluster emission from hot nuclei. There are two well known phenomena that together suggest a promising route to achieve understanding of the cluster emission process. First, statistical decay is driven by entropy, even if average energetics is not favorable for the process. Secondly, the number of nucleonic states in the diffuse nuclear surface is significantly larger than in the nuclear bulk interior. Therefore, a nucleus, if hot enough, gains entropy when enlarging its surface. Consequences of this tendency is that a hot nucleus will expand, that it may split into several smaller fragments, or do both in sequence. A quantitative study of these effects can be carried out in a microcanonical approach, employing the simple model of the nucleus as a Fermi gas of nucleons interacting via realistic nuclear forces. Such interactions can be written in harmonic approximation as [56,57],

$$E^*_{com} = -E_B \left(1 - \frac{\rho}{\rho_o}\right)^2 . \tag{9}$$

In this expression for the compressional nuclear energy, the quantity $E_B \approx 8\,\mathrm{MeV}$ is the average nucleon binding energy. The total excitation energy of a nucleus is then the sum of (collective) compression and (random) thermal excitation energies,

$$E^* = E^*_{com} + E^*_{therm} = -E_B \left(1 - \frac{\rho}{\rho_o}\right)^2 + a \cdot T^2 , \tag{10}$$

with temperature T and a level density parameter that increases with decreasing matter density,

$$a = a_o \left(\frac{\rho}{\rho_o}\right)^{-\frac{2}{3}} , \tag{11}$$

according to the standard, familiar Fermi gas matter density dependence. Consequently, one defines a density dependent nuclear entropy,

$$S = 2\sqrt{aE^*_{therm}} = 2\sqrt{a(E^* - E_{com})} . \tag{12}$$

This dependence is pictured in Fig. 13 for an A = 200 nucleus and for different total excitations E^* (MeV).

At low excitations, the entropy has a maximum close to normal matter density, $\varrho_{equ} \approx \rho_0$. With increasing E^*, this maximum shifts to lower densities, the nucleus expands. At about a critical excitation energy of $E^*/A = 10\,\text{MeV}$, the entropy curve degenerates. Here, no maximum entropy exists anymore, which increases monotonically with $\rho \to 0$. This implies that at excitations higher than the critical excitation, the nucleus expands continuously until it disassembles completely into nucleons. This expansion against the nuclear attractive interaction consumes energy. Naturally, the temperature of the nucleus decreases in the expansion. It is related to the equilibrium matter density by

$$T = \sqrt{\frac{E^*_{therm}}{a}} = \left(\frac{\rho_{equ}}{\rho_o}\right)^{\frac{1}{3}} a_o^{-\frac{1}{2}} \sqrt{E^* - E_B \left(1 - \frac{\rho_{equ}}{\rho_o}\right)^2} \,. \tag{13}$$

As an interesting consequence, the temperature does not increase monotonically with excitation energy, but has a maximum ($T \approx 6\,\text{MeV}$) and decreases again for higher excitations. This behavior leads to apparent negative heat capacities, $c = \partial E^*/\partial T < 0$, when the total excitation energy E^* is substituted for the thermal energy. In experiments, it is desirable but difficult to distinguish between thermal energy and collective potential energy of expansion. Typically, one measures the total excitation energy, which could be a reason for the puzzling negative apparent heat capacities reported [65] in the literature. The significance of an increased entropy for expanded hot nuclear systems for statistical cluster emission becomes obvious when one compares the excitation energy dependencies of the entropies of a spheri-

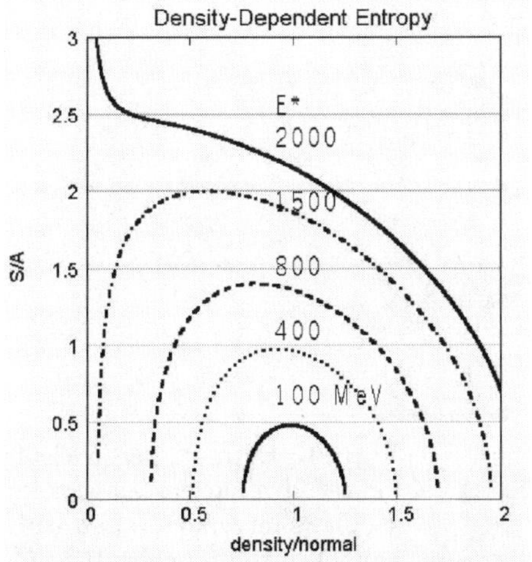

Fig. 13. Entropy vs. relative matter density, ϱ/ϱ_0 for an A = 200 nucleus and excitation energies E^* in MeV

cal hot parent mono-nucleus with that of the di-nuclear saddle shape of the system split into cluster and residue.

Figure 14 illustrates this behavior for the case where a hot expanded ^{16}O cluster emerges from the ^{208}Pb parent nucleus, maintaining nuclear surface contact with the parent [35]. Coulomb contributions to the interaction energy V_{int} are taken for spherical nuclei. It turns out that the barriers for emission of such clusters do not change substantially when expansion is taken into account. However, at total excitation energies of the order of $E^*/A = (4$–$5)$ MeV, the di-nuclear saddle-point configuration attains a higher entropy than the parent mono-nucleus,

$$S_{din}(A_f, Z_f, A_{res}, Z, E^*) = S\left(A_f, Z_f, \varrho\left(\frac{(E^* - V_{int})}{A}\right)\right) + S\left(A_{res}, Z_{res}, \varrho\left(\frac{(E^* - V_{int})}{A}\right)\right) . \quad (14)$$

At such energies, even with unfavorable energetics, reactions can nevertheless proceed, provided that there is a sufficiently large entropy gain. In the example pictured in Fig. 14, the entropy decreases by app. 20 units. Therefore, cluster emission becomes a dominant decay channel of hot, expanded nuclei. Closer evaluation shows that here cluster emission is more likely even than the evaporation of nucleons or alpha particles. The probability for cluster emission can be calculated from the kinetic formula

$$P_{din}(E^*) \propto e^{S_{din} - S_m} \propto e^{-B_{eff}/T} , \quad (15)$$

i.e., according to the relative weights of mono- and di-nuclear configurations. The method is well known from fission theory. The quantity B_{eff} is an effective barrier, introduced to make contact with traditional methods of analysis

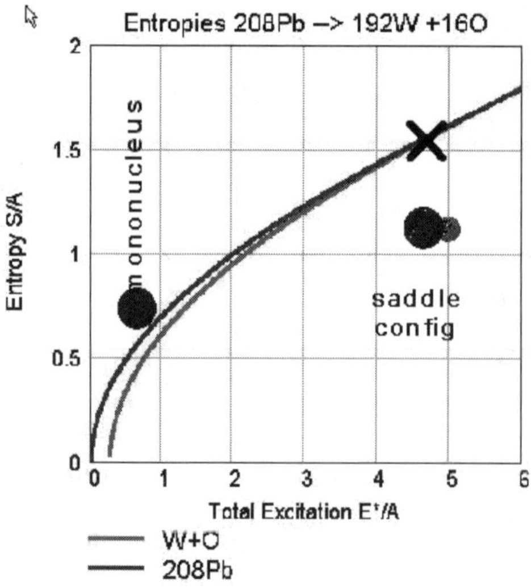

Fig. 14. Entropy per nucleon for a ^{208}Pb nucleus and for its saddle-point configuration with an ^{16}O cluster touching the surface of the parent

of branching ratios. This effective barrier can be shown to drop to zero at about $T = 7\,\text{MeV}$ for the emission of a ^{12}C cluster from a ^{197}Au parent nucleus. From the above discussion, it is obvious that only the emission of complex clusters benefits from the additional entropy gained through expansion and splitting. Therefore, at high excitation energies, cluster emission wins statistically against the emission of protons and neutrons, as well as against other tightly bound particles. Large anomalies in branching ratios may be expected for particles which differ significantly in their intrinsic level schemes. This effect would also explain the observation of incomplete damping of the collisional energy into intrinsic excitation available for nucleon evaporation. cluster emission simply overtakes nucleon emission by large factors at sufficiently high temperatures. One observes in fact that the joint multiplicity distribution $P(m_n, m_{lcp})$ does not expand proportionally (cf. Fig. 5) with bombarding energy but seems to saturate.

4 Summary and Conclusions

It has been shown in this article that reactions between heavy ions at intermediate bombarding energies (E/A = (20–100) MeV) exhibit a complex pattern, much of which appears like a morphing of the dissipative reaction mechanism studied extensively at bombarding energies near the interaction barrier. New dynamical reaction features can operationally be associated with a third entity, a participant zone, in addition to massive remnants of projectile and target nuclei. The theoretical analysis of this reaction type is difficult and still faces serious challenges. cluster emission in such heavy-ion reactions exhibits interesting systematics that have long escaped a comprehensive description. Arguments have been presented why it is essential to view cluster emission within the framework of an overall nuclear reaction mechanism that is rather different from the complete fusion process invoked implicitly in much of the work reported in the literature. In several aspects, cluster emission in heavy-ion reactions resembles a statistical phenomenon, for example, as far as the general shapes of multiplicity, charge, and energy distributions and the reducibility of these distributions are concerned. However, the particle distributions are not consistent with an overall thermal equilibrium. They show higher energies and lead to larger than thermal fluctuations in many observables. For central collisions, in particular, superposition of dynamical and smaller thermal fluctuations become difficult to disentangle. Therefore, more robust data analysis follow reaction trends from peripheral to central collisions. The cluster emission process discussed in this work appears to represent a superposition of two distinct mechanisms, one a more direct and breakup-like phenomenon, the other a statistical emission process. So far, no satisfactory quantitative description is available for the direct cluster emission mechanism, which exhibits general similarities with preequilibrium emission of light particles, e.g., as far as energy spectra are concerned. A coalescence approach in a preliminary realization fails to give a quantitative account of observed cluster distributions, indicating either that cluster correlations build up very fast or that they pre-existed in the original projectile and target nuclei and survive the collision impact. Although it can be observed already in rather peripheral collision, the dynamical cluster emission process becomes dominant in violent central collisions. It would be instructive to compare data for this process to model simulations in terms of QMD or viscous-hydrodynamic calculations. The second, sequential cluster emission process identified experimentally has challenged the traditional statistical-model understanding of nuclear decay. The Coulomb barriers are sufficiently high to prohibit significant cluster emission from nuclei produced with only moderate temperatures, in the reactions of interest. However, when the expansion of a hot nucleus and its large enhancement of surface entropy are taken into account, this emission process becomes not only possible but the dominant decay mechanism. This statistical cluster decay likely involves a rearrangement of the nuclear surface, an interesting process of aggregation of clusters in geometrical confinement. This new access to statistical cluster emission is gained by employing a simple Fermi gas model of interacting nuclei. Not surprisingly, it is found that these interactions are important for the population of final phase space. As interesting consequences, the model predicts the existence of a limiting temperature of approximately $T = 6\,\mathrm{MeV}$ for a heavy nucleus, above which the nucleus is predicted to vaporize, i.e., to disassemble into its nucleonic constituents. Even though the study is based on a simple Fermi gas model, it incorporates the most significant physical effects, which should be robust enough to survive a more sophisticated theoretical treatment. A more explicit characterization of the decay of the massive remnants of reaction fragments is called for, in particular of the projectile remnant, which experimentally is uniquely well defined in dissipative reactions. This could be a new direction promising important and interesting opportunities for the field of nuclear reaction studies.

Acknowledgements

Previously unpublished and partially preliminary experimental results included in this paper have been obtained in a collaboration with W. Gawlikowicz (Univ. of Rochester and Jagellonian University Cracow, Poland), L.G. Sobotka and R.G. Charity (Washington University, St. Louis, MO), and R.T. deSouza (Indiana University, Bloomington, IN). Model calculations have been performed by Z. Sosin and W. Gawlikowicz (Jagellonian University Cracow, Poland).

This work has been supported by the U.S. Department of Energy Grant No. DE-FG02-88ER40414.

References

1. J. Schmelzer, G. Röpke, and R. Mahnke, in *Aggregation Phenomena in Complex Systems* (Wiley–VCH, Weinheim, 1999), p. 263.
2. B. Borderie, J. Phys. G: Nucl. Part. Phys. **28**, R217 (2002).
3. W.U. Schröder and J. Huizenga, in *Treatise on Heavy-Ion Science* (Plenum Press, New York and London, 1984), Vol. 2.
4. B. Djerroud *et al.*, Phys. Rev. **C64**, 034603 (2001).
5. J. Wilczynski, Phys. Lett. **47B**, 484 (1973).
6. R. Dayras, Z.E. Switkowski, and S. Woosley, Nucl. Phys. **A279**, 70 (1977).
7. M.F. Rivet *et al.* (INDRA), Phys. Lett. **B388**, 219 (1996).
8. M. Robel, Ph.D. Thesis, Lawrence Berkeley Lab. preprint **LBL-8181**, (1979).
9. J. Randrup and R. Vandenbosch, Nucl. Phys. **A474**, 219 (1987).
10. J. Bondorf and others, Nucl. Phys. **A333**, 285 (1980).
11. J. Randrup, Nucl. Phys. **A259**, 253 (1976).
12. J. Randrup, Nucl. Phys. **A307**, 319 (1978).
13. J. Randrup, Nucl. Phys. **A327**, 490 (1979).
14. J. Randrup, Nucl. Phys. **A383**, 468 (1982).
15. J. Randrup and B. Remaud, Nucl. Phys. **A514**, 339 (1990).
16. W.U. Schröder *et al.*, in *Nuclear Fission and Heavy-Ion-Induced Reactions* (Harwood, Nucl. Sci. Res. Conf. Ser., ADDRESS, 1987), Vol. 11, p. 255.
17. J. Blocki *et al.*, Ann. Phys. (N.Y.) **113**, 330 (1978).
18. Z. Sosin, Eur. Phys. J. **A11**, 311 (2001).
19. G. Bertsch and S. DasGupta, Phys. Rep. **160**, 189 (1988).
20. W. Bauer, Prog. Part. Nucl. Phys. **30**, 45 (1991).
21. R. Balescu, in *Statistical Dynamics, Matter out of Equilibrium* (Imperial College Press, Univ. Libre de Bruxelles, 2000).
22. W. Bauer and G. Bertsch, Phys. Rev. Lett. **58**, 863 (1987).
23. S. Ayik and C. Gregoire, Phys. Lett. **212B**, 269 (1988).
24. P. Chomaz, G.F. Burgio, and J. Randrup, Phys. Lett. **254B**, 340 (1991).
25. P. Chomaz *et al.*, Phys. Rev. Lett. **73**, 3512 (1994).
26. M. Colonna and P. Chomaz, Phys. Rev. **C49**, 1908 (1994).
27. M. Colonna *et al.*, Nucl. Phys. **A642**, 449 (1998).
28. S. Chattopadhyay, Phys. Rev. **C53**, R1065 (1996).
29. V. Spicka, P. Lipavsky, and K. Morawetz, Phys. Rev. **B55**, 5084 (1997).
30. V. Spicka, P. Lipavsky, and K. Morawetz, Phys. Lett. **240A**, 160 (1998).
31. G. Kortemeyer, F. Daffin, and W. Bauer, Phys. Lett. **374B**, 25 (1996).
32. P. Danielewicz, in *Proc. Int. Workshop Gross Properties Nuclei and Nucl. Excit. XXVII* (GSI, Darmstadt, 1999), Vol. 27, p. 263.
33. B. Li, C. Ko, and W. Bauer, Int. Jour. Mod. Phys. **7**, 147 (1998).
34. K. Morawetz and P. Lipavski, Phys. Rev. **C63**, 61602 (2001).
35. J. Töke and W. Schröder, Phys. Rev. **C**, in press (2002).
36. B. Li, private communication (2002).
37. G. Beauvais, D. Boal, and C. Wong, Phys. Rev. **C35**, 545 (1987).
38. J. Aichelin, Phys. Rep. **202**, 233 (1991).
39. G. Peilert *et al.*, Phys. Rev. **C46**, 1992 (1992).
40. C. Hartnack *et al.*, Eur. Phys. J. **A1**, 151 (1998).
41. J. Lukasik, Ph.D. Thesis, Jagellonian University Cracow (1993).
42. O. Tursunov and O. Zhirov, Phys. Lett. **222**, 110 (1989).

43. J. Aichelin, Univ. de Nantes, Ecole d. Mines, theory group website (2000).
44. A. Ono, H. Horiuchi, T. Maruyama, and A. Ohnishi, Prog. Theor. Phys. **87**, 1185 (1992).
45. H. Feldmeier, Nucl. Phys. **A586**, 493 (1995).
46. D. Gross, Prog. Part. Nucl. Phys. **30**, 155, and references cited (1993).
47. J. Bondorf *et al.*, Phys. Rep. **257**, 113 (1995).
48. L. Moretto and G. Wozniak, Annu. Rev. Nucl. Part. Sci. **43**, 379 (1993), and references therein.
49. L. Moretto, J. Elliott, L. Phair, and G. Wozniak, Phys. Rev. **C66**, 41601R (2002).
50. S. Albergo, S. Costa, E. Constanzo, and A. Rubbino, Nuovo Cimento **89**, 1 (1985).
51. J. Pochadzalla *et al.*, Phys. Rev. **75**, 1040 (1975).
52. D. Gross, Rep. Prog. Phys. **53**, 605 (1990).
53. M. Kleine-Berkenbusch *et al.*, Phys. Rev. Lett. **88**, 22701 (2002).
54. J.B. Elliot *et al.*, Phys. Rev. Lett. **88**, 42701 (2002).
55. M.E. Fisher, Physics **3**, 255 (1967).
56. W.A. Friedman, Phys. Rev. Lett. **60**, 2125 (1988).
57. W.A. Friedman, Phys. Rev. **C42**, 667 (1990).
58. J. Töke, Nucl. Phys. **A681**, 374c (2001).
59. B. Li and W.U. Schröder, in *Isospin Physics in Heavy-Ion Reactions at Intermediate Energies* (Nova Science Publishers, New York, 2001).
60. J. Töke, J. Lu, and W. Schröder, Phys. Rev. In press, (2003).
61. W.U. Schröder, Report DOE/ER/ **79048-1**, (1995).
62. D. Stracener *et al.*, Nucl. Instr. Meth. **A294**, 485 (1990).
63. W. Skulski, J. Tõke, and W. Schröder, Phys. Rev. **C59**, R2130 (1999).
64. R. Planeta *et al.*, Eur. Phys. J. **A11**, 297 (2001).
65. M. Agostino *et al.*, Phys. Lett. **B473**, 219 (2000).
66. R. Nebauer, A. Guertin, R. Puri, Ch. Hartnack, P.B. Gossiaux, and J. Aichelin, Proc. Int. Worksh. Gross Properties Nuclei and Nucl. Excit. XXVII, 263 (1999).
67. R. Nebauer and J. Aichelin, Nucl. Phys. **A681**, 353c (2001).
68. C. Dorso and J. Randrup, Phys. Lett. **215B**, 611 (1988).
69. C. Dorso and J. Randrup, Phys. Lett. **232B**, 29 (1980).
70. T. Døssing *et al.*, Nucl. Phys. **A327**, 490 (1979); ibid. **A433**, 280 (1985).

Correlations in Finite Systems and Their Universal Scaling Properties

Robert Botet and Marek Płoszajczak

Abstract. In this lecture, we shall discuss universal scaling laws of the order-parameter fluctuations in *any finite system* in which the second-order critical behavior can be identified. The relation between the order-parameter, the criticality and the scaling law of fluctuations is established, and the connection between the scaling function and the critical exponents is found. Recent experimental data from fragment production in heavy-ion collisions at intermediate energies, are discussed within this novel framework.

1 Introduction

Many physical systems are strongly fluctuating, and fluctuations of their basic quantities appear to be as large as their average values. In such cases, an understanding of the properties of these fluctuations is indispensable for a comprehension of the precise causes for the observed state of the system. The heavy-ion (HI) collisions at intermediate energies or the ultrarelativistic collisions of leptons, hadrons or nuclei are only few examples of many physical processes where the problem of fluctuations plays a crucial role. Systems produced in these collisions are finite and the multiparticle correlations play an important role in their evolution. In this context, we shall discuss universal scaling laws of the order-parameter fluctuations in *any system*, whether equilibrium or off-equilibrium, with finite or infinite number of constituents, in which a second-order critical behavior can be identified. These laws can be rigorously derived for equilibrium systems and essential elements of this derivation will be presented below. But they are known to hold also for the off-equilibrium critical systems (e.g., the kinetic aggregation or the fragmentation). The relation between the order-parameter, the criticality and the scaling law of fluctuations has been found recently, and the connection between the scaling function and the critical exponents has been established rigorously. This new scaling approach, the so-called Δ-scaling, provides a unique tool to characterize the 'criticality' of small systems. The first-scaling law, which is the extreme fluctuation limit of the Δ-scaling, appears basically in two contexts of the second-order critical system: at the critical point and in the disordered phase. In both cases, the fluctuations of the order parameter are of the same order of magnitude as the average value of this parameter, and the fundamental distinction between these two situations comes from the correlations. At the critical point, the correlation length is infinite and the entire system is correlated. On the contrary, the disordered phase is characterized by short-range correlations; the correlation length ξ is finite, and one can in principle consider that the system is built by uncorrelated blocks of size ξ^d. This is a fundamental idea of the renormalization flow, which leads in this case to the uncorrelated fixed point. For that reason, one can consider a completely uncorrelated system as the generic case of this situation. We shall devote much attention in this lecture to the influence of correlations on the first-scaling law and on the shape of the scaling function. Three features are characteristic of the first-scaling law: the anomalous exponent, the first-scaling itself and the form of the first-scaling function. We shall divide the discussion of those aspects into four cases: (i) the uncorrelated additive order parameter, (ii) the correlated additive order parameter, (iii) the uncorrelated extreme order parameter and (iv) the correlated extreme order parameter. These formal results will be illustrated by simulations in a two-dimensional Potts model and in a three-dimensional percolation network. Finally, the experimental data on the probability distribution of the largest charged fragment in HI collisions at intermediate energies will be discussed.

2 Basic Moments and Generating Functions

In this section we recall some definitions of moments and generating functions used in the general analysis of probability distributions.

Given a normalized probability distribution $\mathrm{P}(s)$, where $s \in \mathsf{R}$, one can define the *moments* of the distribution:

$$\mu_q \equiv \langle |s|^q \rangle = \int_{-\infty}^{\infty} |s|^q \mathrm{P}(s) ds$$

with q any real number for which the above moment is both defined and finite. Since P is a positive function and $\mu_0 \equiv 1$, Hölder inequality implies that the function $q \mapsto (\mu_q)^{1/q}$ is an increasing continuous function of the argument q for $q > 0$. Often, integer values of q are chosen in order to simplify the analysis.

The corresponding *moment generating function* is:

$$\mathcal{M}(u) \equiv \int_{-\infty}^{\infty} \mathrm{e}^{|s|u} \mathrm{P}(s) ds = \sum_q \frac{\mu_q}{q!} u^q \quad .$$

The running argument u may be a complex number. Since $\mathrm{P}(s)$ is a probability distribution, one has: $\mathcal{M}(0) = 1$. All higher order integer moments of $\mathrm{P}(s)$ can be obtained by successive derivatives of $\mathcal{M}$:

$$\langle |s|^q \rangle = \left. \frac{\partial^q \mathcal{M}(u)}{\partial u^q} \right|_{u=0} \quad .$$

If the generating function $\mathcal{M}(u)$ is known for purely imaginary argument $u = i\,\theta$, then the probability distribution $\mathrm{P}(s)$ is given by the Fourier transform of $\mathcal{M}$. The problem of *uniqueness* of the probability distribution defined by the moments μ_q is crucial for the application of generating functions. A sufficient condition has been found by Carleman: If

$$\sum_{q=1}^{\infty} \mu_q^{-1/2q} = \infty \,, \tag{1}$$

then μ_q are the moments of a unique probability distribution. For an extensive discussion of various moments and associated generating functions, the interested reader can refer to [1].

3 Scaling Laws for the Second-Order Critical Phenomena

We describe now scaling laws of the order-parameter distribution for finite parts of a critical system at equilibrium. In this case, precise complete analytical results are known and provide a clear panorama of these particular scaling behaviors. Such features are expected to hold true for out-of-equilibrium critical systems too, but rigorous pieces of evidence are then rare and incomplete. Most of the arguments come then from numerical simulations, and lie beyond the scope of this short introduction to the Δ-scaling laws. More details can be found in [1].

3.1 Scaling Laws in the Landau–Ginzburg Mean-Field Model

Landau–Ginzburg theory is an analytically solvable example of the second-order equilibrium phase transition. The starting point is the homogeneous Landau–Ginzburg free energy density:

$$F(\eta) = \epsilon\eta^2 + b\eta^4 + \cdots$$

where b is a positive constant. Variable η stands for the *intensive* order parameter of the system. It is expected to be 0 in the disordered phase, and not vanishing in the ordered phase. In addition, its proper value in the ordered phase should only depend on the physical state and *not* on the quantity of the material. Note however, that the value of η may fluctuate but these fluctuations are expected to be so small in the infinite system that the actual value of η should coincide with its most probable value.

The reduced driving parameter ϵ governs the critical behavior. In thermal phase transitions, $\epsilon = a(T/T_c - 1)$, where T is the temperature, T_c the critical temperature and a a positive constant. The sign of ϵ characterizes the stability domain of the system. The most probable value of the order parameter η in the disordered phase ($\epsilon > 0$) is implicitly set to 0. This is the same at the critical point ($\epsilon = 0$), but the value $\eta = 0$ is then marginally stable. When ϵ becomes negative, the most probable value of η is finite and this defines the ordered phase. In this phase, $\eta = 0$ is a metastable equilibrium point.

- Distribution of the extensive order parameter

The state of a finite system is not so well defined as in the thermodynamic limit. It is then more convenient to work with the *extensive* order parameter: $s = N\eta$, which is allowed to undergo large fluctuations since the system is small. One should notice the additive structure of this quantity: if s_i is the microscopic order parameter of a site i, s is defined as the sum of local variables: $s \equiv \sum_i s_i$. The probability for the system of size N to get the value s for a given ϵ is [2]:

$$\mathrm{P}_N(s) = \frac{1}{\mathcal{Z}_N} \exp\left[-\beta_T\left(\epsilon\frac{s^2}{N} + b\frac{s^4}{N^3} - \cdots\right)\right] \quad , \tag{2}$$

$k_B T = 1/\beta_T$, and the partition function $\mathcal{Z}_N$ is defined by the normalization of $\mathrm{P}_N(s)$. In the Landau–Ginzburg theory, one assumes that N is so large that the first two terms in the free energy expansion are sufficient to characterize completely the phase of the system. In principle, this condition means that one may study systems for which $|\eta| \ll 1$, *i.e.*, at the critical point or in the disordered phase.

- First-scaling at the critical point

At the critical point ($\epsilon = \epsilon_c \equiv 0$), the leading term in the free energy density is proportional to the fourth power of s. Standard integrations give the partition function $\mathcal{Z}_N$ and the average value of the order parameter $\langle|s|\rangle$:

$$\begin{aligned}\int_{-\infty}^{\infty} \mathrm{P}_N(s)ds \quad &\equiv 1 = \frac{1}{\mathcal{Z}_N}\int_{-\infty}^{\infty} \mathrm{e}^{-\beta_T b s^4/N^3} ds \\ \int_{-\infty}^{\infty} |s|\mathrm{P}_N(s)ds &\equiv \langle|s|\rangle = \frac{1}{\mathcal{Z}_N}\int_{-\infty}^{\infty} |s|\ \mathrm{e}^{-\beta_T b s^4/N^3} ds \quad .\end{aligned}$$

Both $\mathcal{Z}_N$ and $\langle|s|\rangle$ are proportional to $N^{3/4}$, e.g.:

$$\langle|s|\rangle = \frac{\sqrt{\pi}}{(\beta_T b)^{1/4}\Gamma(1/4)} N^{3/4} \quad .$$

Introducing their values in (2), one finds:

$$\langle|s|\rangle \mathrm{P}_N(s) = \frac{4c}{\sqrt{\pi}} \exp\left[-c^2\left(\frac{s}{\langle|s|\rangle}\right)^4\right] \tag{3}$$

with $c = \pi/\Gamma^2(1/4)$. This is a scaling form, called the *first-scaling law* [3]:

$$\langle|s|\rangle \mathrm{P}_N(s) \sim \phi\left(\frac{s}{\langle|s|\rangle}\right)$$

with ϕ the corresponding first-scaling function. Note that the scaling function in the Landau–Ginzburg theory at the second-order critical point, behaves as: $\phi(z_1) \sim \exp(-az_1^4), z_1 = s/\langle|s|\rangle$ and decreases fast (faster than the Gaussian distribution) for s-values larger than the most probable value. The scaling function (3), can be entirely characterized by its moments:

$$\frac{\langle|z_1|^q\rangle}{\langle|z_1|\rangle^q} = \frac{\Gamma((q+1)/4)}{\sqrt{\pi}\ c^{(q-1)/2}} \ .$$

Equivalently, one can study the moments μ_q of the distribution $\mathrm{P}_N(s)$:

$$\frac{\mu_q}{\mu_1^q} \equiv \frac{\langle|s|^q\rangle}{\langle|s|\rangle^q} = \frac{\langle|z|_1^q\rangle}{\langle|z|_1\rangle^q} = \frac{\Gamma((q+1)/4)}{\Gamma(1/4)}\left(\frac{\Gamma(1/4)}{\Gamma(1/2)}\right)^q ,$$

which are always larger than 1 for $q > 1$, since $\Gamma(1/4)/\sqrt{\pi} > 2$. Note that these moments define a unique distribution after Carleman condition (1), since $\mu_q^{-1/2q} \sim q^{-1/8}$, and the sum appearing in (1) is well diverging.

The equivalence between the moments μ_q of s and the moments of the reduced variable z_1 (*i.e.* $\langle|s|^q\rangle/\langle|s|\rangle^q = \langle|z_1|^q\rangle/\langle|z_1|\rangle^q$), is a direct consequence of the first-scaling law for the distribution of s at the critical point. The dependence of $\langle|s|^q\rangle/\langle|s|\rangle^q$ with q, can provide a useful tool to compare experimental scaling curves to the mean-field theory of the second-order critical phenomenon. For example, some special relations involving first four moments are: $\mu_2/\mu_1^2 = \sqrt{2}$ and $\mu_4\mu_1^2/\mu_3^2 = \pi/4$. Obviously, these relations are rigorously valid only in the limit $N \to \infty$. For finite N, the finite-size deviations to the limit scaling function and, hence, to its moments are expected.

- Gaussian first-scaling in the disordered phase

Outside of the critical point in the disordered phase ($\epsilon > 0$), the leading term of the free energy is proportional to s^2, and the probability distribution $\mathrm{P}_N(s)$ is essentially the Gaussian distribution. After similar derivation, and keeping in mind that both $\mathcal{Z}_N$ and $\langle|s|\rangle$ in the disordered phase behave like $N^{1/2}$, e.g.:

$$\langle|s|\rangle = \sqrt{\frac{N}{\pi\beta_T\epsilon}} ,$$

we get the first-scaling form:

$$\langle|s|\rangle\mathrm{P}_N(s) = \frac{1}{\pi}\exp\left[-\frac{1}{\pi}\left(\frac{s}{\langle|s|\rangle}\right)^2\right]$$

with the Gaussian scaling function[1]. Similarly to the analysis at the critical point, one can define a scaling function by its moments:

$$\frac{\mu_q}{\mu_1^q} = \frac{\langle|z_1|^q\rangle}{\langle|z_1|\rangle^q} = \pi^{(q-1)/2}\Gamma((q+1)/2)$$

[1] Note that this scaling function has been written here for the normalization $\langle|s|\rangle = 1$, fixing then the numerical value $1/\pi$ in the exponential term.

which are *always* larger in the disordered phase than at the critical point[2]. In particular, one has: $\mu_2/\mu_1^2 = \sqrt{2}(\pi/2)^{3/2}$, and $\mu_4\mu_1^2/\mu_3^2 = 3/4$.

- Second-scaling in the ordered phase

The low temperature regime ($\epsilon < 0$) cannot be analyzed in a similar way. One has first to define the most probable value of the order parameter which is positive and proportional to the total size N of the system:

$$m^* = \sqrt{-\frac{\epsilon}{2b}N} \ .$$

Developing $\mathrm{P}_N(m)$ of (2) around this point leads to the expression:

$$m^{*1/2}\mathrm{P}_N(m) \simeq \left(-2\frac{\epsilon^3}{b\pi^2}\right)^{1/4} \exp\left(\epsilon\sqrt{-2\frac{\epsilon}{b}}\,\frac{(m-m^*)^2}{m^*}\right) \ , \tag{4}$$

which is different from the first-scaling form. In this case, the average value of the order parameter $\langle m\rangle$ is close to its most probable value m^* and (4) can be rewritten in the scaling form:

$$\langle m\rangle^{1/2}\mathrm{P}_N(m) \sim \exp\left(-a\frac{(m-m^*)^2}{\langle m\rangle}\right)$$

with a a positive constant. This scaling form is called the *second-scaling law* [3].

3.2 Scaling Laws in the Second-Order Phase Transition from the Similarity Laws

The thermodynamic system at the second-order critical point is a typical example of a large ensemble of strongly correlated random variables. The limit distribution of an order parameter or of any critical observable, is then expected to depend strongly on the appropriate correlation functions, hence on specific features of interactions in the considered system. The concept of *scale-invariance* of the singular part of thermodynamic quantities [4] can be used efficiently to investigate the problem of asymptotic features of the fluctuations.

- Anomalous dimension

The *finite-size scaling* analysis allows to link the finite-system behavior to its thermodynamic limit. Near the critical point of an infinite system, the coherence length ξ can be defined, for example, as a typical length on which

[2] This is a consequence of a slower decrease at large z_1 of the scaling function in the disordered phase as compared to the behavior of the scaling function at the critical point.

a decay of the covariance function of the order parameter is significant. But we know that *at* the critical point, ξ must be infinite since one cannot define any length scale in the system and, hence, the covariance function should decrease as a power-law of the distance. One writes: $\sim 1/\epsilon^{\nu}$, with a positive critical exponent ν. The quantity ξ^d defines a coherence volume in d-dimensional Euclidean space. Let us denote by η the intensive order parameter of the system. Because of the infinite-range collective interactions, this quantity behaves singularly with ϵ near the critical point. The fundamental finite-size scaling of the order parameter close to $\epsilon = 0$ in a system of size N is:

$$\langle |\eta| \rangle \simeq \epsilon^{\beta} f(N/\xi^d) ,$$

where f is the cross-over function. In order to remove the singular behavior[3] with ϵ, the only consistent choice for the function f close to the origin is a power-law with the exponent $-\beta/\nu d$. This leads to a power-law for $\langle |N\eta| \rangle$ *vs* the size of the system when the condition $N \ll \xi^d$ is realized:

$$\langle |N\eta| \rangle \sim N/N^{\beta/\nu d} \quad .$$

The anomalous exponent g for the order parameter is introduced as:

$$\langle |s| \rangle \sim N^g \quad .$$

In the case of an equilibrium second-order critical phenomenon, one obtains in this way:

$$g = 1 - \beta/\nu d \quad .$$

Since all three exponents β, ν, d are positive, g must be smaller than 1. Other constraints come from the thermodynamic inequalities. Fisher inequality states that:

$$\gamma \leq (2 - \eta)\nu \ , \tag{5}$$

with the positive exponent γ which describes a divergence of the isothermal susceptibility, and the critical exponent η of the correlation function[4]. In addition, the Buckingham-Gunton inequality states that:

$$2 - \eta \leq \frac{d\nu}{2\beta + \gamma} \quad . \tag{6}$$

Consequently, for any thermodynamical critical system:

$$\frac{1}{2} \leq g \leq 1 \quad .$$

[3] Since N is finite, the *extensive* order parameter $s \equiv N\eta$ cannot be singular at the transition.

[4] We keep here the standard notations for critical exponents. The critical exponent η appearing in (5), (6) should not be confused with the value of the intensive order parameter.

Same arguments can be extended to any non-singular quantity. For example, if a certain quantity depends smoothly on ϵ near the critical point (*i.e.*, it has neither a zero nor a pole) in the thermodynamic limit $N = \infty$, then it cannot depend on N in the finite system when $\epsilon = 0$.

- First-scaling at the second-order critical point

Let N be a *fixed* size of a system[5]. The state of the system is generally characterized by the thermodynamic potential per site: $G = F - h\eta$. In this expression, F is the free energy per site and h is the field conjugated to the order parameter. The function G depends on two parameters ϵ and h, and is a function of a variable η. The free energy F is related to the free-field Hamiltonian of the system and to its entropy. Gibbs probability distribution of the extensive variable s at the temperature T is:

$$\mathrm{P}_{(N,\epsilon,h)}(s) = \frac{1}{\mathcal{Z}_N} \exp\left(-\beta_T N G(\epsilon, h)\right) \ ,$$

with normalization given by the partition function $\mathcal{Z}_N$. This leads to a relation between the moments of the variable s and the thermodynamic potential:

$$\langle s^q \rangle_{(N,\epsilon,h)} = \frac{1}{\mathcal{Z}_N} \left(\frac{1}{N\beta_T}\right)^q \int \frac{\partial^q \mathrm{P}_{(N,\epsilon,h)}(m)}{\partial h^q} dm \ ,$$

where q takes integer values. Fundamental similarity-law assumption tells that the singular part of the thermodynamic potential $G(\epsilon, h)$ in a second-order critical system scales near the critical point as the homogeneous function [5]:

$$\mathrm{G}(\lambda\epsilon, \lambda^{2-\alpha-\beta} h) = \lambda^{2-\alpha} \mathrm{G}(\epsilon, h)$$

for any positive scale parameter λ. For systems with two relevant scaling fields, as it is generally the case, two positive critical exponents α and β characterize completely the *criticality class* of the thermodynamic system. In this framework, the q-th moment of the variable s scales as:

$$\langle |s|^q \rangle_{(\infty, \lambda\epsilon, \lambda^{2-\alpha-\beta} h)} = \lambda^{q\beta} \langle |s|^q \rangle_{(\infty,\epsilon,h)} \ .$$

This relation can be extended easily to the case where q takes non-integer values. Such a scaling has deep consequences. For instance, taking $h = 0$ and $\lambda = 1/\epsilon$, one finds that all normalized moments $\langle |s|^q \rangle / \langle |s| \rangle^q \equiv c_q$ are independent of ϵ near the zero-field transition. Following the finite-size scaling analysis, these moments cannot depend on the system size N. Then, the moment generating function $\mathcal{M}(u)$ of the probability distribution $\mathrm{P}_{(N,0,0)}(s)$ at the critical point can be written as:

$$\mathcal{M}_{(N,0,0)}(u) = \int_{-\infty}^{\infty} \mathrm{P}_{(N,0,0)}(s) \mathrm{e}^{|s|u} ds = \sum_{q=0}^{\infty} c_q \frac{(u \langle |s| \rangle_{(N,0,0)})^q}{q!}$$

[5] N can be either finite or infinite.

with a sequence $c_q \equiv \mu_q/\mu_1^q$ of positive constants. Hence, the moment generating function does not depend on u and N separately, but only on the reduced variable $u\langle|s|\rangle_{(N,0,0)}$. Taking the inverse Fourier transform of $\mathcal{M}$, one obtains:

$$\langle|s|\rangle \mathrm{P}_{(N,0,0)}(s) = \phi(z_1) \quad , \qquad z_1 = \frac{s}{\langle|s|\rangle} \tag{7}$$

at the critical point. In contrast to $\langle|s|\rangle$, the scaling function ϕ does not depend on the size N. The first-scaling law (7) is a characteristic of extremely large fluctuations:

$$\frac{\langle s^2\rangle - \langle|s|\rangle^2}{\langle|s|\rangle} \sim \langle|s|\rangle \to \infty \quad .$$

The scaling limit is defined by the asymptotic behavior of $\langle|s|\rangle\mathrm{P}(s)$ when both s and $\langle|s|\rangle$ tend to infinity, but the value of the scaling variable $z_1 = s/\langle|s|\rangle$ remains fixed. The quantity $\langle|s|\rangle$ plays then a role of the scale factor for the order parameter, replacing N in many phenomenological applications. The scaling function ϕ in (7), is a Fourier transform of the generating function $\mathcal{M}$, which usually is not exactly known except for very simple cases. Nevertheless, it has been demonstrated that the tail of the scaling function for large values of the scaled order parameter is related to the critical exponent δ characterizing the singular behavior of the order parameter with the small field [6]: $\langle|s|\rangle_{(\infty,0,h)} \sim h^{1/\delta}$. The general form in the limit $z_1 \gg 1$ and $s \to \infty$ is:

$$\phi(z_1) \simeq \exp(-c z_1^{\delta+1})$$

with c a positive constant. Since one knows that: $\alpha + 2\beta + \gamma = 2$, and: $\nu d = 2 - \alpha$ (the hyperscaling relation), one can rewrite (8) as:

$$\phi(z_1) \simeq \exp(-c z_1^{1/(1-g)})$$

relating the decrease of the scaling function at large z_1 to the value of the anomalous exponent g. We have seen previously that the basic thermodynamics implies $1/2 \le g \le 1$. Consequently, the scaling function ϕ decreases *always faster* than predicted by a Gaussian law[6] for large arguments.

- Gaussian first-scaling in the disordered phase and the second-scaling in the ordered phase

Let us suppose that the system is *not* at the critical point or that one studies a random variable s whose average value is not singular at the critical

[6] The value of critical exponent δ is generally quite large: it is $\delta = 3$ in the mean-field theory of second-order phase transitions, $\delta \simeq 5$ in a three-dimensional Ising model, and $\delta = 15$ in a two-dimensional Ising model. Even larger values are known.

point. Then the free energy per site can be written as a regular function of this variable:

$$F \simeq a + bs^2 + \cdots \quad , \tag{8}$$

with the non-vanishing parameter b. Previous analysis in the Landau–Ginzburg model outside of the critical point can now be applied without any modification. The basic results can be expressed as follows: if the system is in the disordered phase, then the average order parameter $\langle |s| \rangle$ is proportional to the square root of the size of the system, and the first-scaling law holds with the Gaussian scaling function. In the ordered phase, the average value of the order parameter scales proportionally to the size of the system, and the second-scaling holds. All these features follow from the analytical structure (8) of the free energy thermodynamic function.

3.3 Example 1: The Scaling Law for a Potts Model

As an illustration of the general analysis for thermodynamic systems at the critical point, we present now briefly results obtained in the 2-dimensional, 3-states ferromagnetic Potts model[7]. The Hamiltonian is:

$$\mathcal{H} = J \sum_{\{i,j\}} \delta(s_i, s_j) \, , \tag{9}$$

where δ is the Kronecker symbol, and J is the positive coupling constant to ensure the ferromagnetic interaction. The sum in (9) is restricted to nearest-neighbor spins $\{i, j\}$. Some exact results are known in Potts model for these precise parameters. For example, the infinite system undergoes a second-order phase transition at a critical temperature $(\beta_T J)_c = \ln(1+\sqrt{3})$, and the value of the critical exponent δ is: $\delta = 14$. The order parameter is defined as:

$$s = \frac{3\,\mathrm{Max}_k\{n_k/N\} - 1}{2} \, ,$$

where n_k counts the total number of sites in the state k. The ratio n_k/N corresponds then to a 'magnetization' in the state k. The first-scaling law at a critical temperature is shown in Fig. 1, where the order parameter distributions for three system sizes $N = 32^2, 48^2$ and 64^2, are plotted in the shifted scaled variable: $z_1 \equiv (s - \langle s \rangle)/\langle s \rangle$. Even for such small system sizes, the first-scaling is nearly perfect. Note also a sharp decrease of the scaling function ($\sim \exp(-a z_1^{15})$) for large values of z_1.

[7] In this model, at each node i of the regular square lattice, a spin s_i is attached which can take three values: $k = 0, 1$ or 2.

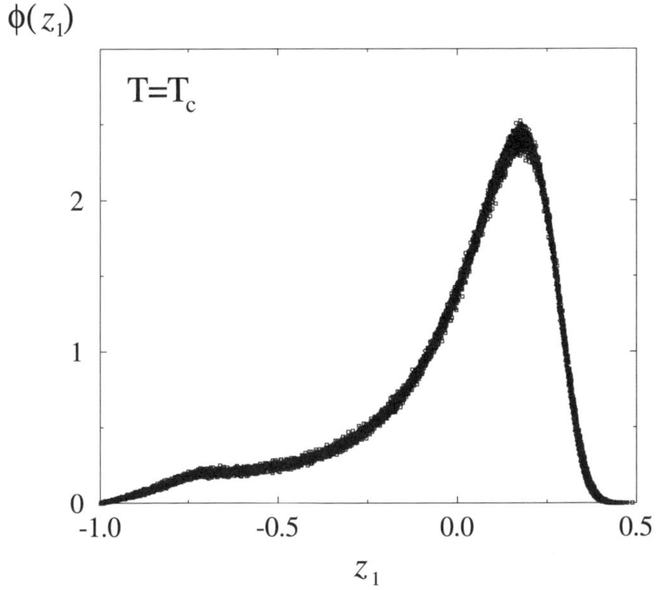

Fig. 1. Scaled s-distributions for a 2-dimensional, 3-states Potts model at the critical temperature. Three system sizes are plotted: $N = 32^2$ (*stars*), $N = 48^2$ (*circles*) and $N = 64^2$ (*squares*)

3.4 Example 2: The Scaling Law for a Percolation Model

The percolation model is a typical example of the equilibrium critical system for which the order parameter is *non-extensive* (here: in the disordered phase). We illustrate the first-scaling behavior for a 3-dimensional bond percolation on a cubic lattice at the critical concentration threshold. For a given value of the occupation probability, one computes sizes of all clusters[8] in the simulated configuration. An actual value of the order parameter is then defined as:

$$s_{max} \equiv \mathrm{Max}_k\{s_k\} ,$$

where s_k is the size of k-th cluster. When the occupation probability p becomes larger than the threshold value $p_c = 0.2488$, the size of the largest cluster becomes comparable to the system size. This corresponds to the appearance of an infinite cluster spanning the lattice. The mathematical structure of s_{max} is quite different from the additive examples discussed previously. This point will be discussed in the following section. Nevertheless, the first-scaling law holds at the critical point, as the general theory predicts. Figure 2 shows how s_{max}-distributions for different system sizes at the percolation threshold collapse into a single curve in the variables of the first-scaling.

[8] A cluster is defined as an ensemble of sites connected by active bonds.

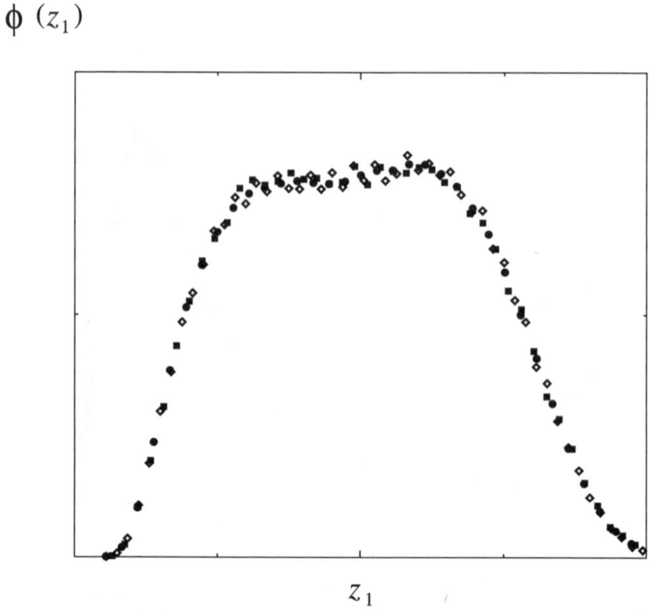

Fig. 2. Scaled s_{max}-distributions for a 3-dimensional bond-percolation model at the critical threshold, as functions of the shifted scaled order parameter $z_1 \equiv (s_{max} - \langle s_{max} \rangle)/\langle s_{max} \rangle$. Three network sizes are plotted: $N = 14^3$ (*stars*), $N = 20^3$ (*circles*) and $N = 32^3$ (*squares*)

The anomalous exponent can be estimated from an overall mean size distribution of clusters. At the percolation threshold, and only there, the cluster size distribution follows a power-law:

$$n_s \propto N/s^\tau$$

with the exponent $\tau \simeq 2.2$, where n_s is the average number of clusters of size s. The N-dependence of the prefactor comes from normalization: $\sum_s sn_s = N$, taking into account $\tau > 2$. This leads to the anomalous scaling:

$$\langle s_{max} \rangle \sim N^g$$

and the anomalous exponent:

$$g = \frac{1}{\tau - 1} \quad .$$

Thermodynamic equalities, which have been established by Fisher for those infinite critical systems in which a notion of the cluster size is relevant (e.g. the Fisher droplet model [7] or the percolation model), tell that the exponent τ should be related to other critical exponents:

$$\tau = 2 + \frac{\beta}{\beta + \gamma} \quad . \tag{10}$$

Since both β and γ have to be positive, eq. (10) implies that τ has to be restricted to an interval $2 \le \tau \le 3$. This implies in turn that: $1/2 \le g \le 1$, as we have already noticed in the general theory of the at-equilibrium second-order critical phenomena.

4 On the Shape of First-Scaling Functions

Discussion of this section will be organized into separate subsections corresponding to (i) the uncorrelated additive order parameter, (ii) the correlated additive order parameter, (iii) the uncorrelated extreme order parameter and (iv) the correlated extreme order parameter.

4.1 The Uncorrelated Additive Order Parameter

This case has been discussed above in the disordered phase of the Landau–Ginzburg model. Formal results can be expressed as follows. Consider the extensive order parameter: $s = \sum_i s_i$, where N microscopic random variables s_i are independent and distributed according to the same probability law $f(s)$. Under general conditions, the Central Limit Theorem applies and leads to the Gaussian distribution of s. It is instructive to give an explicit general condition here, in order to compare with the extreme value case discussed later. If the common probability distribution $f(s)$ is such that:

$$\int_{-\infty}^{\infty} (1 - F(s))ds < \infty$$

$$\lim_{x\to\infty} \frac{x^2(1 - F(x))}{\int_x^{\infty} s^2 f(s)ds} = 0$$

with F the cumulative distribution, then the average value $\langle s \rangle$ exists and is finite, and the probability density $\mathrm{P}(m)$ of the random variable $m \equiv (\sum_{i=1,\ldots,N} s_i - N\langle s \rangle)^2$ is such that:

$$\lim_{N\to\infty} \langle m \rangle \mathrm{P}_N(m) = \exp\left[-\frac{m}{\langle m \rangle}\right] \quad .$$

Note that this limit scaling has been written for the variable m which is essentially the squared (shifted) sum of the random microscopic variables s_i. If rewritten in terms of the variable $|\sum_{i=1,\ldots,N} s_i - N\langle s \rangle|$ for example, this would lead to the usual Gaussian limit function.

As a particular example, let us consider uncorrelated variables s_i which can take only three different values, say -1, 0 or 1, with equal probability $\frac{1}{3}$. The probability $\mathrm{P}_N(s)$ to get the value s, satisfies the recurrence relations:

$$\mathrm{P}_N(s) = \frac{\mathrm{P}_{N-1}(s+1) + \mathrm{P}_{N-1}(s) + \mathrm{P}_{N-1}(s-1)}{3} \quad ,$$

so that the moment generating function $\mathcal{M}_N$ is:

$$\mathcal{M}_N(u) = \left(\frac{e^{-u} + 1 + e^u}{3}\right)^N .$$

All moments of $P_N(s)$ can be readily calculated from $\mathcal{M}_N(u)$. Developing the moment generating function at the second-order near $u = 0$, one has: $\mathcal{M}_N(u) \sim \exp(Nu^2/3)$. A Fourier transform gives then the probability distribution:

$$P_N(s) \propto e^{-3s^2/4N} .$$

In the leading order of N, the average value of $|s|$ is[9]:

$$\langle |s| \rangle \simeq \sqrt{4N/3\pi}$$

and the first-scaling law follows:

$$\langle |s| \rangle P_N(s) \sim \exp\left[-\frac{1}{\pi}\left(\frac{s}{\langle |s| \rangle}\right)^2\right] .$$

One recovers here the same Gaussian scaling function as derived in the Landau–Ginzburg theory. Note also that the average value of s ($\langle s \rangle = 0$) is the most probable value for this variable.

4.2 The Arc-Sine Law

As an example of a correlated additive variable, let us consider a famous example linked to the one-dimensional diffusion problem [8]. Let the additive variable s be the sum: $s = \sum_i s_i$, over N microscopic variables s_i. As in the previous example, each random variable s_i can take three values, say -1, 0 or 1, but now these values are distributed according to the following rule: let x_j be a random variable taking the values $+1$ or -1 with an equal probability. The sum $z_i = \sum_{j=1}^{i} x_j$ represents the location of a 1-dimensional random walker, after i steps, starting from the origin. Then s_i equals -1 if z_i is negative, $s_i = 0$ if $z_i = 0$, and $s_i = 1$ if $z_i > 0$. The successive random variables s_i are indeed correlated since $z_{i+1} = z_i + x_{i+1}$. The limit distribution for this random variable s has been found by Lévy[10]:

$$P_N(s) \sim \frac{1}{\sqrt{1 - s^2/N^2}} .$$

[9] $g = 1/2$ in this case.

[10] This limit distribution is called the 'Arc-sine' law because the cumulative function associated to $P_N(s)$ is the arc-sine function.

For our purpose, one has to calculate an average value of $|s|$:[11]

$$\langle |s| \rangle = N/2$$

and the first-scaling law follows with the scaling function in the form:

$$\langle |s| \rangle \mathrm{P}_N(s) \sim \frac{1}{\sqrt{1 - \frac{1}{4}(s/\langle |s| \rangle)^2}} \quad . \tag{11}$$

In this example, average value of the variable ($\langle s \rangle = 0$) is equal to its least probable value. Comparing this example and the precedent one, one realizes that for two microscopic variables having a similar mathematical structure[12] and for an additive macroscopic variable defined in the same way ($s = \sum_i s_i$) in both examples, the first-scaling is found in both cases but (i) the value of the anomalous exponent changes from $\frac{1}{2}$ in the uncorrelated case, to 1 in the correlated case, and (ii) the first-scaling function changes from a Gaussian in the uncorrelated case to a non-standard form (11) in the correlated case.

4.3 The Uncorrelated Extreme Order Parameter

Let us consider the same set of N microscopic independent real random variables s_i, and the non-extensive global random variable $s_{max} = \mathrm{Max}_{i=1,\ldots,N}\{s_i\}$. Von Mises condition states that if the common probability distribution $f(s)$ is such that:

$$\int_{-\infty}^{\infty} (1 - F(s))ds < \infty$$

$$\lim_{x \to \infty} \frac{f(x) \int_x^{\infty} (1 - F(s))ds}{(1 - F(x))^2} = 1$$

with F the cumulative distribution, then there exist two sequences of positive numbers a_N and b_N such that:

$$\lim_{N \to \infty} a_N \mathrm{P}_N(s_{max}) = \phi\left(\frac{s_{max} - b_N}{a_N}\right)$$

with ϕ the Gumbel distribution, whose general form is:

$$\phi(z) = \frac{1}{\sigma} \exp\left[-\frac{(z - \mu)}{\sigma} - \mathrm{e}^{-(z-\mu)/\sigma}\right] \quad .$$

The Gumbel distribution for the extreme value problem plays a similar fundamental role as the Gaussian distribution for the problem of the sum of independent random variables. An equivalent problem for the global variable $s_{min} = \mathrm{Min}_{i=1,\ldots,N}\{s_i\}$ gives similar results, except that the value of $z - \mu$ in the Gumbel distribution is replaced by $\mu - z$.

[11] $g = 1$ in this case.

[12] In both examples, they can take only three different values -1, 0 or $+1$.

4.4 The Hierarchically Correlated Extreme Order Parameter

As for the Central Limit Theorem, things are quite different when the microscopic random variables are correlated. An explicit example is given by the zero-temperature Derrida–Spohn model [9]: consider the Cayley tree rooted at $\mathcal{O}$ and random energies ϵ_i associated with every bond of the tree. The random variables ϵ_i are independent and with the same distribution $\rho(\epsilon)$. A path of size n goes from the root $\mathcal{O}$ to any of the 2^n nodes at level n. There are then $N = 2^n$ possible paths of size n, and the energy of any of these paths is given by:

$$E = \sum_{i=1,\ldots,n} \epsilon_i \quad .$$

The N random variables E_1, E_2, ..., E_{2^n} are correlated in a hierarchical way and the 2-point correlation between the energies of any two paths is proportional to the number of bonds in common. The distribution $\mathrm{P}_N(E_{min})$ of the minimum energy $E_{min} = \mathrm{Min}\{E_i\}$ is [10]:

$$\mathrm{P}_N(E_{min}) \sim q^2\left(E_{min} + a\ln N\right) \quad .$$

The scaling function q depends in a known way on the distribution $\rho(\epsilon)$. In particular, for $\rho(\epsilon) = \exp(-|\epsilon|)/2$, one finds exponential tails at both ends of the function $q(z)$ $(z \to \pm\infty)$, instead of an exponential tail for $z \to -\infty$, and the super exponential tail for $z \to \infty$ in the uncorrelated case. In this case, correlations kill a characteristic super-exponential tail of the Gumbel distribution.

4.5 The Δ-Scaling for the Extreme Distribution and the Shape of the Scaling Function

A classical theorem proved by Gnedenko [11] about the distribution of the extreme random variable $s_{max} = \mathrm{Max}_{i=1,\ldots,N}\{s_i\}$, with all s_i independent and of same probability distribution $f(s)$, can be stated as follows:

If the distribution f is defined and positive on R, its first derivative exists everywhere, and:

$$\int_{-\infty}^{\infty} (1 - F(s))ds < \infty \tag{12}$$

then, there exist two sequences of positive numbers a_N, and b_N, such that $\lim_{N\to\infty} b_N = \infty$ and:

$$a_N \mathrm{P}_N(s_{max}) \sim \phi\left(\frac{s_{max} - b_N}{a_N}\right) \quad .$$

Moreover, if the derivative of a_N with respect to b_N is such that:

$$\lim_{N\to\infty} \frac{da_N}{db_N} = 0 \ , \tag{13}$$

then the limit scaling function ϕ is the Gumbel distribution:

$$\phi_G(z) \propto \mathrm{e}^{-z-\mathrm{e}^{-z}} , \tag{14}$$

while if:

$$\lim_{N\to\infty} \frac{da_N}{db_N} = c \tag{15}$$

with c a positive number, then the limit scaling function ϕ is the Fréchet distribution:

$$\phi_F(z) \propto \frac{1}{z^{1+1/c}} \mathrm{e}^{-z^{-1/c}} . \tag{16}$$

Important consequences for the scaling laws of the extreme order parameter can be deduced from this theorem. For example, the first scaling law is characterized by the constraint: $a_N \propto b_N$, which is only consistent with the Fréchet case (15). The first-scaling law in the s_{max}-distribution can then only be seen in conjunction with the Fréchet distribution (16). In addition, one knows how to rely the N-dependence of a_N and b_N to the distribution function $f(s)$, namely one can choose [12]:

$$F(b_N) = \mathrm{e}^{-1/N} ,$$
$$a_N = \frac{1}{Nf(b_N)} .$$

From these two relations, one proves that the first-scaling law on s_{max} holds only for the power-law s-distribution:

$$f(s) \sim s^{-\tau} \tag{17}$$

with: $\tau = (c+1)/c$, and c is the constant appearing in (15). Note that, as usual for this kind of scaling or for the Central Limit Theorem, the proper shape of the scaling function is determined uniquely by the large-argument behavior of the distribution of the microscopic variable. Hence, the power-law form (17) is only needed as the asymptotic behavior for the large values of the argument s. From the constraint $c > 0$, one deduces that the exponent τ in (17) should be larger than 1, but the condition (12) is stronger as it leads to: $\tau > 2$.

On the other hand, a more general choice of dependence between a_N and b_N sequences can be written as $a_N \sim b_N^{\Delta}$, with the real parameter Δ. The corresponding scaling is then called the Δ-scaling law [13]. The value $\Delta = 1$ corresponds to the first-scaling and has been discussed above. If value of Δ is strictly smaller than 1, the condition (13) is fulfilled and the scaling function ϕ is the Gumbel distribution. This occurs for example for the generalized Gamma-distribution:

$$f(s) \propto s^{-\Delta} \mathrm{e}^{-u^{1-\Delta}} \tag{18}$$

of the microscopic variables s_i. For (18), relation: $a_N = b_N^{\Delta}/(1-\Delta)$ holds, and the Gumbel distribution (14) is recovered for the Δ-scaling function.

5 Nuclear Multifragmentation

Theoretical description of the fragment production in HI collisions depends on whether the equilibrium has been reached before the system starts fragmenting. If the equilibrium is attained, then the thermodynamic models using different statistical ensembles can be applied to understand basic physics of this process. Otherwise, the description of this process is restricted to a simulation of kinetic equations. Possibility of the finite system critical behavior associated with the transition from the particle evaporation regime at low excitation energies to the explosion of the hot source at about 5–10 MeV/nucleon cannot be excluded. In this complicated process, tools of the theory of universal fluctuations provides a model-independent way of characterizing different phases of the multifragmentation [14,15].

Below, we shall present selected results from the recent systematic studies of the probability distributions of the largest fragment as a function of the system size. There are two generic classes of dynamical critical phenomena, with their appropriate 'relevant observables'. The first one is the sequential cluster fragmentation, where the average cluster size decreases and the cluster multiplicity increases during the process. The cluster (particle) multiplicity is an order parameter for this class of critical phenomena. The second one is the sequential cluster aggregation, where the average cluster size increases and the cluster multiplicity decreases during the process. The size of the largest cluster is here an order parameter. The second class contains for example the liquid-gas phase transition (Fisher model), the percolation transition or the sol-gel phase transition in the irreversible kinetic aggregation. The discussion of Δ-scaling of this section, of which both first-scaling law and second-scaling law are the particular limits, make a reference to the scaling of probability distributions as a function of the system size. Actually in many phenomenological application, the precise meaning of the system size is not necessary and one can use for this purpose the average value of the observable[13], *i.e.* the Δ-scaling of the normalized probability distribution $\mathrm{P}_{\langle m\rangle}(m)$ of the variable m can be written as:

$$\begin{aligned} &\langle m\rangle^{\Delta}\mathrm{P}_{N(\langle m\rangle)}(m) \equiv \langle m\rangle^{\Delta}\mathrm{P}_{\langle m\rangle}(m) = \phi(z_{(\Delta)}) \qquad 0 < \Delta \leq 1 \\ &z_{(\Delta)} = (m - m^*)/\langle m\rangle^{\Delta} \end{aligned} \tag{19}$$

where $\langle m\rangle$ and m^* are the average and the most probable values of m respectively and the system size N is assumed to be a monotonous single-valued function of $\langle m\rangle$.

What makes the study of nuclear fragmentation exciting is that the order parameters for these two different scenarios are measurable with high

[13] The exact meaning of the system size is important if we want to find the relation between scaling parameter Δ and the anomalous dimension.

precision event-by-event. The 'size' of the cluster can be identified with the number of protons of the nuclear fragment (*i.e.* the fragment charge) detected in the final state. The measured fragment multiplicity corresponds to the charged fragment multiplicity. Figure 3 shows the Δ-scaling features of P$[Z_{max}]$-distributions for central Xe+Sn collisions at different bombarding energies: 25 MeV $\leq E_{lab}/A \leq$ 100 MeV. The collision violence is monitored by ranging events in bins of transverse energy E_{t12} of light charged particles ($Z = 1, 2$) normalized by the available energy in the collision E_{avail}. This quantity has been shown to be a good and unbiased indicator of the violence and hence of the centrality of HI collisions in this energy domain. An approximate correspondence between E_{t12}/E_{avail} and the 'impact parameter' b in the collision can then be established. In this scale, the data shown in Fig. 3 corresponds to most central collisions with $b < 0.1 b_{max}$. On the r.h.s. of Fig. 3, Z_{max}-distributions for central collisions at $E_{lab}/A \geq 39$ MeV are plotted in the scaling variables of the Δ-scaling (19) with $\Delta = 0.89$. The scaled distributions for different bombarding energies can be compressed into a single curve and the form of the scaling function is closely reproduced by the Gumbel distribution (the dashed line in Fig. 3b) which, as discussed before, is expected for an uncorrelated extreme order parameter in the disordered phase. Data at lower bombarding energies (see Fig. 3a): $E_{lab}/A = 25, 32, 39$ MeV cannot be superposed in these variables. At $E_{lab}/A = 25$ MeV, Z_{max}-distribution is perfectly Gaussian (see the dashed curve in Fig. 3a), as expected in the ordered phase. On the other hand, at $E_{lab}/A = 32, 39$ MeV, strong deviations from the Gaussian form are seen. Moreover, these data do not collapse in the scaling variables with $\Delta \sim 0.5$, as expected in the ordered phase. Actually, at these bombarding energies the system produced in central Xe+Sn collisions is already in the disordered phase of the fragmentation. Similar scaling

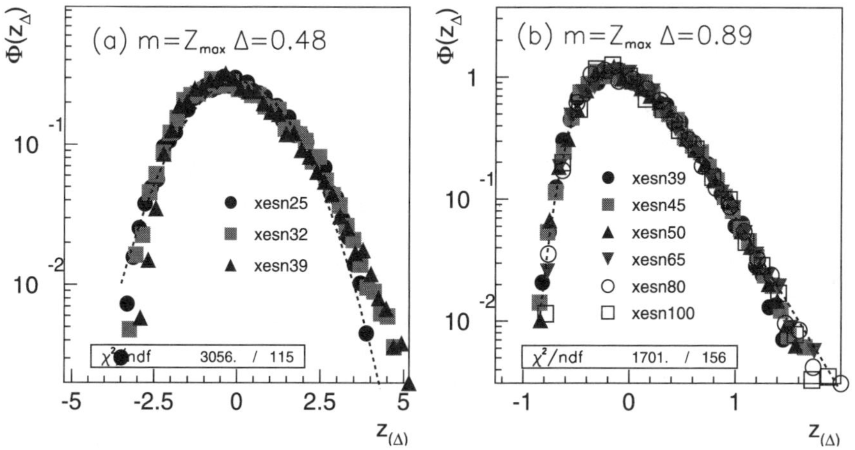

Fig. 3. Z_{max}-distribution for central Xe+Sn collisions (courtesy of J.D. Frankland and INDRA Coll.)

features have been observed in Ni+Ni and Au+Au collisions [15], though the transition from the second-scaling regime to the first scaling regime happens at lower bombarding energy in Au+Au reaction and at higher bombarding energy in Ni+Ni reaction than in Xe+Sn reaction. Also the deviations from $\Delta = 1$ in the disordered phase and from $\Delta = 1/2$ in the ordered phase are more important in light systems, like Ni+Ni than in heavy system like Au+Au.

The scaling pattern of $P(Z_{max})$ in semi-peripheral collisions follows a similar evolution with the bombarding energy (the excitation energy) as seen in central collisions [14,15]. In particular, the Z_{max}-distribution along the first-scaling branch give nearly same scaling function of a Gumbel form, as in the central collisions. This underlines again the universality of a scaling function. Moreover, the scaling function in the second-scaling branch at lower bombarding energies is essentially identical in semi-central and central collisions [14].

The pattern of charged fragment multiplicity distributions $P(n)$ does not show any evolution with the bombarding energy (see Fig. 4). The data is perfectly compressible in the variables of the second-scaling and the scaling function has a Gaussian form. The scaling features of experimental Z_{max}- and n-distributions are complementary and allow to affirm that the fragment production in central HI reactions follows the aggregation scenario.

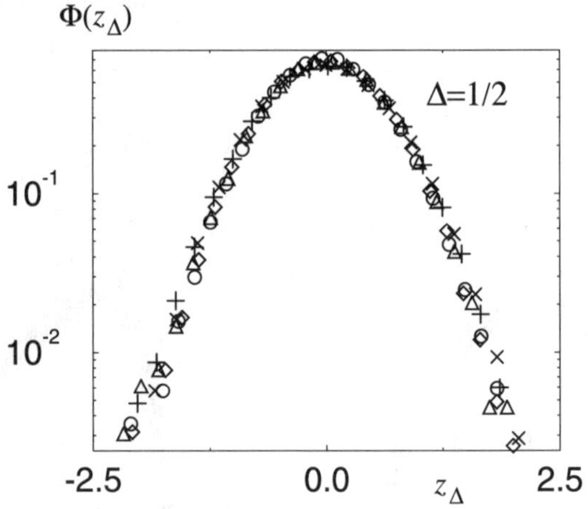

Fig. 4. Multiplicity distributions of charged fragments for central Xe + Sn collisions at $E_{lab}/A = 25$ (*asterisks*), 32 (*crosses*), 39 (*triangles*), 45 (*diamonds*) and 50 (*circles*) MeV are plotted in the second-scaling variables at $E_{lab}/A = 25$ (*asterisks*), 32 (*crosses*), 39 (*triangles*), 45 (*diamonds*) and 50 (*circles*) MeV (from [14])

6 Conclusion

Strongly fluctuating finite systems are frequent in Nature and their understanding is impossible without an understanding of the pattern of the fluctuations. A 'classical' example are the second-order critical systems. Many of those strongly fluctuating systems exhibit also strong long-range correlations. Recently, a general understanding of such finite systems has improved largely as a result of the progress in understanding of the universal scaling laws obeyed by the fluctuations [1]. This vast subject has been addressed in this lecture mainly in the context of the second-oder critical systems. We have discussed in particular the salient features of the scaling function, both for extensive and non-extensive global variables, depending on whether the microscopic variable is correlated or not. The fluctuating system is characterized by the anomalous exponent, the scaling law itself and the form of the scaling function. These features are robust and can be studied even in small systems, as has been illustrated in this lecture by numerical simulations in Potts model, percolation model, and by the experimental data in the domain of nuclear multifragmentation. The tools of the universal scaling approach can be also applied in the classical turbulence where recent studies have disclosed many universal features [16].

An application of the theory of universal fluctuations to symmetric HI reactions in the Fermi energy domain shows that the fragment production is governed by the aggregation scenario with $\langle Z_{max} \rangle$ as the order parameter. Change of a pattern of Z_{max}-fluctuations from the weak fluctuation regime at low energies to the strong fluctuation regime at high energies is compatible with the transition from the ordered phase to the disordered phase in a system with the non-extensive order parameter. In future, more extensive studies are necessary to distinguish between the scenario of finite-system critical behavior and the scenario of the cross-over phenomenon. The conceptual basis is however ready to assure a model *and* assumption independent answer to this question.

References

1. R. Botet, M. Płoszajczak: *Universal Fluctuations: Phenomenology of Hadronic Matter* (World Scientific, Singapore 2002)
2. J.E. Mayer, M.G. Mayer: *Statistical Mechanics* (John Wiley & Sons, London 1957)
3. R. Botet, M. Płoszajczak, V. Latora: Phys. Rev. Lett. **78**, 4593 (1997)
4. B. Widom: Physica **73**, 107 (1974)
5. B. Widom: J. Chem. Phys. **43**, 3898 (1965)
   F.J. Wegner: J. Phys. C **7**, 2098 (1974)
6. R. Botet, M. Płoszajczak: Phys. Rev. E **62**, 1825 (2000)
7. M. Fisher: Physics **3**, 355 (1967)
8. P. Lévy: *Processus stochastiques et mouvement Brownien* (Gauthier, Paris 1948)

9. B. Derrida, H. Spohn: J. Stat. Phys. **51**, 817 (1988)
10. D.S. Dean, S.N. Majumdar: Phys. Rev. E **64**, 046121 (2001)
11. B.V. Gnedenko: Ann. Math. **44**, 423 (1943)
12. J. Pickands: Ann. Probab. **14**, 996 (1986)
13. R. Botet, M. Płoszajczak: 'Universal features of fluctuations'. In: *Proceeding of the IXth International Workshop on Multiparticle Production*, Villa Gualino, Torino, Italy, ed. by A. Giovannini, R. Ugoccioni, Nucl. Phys. (Proc. Suppl.) B **92**, 101 (2001)
14. R. Botet et al.: Phys. Rev. Lett. **86**, 3514 (2001)
15. J.D. Frankland et al. (INDRA Coll.): 'Fragment production in central heavy-ion collisions: reconciling the dominance of dynamics with observed phase transition signals through universal fluctuations'. In: *Proceedings XLth Int. Winter Meeting on Nuclear Physics*, Bormio (Italy), January 21-25, 2002; nucl-ex/0202026
J.D. Frankland et al. (INDRA Coll.): 'Universal fluctuations: a new approach to the study of 'phase transitions' in intermediate energy heavy ion collisions'. In: *Proceedings IWM2001*, LNS-Catania (Italy), November 28–December 1, 2001; nucl-ex/0201020
16. S.T. Bramwell, P.C.W. Holdsworth and J.-F. Pinton: Nature **396**, 552 (1998)

Fluctuations and Instabilities in Nuclear Dynamics: from Multifragmentation to Neutron Stars

Massimo Di Toro

Abstract. The dynamics of nuclear phase transitions, from liquid-gas at low energy to the hadron-deconfined matter at high baryon density, is discussed. Microscopic simulations of Heavy Ion Collisions (HIC) with stochastic mean field approaches are presented. Focus is put on the isospin dynamics, in connection to the new possibilities provided by radioactive beam facilities. Particular attention is devoted to our knowledge of the elusive isovector part of the in-medium nuclear interaction. A selection of sensitive observables from reaction measurements is finally suggested.

1 Introduction

In the first part of the report the nuclear liquid-gas phase transition will be discussed with particular attention to the dynamics of instabilities leading to fragment formation in heavy ion collisions: the spinodal decomposition mechanism. The conditions for an event reduction in order to apply statistical methods will be analyzed. Focus will be put on two recent experimental signatures of the mixed phase existence: i) Negative heat capacity from kinetic energy abnormal fluctuations; ii) "Fossil relics" of the spinodal dynamics from correlations measurements.

The dynamics of a new important dissipation process in the Fermi energy domain, the "neck-rupture" mechanisms for semicentral collisions, will be also presented.

In the second part properties of the elusive symmetry term, from ^{208}Pb to neutron stars, will be discussed, in a non/relativistic and relativistic frame for the effective in medium nuclear interactions. In particular the asymmetric nuclear matter will be dynamically tested at low densities (isospin distillation and isospin effects on neck fragmentation) as well as at high baryon densities (new exotic collective modes, relativistic effects on collective flows, up to the possibility of a transition to a mixed phase with deconfined quark matter).

In conclusion a selection of *Eleven Observables* in HIC's from low to relativistic energies, particularly sensitive to the Isospin Dynamics, will be suggested.

2 The Nuclear Liquid-Gas Phase Transition and Phase Coexistence

The short-range nature of the nuclear interaction, responsible of the stability of matter, and its similarity with Van der Waals interaction, suggests that nuclear matter inside the nucleus resembles a real gas which is in a liquid phase and could undergo a first-order phase-transition below a given critical temperature. Indeed the nuclear Equations of State (EOS) obtained from realistic non-relativistic and relativistic effective interactions, in mean-field and/or correlated schemes, are all predicting a liquid-gas phase transition in dilute matter with a critical point around 20 MeV temperature.

In macroscopic world, phase transitions manifest themselves by abrupt transformations of the matter properties. In finite systems, rapid modifications should remain. This is the origin of the large interest in a microscopic study of the dynamics of phase transitions in nuclear collisions: we expect a great sensitivity to detailed properties of the in-medium interaction.

In the past years, several such modifications have been identified in dissipative heavy-ion collisions around the Fermi energy. It has been recently claimed that more than ten independent experimental evidences have been already observed [1], ranging from the disappearance of the hot Giant Dipole Resonance to the instantaneous fragment production up to vaporization events.

A key issue in favor of a first order phase transition is the detection of a coexistence region. In this context the reaction dynamics plays an essential role and we can distinguish two main experimental scenarios and/or event sorting selection:

- select event classes for which the dynamics can be reduced to the control of few collective variables (mass, excitation energy, freeze-out volume ...) all the other degrees of freedom being statistically distributed
- look for experimental situations presenting clear dynamical signals testing specific transport properties (spinodal instabilities, radial flow and compressibility ...)

The first issue can be solved by sorting the final state of the collision by well chosen global variables which uniquely identify classes of events and then apply statistical methods. This analysis supposes that the reaction is complex enough so that only few global variables have a non trivial dynamics while the remaining available phase space is randomly populate. In the second scenario we expect to see events with fragment properties (mass, charge, velocity distributions ...) largely outside any possible statistical prediction. Of course the two selections correspond to different lifetimes of the fragmenting source. We can expect the first picture to be dominant for fragmentation of "spectator" sources in semicentral collisions and the second more effective in presence of a fast expansion in central events.

2.1 Direct Observables of the Phase Transition

Thermodynamics. *The caloric curve*
A temperature saturating over a broad range of excitation energy was presented as an evidence for a phase transition in nuclei [2]. For each system, a caloric curve has been constructed emphasizing the consistency of all thermometers. The flattening of the caloric curve depends upon the system charge showing that Coulomb instabilities might trigger the phase transition [3]. It should be however stressed that the order parameter of the liquid-gas phase transition is the density i.e. the volume. Then the "caloric curve" is actually bi-dimensional depending both on excitation energy and volume (or pressure). It would therefore be necessary to experimentally measure an exclusive caloric curve constrained by the volume of the system. In actual nuclear reactions, the volume is determined by the dynamics and cannot be directly controlled, apart difficult selections based on particle interferometry.

Negative heat capacity
In a constant energy ensemble (microcanonical) for finite systems negative heat-capacity defines a first order phase transition [4]. The heat capacity of the fragmenting system, selected with a given total energy, can be deduced from the comparison of the kinetic heat-capacity with the event-by-event fluctuations of the kinetic energy [1,5]. Such a measurement was performed in different systems covering a broad range in excitation energy [1,6,7]. The existence of a negative heat-capacity was observed in the excitation-energy range between 3 and 6 MeV, providing a very direct indication of a first order liquid-gas transition.

Dynamics. *Spinodal Decomposition*
A particular aspect of nuclear EOS is the existence of the spinodal region inside the liquid-gas coexistence. The spinodal region is characterized by an increasing pressure when the matter density decreases. This property generates a mechanical instability through the amplification of density fluctuation that will end up in a direct cluster formation. This dynamical mechanism will be reminiscent of the space/time structure of the most unstable modes, with favored wavelenghts (fragment sizes) and time scales [8–10]. The production of fragments through spinodal decomposition of excited nuclear systems has been examined in depth applying, in particular, stochastic mean-field approaches to describe the dynamics of heavy-ion reactions. It was found that, because of the range of the nuclear interaction, spinodal decomposition favors the breaking of the system in close size fragments of charge $Z \approx 12 - 15$ on times of the order of 100 fm/c. However, because of the coupling induced by non linear effects and particle decays, which follow the early growth of instabilities, the original partitions are strongly modified by the subsequent dynamics leaving a small fraction of the events with the original correlation [11]. This "fossil" signal has been indeed recently observed in correlation

measurements in the appropriate dynamical scenario, central collisions in the Fermi energy range [12].

We like to note that this fully dynamical mechanism will be very sensitive to properties of the effective nuclear interactions. This point will be largely exploited in the rest of the report, for isospin effects on fragment production, in connection to the search for elusive isovector channel in regions far away from normal conditions.

2.2 Dynamical Fragmentation of a Neck in a Di-nuclear System

In semi-central collisions around Fermi energies, fragments emitted at mid-rapidity might come from the decay of a neck transiently formed in dissipative binary collisions. Due to the time matching requirements we can also expect a rise and fall of the effect with increasing beam energies [13]. Indeed such process represents a dissipative reaction mechanism of transition from the low energy Deep-Inelastic to the high energy Fireball (Participant-Spectator) picture.

Experimentally this mid-rapidity production of fragments has been well observed [14–16]. Some data are even nicely showing that in the same events we can have a coexistence of fragments produced from the neck and from the excited spectators (projectile-like and/or target-like) [17]. Due to the instability origin we can actually have a wide variety of emitting "sources", from mid-rapidities as well as from rapidities closer to the projectile and/or target side. We observe also limiting cases of anisotropic fast-fission of the spectator-like fragments, which are evidences of dynamically-aligned fission mechanisms [18].

We expect the relative observables, cross sections and fragment properties, to be largely sensitive to transport effects of the interaction. This point will be also largely discussed in the isospin context.

3 The Isospin Degree of Freedom

The isospin degree of freedom in the final state of heavy-ion collisions will provide an additional signal to detect the liquid-gas phase transition.

Indeed, theory of phase transitions in two-fluid systems predict different N/Z concentration in each phase: the liquid phase drives the system toward symmetric matter while the gas phase absorbs the isospin of the heated system, *isospin distillation* [19–24]. In spinodal decomposition, the isovector part of the nucleon-nucleon interaction at sub-saturation densities will favor particular N/Z concentrations in the fragments. The experimental determination of the favored isospin would shed light on the properties of the symmetry term of the nuclear EOS in the low density regime. For semi-central collisions the dynamics of the participant zone appears also to be quite affected by the symmetry term [25–28].

At high density we will discuss new collective modes, collective flows and finally a quite speculative *earlier* possible transition to a mixed phase with deconfined matter.

The EOS symmetry term
The behavior of the symmetry term of the nuclear EOS is poorly known in regions far from normal density. We like to remark that the *Symmetry Pressure/Compressibility* is of relevant importance for structure properties, being linked to the thickness of the neutron skin in n-rich (stable and/or unstable) nuclei, see the discussion in [29–31]. In neutron stars the high density behavior of $E_{sym}(\rho_B)$ is largely affecting the mass/radius relation as well as the proton fraction and then the cooling mechanisms, apart the possibility of a transition to a quark core [32].

In the following we will compare results obtained with forces that have *the same saturation properties for symmetric NM*. We will refer to a "asy-stiff/superstiff" EOS when we are considering a potential symmetry term with a linear/parabolic increase with nuclear density and to a "asy-soft" EOS when it shows a saturation and eventually a decrease above normal density [26].

In Figs. 1, 2 we report, for a ^{124}Sn asymmetry $(N - Z)/A = 0.2$, the density dependence of the symmetry contribution to the mean-field potential (left) and of the chemical potentials (right) for neutrons (top curves) and protons (bottom curves), for the different effective interactions in the isovector channel. From the behavior in the low densities region we expect that when the inhomogenities develop both neutrons and protons have the tendency to move from lower to higher density regions, in phase: the system is

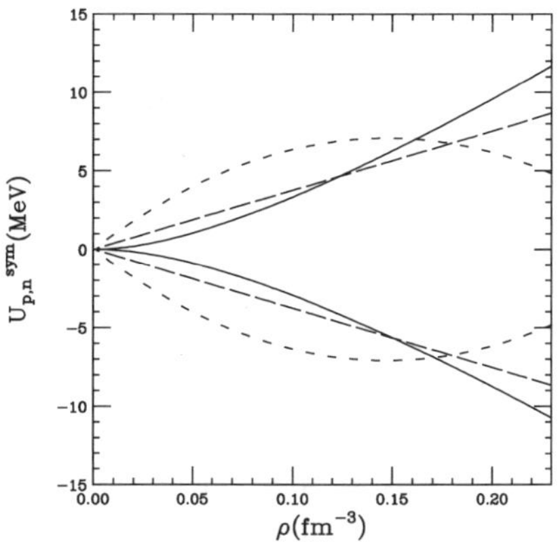

Fig. 1. Symmetry contribution to the mean field at $I = 0.2$ for neutrons and protons: *dashed lines* "asy-soft", *solid lines* "asy-stiff", *long dashed lines* "asy-superstiff"

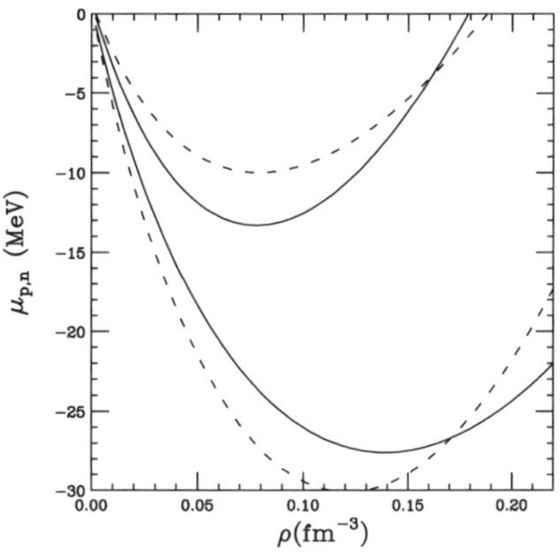

Fig. 2. Density dependence of proton and neutron chemical potential for asy-superstiff (*solid lines*) and asy-soft (*dashed line*) EOS

unstable against *isoscalar-like* fluctuations and not isovector, see later. Since the variations of the two chemical potentials are different (larger for protons) we expect a lower asymmetry in the liquid phase. In the case of a contact between more dilute and "normal" density regions, we see from Fig. 2 that in this range the neutrons have the tendency to move towards the dilute part producing a n-enrichment while the protons will migrate to the higher density regions. This mechanism is present in the "neck fragmentation", [13,27,28]: the neck $IMF's$ will be always more n-rich compared to the fragments produced in the case of bulk fragmentation.

4 Isospin Distillation in Dilute Nuclear Matter

For charge asymmetric systems we expect a qualitative new feature in the liquid-gas phase transition, the onset of chemical instabilities that will show up in a novel structure of the unstable modes [23,24]. Experimentally this will be revealed through the *Isospin Fractionation* or *Distillation* effect in multifragmentation events.

We have now a new degree of freedom, the concentration, and in the spinodal region the fluctuations against which a binary system becomes unstable depend on the nature of the interaction between the two components of the mixture.

In the framework of Landau theory for two-component Fermi liquids the spinodal border is determined by studying the stability of collective modes described by two coupled Landau–Vlasov equations for protons and neutrons. In terms of the appropriate Landau parameters the stability condition can be expressed as [23],

$$(1 + F_0^{nn})(1 + F_0^{pp}) - F_0^{np} F_0^{pn} > 0 \,. \tag{1}$$

It is possible to show that this condition is equivalent to the following thermodynamical condition

$$\left(\frac{\partial P}{\partial \rho}\right)_{T,y} \left(\frac{\partial \mu_p}{\partial y}\right)_{T,P} > 0 \,, \tag{2}$$

where y is the proton fraction and μ_p the proton chemical potential. In Fig. 3 we show the spinodal lines obtained from Eq. (1) (*continuous line with circles*) which for asymmetric nuclear matter is seen to contain the lines corresponding to "mechanical instability", $\left(\frac{\partial P}{\partial \rho}\right)_{T,y} < 0$ (*crosses*). The calculations are performed with non-relativistic Skyrme-like forces [24], but very similar results can be obtained with relativistic mean field approaches [20].

We want to stress, however, that by just looking at the above stability conditions we cannot determine the nature of the fluctuations against which a binary system becomes chemically unstable. Indeed, the thermodynamical condition in Eq. (2) cannot distinguish between two very different

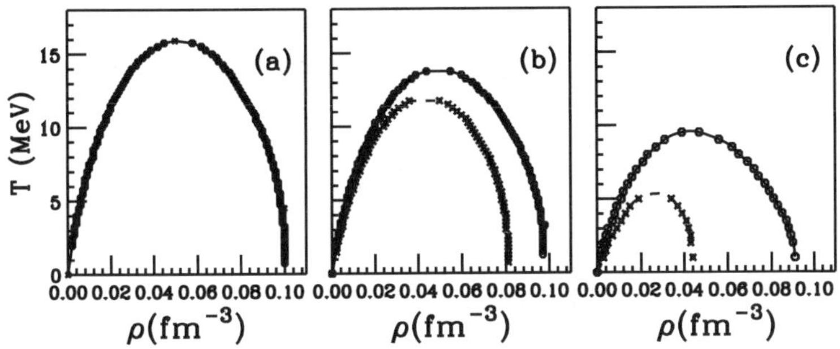

Fig. 3. Spinodal line (chemical plus mechanical) of asymmetric nuclear matter: *open circles*. Mechanical instability line: *crosses*. Three proton fractions: (**a**) $y = 0.5$, (**b**) $y = 0.25$, (**c**) $y = 0.1$

situations which can be encountered in nature: an attractive interaction between the two components of the mixture ($F_0^{np}, F_0^{pn} < 0$), as is the case of nuclear matter, or a repulsive interaction between the two species. For the dilute asymmetric nuclear matter case because of the attractive interaction between protons and neutrons at low density the phase transition is thus due to isoscalar-like fluctuations that induce chemical instabilities while the system is never unstable against isovector fluctuations. Of course the same attractive interaction is also at the origin of phase transitions in symmetric nuclear matter. However, in the asymmetric case isoscalar fluctuations lead to a more symmetric high density phase everywhere under the instability line defined by the Eqs. (1),(2) [23,24] and therefore to a more neutron-rich gas (*isospin distillation*).

Since during the dynamics of a collision the system can deeply enter the instability region it is important to have a more detailed information on the space-time structure of the unstable modes leading to cluster formation. This can be easily achieved following the Landau kinetic approach to the linear response theory. The Landau dispersion relations have indeed imaginary sound velocity solutions with well defined structure in the $(\delta\rho_n, \delta\rho_p)$ components [22,23].

We define density fluctuations as isoscalar-like in the case when proton and neutron densities fluctuate in phase and as isovector-like when the oscillations are out of phase. For the dilute asymmetric nuclear matter because of the attractive force between protons and neutrons at low density the phase transition is *uniquely* driven by isoscalar-like instabilities [24,39].

An intuitive picture is presented in Fig. 4. With increasing asymmetry the direction of the unstable modes (arrows) in the $(\delta\rho_n, \delta\rho_p)$ plane is more and more diverging from the constant concentration value (thick lines), towards a less asymmetric liquid phase. The angle between the two directions, i.e. the amount of isospin distillation, will be proportional to the repulsion of the symmetry term at sub-saturation densities. The measurement of the isospin

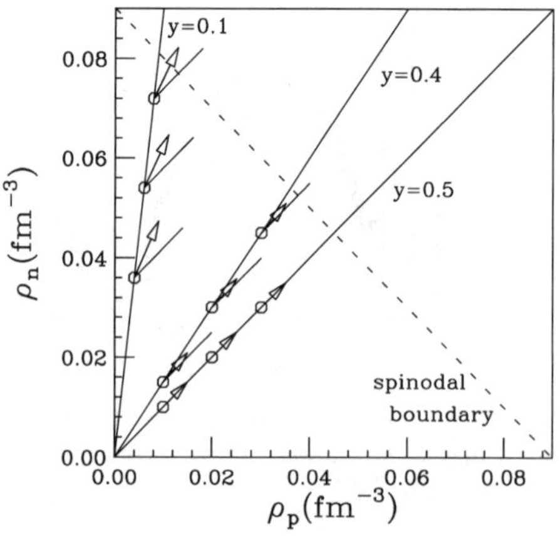

Fig. 4. The mechanism of isospin distillation. The *arrows* show the "direction" of the unstable mode in various points of the spinodal region. The *thick lines* represent a composition corresponding to the initial concentration, the *thin lines* a pure isoscalar mode. Initial proton fractions $y = 0.5,\ y = 0.4$ and $y = 0.1$

content of fragments provides promising new means to access the properties of the nuclear EOS and phase-diagram. Such a study will require a new generation of multidetectors allowing isotopic separation and the availability of unstable neutron and proton rich beams with energies well beyond the Coulomb barrier.

5 Isospin in Stochastic Transport Simulations

Stochastic transport simulations [25,33–35] of fragment production collisions at medium energies are confirming these predictions, see [27,28] where a comparison with recent data [36,37] is also performed. A new code for the solution of microscopic transport equations of Boltzmann–Nordheim–Vlasov (BNV) type has been written where asymmetry effects are suitably accounted for and the dynamics of fluctuations is included. A density dependent symmetry term is used in the ground state construction of the initial conditions, isospin effects on nucleon cross section and Pauli blocking are consistently evaluated. The transport equations are solved following a test particle evolution on a lattice. A parameterization of free NN cross sections is used, with energy and angular dependence.

We consistently include isospin effects also in the treatment of the stochastic term of the transport equation. Indeed the evolution under the influence of fluctuations is described by a transport equation with a stochastic fluctuating term, the so-called Boltzmann–Langevin equation (BLE) [8]. In our simulations we follow two methods to include fluctuations, the stochastic mean field [35] and the corresponding simplified approach used in [9–11]. The

Stochastic Mean Field method is based on a fully self-consistent treatment of the fluctuations during the time evolution. We notice that the variance around a given trajectory of the system $\sigma^2 = \langle (f - \bar{f})^2 \rangle$ in each phase space cell obeys the equation of motion:

$$\frac{d}{dt}\sigma^2 = -\frac{2}{\tau(t)}\sigma^2 + 2\ D(t). \tag{3}$$

with $2\ D(t)$ correlation function of the fluctuating term and $\tau(t) = 1/(w^+ + w^-)$, where w^+ and w^- are the transition probabilities into and out of the phase space cell. The statistical value $\sigma_0^2 = f_0(t_{eq})(1 - f_0(t_{eq}))$ at equilibrium suggests an ansatz for the correlation function of the fluctuation term of the form:

$$2D(t) = (1 - \bar{f})w^+ + \bar{f}w^-, \tag{4}$$

i.e. the magnitude of the fluctuations is given by the total number of collisions (fluctuation-dissipation theorem) [38]. Then we have for the time evolution of the quantity $\Delta = \sigma^2 - \bar{f}(1 - \bar{f})$:

$$\frac{d}{dt}\Delta = -\frac{2}{\tau(t)}\Delta. \tag{5}$$

Since $\Delta = 0$ is a solution of Eq. (5), if the variance is initially locally set equal to its statistical value, it will always follow the evolution of the average distribution function according to the relation $\sigma^2 = \bar{f}(1 - \bar{f})$. From a projection on coordinate space we get local density variances and then, with a Monte Carlo method, density fluctuations at each time step. In this way we can have also the various trajectory branching. The procedure is correct if we assume a local thermal equilibrium, appropriate for the problems discussed here, fragment production in the expansion/separation phase.

The other approach, computationally much easier [9–11], is based on the introduction of density fluctuations by a random sampling of the phase space. The amplitude of the noise is gauged to reproduce the dynamics of the most unstable modes [10]. For each system we have checked the equivalence of the two methods in the description of the collision dynamics, from fast particle emissions to the fragment production. The analysis of the results, presented in the following, is based on events collected with both numerical methods.

5.1 Numerical Results: Heated Nuclear Matter in a Box

Numerical simulations have been performed in order to follow all stages of the fluctuation growth in the fragment formation process. In the numerical approach we consider nuclear matter in a box of size $L = 24\,\text{fm}$, imposing periodic boundary conditions [23].

An initial temperature is introduced by distributing the test particle momenta according to a Fermi distribution. We have followed the space-time evolution of test particles initialized at asymmetry $I = 0.5$, density

$\rho^{(0)} = 0.09\,\mathrm{fm}^{-3}$ and $T = 5\,\mathrm{MeV}$. In such a way we start with a system prepared inside the chemical instability region (see Fig. 3(b). The inital density perturbation was created automatically due to the random choice of the test particle positions.

We report in Fig. 5 the time evolution of neutron (thick histogram) and proton (thin histogram) abundances (a) and asymmetry (b) in various density bins. The dashed line shows the initial uniform density value $\rho^{(0)} = 0.09\,\mathrm{fm}^{-3}$ and the initial asymmetry $I = 0.5$. The drive to higher density regions is clearly different for neutrons and protons: at the end of the dynamical clustering mechanism we have different values of the asymmetry in the liquid and in the gas phase (see 300 fm/c). This result is the same of [23], where the dynamics of mechanical instabilities was studied, showing that the nature of fluctuations associated with chemical *and* mechanical instabilities is essentially the same.

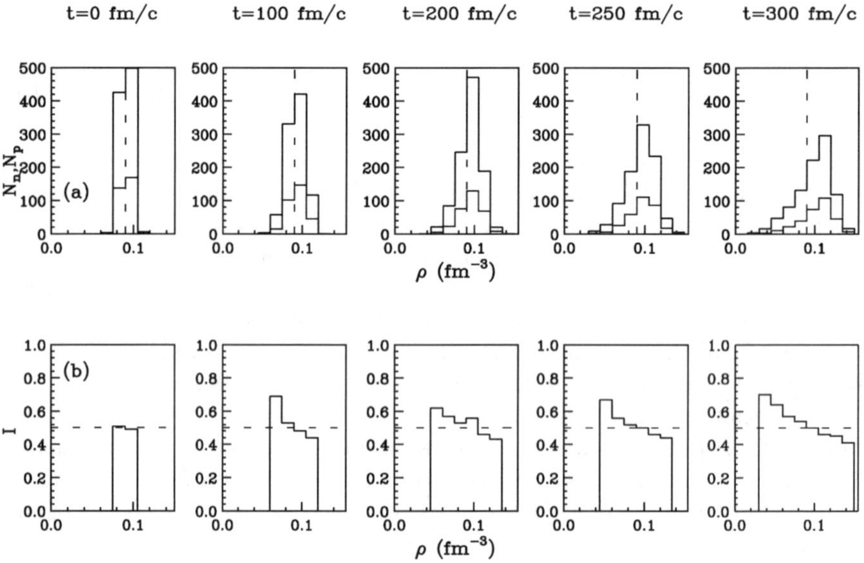

Fig. 5. Time evolution of neutron (*thicks lines*) and proton (*thin lines*) abundances (**a**) and asymmetry (**b**) as a function of density

5.2 Central Heavy Ion Collisions

We have studied the reaction $^{124}\mathrm{Sn} + {}^{124}\mathrm{Sn}$ at $50\,\mathrm{MeV/A}$, where new data are under analysis at $NSCL - MSU$ [36]. In Figs. 6, 7 we show the time evolution of the nucleon density (projected on the reaction plane) for typical events in central (Fig. 6) and semi-central (Fig. 7) collisions.

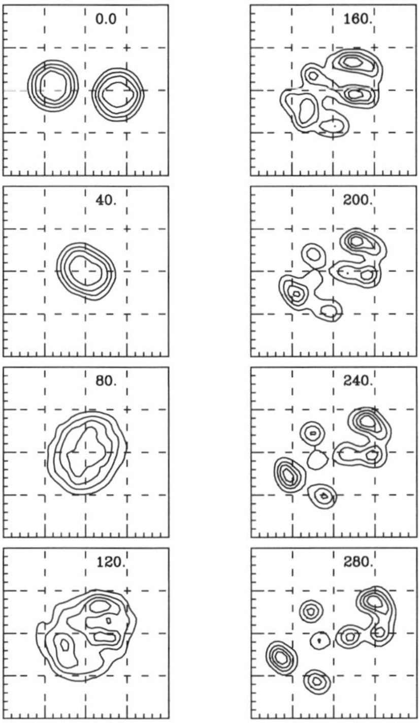

Fig. 6. $^{124}Sn + ^{124}Sn$ $b = 2\,\mathrm{fm}$ collision: time evolution from approaching to freeze-out, iso-density lines (times in fm/c in each panel). Asy-Stiff EOS

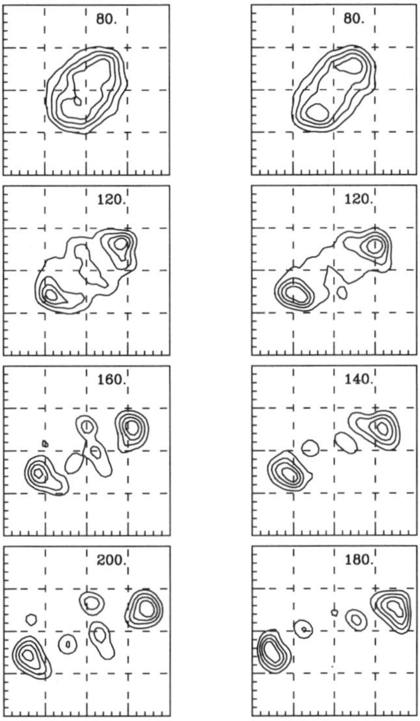

Fig. 7. $^{124}Sn + ^{124}Sn$ $b = 4\,\mathrm{fm}$ (*left*) and $b = 6\,\mathrm{fm}$ (*right*) collision: time evolution from expansion to freeze-out. See previous figure

We will comment first events generated in central collisions ($b = 2\,\mathrm{fm}$) [27,28]. In this case the reaction mechanism corresponds to a bulk fragmentation. We can identify three main stages, each characterized by very particular features of the isospin dynamics, since the system explores different density regions. After the compression phase (until 60–70 fm/c) follows a fast expansion (until 110–120 fm/c) and then during the last stage the systems breaks up into pieces. The values of density and temperature calculated at the beginning of the third stage correspond to situations inside the unstable region of the NM phase diagram. Volume instabilities have time to develop and we expect a kinetic mechanism of formation of the liquid phase of the spinodal decomposition type. At 260 fm/c, we observe a saturation in the time dependence of the average number of produced fragments, hence this time would correspond to the *freeze-out* configuration.

The results obtained with the *asy-stiff* symmetry term are shown in Fig. 8, while the *asy-soft* results are presented in Fig. 9. We have performed a separate investigation in a "central" region having a linear dimension of 20 fm centered around the CM of the system. Since it corresponds to the active volume in which fragmentation takes place, in this way we can obtain more detailed information about the process. We define as belonging to the gas phase the regions with density below $0.03\,\mathrm{fm}^{-3}$.

Each figure is organized in this way:

Left column, from top to bottom: Time evolution of: *Mass* in the liquid (up) and gas (down) phase; *Asymmetry* $I = (N-Z)/(N+Z)$ in the "central" gas (squares joined by solid line), and total gas (squares joined by dashes), "central" liquid (circles joined by solid line) and IMF (clusters with $3 < Z < 23$, stars). The horizontal line shows the initial average asymmetry; *Mean Fragment Multiplicity* $Z \geq 3$.

Right column, properties of the "primary" fragments at the *Freeze-Out Configuration*, from top to bottom: *Charge Distribution*, *Asymmetry Distribution* and *Fragment Multiplicity Distribution* (normalized to 1).

We observe a quite different behavior, concerning expecially the IMF isotopic content, depending on the used EOS. More asymmetric fragments are formed in the stiff case. This can be nicely understood by looking at the behavior of the chemical potentials (Fig. 2).

5.3 Semi-peripheral Collisions

To illustrate the dynamics of semi-peripheral collisions we have considered the reaction $^{124}\mathrm{Sn} + ^{64}\mathrm{Ni}$ at $35\,\mathrm{MeV/A}$, $b = 6\,\mathrm{fm}$. This experiment has been recently performed at LNS by the $REVERSE$ collaboration [37]. We observe a quite different fragmentation mechanism with respect to the central collisions. Now in the overlap region a neck structure is developing. During the interaction time it heats and expands but still remains in contact with denser and colder regions of projectile like (PLF) and/or target like (TLF) type and we are dealing essentially with surface instabilities. In some events small fragments may originate from this structure, while in other cases, this region can be re-absorbed by the PLF and/or the TLF.

In Figs. 10, 11 we present, for semi-peripheral reactions, the same kind of analysis as illustrated above (Figs. 8, 9). Now we observe that small IMF's are more neutron-rich than the larger ones. In fact they mostly originate from the neutron rich neck region. However, there is a clear difference between the asy-stiff and asy-soft predictions. In the soft case the effect is quite reduced. This can be again nicely related to the behavior of the symmetry energy (Fig. 1), but around normal density, since we are facing surface rather than volume instabilities. Now the larger value of the symmetry energy in the stiff case is responsible for a larger migration of neutrons towards the less dense neck region, that will become more neutron rich.

6 The QHD Effective Theory

The *QHD* effective field model represents a very successful attempt to describe, in a fully consistent relativistic picture, equilibrium and dynamical

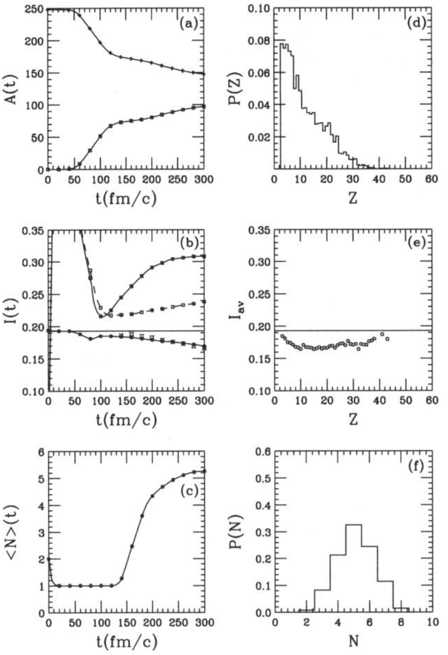

Fig. 8. $^{124}Sn + ^{124}Sn$ $b = 2$ fm collision: time evolution and freeze-out properties. See text. Asy-Stiff EOS

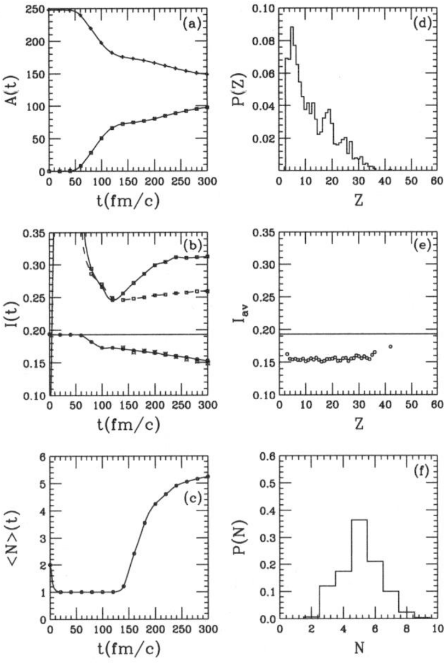

Fig. 9. $^{124}Sn + ^{124}Sn$ $b = 2$ fm collision: time evolution and freeze-out properties. See text. Asy-Soft EOS

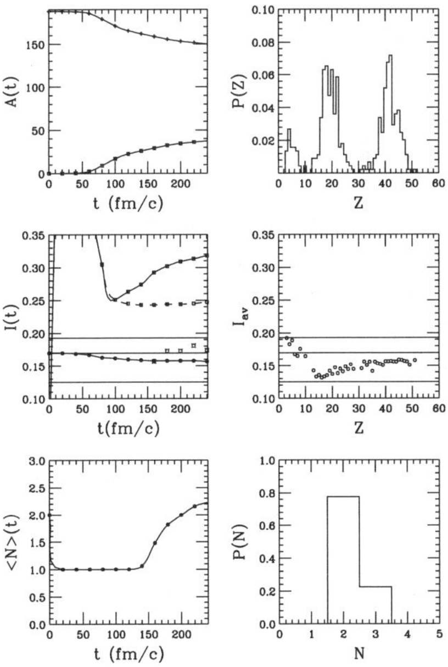

Fig. 10. $^{124}Sn + ^{64}Ni$ $b = 6$ fm collision: time evolution and freeze-out properties. See text. Asy-Stiff EOS

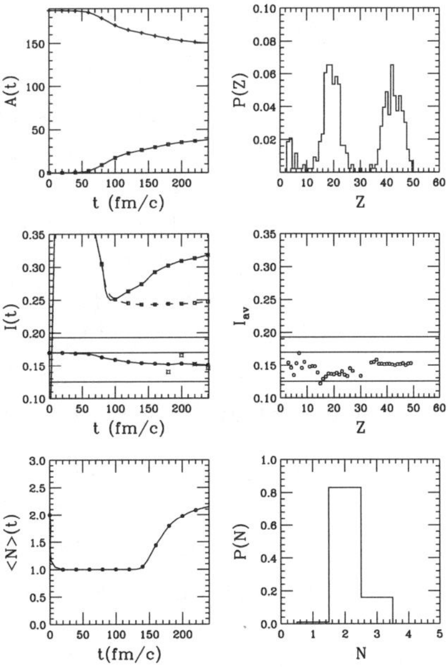

Fig. 11. $^{124}Sn + ^{64}Ni$ $b = 6$ fm collision: time evolution and freeze-out properties. See text. Asy-Soft EOS

properties of nuclear systems at the hadronic level [40–42]. Very nice results have been obtained for the nuclear structure of finite nuclei, for the NM Equation of State and liquid-gas phase transitions and for the dynamics of nuclear collisions.

In this report we will mainly focus our attention on the dynamical response of asymmetric nuclear matter. One of the main points of our discussion is the relevance of the coupling to a scalar isovector channel, the virtual $\delta[a_0(980)]$ meson, not considered in the usual Relativistic Mean Field (RMF) studies. Contributions to this channel are mainly coming from correlation effects [46]. Therefore the correct microscopic approach should be to derive the relative coupling constant, in a QHD mean field framework, from Dirac–Brueckner–Hartree–Fock calculations. Several attempts have been recently performed, see [43–45], but the results are up to now not fully model-independent.

An important outcome of our work is to show that the two effective couplings, vector and scalar, in the isovector channel are influencing in a different way the static (symmetry energy) and dynamic (reaction evolution) properties of asymmetric nuclear matter. This will open new possibilities for a phenomenological determination of these fundamental quantities.

We perform the many-body calculations in the quantum phase-space introducing the Wigner transform of the one-body density matrix for the fermion field defined as:

$$[\widehat{F}(x,p)]_{\alpha\beta} = \frac{1}{(2\pi)^4} \int d^4R e^{-ip\cdot R} \left\langle : \bar{\psi}_\beta \left(x + \frac{R}{2}\right) \psi_\alpha \left(x - \frac{R}{2}\right) : \right\rangle ,$$

where α and β are double indices for spin and isospin. The brackets denote statistical averaging and the colons denote normal ordering. The Wigner function is a matrix in spin and isospin spaces; in the case of asymmetric NM (ANM) it is useful to decompose it into neutron and proton components. From the Dirac equation we obtain a kinetic equation for the Wigner function, that can be eventually solved for the collective response [47] as well as for the relativistic reaction dynamics [48,49].

6.1 Equilibrium Properties: the Nuclear Equation of State

The energy density and pressure for symmetric and asymmetric nuclear matter and the n,p effective masses can be self-consistently calculated just in terms of the four boson coupling constants, $f_i \equiv (\frac{g_i^2}{m_i^2})$, $i = \sigma, \omega, \rho, \delta$, and the two parameters of the σ self-interacting terms, $A \equiv \frac{a}{g_\sigma^3}$ and $B \equiv \frac{b}{g_\sigma^4}$, see [50,51].

The isoscalar meson parameters are fixed from symmetric nuclear matter properties at $T = 0$. They have quite standard values for these minimal non-linear RMF models.

The symmetry energy in ANM is defined from the expansion of the energy per nucleon $E(\rho_B,\alpha)$ in terms of the asymmetry parameter $\alpha \equiv \frac{N-Z}{A}$:

$$E(\rho_B,\alpha) \equiv \frac{\epsilon(\rho_B,\alpha)}{\rho_B} = E(\rho_B) + E_{sym}(\rho_B)\alpha^2 + O(\alpha^4) + \dots ,$$

and so in general

$$E_{sym} \equiv \frac{1}{2}\frac{\partial^2 E(\rho_B,\alpha)}{\partial\alpha^2}|_{\alpha=0} = \frac{1}{2}\rho_B\frac{\partial^2\epsilon}{\partial\rho_{B3}^2}|_{\rho_{B3}=0} \ . \tag{7}$$

In the Hartree case an explicit expression for the symmetry energy can be easily derived [51]

$$E_{sym}(\rho_B) = \frac{1}{6}\frac{k_F^2}{E_F^*} + \frac{1}{2}\left[f_\rho - f_\delta\left(\frac{M^*}{E_F^*}\right)^2\right]\rho_B \equiv E_{sym}^{kin} + E_{sym}^{pot} , \tag{8}$$

where k_F is the nucleon Fermi momentum corresponding to ρ_B, $E_F^* \equiv \sqrt{(k_F^2 + M^{*2})}$ and M^* is the effective nucleon mass in symmetric NM, $M^* = M_N - g_\sigma\phi$.

We see that, when the δ is included, the observed a_4 value actually assigns the combination $[f_\rho - f_\delta(\frac{M}{E_F})^2]$ of the (ρ,δ) coupling constants. If $f_\delta \neq 0$ we have to increase the ρ-coupling. In our calculations we use the value $a_4 = 32\,\mathrm{MeV}$, while the δ coupling is chosen in agreement with the analysis of [43].

Now the symmetry energy at saturation density is actually built from the balance of scalar (attractive) and vector (repulsive) contributions, with the scalar channel becoming weaker with increasing baryon density [51]. This is indeed the isovector counterpart of the saturation mechanism occurring in the isoscalar channel for the symmetric nuclear matter. From such a scheme we get a further support to the introduction of the δ-coupling in the symmetry energy evaluation.

When the δ "channel" is included the high density behavior is stiffer for the relativistic mechanism discussed before, see also [51].

In Fig. 12 we report the density dependence of the symmetry energy for three different models: one including only the ρ field (RMF-ρ), the other including $(\rho+\delta)$ fields (RMF-$(\rho+\delta)$) and the last including only the ρ field but with a covariant density dependence (RMF-$D\rho$), tuned just to give the same EOS of the RMF-$(\rho+\delta)$ case. The latter is useful for disentangling the effects coming from the difference in $E_{sym}(\rho_B)$ and those coming from the strength of the ρ vector field. In such a way we have a model with an isovector contribution given only by the vector field with $f_\rho = 1.1\,\mathrm{fm}^2$, another with a balance between a vector field with $f_\rho = 3.3\,\mathrm{fm}^2$ and a scalar one with $f_\delta = 2.4\,\mathrm{fm}^2$ and a last one with a coupling $f_\rho(\rho_0) = 1.1\,\mathrm{fm}^2$ and an increasing density dependence shown in Fig. 12 (insert). We stress again that in RMF-$(\rho+\delta)$ the symmetry energy is coming from a balance between a scalar,

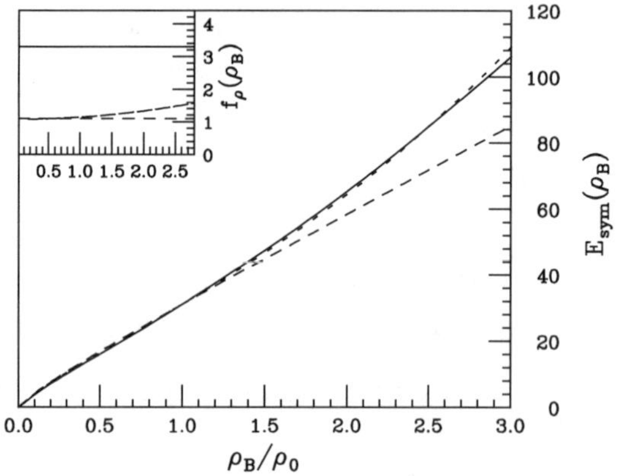

Fig. 12. Total (kinetic + potential) symmetry energy as a function of the baryon density. *Solid*: $RMF - (\rho + \delta)$. *Dashed*: $RMF - \rho$. *Short dashed*: $RMF - D\rho$. In the *insert* the density behavior of the ρ coupling, f_ρ, for the three models is shown

attractive ($\delta - like$), and a vector, repulsive ($\rho - like$), which then is three times larger than in the case RMF-ρ model.

7 Linear Response

We study collective oscillations that propagate in cold nuclear matter due to the mean field dynamics. In some sense we follow a relativistic extension of the method introduced by Landau to study liquid-^{3}He and recently applied to investigate stable and unstable modes in nuclear matter [22,23]. The starting point is the relativistic kinetic transport equation. We look for solutions corresponding to small oscillations of $\hat{F}(x,p)$ around the equilibrium value. Therefore we put

$$\hat{F}(x,p) = \hat{H}(p) + \hat{G}(x,p) , \tag{9}$$

where $\widehat{H}(p)$ is the Wigner function at equilibrium and $\widehat{G}(x,p)$ represent its fluctuations. By Fourier transforming and integrating over four–momentum, in linear approximation we get the set of equations for the scalar and vector fluctuation of each species ($i = 1, 2$ for proton, neutron respectively), see [47].

The normal collective modes are plane waves, characterized by the wave vector ($k^\mu = (k^0, 0, 0, \mid \mathbf{k} \mid)$), they are determined by solving the set of homogeneous linear equations. The solutions correspond only to longitudinal waves and do not depend on k^0 and $|\mathbf{k}|$ separately, but only on the ratio $v_s = k^0/|\mathbf{k}|$.

The sound velocities are given by values of v_s for which the relevant determinant of the set vanishes, i.e. the dispersion relations. In correspondence the neutron/proton structure of the eigenvectors (normal modes) can be derived. It should be remarked that in asymmetric nuclear matter isoscalar and isovector components are mixed in the normal modes. Here this can be argued by the fact that in each linearized equation both proton/neutron densities and currents are appearing. However we remind that one can still identify isovector-like excitations as the modes where neutrons and protons move out of phase, while isoscalar-like modes are characterized by neutrons and protons moving in phase [22,24].

Isovector Modes
One may expect that once a_4 is fixed, the velocity of sound is also fixed. On the other hand our results clearly indicate a different dynamical response with or without the δ-meson channel, for interactions which give *exactly* the same a_4 parameter. In order to get a clear understanding of this effect we will consider the case of symmetric nuclear matter in the Hartree scheme, where the dispersion relations are assuming a transparent analytical form.

Now it is also possible to decouple the collective modes into *pure* isoscalar and isovector oscillations, [22]. After a straightforward rearrangement we have a dispersion relation for the isovector modes:

$$1+\frac{6\,E_F^*}{k_F^2}\left[E_{sym}^{pot}-\frac{f_\rho}{2}\frac{k_F^2}{E_F^{*2}}\left(1-f_\delta\,\frac{M^*}{E_F^{*2}}\,\rho_S\right)\rho_B\right]\varphi(s)=0\;, \tag{10}$$

where the potential part of the symmetry energy explicitly appears in the dispersion relations, but *joined to an important correction term* which shows a different f_ρ, f_δ structure with respect to that of E_{sym}^{pot}, Eq. (8). *We can easily have interactions with the same a_4 value at normal density but with very different isovector response.* E.g. when we include the δ channel we know that we have to increase the f_ρ coupling in order to have the same a_4, but now the "restoring force" (coefficient of the Lindhard function in the Equation (10)) will be strongly reduced.

Isoscalar Modes
As already remarked, we like to note that for symmetric NM there is a tight analogy between the isoscalar and the isovector response in the RMF approach. In the isoscalar degree of freedom the compressibility will play the same role of the symmetry energy in the dispersion relation equations. Also in this case we will have an important correction term coming from the interplay of the scalar and vector channel.

The isoscalar equivalent of the Eq. (10) takes the form:

$$1+\frac{E_F^*}{3\,k_F^2}\left[K_{NM}^{pot}-9\,f_\omega\frac{k_F^2}{E_F^{*2}}\left(1-f_\sigma\,\frac{M^*}{E_F^{*2}}\,\rho_S\right)\rho_B\right]\varphi(s)=0\;, \tag{11}$$

where the K_{NM}^{pot} is the potential part of the nuclear matter compressibility that in the Hartree scheme has the simple structure:

$$K_{NM}(\rho_B)=\frac{3\,k_F^2}{E_F^*}+9\left[f_\omega-f_\sigma\left(\frac{M^*}{E_F^*}\right)^2\right]\rho_B\equiv K_{NM}^{kin}+K_{NM}^{pot}\;. \tag{12}$$

By means of such an analogy, the previous discussion can be extended to isoscalar oscillations with the role of E_{sym} now "played" by the compressibility. In this case however one always takes into account both the scalar and vector channel in any RMF models. However the coupling constant f_ω can assume very different values depending on the required value for effective masses M_0^*. This is easy to understand since in the RMF limit the saturation binding energy has the simple form

$$E/A(0) = E_F^* + f_\omega \rho_B(0) - M_N ,$$

where M_N is the bare nucleon mass. So we see that in order to have the same saturation values of $\rho_B(0)$, $E/A(0)$ when we decrease M_0^* we have to increase f_ω. We derive then at the natural conclusion that if two EOS have different effective masses even if the compressibility is equal the dynamical behavior is expected to be different. This is a very general feature present also in non-relativistic approaches.

7.1 Exotic High Baryon Density Modes

For asymmetric matter we have found that, in all the calculation schemes, with increasing baryon density the isovector modes disappear and the isoscalar-like excitations become dominant: we call such densities ρ_B^{cross}. Around this transition density we expect to have an almost *pure neutron wave* propagation of the sound. Some results are shown in Fig. 13. It should be noticed that the frequency of the isoscalar-like modes is essentially related to the compressibility of the system at the considered density. In Fig. 13(a) we display the sound velocity obtained in Hartree and Hartree–Fock calculations at $\rho_B = 3.5\ \rho_0$, as a function of the asymmetry α, [47].

At $\alpha = 0$ the two Hartree models have exactly the same isoscalar mean fields, but for asymmetric nuclear matter the different behavior of the symmetry energy leads to a different compressibility. The case $NLH - (\rho + \delta)$ which has the stiffer E_{sym} (resulting in a greater incompressibility for $\alpha > 0$) with respect to $NLH - (\rho)$ shows also a greater increase of v_s/v_{Fn} with density.

Some differences are observed also in the chemical composition of the mode (Fig. 13(b)). The black spots show the behavior of ρ_p/ρ_n vs. α. Note the *pure neutron wave* structure of the propagating sound, since the oscillations of protons appear strongly damped ($\delta\rho_p/\delta\rho_n \ll \rho_p/\rho_n$); unfortunately this is an effect not experimentally accessible (at present).

7.2 Isospin Distillation

We have also investigated the response of the system in the region of spinodal instability associated with the liquid–gas phase transition, which occurs at low densities. It is known that in this region an isoscalar unstable mode can be found, with imaginary sound velocity, that gives rise to an exponential

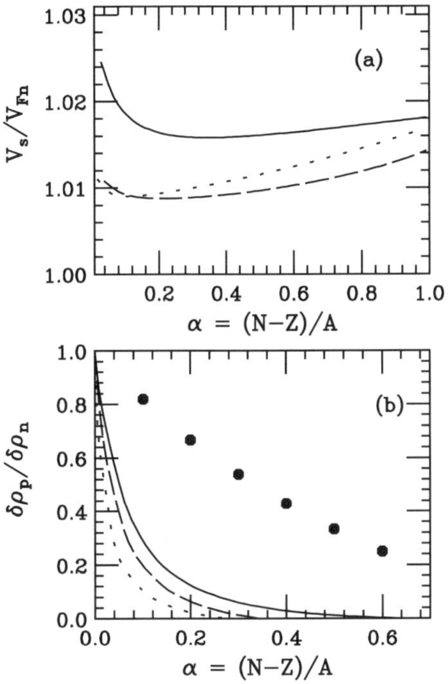

Fig. 13. Sound velocity (**a**) and chemical composition (**b**) of isoscalar-like modes at $\rho_B = 3.5\rho_0$. *Solid line*: $NLHF$. *Long Dashed line*: $NLH - \rho$. *Dotted line*: $NLH - (\rho + \delta)$. The *full circles* in panel (**b**) represent the behavior of ρ_p/ρ_n vs. α

growth of the fluctuations. The latter can represent a dynamical mechanism for the multi–fragmentation process observed in heavy–ion collisions. We have found this kind of solution in the present approach. In Fig. 14 we show the ratio $\delta\rho_p/\delta\rho_n$ as function of the initial asymmetry for such a collective mode. For all the interactions this ratio is different from the corresponding ρ_p/ρ_n of the initial asymmetry α. This is exactly the chemical effect associated with the new instabilities in dilute asymmetric matter discussed before, [24,39].

In particular it is found that, when isoscalar-like modes become unstable, the ratio $\delta\rho_p/\delta\rho_n$ becomes *larger* that the ratio ρ_p/ρ_n (at variance with the stable modes at high densities, see Fig. 13). Hence proton oscillations are relatively larger than neutron oscillations leading to a more symmetric liquid phase and to a more neutron rich gas phase, during the disassembly of the system, see also the discussion in the introduction. This is the so–called isospin distillation effect in fragmentation, and signatures of this effect could be searched by looking at the ratio N/Z of fragments produced in dissipative heavy ion collisions [27,28].

In Fig. 14 we observe that Hartree results (with and the without δ meson) are very similar and, indeed, at low density the symmetry energy behavior is nearly the same in the two cases. We can conclude that there are essentially no effects of the scalar isovector channel on isospin distillation in the spinodal decomposition.

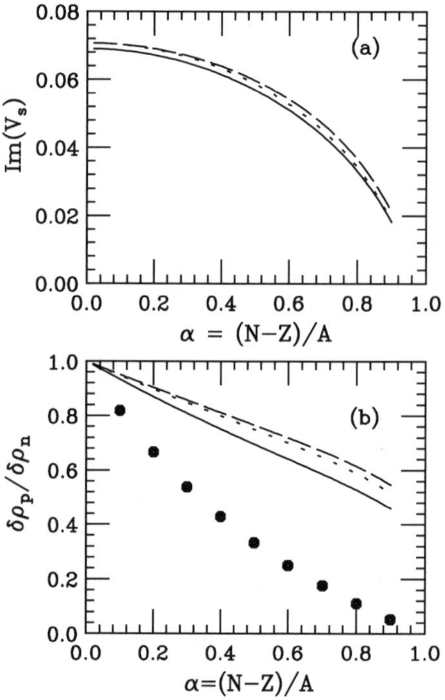

Fig. 14. Isoscalar–like unstable modes at $\rho_B = 0.4\,\rho_0$: Imaginary sound velocity (**a**) and ratio of proton and neutron amplitudes (**b**) as a function of the asymmetry α. *Solid line*: $NLHF$; *Dotted line*: $NLH - (\rho + \delta)$; *Long Dashed line*: $NLH - \rho$.The *full circles* in panel (**b**) represent the behavior of $\delta\rho_p/\delta\rho_n$ vs. α

8 Symmetry Effects at High Baryon Density: Collective Flows

It is quite desirable to get information on the symmetry energy at higher density, where furthermore we cannot have complementary investigations from nuclear structure like in the case of the low density behavior. Heavy Ion Collisions (HIC) provide a unique way to create asymmetric matter at high density in terrestrial laboratories.

The isospin dependence of collective flows has been already discussed in a non-relativistic framework [34,52]. The main new result shown here, in a Relativistic Mean Field (RMF) scheme, is the importance at higher energies of the microscopic covariant structure of the effective interaction in the isovector channel: effective forces with very similar symmetry terms can give rise to different flows in relativistic heavy ion collisions [49]. A full description of the isovector channel in a relativistic framework in principle should rely on the balance between a scalar ($\delta - like$, attractive) and a vector ($\rho - like$, repulsive) contributions, see before. This is a quite controversial point. In relativistic $HIC's$, due to the large counter streaming nuclear currents, one may directly exploit the different Lorentz nature of a scalar and a vector field [49].

For the description of heavy ion collisions we solve the covariant transport equation of the Boltzmann type within the Relativistic Landau–Vlasov (RLV) method [48] (for the Vlasov part) and applying a Monte-Carlo procedure for the collision term, including inelastic processes involving the production/absorption of nucleon resonances, [53].

Typical results for the ^{132}Sn + ^{132}Sn reaction at 1.5 AGeV (semicentral collisions) are shown in Figs. 15, 16. In Fig. 15 we report the differential flow $F^{pn}(y) \equiv 1/N(y)\sum_i p_{x_i}\tau_i$ where $N(y)$ is the total number of free nucleons at the rapidity y, p_{x_i} is the transverse momentum of particle i in the reaction plane, and τ_i is $+1$ and -1 for protons and neutrons, respectively. The $RMF-(\rho+\delta)$ case (full circles and solid line) presents a stiffer behavior relative to the $RMF-\rho$ (open circles) model, as expected from the more repulsive symmetry energy $E_{sym}(\rho_B)$ at high baryon densities [49,51]. We have however repeated the calculation using the $RMF-D\rho$ interaction, i.e. with only a ρ contribution *but* tuned to reproduce the same EOS of the $RMF-(\rho+\delta)$ case. The results, short-dashed curve of Fig. 15, are very similar to the ones of the $RMF-\rho$ interaction. Therefore we can explain the large flow effect as mainly due to the different strengths of the vector-isovector field between $RMF-(\rho+\delta)$ and $RMF-\rho, D\rho$ in the relativistic dynamics. In fact if a source is moving the vector field is enhanced (essentially by the local γ Lorentz factor) relative to the scalar one. Keeping in mind that $RMF-(\rho+\delta)$ has a three times larger ρ field it is clear that dynamically

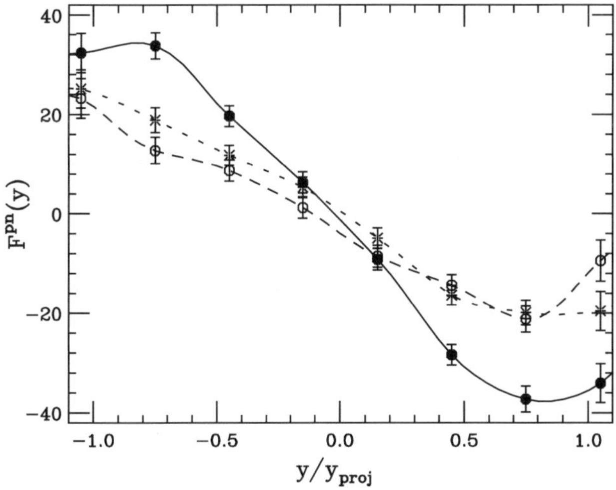

Fig. 15. Proton–neutron differential collective flow in the ^{132}Sn+^{132}Sn reaction at 1.5 AGeV $b = 6$ fm for the three different model for the isovector mean fields. *Full circles* and *solid line*: $RMF-(\rho+\delta)$. *Open circles* and *dashed line*: $RMF-\rho$. *Stars* and *short dashed line*: $RMF-D\rho$

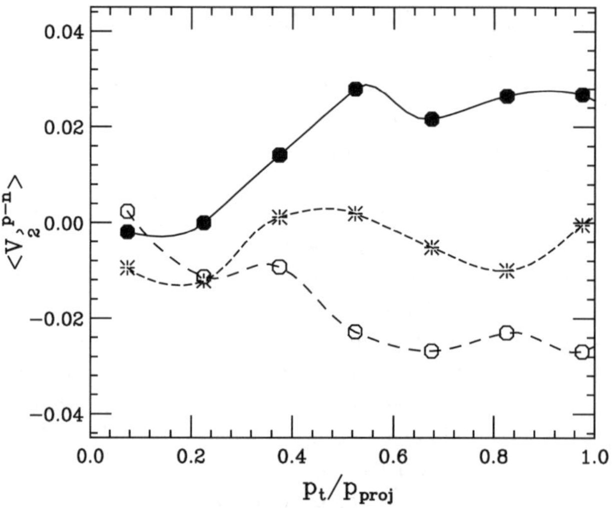

Fig. 16. Difference between neutron and proton elliptic flow as a function of the transverse momentum in the ^{132}Sn + ^{132}Sn reaction at 1.5 AGeV $b = 6$ fm in the rapidity range $-0.3 \leq y/y_{proj} \leq 0.3$. *Full circles* and *solid line*: $RMF - (\rho + \delta)$. *Open circles* and *dashed line*: $RMF - \rho$. *Stars* and *short dashed line*: $RMF - D\rho$

the vector-isovector mean field acting during the HIC is much greater than the one of the $RMF - \rho, D\rho$ cases.

In Fig. 16 we report the elliptic flow $v_2(y, p_t)$, $v_2 = \langle (p_x^2 - p_y^2)/p_t^2 \rangle$ where $p_t = \sqrt{p_x^2 + p_y^2}$ is the transverse momentum [54]. A negative value of v_2 corresponds to the emission of matter perpendicular to the reaction plane, *squeeze-out* flow. The p_t-dependence of v_2 is very sensitive to the high density behavior of the EOS since highly energetic particles ($p_t \geq 0.5$) originate from the initial compressed and out-of-equilibrium phase of the collision. We focus on the proton-neutron difference of the elliptic flow. From Fig. 16 we see that in the $(\rho + \delta)$ dynamics the high-p_t neutrons show a much larger *squeeze-out*. This is fully consistent with an early emission (more spectator shadowing) due to the larger repulsive ρ-field. The v_2 observable, which is a good *chronometer* of the reaction dynamics, appears to be particularly sensitive to the Lorentz structure of the effective interaction.

π^-/π^+ *ratios.* Using the same relativistic transport code we have evaluated the π^- vs. π^+ production for central Au + Au collisions at different energies, see Fig. 17. As expected the larger repulsion seen by neutrons at high densities in the $RMF - (\rho + \delta)$ will show up in a reduced n/p ratio, smaller $\Delta^{0,-}$ density and finally a reduced π^- production. However now reabsorption effects are important and actually the effect appears to be decreasing at higher energies, [55].

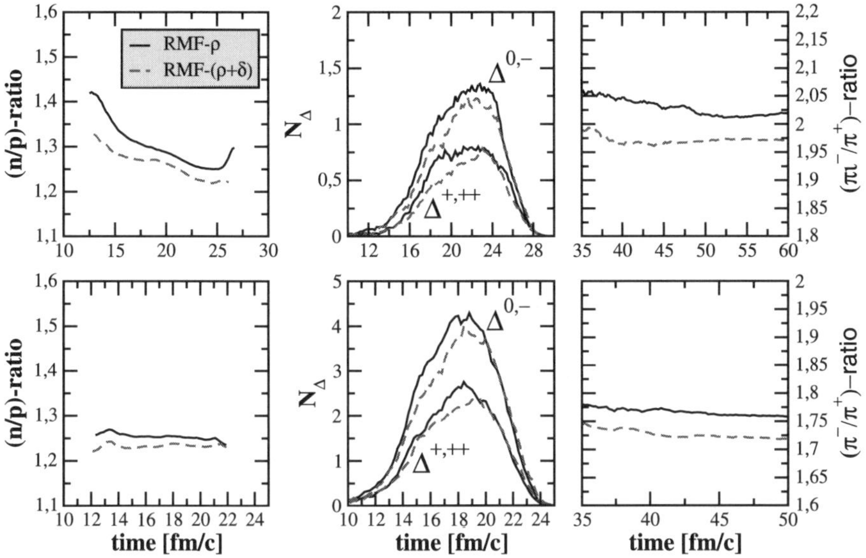

Fig. 17. Central Au + Au collisions at 0.6 AGeV (*top*) and 1.0 AGeV (*bottom*). Time evolution of n/p ratio and Δ resonance production in high density regions ($\rho/\rho_0 \geq 2.0$) (*first two columns*) and of the total π^-/π^+ ratio (*right*). *solid lines*: $RMF-\rho$. *Dashed lines*: $RMF-(\rho+\delta)$

9 Isospin and Deconfinement at High Baryon Density

It is relatively easy to show that at high baryon density and low temperature we can expect a transition from hadronic matter to deconfined quark matter. The procedure is straightforward:
(i) Start from two "reasonable" model Equations of State (EOS), one for the hadronic phase, which can reproduce saturation properties, one for the quark phase, which can reproduce the hadron spectrum.
(ii) Construct the phase separation boundary surface from the Gibbs phase rule.

For symmetric matter the baryon density ρ_{tr} corresponding to the transition to the coexistence region is relatively high, as expected, ranging from 4 to 8 times the saturation value ρ_0, depending on the stiffness of the hadronic EOS at high densities. The new feature we would like to focus on in this report is the isospin dependence of such boundary location. We can foresee an interesting asymmetry effect, in the appealing direction of a decrease of ρ_{tr}, since the hadronic EOS becomes more repulsive [56,57].

The proton fraction Z/A dependence of the ρ_{tr} is reported in Fig. 18 with the bag constant value $B^{1/4} = 150\,\text{MeV}$ and $\alpha_s = 0$ for the quark phase and various choices for the hadronic EOS: *Dotted line* $GM3$ parameterization [58]; *Dashed line* $RMF-\rho$ parameterization [51]; *Solid line* $RMF-(\rho+\delta)$ param-

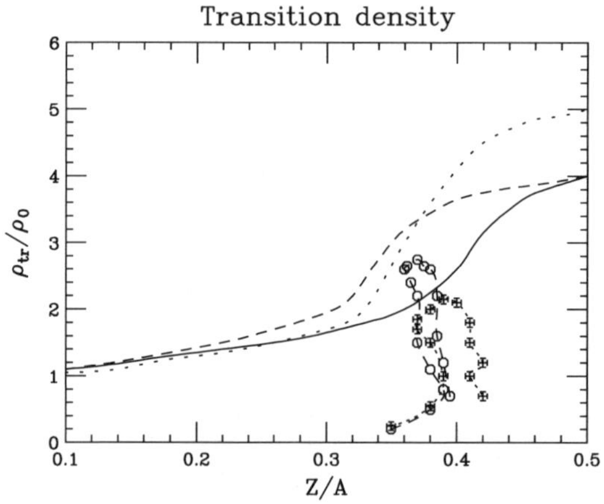

Fig. 18. Variation of the transition density with proton fraction followed in the interaction zone during a semicentral ^{132}Sn + ^{132}Sn collision at 1 AGeV (*circles*) and 300 AMeV (*crosses*)

eterization [51]. $GM3$ and $RMF - \rho$ have the same source of the interaction symmetry term (only the ρ-meson).

The effects of the asymmetry appears now quite dramatic: we see a ρ_{trans} as low as $2\rho_0$ for proton fractions between 0.3 and 0.4, conditions that with some confidence we could "locally" reach in a heavy ion collision at *intermediate* energy using exotic very asymmetric beams.

Using our Relativistic Transport Code, with the $RMF - (\rho + \delta)$ effective interaction, we have performed some simulations of the ^{132}Sn+^{132}Sn (average $Z/A = 0.38$) collision at various energies, for a semicentral impact parameter, $b = 6$ fm, just to optimize the neutron skin effect. In Fig. 18 the paths in the $(\rho, Z/A)$ plane followed in the c.m. region during the collision are reported, at two energies 300 AMeV (crosses) and 1 AGeV (circles). We see that already at 300 AMeV we are reaching the border of the mixed phase, and we are well inside it at 1 AGeV.

In conclusion we support the possibility of observing precursor signals of the phase transition to a deconfined matter in violent collision (central and semicentral) of exotic (radioactive) heavy ions in the energy range of few hundred MeV per nucleon. A possible signature could be revealed through an earlier "softening" of the hadronic EOS for larger asymmetries.

10 Outlook: The Eleven Observables

As a conclusion of our report we like to suggest a selection of *Eleven Observables*, from low to relativistic energies, that we expect particularly sensitive to the microscopic structure of the *in-medium* interaction in the isovector channel, i.e. to the symmetry energy and its "fine structure":

1. Competition of Reaction Mechanisms. Interplay of low-energy dissipative mechanisms, e.g. [25], fusion (incomplete) vs. deep-inelastic vs. neck fragmentation: a stiff symmetry term leads to a more repulsive dynamics.

2. N/Z of fast nucleon emission. Symmetry repulsion of the neutron/proton mean field in various density regions.

3. Neutron/Proton correlation functions. Time-space structure of the fast particle emission and its relation to the baryon density of the source, see the recent [59].

4. Fragment Multiplicities. A more efficient use of protons in forming primary fragments is expected in the asy-stiff case.

5. Isospin Distillation (Fractionation). Isospin content of the Intermediate Mass Fragments in central collisions. Test of the symmetry term in dilute matter.

6. Isospin content of Neck-Fragments. Test of the symmetry term around ρ_0.

7. Fast Fission Multiplicity. The rate of "aligned" fission events of the Projectile-Like/Target-Like Fragments reflects the symmetry repulsion in semicentral collisions.

8. Isospin Diffusion. Measure of charge equilibration in the "spectator" region in semicentral collisions, test of symmetry repulsion.

9. Neutron-Proton Collective Flows. Together with light isobar flows. Check of symmetry transport effects. Test of the momentum dependence (relativistic structure) of the interaction in the isovector channel. Measurements also for different p_t selections.

10. π^-/π^+ *Yields.* Since π^- are mostly produced in nn collisions we can expect a reduction for highly repulsive symmetry terms at high baryon density, see [55,60,61].

11. Deconfinement Precursors. Signals of a mixed phase formation (quark-bubbles) in high baryon density regions reached with asymmetric HIC at intermediate energies.

From the points $3, 4, 5, 8, 9$ in our simulations we presently get some indications for *asy-stiff* behaviors, i.e. increasing repulsive density dependence of the symmetry term, but not more fundamental details. Moreover all the available data are obtained with stable beams, i.e. within low asymmetries.

Acknowledgements

This report is deeply related to ideas and results partially reached in a very pleasant and fruitful collaboration with very nice people: V. Baran, M. Colonna, A. Drago, T. Gaitanos, V. Greco, A. Lavagno, B. Liu, F. Matera, M. Zielinzka-Pfabe', J. Rizzo and H.H. Wolter. I have learned a lot from all of them.

References

1. Ph. Chomaz: 'The nuclear liquid-gas phase transition and phase coexistence'. In: *Nuclear Physics in the* 21st *Century*, INPC 2001 Berkeley, Eds. E. Norman et al., AIP Conf.Proc. **610**, 167 (2002)
2. J. Pochodzalla et al. (ALADIN coll.): Phys. Rev. Lett. **75**, 1040 (1995)
3. J.B. Natowitz et al.: Phys. Rev. **C66**, 031601 (2002) and refs. therein
4. M.S.S. Challa and J.H. Hetherington: Phys. Rev. **A38**, 6324 (1988)
5. Ph. Chomaz, V. Duflot and F. Gulminelli: Phys. Rev. Lett. **85**, 3587 (2000)
6. M. D'Agostino et al.: Phys. Lett. **B473**, 219 (2000)
7. M. D'Agostino et al.: Nucl. Phys. **A699**, 795 (2002)
8. M. Colonna and Ph. Chomaz: Phys. Rev. **C49**, 1908 (1994); Ph. Chomaz, M. Colonna, A. Guarnera and J. Randrup: Phys. Rev. Lett. **73**, 3512 (1994)
9. M. Colonna, G.F. Burgio, Ph. Chomaz, M. Di Toro, and J. Randrup: Phys. Rev. **C47**, 1395 (1993)
10. M. Colonna, M. Di Toro and A. Guarnera: Nucl. Phys. **A580**, 312 (1994)
11. A. Guarnera, M. Colonna and Ph. Chomaz: Phys. Lett. **B373**, 267 (1996); M. Colonna: Nucl. Phys. **A630**, 136c (1998)
12. B. Borderie et al.: Phys. Rev. Lett. **86**, 3252 (2001); G. Tabacaru et al.: arXiv:nucl-ex/0212018, sub.EPJ **A**
13. M. Colonna, M. Di Toro and A. Guarnera: Nucl. Phys. **A589**, 160 (1995); J.D. Frankland: Thesis Orsay 1998;
14. J. Toke et al.: Phys. Rev. Lett. **77**, 3514 (1996) and refs. therein
15. D. Durand: Nucl. Phys. **A654**, 273c (1998)
16. E. Plagnol et al. (INDRA collab.): Phys. Rev. **C61**, 014606 (2000)
17. P. Milazzo et al.: Phys. Lett. **B509**, 204 (2001)
18. F. Bocage, J. Colin et al.: Nucl. Phys. **A676**, 391 (2000)
19. *Isospin Physics in Heavy-ion Collisions at Intermediate Energies*, Eds. Bao-An Li and W. Udo Schroeder, Nova Science Publishers (2001, New York)
20. H. Mueller and B.D. Serot: Phys. Rev. **C52**, 2072 (1995)
21. Bao-An Li and C.M. Ko: Nucl. Phys. **A618**, 498 (1997)
22. M. Colonna, M. Di Toro and A. Larionov: Phys. Lett. **B428**, 1 (1998)
23. V. Baran, A. Larionov, M. Colonna and M. Di Toro: Nucl. Phys. **A632**, 287 (1998)
24. V. Baran, M. Colonna, M. Di Toro and V. Greco: Phys. Rev. Lett. **86**, (2001) 4492
25. M. Colonna, M. Di Toro, G. Fabbri and S. Maccarone: Phys. Rev. **C57**, 1410 (1998)
26. M. Di Toro et al.: Progr. Part. Nucl. Phys. **42**, 125–136 (1999)
27. M. Di Toro et al.: Nucl. Phys. **A681**, 426c (2001)
28. V. Baran et al.: Nucl. Phys. **A703**, 603 (2002)
29. B.A. Brown: Phys. Rev. Lett. **85**, 5296 (2000).
30. C.J. Horowitz and J. Piekarevicz: Phys. Rev. Lett. **86**, 5647 (2001)
31. R.J. Furnstahl: Nucl. Phys. **A706**, 85 (2002)
32. J.M. Lattimer and M. Prakash: Astro. Phys. Jour. **550**, 426 (2001)
33. V. Greco, A. Guarnera, M. Colonna and M. DiToro: Phys. Rev. **C59**, 810 (1999)
34. L. Scalone, M. Colonna and M. Di Toro: Phys. Lett. **B461**, 9 (1999).

35. M. Colonna et al.: Nucl.Phys. **A642**, 449 (1998)
36. M.B. Tsang et al.: Phys. Rev. Lett. **86**, 5023 (2001)
37. E. Geraci, A. Pagano et al. (Chimera Collab.), private communication
38. J. Randrup and B. Remaud: Nucl. Phys. **A514**, 339 (1990)
39. Ph. Chomaz and J. Magueron: arXiv:nucl-th/0212082
40. J.D. Walecka: Ann. Phys. (N.Y.) **83**, 491 (1974)
41. B.D. Serot and J.D. Walecka in *Advances in Nuclear Physics*, edited by J.M. Negele and E. Vogt, Plenum, New York, 1986 Vol. 16
42. B.D. Serot and J.D. Walecka J.D.: Int. J. Mod. Phys. **E6**, 515 (1997)
43. F. Hofmann, C.M. Keil, H. Lenske: Phys. Rev. **C64**, 034314 (2001)
44. E. Schiller and H. Muether: Eur. Phys. J. **A11**, 15 (2001)
45. Z. Ma and L. Liu: Phys. Rev. **C66**, 02434 (2002)
46. V. Greco et al.: Phys. Rev. **C63**, 035202 (2001)
47. V. Greco, M. Colonna, M. Di Toro, F. Matera: Phys. Rev. **C67**, 015203 (2003)
48. Ch. Fuchs and H.H. Wolter: Nucl. Phys. **A589**, 732 (1995)
49. V. Greco, M. Colonna, M. Di Toro, T. Gaitanos, H.H. Wolter and V. Baran: *Relativistic effects in the search for high density symmetry term*, arXiv:nucl-th/0112102, and V. Greco, Ph. D. Thesis 2002.
50. V. Greco, M. Colonna, M. Di Toro, G. Fabbri and F. Matera: Phys. Rev. **C64**, 045203 (2001)
51. B. Liu, V. Greco, V. Baran, M. Colonna and M. Di Toro: Phys. Rev. **C65**, 045203 (2002)
52. Bao-An Li: Phys. Rev. Lett. **85**, 4221 (2000)
53. S. Huber, J. Aichelin: Nucl. Phys. **A573**, 587 (1994)
54. P. Danielewicz: Nucl. Phys. **A673**, 375 (2000)
55. M. Di Toro et al.: ISPUN02 (Halong Bay, Vietnam) Contribution, arXiv:nucl-th/0301033, Nucl. Phys. **A** to appear
56. H. Mueller: Nucl.Phys. **A618**, 349 (1997)
57. M. Di Toro, A. Drago, V. Greco and A. Lavagno: arXiv:nucl-th/0210052
58. N.K. Glendenning and S.A. Moskowski: Phys. Rev. Lett. **67**, 2414 (1991)
59. L.W. Chen, V. Greco, C.M. Ko and B.A. Li: arXiv:nucl-th/0211002
60. V.S. Uma Mashevari et al.: Phys. Rev. **C57**, 922 (1998)
61. B.A. Li: Phys. Rev. **C67**, 017601 (2003)

Index